Values and Valuing in Mathematics Education

Yüksel Dede · Gosia Marschall · Philip Clarkson
Editors

Values and Valuing in Mathematics Education

Moving Forward into Practice

Springer

Editors
Yüksel Dede
Department of Mathematics Education
Gazi Education Faculty
Gazi University
Ankara, Türkiye

Gosia Marschall
Faculty of Education
University of Cambridge
Cambridge, UK

Philip Clarkson
Australian Catholic University
Melbourne, VIC, Australia

University of Melbourne
Parkville, VIC, Australia

ISBN 978-981-99-9453-3 ISBN 978-981-99-9454-0 (eBook)
https://doi.org/10.1007/978-981-99-9454-0

This Springer imprint is published by the registered company Springer Nature Singapore Pte Ltd.
The registered company address is: 152 Beach Road, #21-01/04 Gateway East, Singapore 189721, Singapore

Contents

Contributors

Nor Azura Abdullah Universiti Brunei Darussalam, Gadong, Brunei Darussalam

Yasmine Abtahi University of South-East Norway, Notodden, Norway

Veysel Akçakın Uşak University, Uşak, Türkiye

Fatma Nur Aktaş Kahramanmaras Sutcu Imam University, Onikişubat, Turkey

Annica Andersson University of South-Eastern Norway, Notodden, Norway

Takuya Baba Hiroshima University, Higashihiroshima, Japan

Richard Barwell University of Ottawa, Ottawa, Canada

Yan Cao Chengdu Qingyang Institute of Education Sciences, Chengdu, China

Hui Min Chia Faculty of Education, University of Macau, Macau, China

Philip Clarkson Australian Catholic University/University of Melbourne, Melbourne, Australia

Ernest Kofi Davis University of Cape Coast, Cape Coast, Ghana

Anette De Ron Stockholm University, Stockholm, Sweden

Yüksel Dede Gazi University, Ankara, Türkiye

Noor Elebiary University of Howler, Kurdistan, Iraq

Hiroto Fukuda Okayama University of Science, Okayama, Japan

Yuichiro Hattori Okayama University, Okayama, Japan

Kjellrun Hiis Hauge Western Norway University of Applied Sciences, Bergen, Norway

Julia L. Hill Faculty of Education, The University of Melbourne, Melbourne, Australia

Jodie Hunter Massey University, Palmerston North, New Zealand

Penelope Kalogeropoulos Monash University, Melbourne, Australia

Gürcan Kaya Afyon Kocatepe University, Afyonkarahisar, Türkiye

Margaret L. Kern Faculty of Education, The University of Melbourne, Melbourne, Australia

Hee-jeong Kim Korea University, Seoul, Korea

Leena Kim Graduate School of KNUE, Cheongju-si, South Korea

SeungMin Kim Seoul Jungbu District Office of Education, Seoul-si, South Korea

Won Kim Korea University, Seoul, Korea

Huk Yuen LAW Faculty of Education, The Chinese University of Hong Kong, Hong Kong, China

Gima Lee Graduate School of Education, Korea University, Seoul, Korea

Angela Liyanage Monash University, Melbourne, Australia

Gosia Marschall University of Cambridge, Cambridge, UK

Ahmad Muttaqin UIN Prof. K.H. Saifuddin Zuhri Purwokerto, Purwokerto, Indonesia

Nagisa Nakawa Kanto Gakuin University, Yokohama, Japan

Ifada Novikasari UIN Prof. K.H. Saifuddin Zuhri Purwokerto, Purwokerto, Indonesia

Lisa Österling Stockholm University, Stockholm, Sweden

JeongSuk Pang Korea National University of Education, Cheongju-si, South Korea

James Anthony Russo Monash University, Melbourne, Australia

Wee Tiong Seah Faculty of Education, The University of Melbourne, Melbourne, Australia

Isao Shimada Formerly of Nippon Sport Science University, Tokyo, Japan

Ciera Street California Polytechnic State University, San Luis Obispo, California, USA;
Colorado State University, Fort Collins, Colorado, USA

Jan van Driel Faculty of Education, The University of Melbourne, Melbourne, Australia

Miho Yamazaki Teikyo University, Itabashi, Japan

Qiaoping Zhang The Education University of Hong Kong, Hong Kong SAR, China

Ying Zhang Chengdu Qingyang Institute of Education Sciences, Chengdu, China

Juan Zhong Chengdu Jinsha Elementary School, Chengdu, China

Part I
Theoretical and Reflective Perspectives

Chapter 1
An Overview of Values in Mathematics Education

Yüksel Dede, Gosia Marschall, and Philip Clarkson

1.1 Introduction

> To Madame M. V. Kiselyov
>
> Moscow, October 27, 1888
>
> You are right in demanding that an artist should take an intelligent attitude to his work, but you confuse two things: solving a problem and stating a problem correctly. It is only the second that is obligatory for the artist (Chekhov, 1888).

According to Abbott (2014), in the above prose, Chekhov's aim is to emphasise that the purpose of a creative activity of an artist is to intelligently give the 'recipient' (e.g., the reader) something to think about. On par with Chekhov, the critical and creative aim of this book is to help raise important questions about values and valuing in mathematics education, which are to help expand the possibilities of what teachers might do in their classroom practice. While this book does not provide prescriptive solutions for any particular context, specific classroom or social, cultural, ethnic setting, it articulates some of the multiple problems that we face when we take seriously the issues of values in mathematics education and their implications for practice. Sometimes, as educators, we can become rather arrogant in the way we put forward ideas on how to teach best, often assuming that our colleagues in other classrooms need to follow carefully the recipes we devised. But teaching students

Y. Dede (✉)
Gazi University, Ankara, Türkiye
e-mail: ydede@gazi.edu.tr

G. Marschall
University of Cambridge, Cambridge, UK
e-mail: mzm21@cam.ac.uk

P. Clarkson
Australian Catholic University/University of Melbourne, Melbourne, Australia
e-mail: Philip.Clarkson@acu.edu.au

Y. Dede et al. (eds.), *Values and Valuing in Mathematics Education*,
https://doi.org/10.1007/978-981-99-9454-0_1

is far more complex than following a simple recipe. This is why in this book we (the editors and the authors) try to look beyond offering simple recipes for success, and instead grapple with a wider question of what should be at heart of school mathematics teaching and learning.

The field of values in mathematics education takes the view that if students were able to appreciate the relevance of mathematics and how it works, many of them would end up valuing mathematics much more and, as a consequence, would develop a feeling of empowerment through their mathematics education. That would be wonderful for all concerned (the teachers, the students, the wider society). However, as we know, many students never really develop such an understanding of and appreciation for mathematics. Could this mean that the student's own mathematical education lets them down? Although many students across the world study mathematics as a compulsory subject for many years, most leave school with the view that their mathematics adds little to their daily lives (with the exception of some practical notions such as dealing with money or attending to measurements). We contend that, if students were given an opportunity to engage with mathematics through the lens of values, they would be in a better position to grasp the interconnected ideas that lay at the centre of mathematics and its connection to real world (which we see as going beyond mere application of mathematical algorithms), and appreciate powerful ways in which mathematics can enrich people's lives in a myriad of ways.

At International Congress on Mathematical Education [ICME] 5, in 1984, in Adelaide, Australia, Ubiratan D'Ambrosio gave what has become regarded as the key plenary talk of all times. In this talk, he introduced the world-wide mathematics education community to ethnomathematics, in which he included the notion that values should become a key part of what mathematics teaching was about. Reflecting on his 1984 address, in 2004 ICME Congress, D'Ambrosio suggested further that, in part, mathematics education was "to promote citizenship, transmitting values and understanding rights and responsibilities in society" (D'Ambrosio, p. ix, 2006). On both occasions, he was emphasising the important role of societal values in the teaching and learning of mathematics. Several years later, others began to develop the implication of values further, making a distinction between societal and mathematical values, putting forward an argument that both were critical in the teaching and learning of mathematics (Bishop, 1988; Ernest, 1991).

Alan Bishop, one of the pioneers in the values field, was first to flesh out how both mathematical values and societal values play a central role in mathematics education. His attention to values was drawn during the classes run by Benjamin Bloom, during Bishop's master's study at Harvard. At that time, Bloom and colleagues, who had previously completed their book on cognition (Bloom et al., 1956), were about to publish their subsequent volume on affect, which featured values as a pivotal component (Krathwohl et al., 1964). In Bishop's earliest writing, values were seen as a core aspect of the choices mathematics teachers made at critical moments in the ongoing classroom dynamic, when they had to choose as to which teaching move to make next (Bishop & Whitfield, 1972). In the main discussions of these works, Bishop dealt with what has become known as societal values and mathematical pedagogical values. This was swiftly followed by an inclusion of values embedded within mathematics itself (Clarkson, 2008, 2019). Of course, much has happened in research dealing with mathematics values since those early days, and research on values in mathematics education has become much richer and more profound today. Today, values are investigated by researchers worldwide in a wide range of areas, from determining the values prominent (or not) in mathematics classrooms, through to the teachers' and students' values alignment in ways of promoting mathematical wellbeing.

Values and valuing play a key role in many aspects of education, such as assessment, planning, classroom interactions (James & Pedder, 2006), choosing tasks (Levenson, 2022), and general wellbeing (Kim et al., 2020). This means that what one values and finds important in the learning and teaching of mathematics operates within the intersection of all social, cognitive, and affective aspects of school pedagogy (see Allchin, 1999; Seah, 2019a), making values a significant holistic factor in education. In the context of mathematics education, the extent to which students choose to engage (or not) with mathematics frequently depends on their inclination to embrace the curricular trends and expectations (Bishop et al., 2006). Seah (2019a) argues that such student engagement can be achieved by teachers implementing the values and valuing approach in their pedagogy; or, put differently, that an emphasis on explicitly teaching values in the mathematics classroom could help enhance students' meaningful engagement with the subject, consequently extending their learning of mathematics beyond the simple acquisition of competencies.

The growing appreciation for values in the context of education has led to an increased research interest in values and valuing in mathematics teaching and learning. This has resulted in the publishing of an extensive number of professional and research articles, including the theme of a special issue of the 'ZDM-Mathematics Education journal in 2012. These ideas have also been explored at conferences such as the setting up of Discussion Groups in the 2012 and 2016 ICME, and at the 2015 International Group for the Psychology of Mathematics Education [PME] conference. An Invited Lecture at the 2016 ICME was also dedicated to this topic. While many of the earlier contributions in values research appeared to be predominantly exploratory, recently Seah (2019b) signalled that the field has begun to move towards the intervention and application phase. Consequently, a range of ideas have been explored to date, such as fostering mathematical well-being (Clarkson et al., 2010, 2023; Hill et al., 2021) or engagement (Kalogeropoulos, 2016), teacher noticing (Aktaş et al., 2019), and the pursuit of values alignment (Kalogeropoulos & Clarkson, 2019; Kalogeropoulos et al., 2021).

This book, in many ways, can be seen as a follow up volume to an earlier edited book—*Values and Valuing in Mathematics Education: Scanning and Scoping the Territory* (Clarkson et al., 2019). While the first book brought together contributions from a range of authors that explored the territory and field of values in more general terms, this book continues that task, by shifting the emphasis onto implications for practice. This volume brings together authors' writings situated in many different countries. It draws together studies from preschool through primary and secondary schools, through to higher education level. The works presented here draw on a variety of theoretical perspectives; what they all have in common, however, is that all of them have engaged with the idea of how, to a greater or lesser extent, their own work may influence aspects of practice. In some chapters, there is quite clear advice regarding what teachers could do in their interactions with students, with regards to mathematical values learning by their students; in other chapters, which are more concerned with students, historical developments of the area of research, or of theory, authors use a more speculative approach when discussing practice. Regardless of the structure, however, we hope that the work presented across these chapters will influence how we might interact with students, colleagues in schools and tertiary institutions, government officials who develop curricula documents, politicians, and the general public. Consequently, we hope that these values enriched discourses will lead to enabling our students to experience, understand, appreciate and use their mathematics in ways which will impact their lives and help them live with dignity in their own societies.

1.2 Formation of the Book

The aim of this book was to encourage elaborations on potential teaching strategies that could enhance students' and teachers' exploration of their values, when doing mathematics in a variety of contexts and at different educational levels. We hope that this volume may challenge our colleagues (those in schools and beyond) to think more deeply about the impact of their own mathematics values in their professional work, as well as about how articulating such values could deepen not only their own work but also their students' appreciation for these. We also hope that this volume may provide critical and creative ideas for both beginning and experienced mathematics teachers on how to engage with teaching values in their mathematics classrooms effectively. Finally, we hope that this volume may provide education stakeholders (such as researchers, curriculum developers, and policymakers) with a richer understanding of and perspective on the role of values in mathematics education, which might have an impact on richer, more meaningful and holistic ways of conceptualising classrooms and learning across different contexts.

1.3 Chapter Outlines

In our initial call for contributions for this volume we had anticipated responses from colleagues who work in a variety of different contexts across the globe, and who would be keen to engage with a wide range of issues pertaining to values and mathematics education. Among these, we also expected contributions which would deal with specific issues that had already been emphasised in literature as particularly important, such as, for example, mathematical wellbeing. What we, perhaps, did not anticipate at that stage, however, was how broad the wealth of experience and interests in the field actually was, and how challenging it would be for us to structure a coherent classification of the contributions in the publication process of this volume. Having tackled this complex task, we have arrived at the following structure of this volume according to the following subheadings:

- Theoretical and Reflective Perspectives
- Values Alignment and Classroom Practices
- Utilising the Values Perspective in Promoting and Sustaining Student Mathematical Wellbeing
- Applying the Values Perspective to Teaching Problem Solving
- Values and Socio-cultural Contexts

Despite the breadth of the perspectives across the aforementioned sections, however, we can reassure the reader that, in keeping with the goal of shifting from theory to practice, all chapters of the book engage with such a shift, even if only in speculative terms. In what follows, incredibly grateful for the contributions from all authors, we are excited to present the reader with a rich and creative landscape of innovative research in values in mathematics education.

1.3.1 Theoretical and Reflective Perspectives

We start this volume with chapters which give a general perspective on value studies in mathematics education, from historical, philosophical and theoretical viewpoints.

In Chap. 2, Clarkson, sitting in his armchair and reflecting back on his own practice and on the field more generally, provides some historical accounts of the field development, while pointing to gaps and opportunities (both those that were missed and those that still arise) in the value-focused teaching and learning of mathematics. In that, he shows us that, despite rigidity and prescriptive nature of curricular designs, we can (almost) always find a wriggle room to privilege students' choosing of particular-value enactment; choosing which would consequently allow mathematical learning to flourish. In that, considering values and related pedagogies, Clarkson pays particular attention to the value of *mystery,* which he is concerned is often forgotten, or under played, in the school context of mathematics learning.

Similarly to Clarkson, who briefly alludes to an issue of conceptualisation of values in the field of mathematics education, taking a comprehensive view (in Chap. 3), Chia and Zhang draw our attention to this conceptualisation, across studies in the recent decade. Following the synthesis of 67 studies, they conclude that values definitions and conceptualisations fall broadly into cognitive, discursive and enactive perspectives—an aspect that points to the vast complexity of the concept, and one which has implications for our engagement with it in both practice and further research. In Chap. 4, Barwell and colleagues, extend the aspect of conceptualisation further, pointing towards values' orientations, which they divide into those that are individualist or collectivist. Although this conceptualisation, as they argue, might only contribute further to emphasising the complexity of the concept, it allows us to capture and think about aspects in mathematics education, which are crucial in way of helping students understand and cope with social and ecological challenges of today's complex world. This might lead us to think about student active engagement with and meaning making about the mathematics that they learn and about world that they live in. And it is exactly the aspect of meaning making in the process of mathematics learning that Law touches upon in Chap. 5, emphasising that it is an aspect which is not only to be valued by teachers in mathematics classrooms, but the pursuit of which should necessarily be seen by teachers as their ethical responsibility towards students and their learning.

1.3.2 Values Alignment and Classroom Practices

Classroom practices that utilise approaches focusing on values alignment have been of interest since the field of values started to catch traction. The chapters in this section emphasize the importance of its exploration in five different cultural contexts: Chaps. 6 and 10 provide some results for applying values alignment in mathematics classrooms; Chaps. 7, 8, and 9 attend to an assessment of and provide suggestions for determining the current situation in values alignment in mathematics classrooms.

In this sense, Aktaş, in the Turkish context (Chap. 6), draws attention to the complex and interactive nature of inclusive education pedagogies and processes as well as general classroom practices; she proposes a practical professional development model for how mathematics teachers working in such inclusively educated mathematics classrooms can implement values alignment practices and pedagogies. In Chap. 10, Azura Abdullah examines the dominant values in Brunei mathematics Lesson Study classrooms, based on the framework proposed by Bishop (1988), emphasizes the importance of understanding these values, and reveals the possible factors underlying these values to improve mathematics education.

On the other hand, Chaps. 7, 8, and 9, all of which make use of a Values Alignment Study [VAS] questionnaire, provide valuable insights into what values alignment looks like and how it works in mathematics classrooms in different cultural contexts. These chapters reveal that students' expectations from mathematics teachers, in mathematics classrooms in different cultural contexts, are generally similar. Specifically, in Chap. 7, Pang and colleagues show that Korean students' values generally focus on problem, understanding, and review, as well as that students' values regarding learning mathematics overlap with perceived teacher values regarding teaching mathematics. Similarly, in Chap. 8, Kalogeropoulos and colleagues also reveal some similarities between student and teacher values, while at the same time concluding that Australian students expect their teachers to emphasize the process-oriented aspects of learning mathematics, as much as reasoning and collaboration. This expectation of Australian students coincides with the values of problem, understanding, and review that Korean students attach importance to when learning mathematics. In Chap. 9, Davis and colleagues compare the values of primary school students in two different cultural contexts (Australia and Ghana) and determine that students in both countries generally value mathematics positively. Another perhaps more striking emphasis in this chapter is that, despite the stereotype that girls are less interested in mathematics than boys and do not want to pursue a job that involves mathematics, Ghanaian girls want to work in a position that requires more mathematics than Ghanaian boys.

1.3.3 Utilising the Values Perspective in Promoting and Sustaining Student Mathematical Wellbeing

The interaction between values and student wellbeing has been an issue of study for more than ten years. More specifically, one way of conceptualising student mathematical wellbeing has been through the lens of value alignment. The chapters in this section engage with these aspects, pointing to different findings in relation to student's valuing and mathematical wellbeing across three different cultural contexts. For example, in Chap. 11, Kim and colleagues illustrate Korean students' valuing of cognitive aspects of learning, which revolve around computational thinking (for younger students) and around perseverance and a growth mindset (for older students), while regarding positive emotions such as fun, excitement, interest, and happiness to be of less value. In contrast, in Chap. 13, Hill and colleagues illustrate how Australian Year 8 students value both the cognitive challenges and the positive emotions in mathematic learning, although fulfilling of these values was not achieved to the same degree (with students being least successful in fulfilling their ultimate values related to positive emotions). In addition to these, however, the Australian students in the study of Hill and colleagues, indicate high value in relationships, focusing slightly more on social interactions than the Korean students in the study of Kim and colleagues. These findings reflect both studies' concerns about the current state of affairs in students' sense of fulfilment and happiness in mathematics classrooms, recognising that the aspect of inability to fulfil specific values hinders students' mathematical wellbeing.

A similar aspect is emphasised by Zhong and colleagues (Chap. 12), who affirm the necessity for students to fulfil all seven values (discussed in Hill's framework) in order to achieve mathematical wellbeing. Having established that, the authors then reach beyond the exploration of value fulfilment and student mathematical wellbeing, and elaborate on practices which teachers can employ to help foster such wellbeing.

1.3.4 Applying the Values Perspective to Teaching Problem Solving

Although problem-solving has been a topic in mathematics teaching and learning for decades, it has only recently been linked explicitly to exploring values in mathematics. The chapters here offer several ideas that advance this connection. In this regard, in Chap. 14, Baba and his colleagues assess socially open-ended problems, which they bring in from their previous studies (see Baba, 2007; Baba & Shimada, 2019), analysing normative and socio-critical mathematical modelling tasks against Bishop's (1988) six universal activities. They propose an important and useful link between the context of mathematical modelling, mathematical problem solving, and values in mathematics education. In Chap. 15, Yamazaki compares the mathematical values of students studying the same Japanese mathematics curriculum in Japan and Australia in the context of problem-solving. This comparison reveals that even though students in both countries study the same curriculum, their choices in the problem-solving process and the values underlying these choices differ depending on culture.

1.3.5 Values and Socio-cultural Contexts

Socio-cultural contexts influence education in general and mathematics teaching in particular. The chapters in this section further explore these contexts on mathematics teaching/learning through mathematics values.

The context for the Chap. 16 is pre-school children in Japan. Here, Nakawa sets out to explore social values that the children act on when engaged in mathematical activities. Interestingly, a number of these social values are precursors to mathematical values (e.g., fairness, rule-making), and pedagogical mathematical values, such as 'sympathy for others', which will hopefully help with working cooperatively in groups in school years. In Chap. 17, Novikasari and colleagues also examine social values, but this time in the context of how the different values are embedded among three ethnic groups of Indonesia. The university preservice teaching students were asked to develop mathematical modules of work that drew on their social values that were foundational to their own ethnic group.

Culturally speaking, it can be much easier for values to be shared when the teacher and students come from similar backgrounds, and when the classroom is located within their specific space. If this is not the case, educational context poses challenges to learners. This is an issue which street touches upon in Chap. 19, where she examines the experience of two women of colour on their undergraduate mathematics studies in the United States of America, which was highly gendered and racialized. In her analysis of these experience, Street suggests five areas of values misalignment based on sociohistorical Western mathematical values (meaning of mathematics, mathematics as fast-paced, innateness of mathematical ability, mathematics is competitive, and mathematics as a gatekeeper), and concludes with discussion of possible change in practice that might lead to support more equitable mathematics environments for women of colour.

Facing a similar issue, in Chap. 18, Hunter studies a classroom located in New Zealand, but one run by a 'European New Zealand' teacher and with a student group which includes a mix of Pacific Islanders. In this study, she illustrates how the teacher succeeds in making the students feel positively in her classroom environment by using a culturally sustaining mathematics pedagogy, based on prompts and actions that make implicit and explicit connections to mathematics educational values including family, respect, collaboration, reciprocity, inclusion, and belonging (all crucial values embedded in Pacific Island cultures).

Over the last 10 years, more than 20 countries have participated in the international *What I Find Important (in my mathematics learning) [WIFI]*[1] survey. Although several chapters in other sections of this book touch upon the WIFI study, the final two chapters in this section deal with the WIFI results explicitly, focusing particularly on 'the magic pill' item from Section C of the survey.

In Chap. 20, set in Turkey, Dede and colleagues examine the results from the WIFI survey of grade 9 students who reported that they 'felt good when doing mathematics'. Here, as the authors report, the Turkish students privilege ability, effort, mathematics concepts, fun, teacher, and materials that enhance thinking. In Chap. 21, Österling and colleagues, study the idea of the 'magic pill' with Swedish students from Grade 5 and 9. Although these students seem to privilege similar aspects, such as ability, numbers, calculation processes and being fast, a number of students also include ingredients such as 'unicorns' in their magic pill. This makes the authors wonder whether there should be a change in teaching practice, one which reduces the emphasis on exclusionary practices, such as encouraging students to complete their mathematical problems 'fast', but includes "an ounce of joy and happiness" (in one of the participant's words) in their teaching.

References

Abbott, N. (2014). *Writing advice from Anton Chekhov: Part Three—Ambiguity.* https://nealabbott.wordpress.com/2014/08/25/writing-advice-from-anton-chekhov-part-three-ambiguity/

Aktaş, F. N., Yakıcı-Topbas, E. S., & Dede, Y. (2019). The elementary mathematics teacher's values underlying teacher noticing: The context of polygons. In P. Clarkson, W. T. Seah, & J. S. Pang (Eds.), *Values and valuing in mathematics education: Scanning and scoping the territory* (pp. 209–222). Springer.

Allchin, D. (1999). Values in science: An educational perspective. *Science and Education, 8*, 1–12.

Baba, T. & Shimada, I. (2019) Socially open-ended problems for enriching student learning with mathematical models and social values. In P. Clarkson, W. T. Seah & J. Pang (Eds.), *Values and valuing in mathematics education: Scanning and scoping the territory.* Springer Nature Switzerland AG.

Baba, T. (2007). Japanese education and lesson study: An overview. In M. Isoda, M. Stephens, Y. Ohara, & T. Miyakawa (Eds.), *Japanese lesson study in mathematics: Its impact, diversity and potential for educational improvement.* World Scientific Publishing.

Bishop, A. J. (1988). *Mathematical enculturation: A cultural perspective on mathematics education.* Kluwer.

Bishop, A. J., & Whitfield, R. C. (1972). *Situations in teaching.* McGraw-Hill.

Bishop, A. J., Clarke, B., Corrigan, D., & Gunstone, D. (2006). Values in mathematics and science education: Researcher's and teacher's views on the similarities and differences. *For the Learning of Mathematics, 26*(1), 7–11.

---

[1] Information on this survey can be found at https://thirdwavelab.education.unimelb.edu.au/

Bloom, B. S., Engelhart, M. D., Furst, E. J., Hill, W. H., & Krathwohl, D. R. (1956). *Taxonomy of educational objectives: The classification of educational goals* (Vol. I: Cognitive Domain). David McKay Company.

Chekhov, A. (1888). *Letters on art and artists.* https://www.anthologiablog.com/post/chekhov-letters

Clarkson, P. (2019). A conversation with Alan Bishop. In P. Clarkson, W. T. Seah, & J. Pang (Eds.), *Values and valuing in mathematics education: Scanning and scoping the territory* (pp. 11–24). Springer.

Clarkson, P., Bishop, A., & Seah, W. T. (2010). Mathematics education and student values: The cultivation of mathematical well-being. In T. Lovat & R. Toomey (Eds.), *International handbook on values education and student well-being* (pp. 111–136). Springer.

Clarkson, P., Bishop, A., & Seah, W. T. (2023). Mathematics education and student values: The cultivation of mathematical well-being revisited. In T. Lovat & R. Toomey (Eds.), *International handbook on values education and student well-being* (2nd edn). Springer.

Clarkson, P. (2008). In conversation with Alan Bishop. In P. Clarkson & N. Presmeg (Eds.), *Critical issues in mathematics education: Major contributions of Alan Bishop* (pp. 13–28, 69–70, 107, 149–150, 189–190, 229–230). Springer.

Clarkson, P., Seah, W.T., & Pang, J. (Eds.) (2019). *Values and valuing in mathematics education: Scanning and scoping the territory.* Springer.

D'Ambrosio, U. (2006). Preface. In F. Favilli (Ed.), *Ethnomathematics and mathematics education,* (pp. v–x). Tipografia Editrice Pisana.

de Abreu, G., Bishop, A. J., & Presmeg, N. C. (Eds.). (2002). *Transitions between contexts of mathematical practices.* Kluwer.

Ernest, P. (1991). *The philosophy of mathematics education.* The Falmer Press.

Hill, J. L., Kern, M. L., Seah, W. T., & van Driel, J. (2021). Feeling good and functioning well in mathematics education: Exploring student's conceptions of mathematical well-being and values. *ECNU Review of Education, 4*(2), 349–375.

James, M., & Pedder, D. (2006). Beyond method: Assessment and learning practices and values. *The Curriculum Journal, 17*(2), 109–138.

Kalogeropoulos, P., & Clarkson, P. (2019). The role of value alignment in levels of engagement of mathematics learning. In P. Clarkson, W. T. Seah, & J. S. Pang (Eds.), *Values and valuing in mathematics education: Scanning and scoping the territory* (pp. 115–127). Springer.

Kalogeropoulos, P., Russo, J. A., & Clarkson, P. (2021). Exploring educator values alignment strategies in an intervention context: The emergence of the beacon strategy. *ECNU Review of Education, 4*(2), 327–348.

Kalogeropoulos, P. (2016). *The role of value alignment in engagement and (dis)engagement in mathematics learning.* (PhD), Monash University, Australia.

Kim, W. J., Trung, N. X., Hung, L., & Trung, N. N. (2020). Relationship between cultural values and well-being: Analysis from some East Asian countries. *Journal for Cultural Research, 24*(4), 334–350.

Krathwohl, D. R., Bloom, B. S., & Masia, B. B. (1964). *Taxonomy of educational objectives: The classification of educational goals* (Vol. Handbook II: The Affective Domain). David McKay Company.

Levenson, E. S. (2022). Exploring the relationship between teacher's values and their choice of tasks: The case of occasioning mathematical creativity. *Educational Studies in Mathematics, 109*, 469–489.

Seah, W. T. (2019a). A commentary: Extending the research discourse from interests and beliefs to values. In M. S. Hannula, G. C. Leder, F. Morselli, M. Vollstedt, & Q. Zhang (Eds.), *Affect and mathematics education: Fresh perspectives on motivation, engagement, and identity* (pp. 219–230). Springer.

Seah, W. T. (2019b). Values in mathematics education: Its conative nature, and how it can be developed. *Research in Mathematics Education, 22*(2), 99–121.

Chapter 2
Mystery: One of Six Mathematics Values

Philip Clarkson

2.1 Introduction

I was surprised by how the well-respected science educator Jonathan Osborne began his presidential address at the 2007 National Association of Research in Science Teaching (NARST) conference by paraphrasing Samuel Coleridge's famous poem *The Rime of the Ancient Mariner;* "Science education to me is, 'data, data everywhere and not a thought to think'" (quoted by Carter, 2020). Osborne continued that we should spend "a bit more time in our armchairs, more time picking over and thinking about what we do … to develop better theories about our *goals* and *values* in science education." In this chapter I want to get into my armchair using some critical self-reflection notions on my own practice of teaching and research (Brandenburg et al., 2007; Fook, 2007), putting to one side the mountain of data we have in how we might tweak the teaching of fractions, or develop the notions of multiplicativity, or (well there are just so many other topics I could choose), so students perform at some higher level in assessment driven curricula. Rather I want to self-reflect more deeply about one foundational aspect of mathematics, *mathematics*[1] *values*. How might these help us gain more insight as to why and how we might teach mathematics to children for the rest of this century so that they gain a deep understanding of this key cultural artifact, mathematics. In turn, they may be able to participate insightfully and with dignity in their future society and hopefully help it avoid the possible devastation into which we are at present looking.

[1] Bishop (1988) refers to six *mathematical values* in his seminal book. However, in this chapter I prefer to use the term *mathematics values* to emphasize that these six values form part of the core of mathematics. In much of the later literature the term mathematical values has more often been used to refer to pedagogical values when teaching mathematics. In this chapter I want to make a clear distinction between the two.

P. Clarkson (✉)
Australian Catholic University/University of Melbourne, Melbourne, Australia
e-mail: Philip.Clarkson@acu.edu.au

Y. Dede et al. (eds.), *Values and Valuing in Mathematics Education*,
https://doi.org/10.1007/978-981-99-9454-0_2

Of course, in taking this approach of self-reflection in and of itself brings various limitations to my argument. Many of the various anecdotes and experiences I draw on are clearly personal and hence cannot be claimed to be those of a defined group of educators. However, I suspect many colleagues will nod in agreement with at least some of what I describe as experiences they too have lived. But taken as a whole, I believe this narrative has an argument that raises questions which are worthy of consideration.

People want certainty, it makes them feel safe. Mathematics is assumed by most to give straight forward unequivocal answers; the answer is either right or wrong and there is no grey area between. But this is a misreading of mathematics, although it has often built its public reputation on this (mis)understanding. Mathematics values help us see why certainty is only one of several different aspects of mathematics important to consider. Indeed, one of the core mathematics values identified by Bishop (1988) was 'rationalism', which does speak of certainty. But there are another five values one of which is mystery. 'Mystery' does not have such an air of certitude about it. Mystery suggests that maybe there is something at a deep level in mathematics, which is unfathomable, that is only grasped at but never really caught hold of. That, at times, there might not just be one right answer. Maybe there are times when an unequivocal answer is just not available. And with that, sometimes certainty needs to be held at bay and not gasped as if it is the totality. There is also 'progress', which in Bishop's argument means in part that a core aspect of mathematics is looking for ways of bridging out to new unknown mathematical ideas; or 'openness' with which he suggested that mathematical ideas are always negotiable and need to be argued for and justified, not just taken as undisputed facts.

These values do not directly reflect the mathematics that is normally taught in school. And yet, it is these values, and in particular the mystery value, that Bishop identified as making all the difference to an appreciation of just what doing mathematics indeed is or can be. The network that forms mathematics is in part held together by mathematics values. This is vastly different to learning off by rote disconnected discreet 'facts and figures' with little attempt at introducing students to its network of ideas, as so often happens in much of mathematics teaching in schools. It maybe that a re-examination of mathematics values may be one way in which the western mathematics[2] curriculum can be reordered for the rest of this century. In doing so maybe the way some of the big societal issues to which mathematics can give some deep insights might be discussed in the public domain more thoughtfully and incisively.

I start this chapter by thinking about Bishop's mathematics values, their lack of prominence in my own past school teaching, and in our curriculum documents, before looking at some teaching possibilities that might bring mathematics values into everyday teaching practice. The chapter then moves to a lengthy consideration of the mystery value. I turn my attention there since there is a lack of research interest

---

[2] Although Bishop (1988) argued there were six activities that characterised mathematics in all cultures, he was far more circumspect when it came to values embedded in mathematics only claiming the six he identified as pertaining to western mathematics. This point is explored more in Paraide et al. (2022).

by-and-large not only in the six mathematics values but particularly in mystery. Following this, I develop some teaching ideas that might be useful in explicitly exploring the value of mystery with students. The chapter ends by reiterating that all six values should be a critical component in rethinking a framework that guides the development of school mathematics. And that such rethinking is needed if present day students, when they reach adulthood, are to more deeply appreciate the crucial role that mathematics plays in dealing with the multiple critical societal issues they will face.

2.2 Bishop's Six Mathematics Values

One key source document that informed Bishop's work on mathematics values was Raths, Harmin and Simon (1987). These authors suggested that 'valuing' could be thought of as seven sequential interlinking processes: (1) choosing freely, (2) choosing from alternatives, (3) choosing after thoughtful consideration of the consequences of each alternative, (4) prizing and cherishing, (5) affirming, (6) acting upon choices, and (7) repeating.

It is interesting that the first three points in this sequence are about choosing. That implies that students have some knowledge of the possibilities of choosing to enact specific values, and agency in making their choice, not just accepting their teacher's choices. This implies that students have thought, at least to some extent, about why such a choice can be made. I am not sure that many traditional mathematics classrooms offer a context in which such 'choosing' could be privileged. But, as I argue below, if teachers wish to prioritize 'students' choosing', there is nearly always some 'wriggle room' in the official curriculum to enable this, if it is looked for.

Moving to the last four points in Raths et al. (1987) model, assuming the choosing process of values is possible by students in their classroom context, the teacher needs to be aware of when a student is choosing to enact a specific value, and when this choice is repeated (point 7). Repeating to choose to enact a value probably points to the student prizing and cherishing a value of their choice (point 4). One obvious strategy for the teacher is to explicitly affirm such choices. That will also help with point 5, although in time it is hoped that students will not need the external affirmation. Point 6 will need teaching strategies that allow students to enact their choice of values (see later section on possible teaching strategies).

In all the above at some point students will hopefully come to an understanding that their choices of how they think about mathematics includes mathematics values. Hopefully they will come to explicitly understand just what mathematics values they are choosing to embrace and act on when doing mathematics.

It was considering Rath et al. (1987) seven processes, particularly the last four, that led me to think of valuing as 'beliefs in action' (Clarkson & Bishop, 2000). This notion also stemmed from our wish to clearly distinguish valuing from beliefs. We were struck by our earlier examination of the literature in preparing for the

VAMP[3] project that results from mathematics research literature examining beliefs were not consistent. More specifically, many pure Likert survey-based studies that we examined at the time seemed to sometimes suggest association between students' beliefs and their achievement, but at other times suggested none. However, at the same time, studies which involved some type of observation, or asked students or teachers what they would actually do in classroom situations, suggested some type of direct associations between beliefs and mathematics achievement. The key point seemed to be that once behaviour, either directly observed or specifically inferred, was part of a study then the association became more definite and implied that something more than belief was present. We aligned this distinction to a difference between belief and values. In other words, it seemed that valuing, as distinct from belief, did involve some actual behaviour; and that seemed to be an important distinction to us. Students are very attentive to what teachers and their peers actually do, much more so than what they say. Using this insight, which involves the teacher's and fellow students' behaviours, values become a powerful factor in understanding classroom dynamics.

Considered together, Bishop's six values give a picture of what could be seen as part of the core of school mathematics. Various descriptions for the six values can be found in the literature. Most predominant descriptions that Bishop (1988, 1991, 2001) and others have used include:

> *Rationalism* involves reason or logical thinking. Hence this value concerns the relationship between ideas and is related to explaining, arguing, showing and using mathematical proof.
>
> *Objectism* (not objectivism as sometimes seen in the literature) consists of concretising or materialism. This relates to the genesis of mathematical ideas and the naming of those ideas. Hence it involves teachers and students showing others their working using diagrams and / or learning materials to concretise mathematical ideas.
>
> *Control* consists of rules or mastery. Being able to describe phenomena in mathematical terms and then checking these, as if they were objects, against logical arguments, rules and laws. Hence understanding the process of routine calculations and then checking and justifying answers are included.
>
> *Progress* includes generalisation or questioning. Thus, searches for the unknown and making it known in mathematical terms are included here. Applying known solution strategies to novel situations and exploring ideas beyond given examples are aspects of this value.
>
> *Openness* involves demonstration of facts. So, claims that mathematical ideas are always open for dispute, verification, or falsification, gives meaning to openness. This is not just opinion since such disputation must use the processes of mathematics. This can lead to students presenting and defending mathematical ideas publicly.
>
> *Mystery* suggests there are aspects of the unknown at the heart of mathematics, which can be either negative or positive. Although much mathematics is used and is invented in specific contexts, we do not always know how other mathematical aspects arose or whether they were invented by one person at a specific place and time, or whether there were multiple moments of realisation (e.g., Pythagoras Theorem). This also implies that there is always more mathematics to develop, starting from mathematics that is known or developing mathematics to deal with non-mathematical situations. But there is no certainty in this process. There is

[3] Values and Mathematics Project (VAMP) was a project that Bishop and Clarkson developed in the mid 1990s that ran until the early 2000s. Wee Tiong Seah became a researcher for us on this project from an early point in its life.

no guarantee that a mathematical solution can be developed, or a new line of exploration will be fruitful. Mystery also suggests exploration of one's mathematical imagination on the wonder, beauty, and awe of mathematical ideas, which too is a core aspect of mathematics. This implies that to appreciate such wonder you need to have some understanding of the breadth and depth of the notions and their networked connectivity that takes you beyond the normal into places you least expect.

Care needs to be exercised in using some of the value terms such as control. In this discussion this value refers to control that is gained through mathematics. Likewise progress and openness here are described as mathematics value processes. At times in conference discussions, I have wondered whether colleagues have gradually slipped into the everyday sense of these terms. For example, they seem at times to be discussing classroom control pertaining to students' general behaviour in the classroom, rather than ideas emanating and centred on their interaction with mathematical ideas. Openness here is thought of in terms of how mathematics is open and the need for verification, not just having an 'open discussion' with a colleague when all sorts of ideas and opinions are shared, few being actually based in mathematics at all.

2.3 Reflecting on My Mathematics School Teaching

Recently, when cleaning out archive boxes of past teaching materials, I happened across a selection of year 10 and 11 mathematics examination papers that I devised in my first couple of years of teaching in secondary schools. In reading through them, apart from wondering how to do some of the problems that I had taught with such ease many years ago (fortunately I had also included in the folder my answer sheets), I noticed that virtually all items were manipulation of symbols with the few 'written' items having little resemblance to the students' own lives. Apart from throwing dice for playing various board games (this was a long time before games on screens), others dealt with the cost to councils of constructing footpaths and roadways (students do not pay council rates), and the cost of car trips for holidays (except no students would have a car licence for some years and few went on holidays given the low SES profile of the school in which I taught). But one item in particular caught my attention: "A tall tower stands at one corner of a horizontal rectangular field. The angles of elevation ..." This must have been copied and then copied again from some English source. We do not talk about 'fields' in Australia, we talk about paddocks. And why would you have a tower in a corner of a paddock? And having a rectangular paddock; very few are actually rectangular in Australia, nor are many horizontal. Maybe the reference was to a rugby or soccer field; in Australia that would be a pitch and few, if any, students played those games in the schools in which I taught. I doubt whether my exam papers were very different to most of the mathematics teachers of the time. I doubt whether my students started to think about the deeper connections within mathematics and what held those connections in place if doing mathematics to them was on the whole manipulating symbols, and the few non symbolic problems they worked on did not reflect their lived experience out of school.

Looking back, I hope that, in Hattie's terms at least, I was examining what he terms as necessary 'surface learning' (Hattie et al., 2016). In reality, I think in these examination papers I was interested in whether the students were able to repeat what we had completed together in class. Certainly, the questions asked for knowledge of the skills and concepts covered. Interestingly, the instruction on all papers was: 'Marks may be deducted for badly arranged work, bad writing and the omission of essential working from answers.' This perhaps implies that I was asking for students to make visible their working and hence making visible the underlying rationalism and openness of the process, in the sense of two of Bishop's (1988) values. Although at the time I had no deep knowledge of these, maybe just a hazy feeling that something like that might be important. But really, I suspect I was simply asking them to repeat what they had memorized from my chalkboard work (no electronic whiteboards then), and what they had copied into their own books. In these first couple of years of teaching, although I was growing more and more dissatisfied with what and how I was teaching, which led to my involvement in the Rusden Activity Mathematics Project (RAMP) (Clarkson, 1979; Clarkson & Henry, 1976; Horne et al., 1977), I was years away from a deeper understanding of what might be needed for my students to gain a deep understanding of mathematics.

2.4 Mathematics Values in Curriculum Documents

An examination of our (state of Victoria, Australia) current curriculum documents shows the lack of recognition of mathematics values. The document begins with an Overview statement including the Aims of the mathematics curriculum. The Aims state in part that the curriculum will help develop useful "mathematical and numeracy skills for everyday life and work, (but also will develop students) as active and critical citizens in a technological world" (Victorian Curriculum and Assessment Authority, 2016, p. 4). Then follow the various statements introducing the Strands of the curriculum, such as Number and Algebra, and Statistics and Probability, etc. Each Strand has an introductory cover statement. The first sentence(s) of each reads as if a pure mathematics topic will follow. Any reference to real life situations for that Strand is mentioned only in the last sentence of each statement. There is nothing in any of these introductory statements alluding to mathematics as an aspect of culture, of its history, fundamental undergirding concepts, philosophy, aesthetics, or values, which reflect what is in the Aims. Following the Strand introductions, the main body of the document, the Scope and Sequence, is laid out. This is the part of the document to which teachers look to see what and how they are to teach. Again, in this part of the document nothing is found explicitly reflecting the Aims. Clearly teachers are to make what connections they can.

The disconnect between Aims and the rest of the document is one I noted at a conference back in 1998 when analysing a much earlier version of the curriculum documents (Clarkson & Bishop, 2000). It transpires that little has changed. Yet for students to become critical citizens, they need to understand the deep values embedded in mathematics. They might be able to complete the many mathematical processes that are subsequentially listed in the Scope and Sequence, which form the bulk of the PISA and TIMMS assessment instruments. But such processes are not the whole of mathematics. How these processes interact with cultural ideas alluded to in the Aims, and which are at least in part a matter of values, also needs to be at the heart of school mathematics. No wonder adults in our community continue to remember their school mathematics so negatively and hence see little relevance for it in their daily lives let alone in deciding big societal issues (Clarkson et al., 2001).

2.5 Teaching Mathematics Values

2.5.1 Key Points in Teaching Mathematics Values

It seems to me we have an obligation to understand and then teach foundational ideas of mathematics as well as the processes and techniques. Without the foundational ideas there will be no sound or deep understanding of the subject matter. Foundational ideas form the critical and central nodes around which students build their webs of interconnecting ideas networks. Teaching should also be envisaged as enabling students to appreciate the full scope of possible linkages within their idea webs and networks that they have created. Values are always embedded within such networks and indeed in the linkages, although in traditional views of mathematics teaching, there would be probably an exclusive emphasis on content making up the crucial nodes, and often little emphasis on linkages at all.

Although traditional mathematics teaching has emphasized content and cognition, to some extent rationalism, and maybe control, are inevitably embedded in such teaching either explicitly or more often implicitly. But it is rare to find examples of a focus on the value of mystery in school mathematics. The other three values do sometimes appear. So, a critical issue becomes how do values become an explicit focus for teachers when they are planning their mathematics teaching.

Bishop himself considered briefly what might be seen as useful teaching approaches for the six values he described (Bishop, 1988, 1991). He thought that rationalism and objectism dominated the school curriculum, although often implicitly. He also thought it would be much better if these underlying unifying values were made explicit and hence students (and many teachers) could understand more clearly why and how the 'bits' of mathematics connect. Bishop argued that to make these six values explicit, much more discussion was needed in mathematics classrooms. He argued that the incorporation of small group-work facilitates discussion. Moreover, teaching that used projects, investigations and problem-solving opens up the opportunities for discussion between students and hence the opportunities for teachers to explicitly asked questions that seek to direct discussion and thinking about mathematics values. These strategies also often allow for an envelope of solutions or better answers to be found for a given problem or investigation, which is also an aspect of real-life mathematics that should not be ignored. We explored these and more teaching options including body gestures in Bishop and Clarkson (1998).

Other projects have also suggested important ideas for worthwhile teaching strategies of values. Bereznicki et al. (2008) identified some teaching strategies for general societal values that are worth considering.[4] They suggested that values teaching needed to become part of a teachers' normal practice, and never be seen as an extra, different, or an 'add-on' in teaching. They advocated the need to develop an appropriate metalanguage for creative discussions on values between students and teachers in class. However, learning such a language is not always easy. Using in-school professional learning programs for teachers to learn such a language would be needed for this. Other strategies, not new in themselves but useful when considering the teaching of values include "a range of cooperative learning strategies, critical thinking skills, reflective journal writing, narrative, decision-making, problem-solving skills, and explicit discussion about values (and active) listening" (Bereznicki et al., p. 61).

Another interesting possibility discussed was using the strategies and skills captured by a 'storyfest' (Bereznicki et al., p. 73), which for us would occur within a mathematical context. The notion of 'storyfest' sits well with the idea that students can learn better when ideas are presented, interrogated, and elaborated using narrative forms, both vocalized and written. This notion has been explored in the humanities but rarely with mathematics. The danger here might be to have students only writing and talking about peripheral and trivial aspects of mathematics rather than dealing with core ideas and what these might mean (including what feelings and emotions they may bring to the surface while students engage with the embedded mathematics of the story). The stories associated with Pythagoras and his times is just one possibility (a good starting resource for this is Ferguson, 2011). But investigating and writing about where the theorem led mathematics and science (see perhaps Mlodinow (2002) who starts with Euclid but pivots back to Pythagoras) could be useful. Moreover, there are also other interpretations, not just historical which the above two references mainly are, which could be taken. For example, using Wertheim (1995) as a basis would be a different place to start. These three references take a decided western view of this pivotal theory. But Chinese and Korean formulations started historically from quite a different position. Exploring that and contrasting it with the western story would allow a far more nuanced narrative to emerge (see Park, 2000); important I would think in the multicultural classes that predominate in most Australian classrooms. Here then are some possibilities for students' creating narratives from various viewpoints of a critical and seminal pivot within mathematics, incorporating some of the fundamental mathematics that is crucial to this story. Alternatively exploring

[4] Key researchers involved in these projects later published a seminal handbook in 2010 on values in education and a second edition in 2023 (Lovat et al., 2023) including a chapter in each on mathematics (Clarkson et al, 2023). They also edited a second book exploring the place of values in teacher pre-service education programs (Toomey et al, 2010), which also included a chapter on mathematics (Clarkson et al., 2010). A key idea in that chapter was there was little emphasis, if any, in pre-service education programs on teaching values.

the mathematics embedded in mythical or fairy stories and rewriting them is another possibility (Clarkson, 2006 and for real life projects Clarkson, 2010).

Some data generating techniques used in various research studies investigating values in mathematics teaching could be repurposed as teaching strategies. For example, the pre- and post-discussions with teachers after a classroom observation concerning the possible (pre) or actual (post) observed behaviour that might point to different valuing by students (from the VAMP project), and discussions centred on answers from open-ended items (from the WiFi[5] project) could be explored. Role plays, such as those mentioned below, could be reconfigured to bring explicit discussion of values into classrooms. The strategies identified and classified by Kalogeropoulos and Clarkson (2019) when teaching values, scaffolding, balancing, intervention and refuge, in mathematics classes are also important (also Kalogeropoulos et al., 2020).

Within any teaching strategy, however, language is important. The need for an appropriate and common language that can be shared between teacher and students, and between students to tease out the ideas of mathematics values was identified in the 1990s during the VAMP project (Clarkson et al., 2000). Unfortunately, not much work on language in this area has occurred since and yet it seems that if useful teaching strategies for mathematics values are to be developed then language must be seen as an important element of the process. But this is not an easy issue to grasp. In the bilingual education literature, Cummins developed the notion of BICS (Basic Interpersonal Communicative Skills) competency in students' everyday forms of their languages (in their home language, and in whatever the language of teaching was), and competency in CALP (Cognitive Academic Language Proficiency) the academic languages they needed in schools (again both in their home language and the language of teaching) (Clarkson & Carter, 2017; Cummins, 2008). For bilingual students, once competence is gained in CALP, both in their home language and in the language of teaching, school performance is enhanced. It seems it is the interplay between their languages that promotes the deeper understanding and higher performance, and this is so even in mathematics (Clarkson, 2007). I wonder whether there could be a similar process in play when the learning of mathematics values is the focus. Using students' everyday language to introduce different values intuitively, before moving to use a more sophisticated type of language to explore and deepen the discussion of mathematics values over time might be advised for teaching strategies (see Clarkson, 2009 for a model for bilingual students that might be reconfigured with values as the target).

2.5.2 *Two Schools Considering Teaching Mathematics Values*

Recently I was advising two secondary schools as they sought to revitalize their mathematics teaching. The first school focused on designing a continuous developmental

[5] Information on this survey can be found at https://thirdwavelab.education.unimelb.edu.au/

rubric from year 7 through to year 10. Their work revolved around examining the Scope and Sequence segment of the curriculum (Victorian Curriculum and Assessment Authority, 2016). Initially they were not interested in the Overview, Aims or Strand descriptions (see above). My initial interaction with the group was to remind them of what the Overview and Aims said, and extending this to a discussion of mathematics values and mathematical language. However, staff did not initially develop cells within the first version of their rubric that suggested appropriate language or mathematics/pedagogical values at different points, but they planned to do this in the next iteration. The purpose of the rubric was not just for teachers' planning purposes but was to also be used with students. Sections of the rubric would be pasted into students' workbooks so teachers could more easily initiate conversations focused on the rubric. In time it was hoped that interactions between students would also use the rubric to structure conversations, which would include conversations on their own growth and on what was possible. This notion of a teaching group developing their own planning document, but also a document to be given to students and used by teachers and students to initiate discussion, potentially incorporating how and when mathematics values could be emphasized, seems to me to be a very useful approach.

In the second school, staff decided on a project that involved exploring the possibilities of focussing on affect for both students and teachers, as they also experimented with building in more engagement between students using inquiry-based learning, which included building problem-solving units of work. However, wondering whether embedding shared teacher and student values might be important, they started with staff responding to the survey item 'What do we as a maths faculty value in the maths classroom?' At the same time, year 7 and 8 students completed a survey that included the item 'What do your maths teachers value?' These items were not identical although similar enough that comparing responses would be instructive. The differences in the teachers' and students' responses were striking—there was virtually no overlap between them. Moreover, apart from some responses that could be categorised as 'control' for both groups, and one teacher response that might be classified as 'rationalism', no other of Bishop's mathematics values were noted in the analysis. These results reinforced the idea the staff did indeed need to actively explore how to build shared teacher / student values.

Later in the first year of the project, a professional development session using the JEDI, a four-phase process using the notions of Justifying, Essaying, Declaring, and Identifying (Lowrey, 2021; Seah, 2019) was run with the teachers in this school to help them explore more deeply the values they held. During the introduction to the session, Bishop's six mathematics values were explicitly discussed. Interestingly very few of the 30 plus teachers chose any of them during the session (Clarkson & Seah, 2020). However, the teachers did explore a number of pedagogical values, and this generated a deeper sense of the group working together towards a common goal.

During the second year of this project in the second school, staff decided to teach again the two problem-solving units they had devised and taught the previous year to their year 8 students. At the same time, they built another problem-solving unit for year 7 students. The year 8 teachers in the first year had been surprised but delighted in how students had judged the various solutions to the problems, not just using mathematics but also weaving in elements of aesthetics and real-life living. They also realized that they had started to change their teaching by using incidental moments that occurred in the classroom; like when a student asked, 'I already know the correct answer. So why do you want me to write down all my working?' was an introduction to explore the value of openness. Or 'Yes, I'm finished but not sure whether it is right. Mark's will be right. He's always right' allowed a discussion based around rationalism. The teachers believed the use of their problem-solving units provided far more opportunities for such rich discussions to develop than happens when they used their 'normal' traditional approaches to teaching. Incidentally they were using notions that Bishop had suggested, and from time-to-time they seemed to be implicitly discussing mathematical values, which they recognized later in our discussions. In the second year we tried to build on these beginnings.

2.5.3 Using the 'Wriggle Room' in the Curriculum

What we witness above, is the two schools looking for and discovering their 'wriggle room' in the curriculum. Many of Bishop's values are alluded to in our curriculum, but there is a disconnect between the Aims and the 'Scope and Sequence' section, the section that teachers actually consult. In my own case, colleagues were surprised in one school I taught at when I brought in Dienes multi-based arithmetic (MAB) blocks to help year 7 students learn (Clarkson, 1979). As I pointed out to my teaching colleagues there was nothing in the curriculum to say I could not do that.

But our curriculum is deficient in another way. At one point it indicates that students should become proficient in solving sequence of numbers under various operations. The acronym BOMDAS (or BODMAS) is not mentioned although most teachers will immediately invoke this 'rule'. This 'rule' can be an opportunity in uncovering some of the basic fabric of mathematics which is not even alluded to by the curriculum document. In discussing the notion of 'rule', as well as convention, law, proof, theorem, definition, and the like at appropriate year levels, gives opportunities for introducing the values of rationalism, objectism, control and progress. Whether we should use BOMDAS can be argued, but like several other 'key' points in teaching mathematics it is a convention and could be treated as such (such as the conventions of using differing symbols for number operators, and for the decimal point, depending in which country you live). Conventions, rules, etc. are not plucked out of nothingness. Conventions are not mathematical laws, but they do have history behind them which explains why some people use them. Definitions and theorems are bound up in the logical processing of mathematics ideas. Each should be discussed for what they are. An electronic search of our curriculum documents for these terms showed there are only two mentions of 'rule', one for 'laws', one for 'definition', and three for 'theorem'. But there is wriggle room. Even if the curriculum does not imply that such discussions are needed, there is nothing to stop teachers exploring these, and the mathematics values behind them. If such discussions were initiated from the late years in primary school (e.g., Clarkson, 1984, 1997, 2008b), students may gain a deeper understanding of how mathematics is structured in a rational manner, and that paradoxically might lead to spaces when mystery might seep in. Mathematics might be seen not as a matter of guesswork in trying to get the right answer, and mystery might not be automatically equated to 'I do not know'.

2.6 Mystery[6]

Although Bishop (1988) did not neglect societal or pedagogical values in his treatise, it was the six mathematics values that he thought were crucial to the underlying fabric of western mathematics, a position he continues to hold (Clarkson, 2008a; 2019). Hence, it is somewhat ironic that these six mathematics values, which many authors look back to as conceptualizing the study of values and mathematics, have in the decades since been the least studied (pedagogical mathematical values have been far more prominent in the literature). In particular, his mathematics value of mystery seems to have disappeared altogether for most researchers. For example, except when listing Bishop's three pairs of mathematics values, there was no significant mention of mystery evident during a brief scan of the eleven articles in the special 2012 ZDM journal devoted to *Values in East Asian Mathematics Education—The Third Wave Project*, which has since turned into the ongoing '*WIFI Project*' (Seah & Wong, 2012).

Another example is when I organized Discussion Groups on values in mathematics education at the 2015 PME conference and at the 2016 ICMI conference (Clarkson, 2015; Clarkson et al, 2017). Essentially these were built around preparing for and then acting out a role-play using participants as actors. Some participants played students in a make-believe classroom with each 'student' being asked to emphasize one of Bishop's six mathematics values in their behaviour, with the remainder of the participants playing research observers. Interestingly the observers were able to identify which values matched the different 'students', but they were not always able to be easily articulate their reasoning for their decisions. However, although I made valiant efforts to steer the resulting post role-play discussions to one or more of the six mathematics values, participants reverted to pedagogical values pertinent in teaching mathematics. As for mystery, few comments were made.

Although mystery was noted by most authors in Clarkson et al. (2019) as one of Bishop's six values, there was little emphasis on this value except for a section in the chapter by Clarkson (2019). Most other mentions were made somewhat in passing. For example, Corey and Ninomiya (2019) noted that:

> Japanese mathematics teachers are true to the discipline of mathematics as they emphasize logical reasoning and mathematical thinking [Bishop's rationalism]. Other values … are also

---

[6] Bishop arranged his six values into three pairs, The value mystery was paired with the value openness, each sitting at the end of a continuum. It would be interesting to examine this pair of values together to see what emerges as to teaching strategies. However, I leave that task to others. Like Andersson and Wagner (2018) I will only concentrate on the value of mystery here.

> prevalent in the teaching of Japanese teachers, the most salient being *Mystery* and *Openness*. Japanese teachers emphasize *mystery* by using problems that have surprising results or allow students to find interesting patterns. *Openness* is emphasized as students work together to solve problems and as teachers conduct a whole-class discussion of selected student solutions (*neriage*). (p. 61)

Although they do note mystery and openness here, for the remainder of their chapter mystery is not mentioned in any significant manner.

In the same book, Andersson and Österling (2019, p. 80) described mystery as "the mystique of mathematical ideas and their origin" and note that in the WIFI survey conducted with 740 Swedish students the item 'Mystery of mathematics' was ranked more or less in the middle of the 60 items. Another mention of mystery was by Davis et al., (2019, p. 93) when they described mystery as relating to "valuing of the wonder, fascination and the mystique of mathematical ideas" as well as stories about recent developments in mathematics. They suggested that as students progressed through school, they start to let go of the value of mystery as embodied in stories about mathematics and mathematicians, focusing far more on control and rationalism. Consequently, the question they leave unaddressed is, how do you then engage older students to think about the value of mystery?

Data that at the moment are being codified from Section C of the WIFI survey completed by to 223 Hong Kong students elicited 669 written responses.[7] Sadly about 25% of the responses were categorized as irrelevant. Of the remaining responses many related to the individual learner or specifying some obvious content aspect of mathematics (addition, 'x', etc.). Only seven responses (0.01%) were categorized as 'mystery'. Moreover, some of these, magic, luck, mystery, might have a different meaning for the student than for the researchers, the ever-present difficulty with interpreting survey responses. It appeared that only two of these seven responses relating to mathematics seemed to elicit the sense of mystery discussed later in this article; 'beauty' and 'can feel but cannot see'. Clearly in the view of this group of students, mystery is not something that you readily associate with mathematics.

One exception in which mystery is the specific focus for the writers is an article by Andersson and Wagner (2018). They make the important distinction between mystery in the sense of wonder with a landscape that stretches forever, if you are willing to start the journey into it, with the alternative understanding of uncovering ideas that have been purposefully concealed by someone who already knows, in most cases the teacher or textbook writer. Although the latter is understood by many as mystery, and perhaps most students would agree with this idea, the authors argued that this is a fabrication of what mathematics mystery should be about. In fact, rather than teaching something about the value of mystery in mathematics, most teaching they opined is more of an exercise in who has the knowledge power. They argued that if mystery is thought of as the boundless journey into an intriguing unknown and this was able to be imparted to students, then the students would have a deeper understanding of what mathematics was about. This agrees with Bishop's argument that the values he

[7] Personal communication with Wee Tiong Seah.

described including that of mystery, are part of the crucial foundations of western mathematics.

2.7 Teaching Mathematics Mystery

Teaching in such a manner that students come to understand and experience something of the mathematics 'mystery' value will clearly use some of the strategies discussed above; strategies that could be used to teach any of the mathematics values. However, thinking and reflecting on this value of mystery separately and discussing it explicitly seems important. This importance stems from at least two reasons. First, this value has been virtually neglected in the research literature, even though Bishop (1988) argued all six mathematics values are a set that need to be part of teaching mathematics. They together play a crucial role for a balanced insight into what should be one of the foundational aspects of a new mathematics curriculum, and that includes mystery. Secondly, mystery can add the critical role of uncertainty when thinking through a possible new framework for this curriculum. But as with other values in the set, there needs to be some specific aspects of the teaching strategies that will be peculiar to mystery, as there are with other values.

2.7.1 Teaching Beauty?

When considering possible teaching strategies which will help address the mystery value, we have to necessarily start with a question of 'Can you actually teach mystery?' Thinking about beauty, one component of Bishop's discussion of mystery, and its teaching in the arts in school, might give some insights into how the value of mystery can emerge in the teaching of mathematics. There are many articles to be found on the web analysing, defining, and discussing beauty. For example, Sartwell (2016), after discussing the ongoing issue of whether there is a useful distinction to be made by thinking of beauty in subjective or objective terms, ends his discussion by quoting Scruton: "We call something beautiful when we gain pleasure from contemplating it as an individual object, for its own sake, and in its presented form" (Scruton, 2009, p. 26). Wellington (2018) suggests that beauty is somewhat difficult to grasp, and yet we seem to intuitively know what we think is beautiful:

> … beauty is … almost impossible to describe. … the beauty of an object is like trying to explain why something's funny — when it's put into words, the moment is lost. … Beauty might not be an objective quality in the work of art … It's not something we can teach, and perhaps it's not something you can learn. But … our ability to perceive beauty is often what makes a work of art compelling. It is a feeling that reveals a pure moment of humanity that we share with the maker, transcending time and place.

Sartwell's and Wellington's ideas might be useful in thinking about the mathematics value of mystery in that beauty is a very individual perception of the art form.

Implied then is that the art form must be perceived. A discussion of whether this painting or that sonnet or indeed this sonata is beautiful in some way will not get far if the students have not seen the painting, read the sonnet, or listen to the sonata. To gain a deep appreciation of 'good' literature students must read the text and know how to go about applying some of the technical knowledge to gain understanding and meaning. But teachers of literature hope there is a deeper process involved, which goes beyond the learning of techniques, that is helped by discussing passionately their own feelings, attitudes, emotions that are evoked by this literature. Art and music teachers create opportunities for similar occasions when considering how this piece slots into some accepted genre, as well as what emotions are evoked and why others are not. But when do mathematics teachers create occasions when mathematics itself is thought about, talked about, emoted about, rather than how and why the doing of this or that is better, more efficient, correct, incorrect, appropriate? Interestingly various websites discussing teaching of the arts suggest even more; that a key element is the 'participation' in art forms. Students' participation is seen to be crucial. So, seeing and listening and even discussing is good, but it is also crucial for students to actively participate and then verbalize what they themselves are creating. Do we ever expect students to create mathematics and, even more so, reflect on this act of creation?

This discussion in no way aims to reduce the value of mystery to the value of beauty; it only initiates a discussion on possibilities. I leave it to colleagues to explore other components that Bishop includes in his discussion of mystery that may also give valuable insights into its teaching.

2.7.2 Teaching Mystery

The above gives some ideas regarding teaching mathematics mystery. When actually 'doing' mathematics, students might be encouraged to think about whether this mathematics is in any way surprising, beautiful, awful (in the sense of awe not horrible—is mathematics ever horrible?), or mysterious, and verbalize why this is or is not so. Experiencing mystery, like beauty in art, is intensely personal and it is a choice made by individuals. This implies that the doing of mathematics needs to be for oneself, not an obligation for someone or something else: for teachers, parents, for future careers—one likes doing it; you want to create mathematics when solving problems because that is enjoyable, it gives a sense of fun, it seems authentic. This is not always the case for many students; maybe most at the present time. Few students are able to choose this personal approach, given most students are taught in a traditional manner which does not normally privilege student choice. Hence, students may not experience a sense of mathematics mystery, and it follows probably that they will not have the opportunity to gain the deepest appreciation of mathematics.

However, it is actually hard to envisage how mathematics mystery could be taught in mathematics classrooms as they are so often presently constituted. School mathematics has always emphasised the utilitarian aspects of mathematics, particularly in primary school years. This has often led to the term numeracy being used. The over emphasizing the legitimate utilitarian aspects of mathematics has helped overcome to some degree the stigma attached to school mathematics for being useless for most students, and only used by power brokers as a sieving device to get rid of those students who should not go onto higher levels of education. But this has also meant that the fundamental abstract nature of mathematics ideas has been neglected, which does a disservice not only to the subject but also to teachers and students. This, consequently, has led to a situation in which some aspects of 'numeracy' are not fully understood by teachers and students alike, because many of these aspects are rooted in abstract mathematics ideas which are never examined by the teachers.

I taught at least one undergraduate unit in pre-service primary teaching degree programs for each of some 35 years. Without fail, after we had examined fraction ideas, with many students at the start admitting they had never understood fractions, at least two or three of my university students would approach me after the two to three tutorial sessions still somewhat confused. After reiterating the critical point made in the lecture that we were dealing with a different number system, that certainly contained whole numbers etc., but was actually different, light began to dawn. If these numbers were different, then maybe there were different rules that might apply. And no, not everything that you can do with fractions is easy to demonstrate with wooden blocks and/ or diagrams. That is because we are actually dealing with abstract ideas that do not always conform to the real concrete world. And so on. It seems that, once the students had grasped the idea that mathematics was deeper than numeracy, not only did their understanding increase, but they started to see mysterious possibilities that could be explored and could be exciting. That brought a smile to their faces, as well as mine.

This suggests that there is probably a threshold of cognitive understanding that is necessary for experiencing mathematics mystery, and some of the larger logical network that underpins mathematics. Nevertheless, enhancing the possibilities of students gaining some appreciation of mathematics mystery should not be linked to possible student's cognitive gains; although that might happen of course. But the notion of understanding something of mathematics mystery is an end in and of itself; this goes to the deep understanding of mathematics, not performance in mathematics.

2.7.3 Experiencing Mystery, Yourself

Another important ingredient in teaching mathematics mystery is teachers having experienced something of this themselves if they are to lead students to places where they too might also have such experiences. You read the same about beauty in the arts. For my university undergraduates, learning something more about fractions left them experiencing a 'wow' moment, which was exciting and memorable for them. That seems to be part of the mystery. Bishop notes that he found the fact that when you multiply the three elements of a Pythagorean triple together you will always obtain a multiple of 60 somewhat mysterious (Bishop, 1988).[8] Two of my colleagues, Wee Tiong Seah and Ann Downton, offered the Golden Ratio and the Fibonacci Series respectively as ideas that still held fascination for them after many years of doing and teaching mathematics, not just because of the number pattern, but the way they can be found in nature.

When an odd number is added to an even number you will always obtain an odd number. When I first worked that out, in about year two or three, I wondered whether in some strange way odd numbers were stronger or smarter or something, compared to even numbers, hence the odds always won. Even numbers added together give even numbers as the answer, which seemed obvious. But then I realized that two odd numbers added together always give even numbers; that really was a surprise. When I sorted out what happened to odd and even numbers under multiplication some years later, those rules just seemed to be so obvious, but interestingly increased my feeling that something really deep was going on here. Although I remember one teacher commenting that these 'rules' helped you check whether answers might be wrong, he did not explore why they worked. And certainly, he nor any other teacher in school ever asked me whether I found these notions interesting, exciting, awe inspiring. Clearly none of that belonged in a mathematics class; and yet of course it did, inside me. It was not until university, after I learnt some number theory, that I understood the background. And yet the fascination of my early discovery has never really left. There still seems to be a mysterious quality hanging around this property of numbers that engages me, that still gives we a little bit of a 'wow' feeling.

There are various other number 'oddities' that can capture the imagination and have a sense of mystery about them. The year '2020' was not only an awful (horrible) year because of COVID, but it also consisted of the same two numbers repeated. The following year '2021', not surprisingly, consists of two consecutive numbers, but it also is the product of two consecutive primes, 43 and 47. I think it was the great mathematician Ramanujan who devised the lovely symmetrical set of number sentences: $1 \times 9 + 2 = 11$; $12 \times 9 + 3 = 111$; $123 \times 9 + 4 = 1111$. Item 60 of the student questionnaire of the WIFI survey has another lovely example that is labelled "mystery of maths"; $111\,111\,111 \times 111\,111\,111 = 12\,345\,678\,987\,654\,321$. And in geometry why is it that there are only five Platonic solids, and why are they always made from 3, 4 or 5 sided polygons? What happened to 6 to continue this sequence? I played around for ages, but failed during secondary school, trying to construct one with hexagons; the angles were just not right. Interestingly, this was never part of the mathematics I learnt there.

It is my hope that my lasting fascination with 3D models and the like has rubbed off on some of the education students, both preservice and during professional learning sessions, I have taught over the years. This certainly led me to often using Escher tessellations, posters, etc. and hopefully these produced a lasting source of fascinating, surprising discoveries for students with a feeling of 'how can that be?' Well 'how can that be' is often sorted out with some mathematics; but the feelings of

[8] Well, this reference is 35 years old but in recent conversations when quizzed on this statement, Bishop said he still has a sense of mystery lingering with regards Pythagorean triples.

wonder are just as important and more often than not longer lasting and inspiring. However, not every teacher will have had such experiences. It is a sad observation that out of many mathematics teachers I have worked with, very few have been able to articulate an instance of mathematics that has invoked a sense of mystery. Occasionally in professional learning sessions when we have explored some surprising mathematics together, some teachers after gentle prompting started to remember back and could pinpoint some experiences that did, at the time, seem surprising and had a little 'wow' feeling for them. Considering the starting point for teaching mystery value is to try and help teachers to remember such experiences that they may have had themselves, experiences that brought enjoyment because they were fascinating, and in a way mysterious. This might lead to them creating teaching moments that might invoke similar senses of mystery for their students. This includes pre-service teaching, which should develop moments in the course which students will find intriguing, puzzling, but enjoyable. Perhaps they will be able to use such experiences as a basis for moments in their teaching that will lead their students to experience some mathematical mystery.

2.7.4 Student Problem Creators and Student Problem Solvers

Andersson and Wagner (2018) (see above) described another teaching strategy that could lead to students grappling with mathematics mystery. They suggest that one approach is to encourage students to develop problems for their peers in problem-solving episodes. Andersson and Wagner's (2018) observations suggest that many students are very good at judging what problems will suit their peers by constructing tasks that are within reach but not always obvious. The authors discuss such an episode with year 9 students. The students behaved according to the norms they had learnt in their nine years of schooling; the problem devised and solved was a bit hard and sneaky. They report on how both the student problem developers and the problem solvers reacted to the whole process. Andersson and Wager certainly wanted these students to relax and rejoice in the developing and solving experiences and perhaps gain a sense of wonder in what they were achieving. But the sense the authors got was that sadly the students were enacting the mathematical myth prevalent in their schooling—that the object of the exercise was simply to build barriers in the problem situation to see whether these could be overcome. In effect, the student problem developers were taking the place of their teacher and were now the gate keepers to knowledge to which their peers had to somehow try and find the right key (a mechanical process devoid of emotions and deeper insights). And their peers, the problem solvers, obliged.

I wonder however whether, when using such a useful and engaging approach, teachers should also be encouraged to intervene with discussion starters during and/or after the process that might evoke comment on the excitement experienced by the student problem developers and the problem solvers as gradually the problem is developed and solved. What surprises are each group of students experiencing? Do they have a sense of joy in developing/ solving the puzzle? Is there only one solution and why, or what are the possible solutions to the problem? Is that a surprise? What criteria might be used to judge the most appropriate solution for this context if more than one exists? What happens if the context is changed; will that change the criteria? Will this experience help you think forward to when you can do it all over again? Do you think you will remember this experience as something that might lead you to think beyond the mechanics of the solving process? Would this change lead to more ingenuity being built into future problems? These types of questions do not supplant the more 'obvious' cognitive ones as to what were the mathematical processes, ideas and algorithms employed in solving the task. But perhaps the former question could be prioritized when the students were still experiencing the accompanying emotions, and the cognitive questions asked sometime later.

2.8 The Importance of All Six Values

> Most people can only judge of things by the experiences of ordinary life, but phenomena outside the scope of this are really quite numerous. How insecure it is to investigate natural principles using only the light of common knowledge. (Montefiore, 2022, p. 285)

This quote, from Shen Gua, a high ranking Chinese official of the Song dynasty in the 900's CE, comments on the numerous instances of phenomena that cannot be easily defined. It seems to me to still ring true. In particular, not all aspects of mathematics are obvious from our real-life experiences, including for many students, mathematics values. And yet I have argued that for a deep understanding of mathematics an understanding of mathematics values is needed.

Bishop (1988) argued that all six values should find a place in mathematics teaching and learning. Although some, especially rationalism, control and progress, have been recognized as key in the underlying fabric of western mathematics, these have not always been obvious in the way mathematics has been taught in schools. Bishop listed the six values in *complementary* pairs, but not all the values in the set seem to knit neatly together. For example, rationalism and mystery might seem to be at variance, and yet both are needed with their own distinctive contributions. But an over emphasis on rationalism has led to a misunderstanding of what mathematics can do in societal contexts.

Mathematics is a vast network of interconnected abstract ideas. Yet, most people see it as various facts strewn around without any real coherence, although some of the bits and pieces do have utilitarian value for small pockets of our lives. Such bits of mathematics have only one right answer, even if getting that answer can be very difficult and not always obvious. Since they believe there is only one right answer, mathematics gives a sense of certainty, not often available in our lives. But that is not always what mathematics does do. In many instances, even within the confines of pure mathematics and science, let alone when applied to the real lives of humans, mathematical results can have a region of uncertainty around them. No wonder then that many people do not understand the role mathematics plays in understanding the great issues that confront the world society today, even though most adults in western societies have studied mathematics for 9 to 10 years in school.[9]

A mathematical model is a way of structuring how we can look at an issue, such as the recent COVIID pandemic. It can allow us to take early data, make some plausible assumptions based on that analysis, then look for logical implications based on those assumptions. We can then compare those results with the growing datasets to understand what might be driving the patterns that emerge. This process can be repeated in an iterative manner for as long as is needed. Such a process makes sense of patchy early data but does allow possible outcomes to be established which are updated over time, but in the interim do give 'best possible' actionable initiatives which also will change over time. However, it does not tell us for certain what will happen. Yet, in real life, interim results are often, annoyingly, presented to the public as 'certain' scientific and mathematical outcomes because of the certainty that the public crave, rather than their real nature being noted as a consensus reached often after many twists and turns, and normally still held somewhat tentatively, by the mathematicians/scientists involved in the process. This inevitable messiness within mathematics and science needs to be clearly established as one aspect in the teaching of school mathematics, and I suspect the notions embedded in the mathematics values, including that of mathematics mystery, will help. Students, their parents, and indeed whole communities need to come to a new appreciation of what mathematics can do and what it cannot do.

The COVID pandemic is just the latest of ongoing world-wide issues that need mathematical input into their solutions. There is also the planet's over-population by humans and climate change, both of which David Attenborough has forcefully spoken of for many years. Global warming on average increased by 0.8 degrees between 1926 and 2011 (Attenborough, 2020, p.31). Although we know that the world's temperature has increased and decreased over the millennia, that is a rate of increase which vastly exceeds any that has occurred in the last 10,000 years when it did not vary by more than 1 degree (p. 20). In the late 1930s the world population was about 2.3 billion, having taken many millennia to reach that figure (Attenborough, 2020, p. 13). It has more than tripled in my lifetime. These issues have been in the public's awareness for decades. In 1973 I showed a 16mm version of the feature film *Soylent Green* to the year 7 and 8 students at the very end of the school year. The film story is set in our world which has been decimated with the run-away impact of global warming (interestingly now termed 'climate change') and catastrophic over-population by humans; 1973 was 50 years ago. Science and mathematics were central in the film.

Why after nine or more years of studying school mathematics do significant numbers of the populace believe that mathematician's and statistician's opinions are totally irrelevant in a pandemic? One possibility is that the school mathematics

---

[9] But even when the mathematics is not dealing with great societal issues readers' complaints to newspapers can show a decided lack of understanding. See Ribbans (2022) for an amusing but sobering note.

curriculum, that is the mathematics story that powerful people in western society deem to be crucial to its functioning, is no longer relevant for most students, and hence of little concern to them when adults as they consider social issues. This suggests there needs to be a root and branch rethink of the mathematics curriculum.[10] We will need to start incorporating long forgotten elements into this cultural foundational story of mathematics if many more future members of our societies are to see the relevance of mathematics as appropriate to the functioning of society on many levels.

2.9 Summary Statement

In an earlier publication I and others argued for the reconceptualization of the mathematics curriculum by starting with mathematics and pedagogical values rather than mathematical content (Seah et al., 2016). Here, from my armchair, I have again argued that mathematics values are not optional if mathematics is to be understood and appreciated at a deep level. An appreciation of the mathematics values is one central component that holds the network that is mathematics together and gives it a sense of coherence. The need for such an understanding of just what mathematics is and can do has been thrown into stark relief with the pandemic of 2020 and beyond, let alone the other crucial world issues which we face. Once mathematics is used in the social sphere the certainty that most people crave and believe mathematics gives is a misunderstanding of its nature since this is only part of what mathematics is. A deeper understanding is needed to be taught in school mathematics.

References

Andersson, A., & Wagner, D. (2018). Remythologizing mystery in mathematics: Teaching for open landscapes versus concealment. *Educational Science, 8*(41). https://doi.org/10.3390/educsci8020041

Andersson, A., & Österling, L. (2019). Democratic actions in school mathematics and the dilemma of conflicting values. In P. Clarkson, W. T. Seah, & J. Pang (Eds.), *Values and valuing in mathematics education: Scanning and scoping the territory* (pp. 69–88). Springer.

Attenborough, D. (2020). *A life on our planet*. Witness Books.

Bereznicki, B., Brown, D., Toomey, R., & Weston, J. (2008). *At the heart of what we do: Values education at the centre of schooling*. Curriculum Corporation.

Bishop, A. (1988). *Mathematics enculturation*. Kluwer.

Bishop, A. (1991). Mathematical values in the teaching process. In A. Bishop, S. Mellin-Olsen, & J. Van Dormolen (Eds.), *Mathematical knowledge: Its growth through teaching* (pp. 195–214). Kluwer.

Bishop, A., & Clarkson, P. (1998). What values do you think you are teaching when you teach mathematics? In J. Gough & J. Mousley (Eds.), *Exploring all angles* (pp. 30–38). Mathematics Association of Victoria.

[10] Gal and Geiger (2022) also argue for this but have a different conception of what changes are needed.

Bishop A. J. (2001). Educating student teachers about values in mathematics education. In F. Lin & T. Cooney (Eds.), *Making sense of mathematics teacher education* (pp. 233–246). Springer. https://doi.org/10.1007/978-94-010-0828-0_11

Brandenburg, R., Glasswell, K., Jones, M., & Ryan, J. (2007). *Reflective theory and practice in teacher education*. Springer.

Carter, L. (2020, Dec). *Science education in the time of COVID: Unavoidable challenges*. Keynote paper presented at the Science SIG of the Australian Association of Research in Education.

Clarkson, P. (1979). RAMP: A resource for school-based curriculum development. In B. Atweh, G. Booker, C. Irons, & G. Jones (Eds.), *Research in mathematics education in Australia* (pp. 1–20). Mathematics Education Research Group of Australia.

Clarkson, P. (1984). Hex: A game of strategy. *Australian Mathematics Teacher, 40*(3), 19–20.

Clarkson, P. (1997). Games, strategies and 'whatdayado next'? In D. Clarke, P. Clarkson, M. Horne, L. Lowe, A. MacDonough, & M. Mackinlay (Eds.), *Mathematics: Making connections* (pp. 92–94). Mathematics Association of Victoria.

Clarkson, P. (2006). Rhyming and folk tales: Resources for mathematical thinking. *Australian Mathematics Primary Classroom, 11*(4), 18–23.

Clarkson, P. (2007). Australian Vietnamese students learning mathematics: High ability bilinguals and their use of their languages. *Educational Studies in Mathematics., 64*, 191–215.

Clarkson, P. (2008a). In conversation with Alan Bishop. In P. Clarkson & N. Presmeg (Eds.), *Critical issues in mathematics education: Major contributions of Alan Bishop* (pp. 13–28). Springer.

Clarkson, P. C. (2008b). Exploring the possibilities of using 'ticktacktoe' to think and communicate about mathematics. *Australian Mathematics Teacher, 64*(2), 28–35.

Clarkson, P. (2009). Globalisation and mathematics teaching: The global importance of local language contexts. In J. Zajda & V. Rust (Eds.), *Globalisation, policy and comparative research: Discourses of globalisation* (pp. 135–156). Springer.

Clarkson, P. (2010). Mathematics and water in the garden: Weaving mathematics into the student's lived environment. *Australian Primary Mathematics Classroom, 15*(1), 11–14.

Clarkson, P. (2019). A conversation with Alan Bishop. In P. Clarkson, W. T. Seah, & J. Pang (Eds.), *Values and valuing in mathematics education: Scanning and scoping the territory* (pp. 11–24). Springer.

Clarkson, P., & Bishop, A. (2000). Values and mathematics. In A. Ahmed, J. Kraemer, & K. Williams (Eds.), *Cultural diversity in mathematics education* (pp. 239–244). Horwood Publishing.

Clarkson, P., & Carter, L. (2017). Multilingual contexts: A new positioning for STEM teaching/ learning. In K. Hahl, K. Juuti, J. Lampiselkä, A. Uitto, & J. Lavonen (Eds.), *Cognitive and affective aspects in science education research* (pp. 233–242). Springer.

Clarkson, P., Bishop, A., FitzSimons, G., & Seah, W. (2000). Challenges and constraints in researching values. In J. Bana & A. Chapman (Eds.), *Mathematics education beyond 2000* (pp. 188–195). Mathematics Education Research Group of Australasia.

Clarkson, P., Bishop, A., Seah, W. T., & FitzSimons, G. (2001). An undervalued legacy of mathematics education. In G. FitzSimons, J. O'Donoghue, & D. Coben (Eds.), *Adult and lifelong education in mathematics* (pp. 47–70). Language Australia.

Clarkson, P., Seah, W. T., & Bishop, A. (2010). Mathematics well-being and teacher education. In R. Toomey, T. Lovatt, N. Clement, & K. Dally (Eds.), *Taking values education to teacher education* (pp. 179–194). David Barlow Publishing.

Clarkson, P., & Henry, B. (1976). *Angles*. Heinemann Educational.

Clarkson, P., & Seah, W. T. (2020). Revitalizing a school's mathematics teaching culture: Can affect be the key? Paper accepted for presentation of the 14th International Congress on Mathematical Education, Shanghai. https://www.researchgate.net/publication/353820770

Clarkson, P., Andersson, A., Bishop, A., Kalogeropoulos, P., & Seah, W.T. (2017). Connections between valuing and values: Exploring experiences and rethinking data generating methods. In G. Kaiser (Ed.), *Proceedings of ICME-13 Monographs* (pp. 643–644). Springer.

Clarkson, P., Seah, W. T., & Pang, J. (2019). *Values and valuing in mathematics education: Scanning and scoping the territory*. Springer.

Clarkson, P., Bishop, A., & Seah, W. T. (2023). Mathematics education and student values: The cultivation of mathematical wellbeing revisited. In T. Lovat, R. Toomey & N. Clement (Eds.), *International research handbook on values education and student wellbeing*. Springer.

Clarkson, P. (2015). Connections between valuing and values: Rethinking data generating methods. *PME Newsletter, November/December*, 7–10. http://www.igpme.org/wp-content/uploads/2019/05/15_12_PME_Newsletter.pdf

Corey, D., & Ninomiya, H. (2019). Values of the Japanese mathematics teacher community. In P. Clarkson, W. T. Seah, & J. Pang (Eds.), *Values and valuing in mathematics education: Scanning and scoping the territory* (pp. 53–67). Springer.

Cummins, J. (2008). BICS and CALP: Empirical and theoretical status of the distinction. In B. Street & N. Hornberger (Eds.), *Encyclopedia of language and education* (2nd Ed., Vol. 2, pp. 71–83). Springer Science and Business Media LLC.

Davis, E., Carr, M., & Ampadu, E. (2019). Valuing in mathematics learning amongst Ghanaian students: What does it look like across grade levels? In P. Clarkson, W. T. Seah, & J. Pang (Eds.), *Values and valuing in mathematics education: Scanning and scoping the territory* (pp. 89–102). Springer.

Ferguson, K. (2011). *Pythagoras*. His lives and the legacy of a rational universe.

Fook, J. (2007). Reflective practice and critical reflection. In J. Lishman (Ed.), *Handbook for practical learning in social work and social care: Knowledge and theory* (2nd edn, pp. 363–375). Jessica Kingsley.

Gal, I., & Geiger, V. (2022). Welcome to the era of vague news: A study of the demands of statistical and mathematical products in the COVID-19 pandemic media. *Educational Studies in Mathematics, 111*, 5–28.

Hattie, J., Fisher, D., Frey, N., Gojak, L., Moore, S., & Mellman, W. (2016). *Visible learning for mathematics*. Corwin.

Horne, M., Henry, B., Clarkson, P., & Meehan, A. (1977). *Decimals and per cent*. Heinemann Educational.

Kalogeropoulos, P., & Clarkson, P. (2019). The role of value alignment in levels of engagement of mathematics learning. In P. Clarkson, W. T. Seah, & J. Pang (Eds.), *Values and valuing in mathematics education: Scanning and scoping the territory* (pp. 115–128). Springer.

Kalogeropoulos, P., Russo, J., & Clarkson, P. (2021). Exploring educator values alignment strategies in an intervention context: The emergence of the beacon strategy. *ECNU Review of Education, 4*(2), 327–348.

Lovat, T., Toomey, R., & Clement, N. (2023). *International research handbook on values education and student wellbeing*. Springer.

Lowrey, R. (2021). Teaching mathematics the JEDI way. *Vinculum, 58*(2), 6–7.

Mlodinow, L. (2002). *Euclid's window*. Penguin.

Montefiore, S. (2022). *The world: A family history*. Weidenfeld & Nicolson.

Paraide, P., Owens, K., Muke, C., Clarkson, P. & Owens, C. (2022). *Mathematics education in a neocolonial Country: The case of Papua New Guinea*. Springer.

Park, S.-R. (2000). Introduction of western sciences in China, Japan and Korea: A comparative comment. In W.-S. Horng & F.-L. Lin (Eds.), *History in mathematics education; Challenges for a new millennium* (Vol. 1, pp. 11–17). Department of Mathematics, NTNU.

Raths, L. E., Harmin, M., & Simon, S. B. (1987). Selections from values and teaching. In P. F. Carbone (Ed.), *Value theory and education* (pp. 198–214). Krieger.

Ribbans, E. (2022). On the sixth day of Christmas—Or was it the seventh? Why numbers matter more than ever. *The Guardian, 30 December*. https://www.theguardian.com/profile/elisabethribbans

Sartwell, C. (2016). Beauty. In *Stanford encyclopedia of philosophy*. Accessed on 23 Dec 2020 at https://plato.stanford.edu/entries/beauty/

Scruton, R. (2009). *Beauty*. Oxford University Press.

Seah, W. T. (2019). Values in mathematics education: Its conative nature, and how it can be developed. *Journal of Korean Society of Mathematics Education, 22*(2), 99–121.

Seah, W. T., Andersson, A., Bishop, A., & Clarkson, P. (2016). What would the mathematics curriculum look like if values were the focus? *For the Learning of Mathematics, 36*(1), 14–20.

Seah, W. T., & Wong, N. (2012). Values in East Asian mathematics education—The third wave (Special Edition). *ZDM Mathematics Education, 44*(1).

Toomey, R., Lovatt, T., Clement, N., & Dally, K. (Eds.) (2010). *Taking values education to teacher education.* David Barlow Publishing.

Victoria Curriculum and Assessment Authority (2016). *The Victorian curriculum: Mathematics.* http://victoriancurriculum.vcaa.vic.edu.au/mathematics/introduction/rationale-and-aims

Wellington, R. (2018). Friday essay: In defence of beauty in art. *The Conversation,* March 16. Accessed on 22 Dec 2020 at https://theconversation.com/friday-essay-in-defence-of-beauty-in-art-89921

Wertheim, M. (1995). *Pythagora's trousers.* Times Books.

Chapter 3
Towards a Reconceptualisation of Values Research in Mathematics Education: A Systematic Review

Hui Min Chia and Qiaoping Zhang

3.1 Introduction

Values can be considered a sociocultural product that individuals hold; a process of valuing is one in which individuals embrace convictions of importance and worth personally, and one which can bridge personal affective and cognitive aspects with behaviour (Seah, 2019). Various factors, including cultural and social norms, upbringing, personal experiences, and individual personality traits, shape values. Research on values has been conducted across many disciplines, including anthropology, sociology, education, philosophy, and psychology (Bishop, 1988, 1999; Dahlke, 1958; Feather, 1995; Hofstede, 2001; Kluckhohn, 1959; Kluckhohn & Shrodtbeck, 1961; Kohlberg, 1981; Raths et al., 1978; Rokeach, 1973; Schwartz, 1992), with growing attention over the past decades in the field of mathematics education, across peer-reviewed journals (Daher, 2021), books (Clarkson et al., 2019; Palm & Skott, 2018), special issues (*ECNU Review of Education*, July 2021). Across those, different aspects of values and valuing in mathematics teaching and learning have been discussed. For example, studies have been conducted on teachers' values in mathematics teaching (Akyıldız et al., 2021), students' expectancy values in mathematics (Fielding-Wells et al., 2017), and values in mathematics learning (Dede et al., 2022; Zhang, 2019).

Despite the significance and the prominence of the concept of values (Seah, 2019), there remain some challenges in applying it in research. This can be for a number of different reasons. First, there is the multidimensional nature of values (Bishop et al., 2003; Henderson & Thompson, 2003). For instance, researchers (Bishop,

H. M. Chia (✉)
Faculty of Education, University of Macau, Macau, China
e-mail: hmchia@um.edu.mo

Q. Zhang
The Education University of Hong Kong, Hong Kong SAR, China
e-mail: zqiaoping@eduhk.hk

Y. Dede et al. (eds.), *Values and Valuing in Mathematics Education*,
https://doi.org/10.1007/978-981-99-9454-0_3

2020; Bishop et al., 2003; Seah, 2019) conceptualised values as psychological and sociocultural constructs. Secondly, there is the issue of unclear definition of values across the literature (Kalogeropoulos et al., 2021; Seah, 2019). The implicit nature of values in mathematics teaching and learning (Bishop, 1991; Bishop et al., 2005) has made values look abstract (Kalogeropoulos & Clarkson, 2019) and difficult to define (Swadener & Soedjadi, 1988). Yet, a clear and consistent concept definition is crucial to constructively guiding research (Martinez, 2017). Consequently, there is a need to have a shared conception of values to engage with and discuss values more critically in mathematics education.

About twenty years ago, following a comprehensive review, Bishop et al. (2003) provided suggestions for the conception of values. Over a decade later, Carr (2019) conducted another systematic literature review of value research in mathematics education published between year 2003 to 2018. Although she provided an overview of the recent trend of empirical values research in terms of location, stakeholders, the expansion of values research, and consistency of research findings, she did not cover the methodology issue in her review—particularly how value had been conceptualised with various methodological tools. As a result, given the rapid development of values research in mathematics education (Zhang & Seah, 2021), we felt that a review of the trend in the conception of values as well as methods used for the recent ten years was needed.

Across the field of values, there appears to be a tendency for researchers to operationalise values in different ways and to use various data collection approaches to study them, depending on their differing conceptualisations. This can create challenges when comparing studies, particularly those that use different methodological approaches. Conducting a systematic methodology review can help address these challenges by providing a structured and transparent approach to synthesising the existing literature. By systematically identifying and evaluating the strengths and limitations of different methodological approaches, a methodology review can help identify best practices for measuring and analysing values, as well as areas where further research might be needed. This, in turn, can improve the rigour and quality of future research on values and enable researchers to build upon past findings more systematically and robustly. Hence, in the current study, two main research questions have been posed:

(1) RQ1: What features of definitions are shared by the different conceptualisations of values in mathematics education?
(2) RQ2: What are the different research methods used to investigate values in mathematics education informed by different features of values definitions?

3.2 Conceptual Framework

So far in the literature values have been mostly defined as convictions (Seah & Andersson, 2015), criteria (Schwartz, 1992; Williams, 1979), principles (Chin & Lin, 2000) or standards (Chin & Lin, 2000; Kluckhohn, 1959) that pertain to particular actions, events, goals or objects, indicating their desirability (Feather, 1995; Kluckhohn, 1959; Raths et al., 1978), importance (Chin & Lin, 2000; Seah & Andersson, 2015), preferences (Hofstede, 2001; Raths et al., 1978; Seah & Andersson, 2015), priorities (Schwartz, 1992), or worth (Raths et al., 1978; Seah & Andersson, 2015; Swadener & Soedjadi, 1988). They have been described as enduring beliefs that influence attitudes and actions (Rokeach, 1973). Values represent a kind of "duality" Rokeach (1979, p. 51), meaning that they are implicitly and explicitly held by individuals as well as by groups, structured collectively within a particular culture (Bishop, 1988; Hofstede, 2001; Kluckhohn, 1959). Consequently, they can be altered by cultural, societal, and personal experiences (Seah, 2019; Williams, 1979).

Existing values research in mathematics education has been conducted in relation to different concepts, such as preferences in learning mathematics, well-being, achievement motivation, decision-making, noticing, and teaching practices. For example, studies used students' self-reports to investigate students' preferences in learning (Seah, 2013), which could later be linked to students' mathematical well-being (Clarkson et al., 2011). Students' expectancy values toward mathematics have been associated with achievement motivation and task engagement in the classroom (Eccles et al., 1984; Wigfield & Eccles, 2000). Teachers' values in teaching have been linked with students' learning experiences in the classroom (Bishop et al., 2005).

Another strand of values research focused on the events in the classroom, which related teachers' values to their decision-making. For example, a case study approach with questionnaires, interviews, and observations was used in studies to investigate teachers' pedagogical values (Chin & Lin, 2000), or intended and implemented values (Bishop et al., 2001). In the realm of value alignment strategies, some studies focused on ongoing teacher-led instruction and students' continual self-evaluation of their classroom learning process (e.g., Kalogeropoulos & Clarkson, 2019; Seah & Andersson, 2015). Others still explored the correlation between a teachers' values and their capacity to notice and react to crucial classroom events significantly, which affected their decision-making process (Aktaş et al., 2019).

To put shortly, values research in mathematics education stems from a wide range of ideas and perspectives and engages with a variety of different methodological approaches, which are in need of some synthesis. The systematic literature review presented in this chapter attended to two aspects of this synthesis: conceptualisation and methodology. Considering the conceptual framework of values, this synthesis presents that values comprised three key dimensions in defining values, each with two poles. The subject dimension, the first one, pertains to the topic or entity that the values denote, which can be personal or sociocultural. The second dimension is representation, which deals with how values are portrayed—they can be implicit as beliefs or explicit when demonstrated through actions. The final dimension is stability, which refers to the state of values; they can guide actions and simultaneously remain open to change over time. Considering research methodology, the synthesis indicates that the most prominent quantitative values research to date has employed methods such as questionnaires, while qualitative research has involved interviews and lesson observations.

3.3 Method

In conducting the review, we followed the recommendations of Cooper (2017). The process began with searching the literature, screening, evaluating, and selecting the items. The following sections describe the process of the review in detail.

3.3.1 Literature Search

A literature search in three databases such as EBSCOhost Research Databases (APA Psycinfo, British Education Index, and ERIC), Web of Science and Scopus. The following search terms were used: TI "valu*" AND TI "math*" AND NOT "place value" OR "absolute value" OR "numerical value" OR "valuable". Then, the search term AND NOT was used to refine the search due to the initial query of terms *math** AND *valu** turned out a big volume of results in EBSCOhost Research Databases with 608,991 records. After scanning through the results, it was noted that some of the studies involved mathematics concepts, such as place value and absolute value. Hence, due to the multidimensional definition of value, the search term AND NOT was also applied. The selection of data for this review was done using the Preferred Reporting Items for Systematic Reviews and Meta-Analyses (PRISMA) method as a strategy for retrieving articles. PRISMA is the most commonly used method for conducting systematic reviews and meta-analyses (Page et al., 2021). In this study, data extraction consisted of the following four steps: identification, screening, eligibility and included, as shown in Fig. 3.1.

The literature was limited to research published from January 2012 to December 2022. Such a time frame was selected in order to review the most recent values research in mathematics education in the past eleven years, to identify the trends in conceptualisation and operationalisation of the concept. The search conducted yielded a total of 3002 publications. After the database's auto-filter filtered out duplicate articles and limited to studies to peer-reviewed articles in English only, the total number of articles 1916 remained.

3.3.2 Selection of Articles

The selection criteria included empirical studies on primary or secondary school and tertiary level, values as one of the variables in the study, and specific methodology used. The criteria for rejection of articles were:

1. The concept of value is related to mathematics concepts, such as extreme value.
2. The studies are not related to mathematics education or a mixture of other subjects such as science.

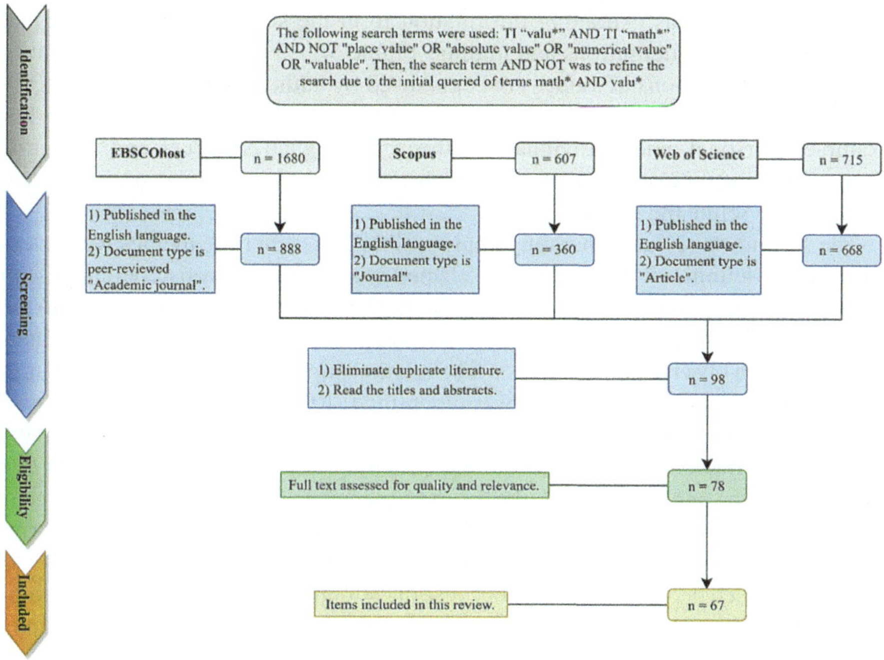

Fig. 3.1 Flowchart of data collection

3. The type of papers was commentary, literature review, meta-analysis, validation of instruments, or theoretical.

After evaluating the title and abstract, the selection of 137 articles for further screening was made. The number of 1779 articles were excluded according to each criterion. Out of the 137 articles, 67 provided definitions of values and a theoretical or conceptual framework in their studies.

3.3.3 Analysis of Studies

All 67 selected studies were imported into qualitative analytic software (a complete list of these studies is included in the Appendix). Thematic synthesis analysis was conducted (Thomas & Harden, 2008) to suit the qualitative characteristics of this review and its research questions. The first stage of analysis was conducted based on the conceptual framework proposed, focusing on the dimensions (themes) emerging from the data. The second stage was mainly inductive, whereby abstract conceptualisation was developed from the defining dimensions and the methodology used in studies. The data analysis began with line-by-line coding and moved on to analytical coding (Thomas & Harden, 2008).

The first stage was extracting information from the study's full text, including aims, methodology, definitions and findings (Table 3.1). This step aimed to identify the defining dimensions of values conceptualised by authors in their studies (RQ1). Line-by-line coding for the themes emerging from the information selected. For example, "personal" and "individual" are the psychological construct, and "cultural" interaction is the sociocultural construct. Constant comparisons between studies were conducted to refine the coding.

The second stage was generating analytical themes. After the themes were identified in the definitions referred to by selected studies, the analysis continued with themes emerging from the methodology. Seven themes were coded according to the methods applied in the study: case study, document analysis, survey, interview, intervention, lesson observations, and task. The relationships between descriptive themes in the definition and methodology were identified (RQ2). Then, the abstracting of related concepts was based on the methodologies into three categories: personal attributes, narratives, and ways of acting (teaching and learning). More than one method could be applied in the study, for example, questionnaires followed by interviews or interviews by lesson observations. However, the coding of the categories depended on the primary method used in data collection. For example, Study A (Table 3.1) coded personal attributes because questionnaires were used to collect individual's psychological aspects and espoused values, even though one of the defining

Table 3.1 The stages of data analysis

Stage 1: extract information	Defining dimensions	Stage 2: analyse methods	Methodology categories
Study A: Value is described as a "belief one holds deeply, even to the point of cherishing, and acts upon" (Philipp, 2007, p. 259) or regarded as beliefs in action (Clarkson et al., 2000)	– Psychology (one) – Espoused (belief) – Enacted (acts upon)	It is part of an international "What I Find Important (in my mathematics learning)" [WIFI] study, which assesses the attributes of mathematics learning and teaching that are valued by students. The participants of this study were a total of 816 Korean students who responded to the WIFI questionnaire	Individual attributes
Study B: According to Clarkson and Bishop (1999), values are beliefs in action Espoused values as values that we want other people to believe we hold, and enacted values as values that we actually practice	– Psychology – Espoused (beliefs) – Enacted (in action/ practice)	Six "excellent" mathematics teachers. Qualitative data were collected through video-recorded lesson observations (three lessons for each teacher) and in-depth interviews with teachers after each observation	The way of acting

features consisted of the enacted dimension. Study B focused on observing classroom practices through lesson observations as the primary data source instead of analysing the narratives to depict the behavioural pattern.

3.4 Findings

In this section, we summarise the analysis results of 67 articles. Among the reviewed articles, more than half of the studies focused on students (n = 41, 62%), followed by in-service teachers (n = 13, 20%), pre-service teachers (n = 5, 8%), teachers and students (n = 5, 8%), document analysis (n = 2, 3%), and pre-service and in-service teachers (n = 1, 2%). Below, we explain the results regarding each research question.

3.4.1 The Defining Features of Value

In examining the definition of values, we identified three main dimensions emerging from studies: subject, representation, and stability. Those dimensions exist in two poles that reflect the multidimensional characteristics of values. The subject dimension was denoted as the topic that was being discussed, either the person or the phenomena. The subject dimension was the psychology pole when the authors mentioned value as "individual's preference", "personal truths", or "personal importance". In conjunction with this dimension, some studies conceptualised values as a sociocultural product. For example, "pedagogical identity", "sociocultural norms and standards", "social identities", and "a component of culture". Some authors conceptualised values as "social practices", "culture-dependent", or "classroom environment". All these were denoted as the sociocultural pole in the subject dimension.

The representation dimension was related to the depiction of values either in symbols or actions. Values identified by authors were espoused through language, writing, and text. This implied that values were the meaning established through symbolic representations. Those definitions posed the keywords of "convictions", "perceptions", "perceived importance", "self- and other focused beliefs", "a concept", "reflect", or "emotional dispositions". The enacted pole was linked to the act of task performing, engagement, responses, teaching and learning activities in the classroom. This pole of the representation dimension focused on classroom practices as the source of values. Concepts involved were "beliefs in action", "behaviour", "choose to engage", "positive response", or "acted on". The actions and behaviours in classrooms reflected the values held by individuals.

The third dimension was associated with the stability of values, which involved the steadiness of change and a lack of change. Here the authors considered values as the independent variable that regulates teaching and learning. For instance, values "regulate", "guide", or "shape" a person's action. The changeable pole involved the conception of values as a process influenced by the sociocultural context.

Values are the dependent variable being "influenced" and "affected" by other factors such as cultures, social contexts, and interventions. According to some researchers (Kalogeropoulos et al., 2021), values can also become "aligned" between teachers and students in the classroom—for example, teachers can try to change or adjust their values so that different values could exist harmoniously in the classroom.

Naturally, across the studies, the three dimensions and their two respective poles often overlapped with each other. For example, most of the studies referred to more than one conception of values in their studies. Unsurprisingly, the conception of values could be intra- and inter-personal. Values could be subjective and change conceptually with longitudinal studies (Lee & Seo, 2021; Putwain et al., 2018; Weidinger et al., 2020) or interventions (Gaspard et al., 2015; O'Meara & Fitzmaurice, 2022; Rosenzweig et al., 2019). The overlapping of dimensions implies the possibilities of multiple combinations of the different characteristics of definitions and methodologies applied. Thus, we created the categories of conceptualising and operationalising values empirically in the next section.

3.4.2 Research Methods for Exploring Values in Mathematics Education

Concerning RQ2, the research methods used to investigate values in mathematics education according to different features of definitions, we identified three categories of studies: values as individual attributes, values as narratives, and values as the ways of acting (see Table 3.2).

The first group of studies was denoted by a definition of values, which emphasised the psychology dimension, a notion of an individual as the personal attributes (n = 49, 73%). These attributes were individual characteristics and independent from the sociocultural context that mainly investigated through self-reports, such as surveys (the espoused dimension). The studies reported the subjective view related to the mathematics subject, specific task, or specific practice in the classroom. Values here were relatively stable and regulated individual preferences, although some longitudinal studies (e.g., Lee & Seo, 2021; Putwain et al., 2018) or studies that applied interventions (e.g., Gaspard et al., 2015; O'Meara & Fitzmaurice, 2022) conceptualised values as changeable.

A large group of studies developed from the expectancy-value theory (e.g., Chatzistamatiou & Dermitzaki, 2014; Lazarides et al., 2022; Phan, 2014; Safavian & Conley, 2016) investigated participant's values as perceptions and interest in mathematics subject or tasks about their decision in the classroom. Some studies (e.g., Karamarkovich & Rutherford, 2021; Peixoto et al., 2017; Putwain et al., 2018) adopted the control-value theory, emphasising values and their relationship with emotions in learning mathematics. Those studies conceptualised values as cognitive and affective variables—particularly the task values based on the perceived importance and usefulness of the task (cognitive), and students' interest in and enjoyment drawn from

Table 3.2 Different conceptual and methodological emphases in researching values

Name of methodology categories	Definition dimensions			Descriptions	Examples
	Subject	Representation	Stability		
Values as personal attributes	Psychology/ sociocultural	Espoused	Stable/ change	Values as the level of agreement with different statements about oneself (e.g., preferences, importance). Data collection through surveys/ interviews	Dede, 2014; Hill et al., 2021; Putwain et al., 2018; Skaalvik et al., 2017
Values as narratives	Sociocultural	Espoused	Stable/ change	Values as narrated using social discourses or represented in documents. Data collection through narratives or document analysis	Kalogeropoulos et al., 2021; Mandt & Afdal, 2020; Weber et al., 2020
Values as the way of acting	Psychology/ sociocultural	Enacted	Stable/ change	Values as acted in particular social contexts. Data collection through observations and mixed methods. Interventions involved participation	Gaspard et al., 2015; Lim & Kor, 2012; Matthews, 2018

engaging with the task (affective). Another group of studies (e.g., Dede, 2014, 2015; Tang et al., 2021; Zhang et al., 2016) conceptualised values in relation to specific classroom teaching and learning practices, viewing personal values as part of the classroom culture. Other still (Hill et al., 2021; Sum et al., 2022) proposed values as a conative concept, building on Seah's (2019) idea that values act as a bridge between cognitive and affective dimensions, leading to behaviour and action. The integration of values with other constructs, such as students' mathematical well-being, was also seen (Hill et al., 2021).

The second group of studies investigated values through phenomena in the form of narratives (n = 7, 10%). Those studies emphasised the relevance of sociocultural context in shaping personal preferences (e.g., Kalogeropoulos et al., 2021), by highlighting students' personal experiences in the classroom. Other studies in this category analysed curriculum documents (Dede et al., 2021) or textbooks (Daher, 2021) as the sources of sociocultural context (which is different from the perspective focusing on individual attributes that place more emphasis on personal perceptions). This group focused on the in-depth stories espoused by a person in a specific context. For example, Kalogeropoulos et al. (2021) researched value alignment strategies used by teachers in the classroom and documented their narratives throughout the intervention process.

The third group of studies conceptualised values as ways of acting (n = 11, 16%). It focused on the enacted aspect of value in the classroom. Unlike the previous two groups, the studies in this group conceptualised values as reflected in teaching practices and students' classroom engagement. The primary analytical approaches here relied on observations of teacher-students interactions inside the classroom, student engagement during the given task (Fielding-Wells et al., 2017; Matthews, 2018), or teachers' teaching practices (Aktas & Argun, 2018; Law et al., 2012; Lim & Kor, 2012); with some studies supplementing these with interviews (e.g., Lim & Kor, 2012), scoring rubrics, field notes and students' work (Fielding-Wells et al., 2017).

3.5 Discussion and Conclusion

The main goal of this chapter was to explore how values had been conceptualised and operationalised in mathematics education research over the last decade; and then to propose a multidimensional model that needs to be considered when using definitions of value. This systematic review can provide a shared language that potentially bridges theoretical and methodological differences between studies.

We proposed the three dimensions of defining values: subject, representation, and stability. The discussion of values as a psychological (subject) dimension in the earlier literature suggested that values are either a cognitive or affective construct. On the one hand, Rokeach (1973) mentioned, "a person has a value is to say that cognitively he knows the correct way to behave or a correct end-state to strive for", which is known as "a cognition about the desirable" (p. 7). On the other hand, Bishop (1999) suggested values are "deep affective qualities" that "appear to survive longer in people's memories than does conceptual and procedural knowledge" (p. 2). But it has also been previously suggested that values could also be conative (Carr, 2019; Fischer & Boer, 2016; Seah, 2019). Cognitive variable involves "the dimension (or domain) of human needs and drives, desires and goals, choices, intentionality, and "will"- that is, the "why" behind human behaviour" (Goldin, 2019, p. 113). Values are connected to motivation and will that regulate a person's or a group's behavioural aspects. Despite recent proposals to view values as a conative variable, this dimension has received less attention from researchers (Carr, 2019). Atweh and Seah (2008) proposed that "values can be understood as commitments that guide an agent's decision-making and actions" (p. 3). This highlights the importance of investigating the sociocultural dimension of values in mathematics education, as mathematics is a knowledge domain shaped by culture (Bishop et al., 2003). Society, school, and classroom values are interrelated with personal values. By investigating the sociocultural dimension of values, we can better understand how values are formed within a particular sociocultural context.

Most studies reviewed in the process of this synthesis have focused on the espoused pole of representation dimension. Those studies involve examining how values are expressed through decisions and preferences. Large-scale studies investigating values across different cultural groups can be precious (Carr, 2019; Dede, 2014, 2015). However, the espoused pole has limitations as it may not always align with actual behaviour in real-life situations. Lee et al. (2022) have argued that the relationship between values and behaviours is non-linear. Additionally, individuals such as teachers might not always be consciously aware of their values (Bishop et al., 2001). Therefore, the enacted pole of representation dimension, which focuses on how values are put into practice and regulated, could offer more comprehensive insights into theorising the relationship between values and decision-making. Although the enacted pole is closely related to the conative variable, it has received less attention in terms of operationalisation in research.

Existing studies on values have primarily focused on the stable pole (stability dimension), emphasising value's persistence and consistency over time. In contrast, the malleability pole has received less attention. Studies that have explored changes in values have been limited to longitudinal studies, interventions, or changes observed in classroom situations. This limited focus on change may be attributed to the challenges associated with documenting changes in values over a long period of time or researching moment-by-moment changes in the classroom. Furthermore, there is the issue of conceptualisation of what actually constitutes change in values and what does not.

In addition to the insights provided by previous reviews (e.g., Bishop et al., 2003; Carr, 2019), our study introduces three categories of methodologies to examine values by making the distinction between their focus on: individual attributes, narratives, and ways of acting. The research which investigated individual attributes focused predominantly on self-report surveys; the narrative studies explored how values are expressed through interviews; and the studies focusing on the way of acting utilised observations of classroom interactions and practices that reflect values. It is important to note that this review did not aim to critique any specific approaches in researching values. Instead, by proposing these three categories, the study hoped to help identify possible limitations in the field. For example, our review reveals that recent studies focused mainly on values as individual attributes and espoused characteristics of values. While findings from such studies can be used to identify students' values, which can subsequently assist teachers in selecting and designing effective teaching strategies in the classroom, an overemphasis on individual attributes can result in a lack of evidence regarding the enacted aspect of values.

Numerous studies have indicated that values play a role in regulating a person's actions. Therefore, incorporating evidence from lesson observations on students' participation and teachers' teaching practices seems crucial to enrich our current knowledge of values in action. One possible challenge in conducting studies focusing on the ways of acting, however, is determining what to exactly observe and how during the lesson. The studies reviewed in this chapter have provided valuable insights into using supplementary data, such as field notes and scoring rubrics, to document teachers' teaching practices and students' participation. Additionally, this review suggests that incorporating other concepts such as teacher noticing (Aktaş et al., 2019), students' well-being (Clarkson et al., 2011; Hill et al., 2021), and expectancy values (Matthews, 2018) might further enhance the understanding of values in action. The findings from the ways of acting research can then contribute to teachers' professional development by informing the implementation of interventions and new teaching practices.

In conclusion, this systematic review highlights the importance of deliberately constructing conceptual and operational definitions of values, while considering the concept's multidimensional nature. We acknowledge that this review has a limitation in depicting the conception of values with the micro aspects of values conception, such as the differences between teacher values and student values. Teachers' values can be primarily link to their teaching practices, while students' values can be connected to their engagement in learning. This review proposes a general framework of definition dimensions derived from existing empirical studies. This framework can serve as a valuable guide for researchers exploring values in mathematics education, by pointing to the complexity of the concept. For example, researchers can identify the dimensions they want to focus on and design their studies accordingly (including the research goals, operationalisation of variables, and the selection of tools for data collection tools). By utilising this, researchers can contribute to conceptual coherence in values research. Furthermore, this study provides a common ground for researchers to engage in meaningful discussions and contribute to the development of the field. Researchers can collectively advance the knowledge and understanding of values in mathematics education by fostering a shared understanding of values and their dimensions.

Acknowledgements The study was partially funded by the Research Grants Council of the Hong Kong Special Administrative Region, China (Grant no. ECS 28609121).

References

Aktas, F. N., & Argun, Z. (2018). Examination of mathematical values in classroom practices: A case study of secondary mathematics teachers. *Education and Science, 43*(193), 121–141. https://doi.org/10.15390/EB.2018.7177

Aktaş, F. N., Yakıcı-Topbaş, E. S., & Dede, Y. (2019). The elementary mathematics teachers' values underlying teacher noticing. In P. Clarkson, W. T. Seah, & J. Pang (Eds.), *Values and valuing in mathematics education: Scanning and scoping the territory* (pp. 209–222). Springer International Publishing. https://doi.org/10.1007/978-3-030-16892-6_15

Akyıldız, P., Aktaş, F. N., Dede, Y., & Hacıömeroğlu, G. (2021). Mathematics teachers' values about teaching mathematics. *Studies in Educational Evaluation, 68*. https://doi.org/10.1016/j.stueduc.2020.100954

Atweh, B., & Seah, W. T. (2008). Theorising values and their study in mathematics education. In P. L. Jeffery (Ed.), *Proceedings of the 2007 AARE conference* (pp. 1–11).
Bishop, A. J. (1991). Mathematical values in the teaching process. In A. J. Bishop, S. Mellin-Olsen, & J. Van Dormolen (Eds.), *Mathematical knowledge: Its growth through teaching* (pp. 193–214). Springer.
Bishop, A. J., Seah, W. T., & Chin, C. (2003). Values in mathematics teaching—The hidden values in persuaders? In A. J. Bishop, M. A. Clements, C. Keitel, J. Kilpatrick, & F. K. S. Leung (Eds.), *Second international handbook of mathematics education* (pp. 717–765). Kluwer Academic Publishers.
Bishop, A. J., FitzSimons, G., Seah, W. T., & Clarkson, P. (2001). Do teachers implement their intended values in mathematics classrooms? In M. v. d. Heuvel-Panhuizen (Ed.), *Proceedings of the 25th conference of the international group for the psychology of mathematics education* (pp. 169–176). Freudenthal Institute.
Bishop, A. J., Clarke, B., Corrigan, D., & Gunstone, D. (2005). Teachers' preferences and practices regarding values in teaching mathematics and science. *Building connections: research, theory and practice (Proceedings of the annual conference of the mathematics education research group of Australasia)* (pp. 153–160).
Bishop, A. J. (1988). *Mathematical enculturation: A cultural perspective on mathematics education*. Kluwer Academic.
Bishop, A. J. (1999). Mathematics teaching and values education—An intersection in need of research. *European Journal of Teacher Education, 25*(1), 77–81. https://doi.org/10.1080/0261976022000012877
Bishop, A. J. (2020). Values in mathematics education. In S. Lerman (Ed.), Encyclopedia of mathematics education (pp. 893–896). Springer. https://doi.org/10.4324/9780367352028-7
Carr, M. E. (2019). Student and/or teacher valuing in mathematics classrooms: Where are we now, and where should we go? In P. Clarkson, W. T. Seah, & J. Pang (Eds.), *Values and valuing in mathematics education: Scanning and scoping the territory* (pp. 25–52). Springer International Publishing. https://doi.org/10.1007/978-3-030-16892-6_3
Chin, C., & Lin, F.-L. (2000). A case study of a mathematics teachers' pedagogical values: Use of a methodological framework of interpretation and reflection. *Proceedings of the National Science Council Part d: Mathematics, Science, Technology, and Education, 10*(2), 90–101.
Clarkson, P., Bishop, A., Seah, W. T., & FitzSimons, G. (2000, December). *Methodology challenges and constraints in the values and mathematics project*. [Paper presentation]. Annual Meeting of the Australian Association for Research in Education, Sydney, Australia.
Clarkson, P., Bishop, A. J., & Seah, W. T. (2011). Mathematics education and student values: The cultivation of mathematical wellbeing. In N. Clement, R. Toomey, & T. Lovat (Eds.), *International research handbook on values education and student wellbeing* (pp. 111–135). https://doi.org/10.1007/978-90-481-8675-4
Clarkson, P., Seah, W. T., & Pang, J. (2019). *Values and valuing in mathematics education: Scanning and scoping the territory*. Springer International Publishing. http://www.springer.com/series/15585
Cooper, H. M. (2017). *Research synthesis and meta-analysis: A step-by-step approach* (5th ed.). SAGE Publications, Inc.
Daher, W. (2021). Values in the geometry and measurement unit of the Palestinian Grade six mathematics book. *Journal of Beliefs and Values, 42*(4), 523–536.
Dahlke, H. O. (1958). Action and values: A cultural approach to education. In *Values in culture and classroom: A study in the sociology of the school* (pp. 3–11). Harper & Brothers Publisher.
Dede, Y. (2014). A comparison of Turkish and German mathematics teachers' values: A gender perspective. *Education and Science, 39*(171), 180–198.
Dede, Y. (2015). Comparing primary and secondary mathematics teachers' preferences regarding values about mathematics teaching in Turkey and Germany. *International Journal of Science and Mathematics Education, 13*, 227–255.

Dede, Y., Akçakın, V., & Kaya, G. (2022). Identifying students' mathematical and mathematics educational values in Turkish culture: A cross-sectional study. *Culture and Education*, 1–33. https://doi.org/10.1080/11356405.2022.2058795

Eccles, J., Adler, T. F., Futterman, R., Goff, S. B., Kaczala, C. M., & Meece, J. L. (1984). Expectancies, values, and academic behaviors. In J. T. Spence (Ed.), *Achievement and achievement motivation* (pp. 75–146). https://doi.org/10.2307/1422172

Feather, N. T. (1995). Values, valences, and choice: The influence of values on the perceived attractiveness and choice of alternatives. *Journal of Personality and Social Psychology, 68*(6), 1135–1151. https://doi.org/10.1037/0022-3514.68.6.1135

Fielding-Wells, J., O'Brien, M., & Makar, K. (2017). Using expectancy-value theory to explore aspects of motivation and engagement in inquiry-based learning in primary mathematics. *Mathematics Education Research Journal, 29*(2), 237–254. https://doi.org/10.1007/s13394-017-0201-y

Fischer, R., & Boer, D. (2016). Values: The dynamic nexus between biology, ecology and culture. *Current Opinion in Psychology, 8*, 155–160. https://doi.org/10.1016/j.copsyc.2015.12.009

Gaspard, H., Dicke, A. L., Flunger, B., Brisson, B. M., Häfner, I., Nagengast, B., & Trautwein, U. (2015). Fostering adolescents' value beliefs for mathematics with a relevance intervention in the classroom. *Developmental Psychology, 51*(9), 1226–1240. https://doi.org/10.1037/dev0000028

Goldin, G. A. (2019). Exploring a conative perspective on mathematical engagement. In S. Chamberlin & B. Sriraman (Eds.), *Affect in mathematical modeling* (pp. 111–129). Springer Nature Switzerland AG. https://doi.org/10.1007/978-3-030-04432-9_8

Henderson, M., & Thompson, D. (2003). *Values at work: The invisible threads between people, performance and profit*. Harper Collins.

Hill, J. L., Kern, M. L., Seah, W. T., & van Driel, J. (2021). Feeling good and functioning well in mathematics education: Exploring students' conceptions of mathematical well-being and values. *ECNU Review of Education, 4*(2), 349–375. https://doi.org/10.1177/2096531120928084

Hofstede, G. (2001). *Culture's consequences: Comparing values, behaviors, institutions, and organizations across nations* (2nd ed.). Sage Publications.

Kalogeropoulos, P., & Clarkson, P. (2019). The role of value alignment in levels of engagement of mathematics learning. In P. Clarkson, W. T. Seah, & J. Pang (Eds.), *Values and valuing in mathematics education: Scanning and scoping the territory* (pp. 115–127). Springer International Publishing.

Kalogeropoulos, P., Russo, J. A., & Clarkson, P. (2021). Exploring educator values alignment strategies in an intervention context: The emergence of the Beacon strategy. *ECNU Review of Education, 4*(2), 327–348. https://doi.org/10.1177/2096531120923127

Karamarkovich, S. M., & Rutherford, T. (2021). Mixed feelings: Profiles of emotions among elementary mathematics students and how they function within a control-value framework. *Contemporary Educational Psychology, 66*, 101996. https://doi.org/10.1016/j.cedpsych.2021.101996

Kluckhohn, C. (1959). Values and value-orientations in the theory of action: An exploration in definition and classification. In T. Parsons & E. A. Shils (Eds.), *Toward a general theory of action* (pp. 388–433). Harvard University Press.

Kluckhohn, F. R., & Shrodtbeck, F. L. (1961). *Variations in value orientations*. Greenwood Press.

Kohlberg, L. (1981). *Essays on moral development: The psychology of moral development* (Vol. 1). Harper & Row Publisher.

Law, H. Y., Wong, N. Y., & Lee, N. Y. L. (2012). A study of espoused values in Hong Kong's mathematics classrooms. *ZDM—International Journal on Mathematics Education, 44*(1), 45–57. https://doi.org/10.1007/s11858-012-0389-y

Lazarides, R., Schiepe-Tiska, A., Heine, J. H., & Buchholz, J. (2022). Expectancy-value profiles in math: How are student-perceived teaching behaviors related to motivational transitions? *Learning and Individual Differences, 98*. https://doi.org/10.1016/j.lindif.2022.102198

Lee, J. A., Bardi, A., Gerrans, P., Sneddon, J., van Herk, H., Evers, U., & Schwartz, S. H. (2022). Are value–behavior relations stronger than previously thought? It depends on value importance. *European Journal of Personality, 36*(2), 133–148. https://doi.org/10.1177/08902070211002965

Lim, C. S., & Kor, L. K. (2012). 'Excellent' primary mathematics teachers' espoused and enacted values of effective lessons. *ZDM—International Journal on Mathematics Education, 44*(1), 59–69. https://doi.org/10.1007/s11858-012-0390-5

Martinez, L. S. (2017). Variables (operationalization). In M. Allen (Ed.), *The SAGE encyclopedia of communication research methods*. SAGE Publications, Inc.

Matthews, J. S. (2018). When am I ever going to use this in the real world? Cognitive flexibility and urban adolescents' negotiation of the value of mathematics. *Journal of Educational Psychology, 110*(5), 726–746. https://doi.org/10.1037/edu0000242

O' Meara, N., & Fitzmaurice, O. (2022). Mathematics teacher educations' missing component: Developing pre-service teachers' appreciation of the utility-value of mathematics. *International Journal of Mathematical Education in Science and Technology*. https://doi.org/10.1080/0020739X.2022.2128455

Page, M. J., McKenzie, J. E., Bossuyt, P. M., Boutron, I., Hoffmann, T. C., Mulrow, C. D., Shamseer, L., Tetzlaff, J. M., Akl, E. A., Brennan, S. E., Chou, R., Glanville, J., Grimshaw, J. M., Hróbjartsson, A., Lalu, M. M., Li, T., Loder, E. W., Mayo-Wilson, E., McDonald, S., … & Moher, D. (2021). The PRISMA 2020 statement: An updated guideline for reporting systematic reviews. *The BMJ, 372*. https://doi.org/10.1136/bmj.n71

Palm, H., & Skott, J. (2018). *Students' and teachers' values, attitudes, feelings and beliefs in mathematics classrooms: Selected papers from the 22nd MAVI conference*.

Peixoto, F., Sanches, C., Mata, L., & Monteiro, V. (2017). How do you feel about math?: Relationships between competence and value appraisals, achievement emotions and academic achievement. *European Journal of Psychology of Education, 32*(3), 385–405. https://doi.org/10.1007/s10212-016-0299-4

Philipp, R. A. (2007). Mathematics teachers' beliefs and affect. In F. K. Lester (Ed.), *Second handbook of research on mathematics teaching and learning* (pp. 257–315).

Putwain, D. W., Pekrun, R., Nicholson, L. J., Symes, W., Becker, S., & Marsh, H. W. (2018). Control-value appraisals, enjoyment, and boredom in mathematics: A longitudinal latent interaction analysis. *American Educational Research Journal, 55*(6), 1339–1368. https://doi.org/10.3102/0002831218786689

Raths, L. E., Harmin, M., & Simon, S. B. (1978). *Values and teaching: Working with values in the classroom* (2nd ed.). C.E. Merrill Pub. Co.

Rokeach, M. (1979). From individual to institutional values: With special reference to the values of science. In M. Rokeach (Ed.), *Understanding human values* (pp. 47–70). The Free Press.

Rokeach, M. (1973). *The nature of human values and value system*. The Free Press.

Rosenzweig, E. Q., Hulleman, C. S., Barron, K. E., Kosovich, J. J., Priniski, S. J., & Wigfield, A. (2019). Promises and pitfalls of adapting utility value interventions for online math courses. *Journal of Experimental Education, 87*(2), 332–352. https://doi.org/10.1080/00220973.2018.1496059

Schwartz, S. H. (1992). Universals in the content and structure of values: Theoretical advances and empirical tests in 20 countries. *Advances in Experimental Social Psychology, 25*, 1–65. https://doi.org/10.1016/S0065-2601(08)60281-6

Seah, W. T., & Andersson, A. (2015). Teacher alignment of values in mathematics classrooms. In *CERME 9-ninth congress of the European society for research in mathematics education* (pp. 3122–3128).

Seah, W. T. (2013). Assessing values in mathematics education. In A. M. Lindmeier & A. Heinze (Eds.), *Proceedings of 37th conference of the international group for the psychology of mathematics education* (Vol. 4, pp. 193–200). PME.

Seah, W. T. (2019). Values in mathematics education: Its conative nature, and how it can be developed. *Research in Mathematics Education, 22*(2), 99–121. https://doi.org/10.7468/jksmed.2019.22.2.99

Sum, E. S. W., Wong, K. L., & Seah, W. T. (2022). Attributes of mathematics learning valued by ethnic minority students in Hong Kong mainstream schools. *Asia Pacific Journal of Education.* https://doi.org/10.1080/02188791.2022.2071215

Swadener, M., & Soedjadi, R. (1988). Values, mathematics education, and the task of developing pupils' personalities: An Indonesian perspective. *Educational Studies in Mathematics, 19*(2), 193–208.

Thomas, J., & Harden, A. (2008). Methods for the thematic synthesis of qualitative research in systematic reviews. *BMC Medical Research Methodology, 8,* 1–10. https://doi.org/10.1186/1471-2288-8-45

Weidinger, A. F., Spinath, B., & Steinmayr, R. (2020). The value of valuing math: Longitudinal links between students' intrinsic, attainment, and utility values and grades in math. *Motivation Science, 6*(4), 413–422. https://doi.org/10.1037/mot0000179

Wigfield, A., & Eccles, J. S. (2000). Expectancy-value theory of achievement motivation. *Contemporary Educational Psychology, 25*(1), 68–81. https://doi.org/10.1006/ceps.1999.1015

Williams, R. M. (1979). Change and stability in values and value systems: A sociological perspective. In M. Rokeach (Ed.), *Understanding human values.* The Free Press.

Zhang, Q. P., & Seah, W. T. (2021). Thematic issue on values and valuing in mathematics education: Revisiting mathematics education from cultural perspectives. *ECNU Review of Education, 4*(2), 225–229.

Zhang, Q., Barkatsas, T., Law, H. Y., Leu, Y. C., Seah, W. T., & Wong, N. Y. (2016). What primary students in the Chinese Mainland, Hong Kong and Taiwan value in mathematics learning: A comparative analysis. *International Journal of Science and Mathematics Education, 14*(5), 907–924. https://doi.org/10.1007/s10763-014-9615-0

Zhang, Q. (2019). Values in mathematics learning: Perspectives of Chinese Mainland primary and secondary students. In P. Clarkson, W. T. Seah, & J. Pang (Eds.), *Values and valuing in mathematics education: Scanning and scoping the territory* (pp. 185–196). Springer International Publishing. https://doi.org/10.1007/978-3-030-16892-6_13

Appendix: Selected Empirical Articles for Review in This Study (N = 67)

Aktas, F. N., & Argun, Z. (2018). Examination of mathematical values in classroom practices: A case study of secondary mathematics teachers. *Education and Science, 43*(193), 121–141. https://doi.org/10.15390/EB.2018.7177

Akyıldız, P., Aktaş, F. N., Dede, Y., & Hacıömeroğlu, G. (2021). Mathematics teachers' values about teaching mathematics. *Studies in Educational Evaluation, 68.* https://doi.org/10.1016/j.stueduc.2020.100954

Asmail, R., Spangenberg, E. D., & Ramdhany, V. (2020). What Grade 8–10 learners from a mathematics-and-science-focus school value as important in the learning of mathematics. *African Journal of Research in Mathematics, Science and Technology Education, 24*(2), 241–252. https://doi.org/10.1080/18117295.2020.1818041

Cambria, J., Brandt, H., Nagengast, B., & Trautwein, U. (2017). Frame of reference effects on values in mathematics: Evidence from German secondary school students. *ZDM—Mathematics Education, 49*(3), 435–447. https://doi.org/10.1007/s11858-017-0841-0

Cavanna, J. M., Herbel-Eisenmann, B., & Seah, W. T. (2015). Investigating teachers' appraisal of unexpected moments and underlying values: An exploratory case in the context of changing mathematics classroom discourse. *Research in Mathematics Education, 17*(3), 163–182. https://doi.org/10.1080/14794802.2015.1112301

Chatzistamatiou, M., & Dermitzaki, I. (2014). Self-regulatory teaching in mathematics: Relations to teachers' motivation, affect and professional commitment. *European Journal of Psychology and Educational Research, 29,* 295–310. https://doi.org/10.1007/s10212-013-0199-9

Daher, W. (2021). Values in the geometry and measurement unit of the Palestinian Grade six mathematics book. *Journal of Beliefs and Values, 42*(4), 523–536. https://doi.org/10.1080/13617672.2021.1878722

Davis, E. K., Seah, W. T., Howard, N., & Wilmot, E. M. (2021). The attributes of mathematics learning which Ghanaian senior high school students value. *Journal of Global Education and Research, 5*(1), 1–14. https://doi.org/10.5038/2577-509x.5.1.1096

Dede, Y. (2013a). Examining the underlying values of Turkish and German mathematics teachers' decision making processes in group studies. *Educational Sciences: Theory & Practice, 13*(1), 690–706.

Dede, Y. (2013b). The effect of German and Turkish mathematics teachers' teaching experience on mathematics education values: A cross-comparative study. *International Journal of Mathematical Education in Science and Technology, 44*(2), 232–252. https://doi.org/10.1080/0020739X.2012.714489

Dede, Y. (2014). A comparison of Turkish and German mathematics teachers' values: A gender perspective. *Education and Science, 39*(171), 180–198.

Dede, Y. (2015). Comparing primary and secondary mathematics teachers' preferences regarding values about mathematics teaching in Turkey and Germany. *International Journal of Science and Mathematics Education, 13*, 227–255.

Dede, Y., Akçakın, V., & Kaya, G. (2021). Mathematical, mathematics educational, and educational values in mathematical modeling tasks. *ECNU Review of Education, 4*(2), 241–260. https://doi.org/10.1177/2096531120928089

Dicke, A. L., Rubach, C., Safavian, N., Karabenick, S. A., & Eccles, J. S. (2021). Less direct than you thought: Do teachers transmit math value to students through their cognitive support for understanding? *Learning and Instruction, 76*. https://doi.org/10.1016/j.learninstruc.2021.101521

Fadda, D., Scalas, L. F., Morin, A. J. S., Marsh, H. W., & Gaspard, H. (2020). Value beliefs about math. *European Journal of Psychological Assessment, 36*(2), 259–268.

Federici, R. A., & Skaalvik, E. M. (2014). Students' perception of instrumental support and effort in mathematics: The mediating role of subjective task values. *Social Psychology of Education, 17*, 527–540. https://doi.org/10.1007/s11218-014-9264-8

Fielding, J., Geiger, V., Miller, J., Bruder, R., Towara, U., & Ratnayake, I. (2022). Using situated expectancy value theory to explore initial teacher education students' motivation to engage with challenging mathematical tasks. *Teaching and Teacher Education, 113*. https://doi.org/10.1016/j.tate.2022.103663

Fielding-Wells, J., O'Brien, M., & Makar, K. (2017). Using expectancy-value theory to explore aspects of motivation and engagement in inquiry-based learning in primary mathematics. *Mathematics Education Research Journal, 29*(2), 237–254. https://doi.org/10.1007/s13394-017-0201-y

Gaspard, H., Dicke, A. L., Flunger, B., Brisson, B. M., Häfner, I., Nagengast, B., & Trautwein, U. (2015a). Fostering adolescents' value beliefs for mathematics with a relevance intervention in the classroom. *Developmental Psychology, 51*(9), 1226–1240. https://doi.org/10.1037/dev0000028

Gaspard, H., Dicke, A., Flunger, B., Schreier, B., Häfner, I., Trautwein, U., & Nagengast, B. (2015b). More value through greater differentiation: Gender differences in value beliefs about math. *American Psychological Association, 107*(3), 663–677.

Guo, J., Marsh, H. W., Parker, P. D., Morin, A. J. S., & Yeung, A. S. (2015). Expectancy-value in mathematics, gender and socioeconomic background as predictors of achievement and aspirations: A multi-cohort study. *Learning and Individual Differences, 37*, 161–168. https://doi.org/10.1016/j.lindif.2015.01.008

Haciomeroglu, G. (2020). Mathematics education values portrayed by elementary student teacher. *Educational Policy Analysis and Strategic Research, 15*(2), 259–270. https://doi.org/10.29329/epasr.2020.251.14

Hanin, V., & Van Nieuwenhoven, C. (2016). The influence of motivational and emotional factors in mathematical learning in secondary education. *Revue Europeenne De Psychologie Appliquee, 66*(3), 127–138. https://doi.org/10.1016/j.erap.2016.04.006

Henschel, S., & Roick, T. (2017). Relationships of mathematics performance, control and value beliefs with cognitive and affective math anxiety. *Learning and Individual Differences, 55*, 97–107. https://doi.org/10.1016/j.lindif.2017.03.009

Hill, J. L., Kern, M. L., Seah, W. T., & van Driel, J. (2021). Feeling good and functioning well in mathematics education: Exploring students' conceptions of mathematical well-being and values. *ECNU Review of Education, 4*(2), 349–375. https://doi.org/10.1177/2096531120928084

Kadroon, T., & Inprasitha, M. (2013). Professional development of mathematics teachers with lesson study and open approach: The process for changing teachers values about teaching mathematics. *Psychology, 04*(02), 101–105. https://doi.org/10.4236/psych.2013.42014

Kalogeropoulos, P., Russo, J. A., & Clarkson, P. (2021). Exploring educator values alignment strategies in an intervention context: The emergence of the Beacon strategy. *ECNU Review of Education, 4*(2), 327–348. https://doi.org/10.1177/2096531120923127

Karamarkovich, S. M., & Rutherford, T. (2021). Mixed feelings: Profiles of emotions among elementary mathematics students and how they function within a control-value framework. *Contemporary Educational Psychology, 66*. https://doi.org/10.1016/j.cedpsych.2021.101996

Krawitz, J., & Schukajlow, S. (2017). Do students value modelling problems, and are they confident they can solve such problems? Value and self-efficacy for modelling, word, and intra-mathematical problems. *ZDM—Mathematics Education, 50*(1–2), 143–157. https://doi.org/10.1007/s11858-017-0893-1

Law, H. Y., Wong, N. Y., & Lee, N. Y. L. (2012). A study of espoused values in Hong Kong's mathematics classrooms. *ZDM—International Journal on Mathematics Education, 44*(1), 45–57. https://doi.org/10.1007/s11858-012-0389-y

Lazarides, R., Buchholz, J., & Rubach, C. (2018). Teacher enthusiasm and self-efficacy, student-perceived mastery goal orientation, and student motivation in mathematics classrooms. *Teaching and Teacher Education, 69*, 1–10. https://doi.org/10.1016/j.tate.2017.08.017

Lazarides, R., Schiepe-Tiska, A., Heine, J. H., & Buchholz, J. (2022). Expectancy-value profiles in math: How are student-perceived teaching behaviors related to motivational transitions? *Learning and Individual Differences, 98*. https://doi.org/10.1016/j.lindif.2022.102198

Lee, Y., & Seo, E. (2021). Longitudinal relations between South Korean adolescents' academic self-efficacy and values in mathematics and English. *British Journal of Educational Psychology, 91*(1), 217–236. https://doi.org/10.1111/bjep.12357

Levenson, E. S. (2022). Exploring the relationship between teachers' values and their choice of tasks: The case of occasioning mathematical creativity. *Educational Studies in Mathematics, 109*, 469–489. https://doi.org/10.1007/s10649-021-10101-9

Liebendörfer, M., & Schukajlow, S. (2020). Quality matters: How reflecting on the utility value of mathematics affects future teachers' interest. *Educational Studies in Mathematics, 105*(2), 199–218. https://doi.org/10.1007/s10649-020-09982-z

Lim, C. S., & Kor, L. K. (2012). 'Excellent' primary mathematics teachers' espoused and enacted values of effective lessons. *ZDM—International Journal on Mathematics Education, 44*(1), 59–69. https://doi.org/10.1007/s11858-012-0390-5

Madosi, T., Spangenberg, E. D., & Ramdhany, V. (2020). The values learners consider as important in the learning of mathematics. *Perspectives in Education, 38*(1), 181–196. https://doi.org/10.18820/2519593x/pie.v38i1.13

Mandt, H. M., & Afdal, G. (2020). On a math mission: The time and space compression of professional identity and values in prospective mathematics teachers' stories. *European Educational Research Journal*, 1–17. https://doi.org/10.1177/1474904120957208

Matthews, J. S. (2018). When am I ever going to use this in the real world? Cognitive flexibility and urban adolescents' negotiation of the value of mathematics. *Journal of Educational Psychology, 110*(5), 726–746.

O' Meara, N., & Fitzmaurice, O. (2022). Mathematics teacher educations' missing component: Developing pre-service teacher's appreciation of the utility-value of mathematics. *International Journal of Mathematical Education in Science and Technology.* https://doi.org/10.1080/0020739X.2022.2128455

Pang, J., & Seah, W. T. (2021). Excellent mathematical performance despite "Negative" affect of students in Korea: The values perspective. *ECNU Review of Education, 4*(2), 285–306. https://doi.org/10.1177/2096531120930726

Parrisius, C., Gaspard, H., Trautwein, U., & Nagengast, B. (2020). The transmission of values from math teachers to their ninth-grade students: Different mechanisms for different value dimensions? *Contemporary Educational Psychology, 62.* https://doi.org/10.1016/j.cedpsych.2020.101891

Peixoto, F., Sanches, C., Mata, L., & Monteiro, V. (2017). How do you feel about math?: Relationships between competence and value appraisals, achievement emotions and academic achievement. *European Journal of Psychology of Education, 32*(3), 385–405. https://doi.org/10.1007/s10212-016-0299-4

Peng, A., & Nyroos, M. (2012). Values in effective mathematics lessons in Sweden: What do they tell us? *Mathematics Enthusiast, 9*(3), 409–429.

Petersen, J. L., & Hyde, J. S. (2017). Trajectories of self-perceived math ability, utility value and interest across middle school as predictors of high school math performance. *Educational Psychology, 37*(4), 438–456. https://doi.org/10.1080/01443410.2015.1076765

Phan, H. P. (2014). Expectancy-value and cognitive process outcomes in mathematics learning: A structural equation analysis. *Higher Education Research and Development, 33*(2), 325–340. https://doi.org/10.1080/07294360.2013.832161

Putwain, D. W., Pekrun, R., Nicholson, L. J., Symes, W., Becker, S., & Marsh, H. W. (2018). Control-value appraisals, enjoyment, and boredom in mathematics: A longitudinal latent interaction analysis. *American Educational Research Journal, 55*(6), 1339–1368. https://doi.org/10.3102/0002831218786689

Putwain, D. W., Schmitz, E. A., Wood, P., & Pekrun, R. (2021). The role of achievement emotions in primary school mathematics: Control–value antecedents and achievement outcomes. *British Journal of Educational Psychology, 91*(1), 347–367. https://doi.org/10.1111/bjep.12367

Reindl, M., Gniewosz, B., & Dresel, M. (2021). Friends' influence on the development of academic values in mathematics: Are there differences between female and male dyads? *European Journal of Psychology of Education, 36*, 781–797.

Rosenzweig, E. Q., Hulleman, C. S., Barron, K. E., Kosovich, J. J., Priniski, S. J., & Wigfield, A. (2019). Promises and pitfalls of adapting utility value interventions for online math courses. *Journal of Experimental Education, 87*(2), 332–352. https://doi.org/10.1080/00220973.2018.1496059

Safavian, N. (2019). What makes them persist? Expectancy-value beliefs and the math participation, performance, and preparedness of Hispanic youth. *AERA Open, 5*(3), 233285841986934. https://doi.org/10.1177/2332858419869342

Safavian, N., & Conley, A. (2016). Expectancy-value beliefs of early-adolescent Hispanic and non-Hispanic youth. *AERA Open, 2*(4), 1–17. https://doi.org/10.1177/2332858416673357

Sahin, S. M., & Özpinar, I. (2020). Reflections of mathematical and mathematics educational values on pedagogical formation students' teaching preferences. *Asian Journal of Education and Training, 6*(1), 80–92. https://doi.org/10.20448/journal.522.2020.61.80.92

Sahin, S. A., Tunca, N., Altinkurt, Y., & Yilmaz, K. (2016). Relationship between professional values and critical thinking disposition of science-technology and mathematics teachers. *Eurasia Journal of Mathematics, Science and Technology Education, 12*(1), 25–40. https://doi.org/10.12973/eurasia.2016.1371a

Scheidemann, B., Gasteiger, H., & Puca, R. M. (2022). Effects of affiliation-, achievement-, and power-related topics in mathematical word problems on students' performance, task- related values, and expectancies. *PLoS ONE, 17*(6), 1–28.

Seah, W. T., & Peng, A. (2012). What students outside Asia value in effective mathematics lessons: A scoping study. *ZDM—International Journal on Mathematics Education, 44*(1), 71–82. https://doi.org/10.1007/s11858-012-0398-x

Skaalvik, E. M., Federici, R. A., Wigfield, A., & Tangen, T. N. (2017). Students' perceptions of mathematics classroom goal structures: Implications for perceived task values and study behavior. *Social Psychology of Education, 20*(3), 543–563. https://doi.org/10.1007/s11218-017-9382-1

Sum, E. S. W., Wong, K. L., & Seah, W. T. (2022). Attributes of mathematics learning valued by ethnic minority students in Hong Kong mainstream schools. *Asia Pacific Journal of Education*, 1–15. https://doi.org/10.1080/02188791.2022.2071215

Tang, H., Seah, W. T., Zhang, Q., & Zhang, W. (2021). The mathematics learning attributes valued by students in Eastern China. *ECNU Review of Education, 4*(2), 261–284. https://doi.org/10.1177/2096531120930240

Tossavainen, T., & Faarinen, E. C. (2019). Swedish fifth and sixth graders' motivational values and the use of ICT in mathematics education. *Eurasia Journal of Mathematics, Science and Technology Education, 15*(12). https://doi.org/10.29333/ejmste/108533

Wang, W., & Yin, H. (2017). Environment matters: Exploring the relationships between the classroom environment and college students' affect in mathematics learning in China. *Asia Pacific Education Review, 18*(3), 321–333. https://doi.org/10.1007/s12564-017-9478-5

Weber, K., Lew, K., & Mejía-Ramos, J. P. (2020). Using expectancy value theory to account for individuals' mathematical justifications. *Cognition and Instruction, 38*(1), 27–56. https://doi.org/10.1080/07370008.2019.1636796

Weidinger, A. F., Spinath, B., & Steinmayr, R. (2020). The value of valuing math: Longitudinal links between students' intrinsic, attainment, and utility values and grades in math. *Motivation Science*. https://doi.org/10.1037/mot0000179

Xu, C., Lem, S., & Onghena, P. (2021). Examining developmental relationships between utility value, interest, and cognitive competence for college statistics students with differential self-perceived mathematics ability. *Learning and Individual Differences, 86*(September 2020), 101980. https://doi.org/10.1016/j.lindif.2021.101980

Yurt, E. (2015). Understanding middle school students' motivation in math class: The expectancy-value model perspective. *International Journal of Education in Mathematics, Science and Technology, 3*(4), 288. https://doi.org/10.18404/ijemst.26938

Yurt, E. (2022). Mathematics self-efficacy as a mediator between task value and math anxiety in secondary school students. *International Journal of Curriculum and Instruction, 14*(2), 1204–1221.

Zhang, Q., Barkatsas, T., Law, H. Y., Leu, Y. C., Seah, W. T., & Wong, N. Y. (2016). What primary students in the Chinese Mainland, Hong Kong and Taiwan value in mathematics learning: A comparative analysis. *International Journal of Science and Mathematics Education, 14*(5), 907–924. https://doi.org/10.1007/s10763-014-9615-0

Chapter 4
Collectivist and Individualist Values in Mathematics Education

Richard Barwell, Yasmine Abtahi, and Kjellrun Hiis Hauge

4.1 Introduction

Western, Eurocentric 'standard' mathematics has long been associated with a set of values that emphasise, among other things, rationality, control, mystery and progress (Bishop, 1988). In this sense, mathematics is seen as working with science to solve important problems, such as finding cures for disease, developing energy sources or ensuring stable food supplies for the growing human population. In this chapter, we argue that this view of mathematics is rather limited, both in terms of its vision of mathematics, and in terms of the role of mathematics in society. We draw on Bishop's work on values in mathematics and mathematics education, as well as on ideas from critical mathematics education (Skovsmose, 2023) and the philosophy of post-normal science (Funtowicz & Ravetz, 1993), to argue for recognising and teaching more expansive, collectivist values in relation to mathematics. In the last part of the chapter, we develop these ideas through a comparison of the values embedded in two mathematics curricula (from Ontario, Canada, and Norway). We do not see individualist and collectivist orientations as a binary distinction. Instead, we think about these ideas as a kind of continuum in which either an individualist orientation or a collectivist orientation may be emphasised more but without the exclusion of the other.

R. Barwell (✉)
University of Ottawa, Ottawa, Canada
e-mail: richard.barwell@uottawa.ca

Y. Abtahi
University of South-East Norway, Notodden, Norway
e-mail: Yasmine.Abtahi@usn.no

K. H. Hauge
Western Norway University of Applied Sciences, Bergen, Norway
e-mail: Kjellrun.Hiis.Hauge@hvl.no

Y. Dede et al. (eds.), *Values and Valuing in Mathematics Education*,
https://doi.org/10.1007/978-981-99-9454-0_4

Critical mathematics education is interested in the role of mathematics education in wider society and offers a contrasting perspective to popular discourses about mathematics (Andersson & Barwell, 2021). Various different strands of critical mathematics education emphasise similar goals. Some key points include:

- mathematics is an important tool for analysing social problems and critiquing policies and practices, particularly with respect to social justice;
- discourses about mathematics circulating in society produce or make possible various subject positions, such as in relation to gender or race, while limiting other possibilities;
- mathematics is deeply embedded in contemporary society, especially through technology, and as such shapes social reality;
- mathematics education needs to educate children to understand how mathematics is used to shape social reality.

The values implied by critical mathematics education include that mathematics is related to social responsibility and that it should be used to further collective well-being, equity, social justice, democracy and so on. Mathematics education should promote critique, questioning and reflection, not just in terms of the technical aspects of mathematical reasoning, but also in terms of how mathematics is used in society and shapes society. We see these values as having a more strongly collectivist orientation, since they are focused on how mathematics is used for collective concerns and through collective processes.

One example that explicitly shows collectivist values in teaching and learning is the four pillars of education proposed by Sobe (2021) with the aim of understanding how to sustain the relations and the space shared by all of us humans, trees, animals, coral reefs and more. With the concept of the commons in mind, the four pillars for learning are described as: 1. learning to study, inquire and co-construct together; 2. learning to collectively mobilise; 3. learning to live in a common world; 4. learning to attend and care. The values that are promoted in these pillars are mostly collectivist. Values such as acknowledging diverse and networked dimensions of knowledge, collaborative capability to empower learners to take action together, tolerating and respecting the rights of other humans and the natural world of which we are a part, and finally to be more cautious about the dangers of focusing on individualism and diminishing empathy. Again, how these pillars can be weighted in curricula or classrooms form a continuum.

In recent years, collectivist values have been explored in the teaching and learning of mathematics. For example, Moses and Cobb (2001), navigating the "Algebra Project" to gain insight into the United States civil rights movement through mathematics, explained that unless pupils become decisive members of their education, public school education itself will remain a manifestation of social inequality. With an even broader perspective, D'Ambrosio asked: "How do we, as mathematicians and mathematics educators, fulfil our commitments to mankind?" (1998, p. 67). With the pursuit of peace, D'Ambrosio and Rosa (2017) located (ethno)mathematics school practices to respect for solidarity, human dignity and cooperation with others. They claimed that in addition to techniques and tools to solve mathematical problems, pupils need to extend their understanding to include how mathematics connects to other disciplines and problems in society and the environment. Similarly, explaining that mathematics education research and practices are neither the cause of nor the solution to intolerance and inequity, Valero et al. (2012) argue that democracy as "the striving for a chance of a dignified life" (p. 2) is an undertaking to which mathematics teaching and learning can contribute. Valero et al. propose the displacement of the core, traditional mathematical concepts and competencies as the centre of the curriculum. Finally, in teaching mathematical concepts to a group of Canadian Indigenous teachers, Abtahi (2022) highlights ethical awareness to increase sensitivity, responsiveness and capabilities in other-oriented and imaginative attempts to experience events as another human has experienced them. These studies show the importance of incorporating collectivist values in the teaching and learning of mathematics in more ideal and theoretical ways. In the next sections, we present key ideas from post-normal science, in order to frame the role of mathematics in society, before relating them to Bishop's work on values in mathematics and mathematics education.

4.2 Values in Mathematics as Part of Post-Normal Science

The philosophy of post-normal science was developed to support science in relation to policy to meet the needs of complex contemporary problems (Funtowicz & Ravetz, 1993). The term 'post-normal' points to Funtowicz and Ravetz's understanding of required science as different from Kuhn's understanding of so-called normal science. The term derives from Kuhn's (1962) seminal account of scientific rationality which he characterises as alternating between 'normal science' and 'revolutionary' periods that lead to paradigm shifts and new "truths". Normal science works within established paradigms and amounts to a kind of puzzle solving. Funtowicz and Ravetz (1993) place scientific activity in a broader social context that seeks to solve complex problems in the world, in particular where single "truths" are not possible to find and puzzle pieces may not fit but may still be relevant for shedding light on the problem and finding solutions. Funtowicz and Ravetz (1993) explain:

> To characterize an issue involving risk and the environment, in what we call 'post- normal science', we can think of it as one where facts are uncertain, values in dispute, stakes high and decisions urgent. In such a case, the term 'problem', with its connotations of an exercise where a defined methodology is likely to lead to a clear solution, is less appropriate. We would be misled if we retained the image of a process where true scientific facts simply determine the correct policy conclusions. (p. 744)

In post-normal situations, science can provide analyses, examine and propose possible response scenarios and suggest projections, but it cannot provide a detailed roadmap because quantifications are built on simplifications and assumptions. This is the kind of situation we face with things like climate change, global pollution, pandemics, economic inequality or the search for peace. A key feature of post-normal situations is that all options involve different degrees of impact or risk for different groups of people, including the option of doing nothing. In such situations, in which science cannot provide a single best solution, *values* come into play and may be highly contested. For example, responses to climate change might include choices between reducing consumption, finding new forms of energy production in order to maintain levels of consumption, or business as usual. The arguments in relation to each of these choices may be based on mathematics and science, but will also include a values dimension, such as whether we place more value on human consumption as a route to happiness and well-being or more value on not interfering with the planetary ecosystem.

Values are important in that decisions affect people and nature in different ways and, because knowledge is uncertain, scientific knowledge may favour certain values in hidden ways, such as, for example, in the data collected, in methodological choices or even in the research question asked. Post-normal science takes these features into account through placing uncertainties at the centre of debate and decision-making, acknowledging that statistics may not suffice in assessing uncertainties, and inviting citizens to contribute to decision-making. The idea behind this latter component is that citizens can contribute with, in particular, local knowledge, context perspectives, reviewing scientific activity and discussing the values dimensions and relevance perspectives of science in post-normal situations.

The distinction between normal and post-normal science relates to increasing degrees of uncertainty and increasingly high stakes. As the stakes get higher, the role of values comes to the fore and the uncertainty together with how it is handled becomes more important. Furthermore, we argue that the values in question shift from individualist values to collectivist values. In normal science, key values are related to the integrity of scientific activity in itself: honesty of reporting, the methodical nature of the work, safety of participants and so on. In post-normal science, the role of values becomes much more collectively oriented, due to the high-stakes nature of possible decisions which affect larger groups of people. Because groups of people may have very different views on what the problem is and what decisions are beneficial, the approach to decision-making is (or should be) more democratic, including dialogue between stakeholders, citizens, experts and decision-makers. Taken together, how values are weighted in normal science and post-normal science forms a continuum from individualist values to collectivist values.

One of the features of post-normal situations is that they are complex non-linear systems, so that choices may have unpredictable impacts on the system and there is no single best solution. A good example was observed in the COVID-19 pandemic, in which countries made choices between lockdown strategies or so-called 'herd immunity' strategies. Science was not sufficient to provide answers on which strategy to choose as knowledge was uncertain, and decisions depended on social values. Each option had a huge impact on society. Moreover, mathematical information was an important part of reviewing these options and in the communication of the political choice. Mathematics was shown to be a useful but limited tool in describing, predicting and communicating the pandemic, and decisions were informed by mathematics-based knowledge, but based on political processes. Decisions and processes have been extensively criticised, for example because uncertainty was not properly communicated, and citizens were not included in decision-making, i.e., not applying principles of post-normal science (Biggeri & Saltelli, 2021).

The involvement of citizens in post-normal science through democratic processes underlines the importance of mathematics and science education in preparing students to participate in such processes. That is, education needs to prepare students to understand and critique mathematical and scientific information, to understand and critique mathematical and scientific data, methods and techniques, and to understand the role of uncertainty, risk and values in decision-making. These activities also require a critical awareness of different values associated with mathematics (Hauge & Barwell, 2017).

4.3 Values in Mathematics: Towards Individualist and Collectivist Orientations

Our next point of departure is Bishop's (1988) framework of three pairs of values connected to mathematics: *ideology* (rationalism, objectivism); *sentiment* (control, progress); and *sociology* (openness, mystery). We propose that each of Bishop's (1988) six categories of values can be understood as either individualist versions or as collectivist versions. We understand individualist values as those values that promote individual development, success, or well-being. We understand collectivist values as those that promote learning collectively, mobilise collective actions in accordance with critical mathematics education and post normal science, and promote care for others. This proposal is explained in the following text and summarised in Table 4.1.

Bishop (1988) emphasizes that the common ideology of mathematics is strongly associated with the process of rational and logical reasoning, which makes *rationalism* a clear value connected to mathematics. Mathematical reasoning includes deductive reasoning, mathematical proofs, etc. We relate this to the individual student valuing particular forms of reasoning. Of course, these values are widely shared in the mathematics community, so in that sense they are collective, but in mathematics education, we assume that mathematics instruction seeks to foster these values in each student. Bishop continues that rational reasoning is highly appreciated in society in general, also outside the field of mathematics, and the applicability of mathematics to problems has been valued in society. This we relate to the rationality of democratic participation and reasoning. Hence, rationalism is about critical reflection and critical mathematical literacy. It involves some of the values of post-normal science and critical mathematics education (see above), such as reasoning through applied mathematics, critically reviewing mathematical information and having a critical sense of the strengths, limits and drawbacks of mathematical reasoning in a given context.

Bishop (1988) continues by denoting the complementary value to rationalism as *objectivism*, which can be understood as the product of rationalism due to the manipulation of mathematical objects through symbol systems and which is detached from

Table 4.1 Values in mathematics education

	Individualist	Collectivist
Ideology	Rationalism	
	E.g., logical reasoning, proof	E.g., reasoning in democratic participation
	Objectivism	
	E.g., symbolic manipulation, application	E.g., argumentation processes, democratic debate
Sentiment	Control	
	E.g., mastery of mathematical techniques	E.g., applying mathematics in societal and environmental problems with uncertain outcomes
	Progress	
	E.g., structured curriculum	E.g., exploration of social problems
Sociology	Openness	
	E.g., critical thinking within mathematics	E.g., ethics and empathy towards relationships
	Mystery	
	E.g., sense of wonder	E.g., awareness of strengths and limitations of mathematics in post-normal situations

the living world. The value of objectivism concerns the worldview that mathematics involves manipulating Platonic objects and thereby obtains a neutral aura. We argue that at an individualist level, education promotes this value, which is important for learning the capacity and power of mathematics and what can be achieved through using mathematical symbol systems. A collectivist dimension of objectivism extends this idea to consider how mathematics is incorporated in broader argumentation processes as part of democratic debate, in line with critical mathematics education and post-normal science. The weighting of objectivism therefore goes from, for example, an appreciation of the construction of mathematical models, to a broader awareness of how mathematical models function in society, what their strengths and drawbacks are, and how they may be effectively used or misused in mathematical communication. An awareness of collectivist dimensions depends on values at an individual level, but collective values can be given more weight in curriculum or in a teaching and learning situation.

The next pair of Bishop's (1988) values, *control* and *progress*, is connected to sentiment, i.e., to feelings and attitudes. Bishop argues that knowledge can be associated with the feeling of control, both related to the mathematics itself and to feeling secure in a changing world. We argue that individualist values in relation to sentiment include students' sense of personal mastery and control when doing mathematics together with valuing students having the methods, facts and solutions at their fingertips. These understandings of sentiment are often cited in media discourses.

Bishop connects *control* to both pure and applied mathematics which indicates that the value can be more collectivist. We understand broader collectivist values in relation to sentiment to be about how mathematics is related to control and mastery of mathematics applied to societal and environmental topics, how it affects people and nature, and how this mastery can contribute to (often problematic) change in society. Mathematics is integral to processes of social control, such as policing, advertising or surveillance, and mathematics is central to the technological processes through which global capitalism functions, such as in supply chains, transportation or finance. These processes involve exploitation of marginalised peoples as well as exploitation of ecosystems (Andersson & Barwell, 2021).

The value *progress* Bishop explains is about understanding more, both concerning how mathematical understanding can develop, either individually or collectively, and how mathematics can contribute to progress, for example through technology. Progress can thus be an individualist value related to individual learning of mathematics through ordered, structured curricula. However, Bishop emphasises that technological progress is a double-edged sword, referring to how so-called technological progress can result in damage or unfavorable effects. Similarly, control through applied mathematics can have positive and negative effects. The collectivist value of progress can thereby involve collective learning, growth and well-being. These kinds of values are related to critical mathematics education activities that analyse social problems through collective exploration, for example, as well as to the role of mathematics education in preparing future citizens to participate in extended peer review of scientific activity in post-normal situations.

Sociology denotes Bishop's (1988) last pair of values and is about relationships to people, institutions and mathematics itself. The first value, *openness*, refers to mathematics being open to scrutiny. The formalisation, abstraction and dehumanisation of mathematics contributes to this openness. The value of *openness* in mathematics at an individualist level we understand as highlighting the importance of critical thinking within mathematics; to understand that there are many ways to tackle a problem or think about mathematics. At the collectivist level, we relate openness to ethics as a concern for others and respectful participation in mathematics-based discussions on societal issues. Mathematics through its role in the creation of social reality affects everyone in society. Students need to explore the impact of mathematics on fellow human beings and other species.

Finally, values relating to *mystery* concern the sense of individual wonder that mathematics and mathematicians can inspire. Bishop argues that this mystery can be associated with aesthetic aspects of mathematics and with wonders of computers, mathematics-based algorithms and computations, which we relate to individualist values. From a broader collectivist perspective, we see mystery as relating to the possibilities and limits of mathematics. Critical mathematics education and post-normal science show that students and other citizens can contribute to change through mathematics. In addition, in post-normal situations, there is much that mathematics cannot do, particularly in the context of chaotic, non-linear systems. Mystery may also be about the uncertainty, indeterminacy and possibilities that arises in such situations.

Although values to support the individual are important, in this chapter, we put more emphasis on the values that promote collective relationships. Our sense of collectivist values involves the webs of relations we are connected to and interdependent with both humans and other species. More specifically, with regard to the learning of mathematics, we explore values that promote the sustaining of social and ecological relations we are part of, by locating our relationships with one another and with more-than-humans at the centre of our practices of teaching and learning.

4.4 Two Mathematics Curriculum Examples

Having set out our proposed conceptualisation of values in mathematics education in relation to individualist and collectivist orientations, in this section, we examine two mathematics curricula using these ideas. Our goal is both to explore and deepen what this conceptualisation might look like, as well as to observe how these orientations might appear in official curricula. Of course, what appears in a published curriculum must be understood as an 'intended' curriculum; what actually happens in mathematics classrooms will be related to many other factors.

We have selected mathematics curricula with which we are familiar: from Norway, and Ontario, Canada. Both curricula were revised in 2020. One of the major changes in both these curricula is a more explicit addition of social and emotional values. Ontario's new curriculum (OME, 2020) adds values such as helping students "develop confidence, cope with challenges, and think critically" (p. 80) and to "acquire a positive attitude towards mathematics, cope with stress and anxiety, persevere and learn from their mistakes [...] and become capable and confident math learners" (p. 62). The Norway curriculum (Ministry of Education and Research, 2017) adds other values such as requiring students to be a reflective and responsible by making "ethical assessments and [being] cognisant of ethical issues" (p. 8) to "contribute to the protection of human dignity and reflect on how they can prevent the violation of human dignity" (p. 7), and to "practise different forms of democratic participation in day-to-day work" (p. 10).

4.4.1 The Norway Curriculum

The Norway mathematics curriculum (Ministry of Education and Research, 2019, p. 2) starts like this:

> Mathematics is an important subject for understanding the patterns and relationships within society and nature through the use of modelling and applications. Mathematics shall help pupils to develop a precise language for reasoning, critical thinking and communication through abstraction and generalisation. Mathematics shall prepare pupils for a society and working life in development by providing them with the competence to explore and solve problems.

The paragraph indicates both individualist and collectivist values. It expresses an *ideology* of mathematics as *rational*, as it can help understand patterns, and *objective*, as it offers a precise language. Further, the paragraph points to *control* through developing *mastery* of mathematics and applications, and it points to *progress* as the students are to explore and solve problems. The expressed aim of critical thinking is related to the value of *openness*. Society is highlighted twice, giving the impression that the values are primarily collectivist. However, the emphasis on abstraction and generalisation also suggests an individualist orientation, in particular related to abstraction and generalisation.

The curriculum continues with, what are called six core elements: exploration and problem solving; modelling and applications; abstraction and generalisation; reasoning and argumentation; representation and communication; and abstraction and generalisation; as well as mathematical fields of knowledge. The description of some of the core elements appear more individualist than what the first paragraph conveys. For example, society is not mentioned in relation to the core element *exploration and problem solving,* and the kind of problems to be solved, within or outside the mathematics domain, is not clarified. Furthermore, it explains that "students must place more emphasis on the strategies and procedures than on the solutions" (p. 2), which can be understood as a means to ensure students' sense of mastery at an individual level. Also, the two core elements *reasoning and argumentation* and *representation and communication* seem to be individualist in similar ways.

As expected, societal perspectives are present in the description of the core element *modelling and applications,* as it includes mathematical models in society. The ideology of mathematics is not noticeably present. More attention is given to the value of *control* through applying mathematics, together with *openness* and *mystery* through critical evaluation and through becoming aware of limitations of models, all of which we understand as collectivist values. Statistics and probability is one of the *mathematical fields of knowledge* and is associated with the collectivist value of ideology. Because it explains that this field "gives the pupils a good foundation for making choices in their own lives, in society and in working life" (p. 4), it implicitly refers to *rationalism* and *objectivism* through *reasoning* and *argumentation.* We consider these expressed values to be both individualist and collectivist since they concern both the individual and society, although mathematics is applied in either case.

The Norway core curriculum (Ministry of Education & Research, 2017), which indicates what formation processes schools must facilitate, includes six core values as a foundation of pedagogical practices:

1. Human dignity
2. Identity and cultural diversity
3. Critical thinking and ethical awareness
4. The joy of creating, engagement and the urge to explore
5. Respect for nature and environmental awareness
6. Democracy and participation

The curriculum explicitly notes that the values are included as "the foundation of our democracy, shall help us to live, learn and work together in a complex world and with an uncertain future" (p. 6). These are overarching values that need to be included in the teaching of all school subjects, including mathematics, and which hold a very strong collectivist orientation. Choosing ethical awareness as an example, the curriculum explains that it "means balancing different considerations is necessary if one is to be a reflecting and responsible human being. The teaching and training must develop the pupils' ability to make ethical assessments and help them to be cognisant of ethical issues" (p. 8). Ethical awareness can be understood as *openness* as it is about empathy and critical thinking. Furthermore, the value is collectivist as it is about being a responsible person and being aware of ethical issues. An even more collectivist orientation is found in relation to the fifth value: Respect for nature and environmental awareness. The curriculum explains: "Children and young people will need to deal with the today's and tomorrow's challenges, and our common future depends on the coming generations and their willingness and ability to protect our world. Global climate changes, pollution and loss of biological diversity are some of the greatest environmental threats in the world" (p. 10). This requires *openness* and inspiration to *master* mathematics as a tool to explore possibilities for *progress*, as collective orientated values. The curriculum also highlights that these issues require collective actions: "These challenges must be solved together" (p. 10).

The main rationale underlying the values is democratic participation. The curriculum mentions "Participating in society means respecting and endorsing fundamental democratic values, such as mutual respect, tolerance, individual freedom of faith and speech, and free elections. Democratic values shall be promoted through active participation throughout the entire learning path" (p. 10). This position implies that mathematics education is obliged to put emphasis on collectivist values.

Of the six core values, critical thinking and democracy are explicitly included in the mathematics subject curriculum. Democracy and citizenship is a multidisciplinary topic in which mathematics is involved together with all other school subjects. The mathematics curriculum notes (Ministry of Education and Research, 2019, p. 4):

> In mathematics, the interdisciplinary theme of democracy and citizenship is about giving students competence in exploring and analysing findings from real datasets and numerical material from nature, society, working life and everyday life. Furthermore, it is about the pupils learning to assess how valid such findings are. Such competence is important to be able to formulate one's own argument and participate in the social debate. The subject should make the students aware of assumptions and premises for mathematical models that form the basis of decisions in their own lives and in society.

We notice the value *objectivism* as a collectivist value since the subject of mathematics is required to assist the students "to formulate one's own argument and participate in social debate". The quotation also carries values related to collectivist *control* and *progress*, in that mathematics should give students competence in "exploring and analysing findings from real datasets and numerical material from nature, society, working life and everyday life".

4.5 The Ontario Mathematics Curriculum

The 2020 Ontario Mathematics Curriculum (OMC) (OME, 2020) for Grades K-8 begins with more than 60 pages of general preamble the covers a wide variety of cross-cutting themes, including English language learners, effective instruction, human rights and assessment. The start of the section focused specifically on mathematics states:

> The Ontario Curriculum, Grades 1–8: Mathematics, 2020 focuses on fundamental mathematics concepts and skills, as well as on making connections between related math concepts, between mathematics and other disciplines, and between mathematics and everyday life. It also supports new learning about mathematical modelling, coding, and financial literacy, and integrates mathematics learning with learning in other STEM (science, technology, engineering, and mathematics) subjects. As well, this curriculum is designed to help students build confidence in approaching mathematics and acquire a positive attitude towards mathematics, cope with stress and anxiety, persevere and learn from their mistakes, work collaboratively with others towards a shared goal, value deep thinking and making connections, and become capable and confident math learners. (p. 62)

This general framing includes both individualist and collectivist values, although the former are more prominent. The emphasis on fundamental mathematics concepts and skills relates to *rationalism* and implies individual learners acquiring the new knowledge. The references to coding and financial mathematics reflect the values of *objectivism and control*, focused on individual application of mathematics. The emphasis is on individual well-being. Collectivist values of *openness* are briefly apparent in the reference to working collaboratively, although even here, the goal appears to be individual learning and mastery.

The curriculum is based on seven principles:

1. A mathematics curriculum is most effective when it values and celebrates the diversity that exists among students.
2. A robust mathematics curriculum is essential for ensuring that all students reach their full potential.
3. A mathematics curriculum provides all students with the foundational mathematics concepts and skills they require to become capable and confident mathematics learners.
4. A progressive mathematics curriculum includes the strategic integration of technology to support and enhance the learning and doing of mathematics.
5. A mathematics curriculum acknowledges that the learning of mathematics is a dynamic, gradual, and continuous process, with each stage building on the last.
6. A mathematics curriculum is integrated with the world beyond the classroom.
7. A mathematics curriculum motivates students to learn and to become lifelong learners.[1]

Implicit in these principles, we notice the expression of many values of rationalism, objectivism; control, progress; and openness, mystery appearing in different

[1] https://www.dcp.edu.gov.on.ca/en/curriculum/elementary-mathematics/context/principles-underlying-the-ontario-mathematics-curriculum.

ways. We can also see individualist values and collectivist values. For example, principle 3 includes an explanation that "Learning in the mathematics curriculum begins with a focus on the fundamental concepts and foundational skills. This leads to an understanding of mathematical structures, operations, processes, and language that provides students with the means necessary for reasoning, justifying conclusions, and expressing and communicating mathematical ideas clearly" (p. 64). This statement encapsulates *rationalism* at an individual level, with a focus on *logical reasoning*. This principle can also be related to *objectivism* and *mastery* through the connection to confidence and capability: the manipulation of mathematics symbols and concepts is linked to individuals becoming more confident.

Principle 2 is related to the value of control. For all students to learn, there needs to be a "robust mathematics curriculum". The curriculum sets out a framework for learning important skills, such as problem solving, coding, and modelling, as well as opportunities to develop critical data literacy skills, information literacy skills, and financial literacy skills. An example of *mystery* as a sense of wonder can be found where the OMC explains the need to provide "opportunities for all students to investigate and experience mathematical situations they might find outside of the classroom and develop an appreciation for the beauty and wide-reaching nature and importance of mathematics" (p. 66).

While several principles appear to suggest a more individualist orientation, there are also hints of a collectivist orientation. The first principle, for example, highlights 'diversity' which could create awareness of multiple ways of doing and knowing mathematics and allow learners to bring these perspectives into their mathematics class. Similarly, principle 6 could create an opportunity for teaching mathematics in relation to broader social issues in line with the collectivist column of Table 1 (although we recognise that this principle does not require such an approach).

One of the major changes in the OMC is that in addition to the overarching principles, there are new strands. For each grade level, the previous curriculum had five mathematical strands: 1. number sense and numeration; 2. geometry; 3. measurement; 4. patterning and algebra; and 5. data management and probability. The new curriculum changes the strands to: 1. social-emotional learning and mathematical processes; 2. number; 3. Algebra; 4. data; 5. spatial sense; and 6. financial literacy.

In promotion of the value of individual wellbeing the strand of Social-Emotional Learning (SEL) involves the skills listed in Table 4.2.

These skills carry certain values, both individualist and collectivist. Values such as expressing feelings, developing resilience, fostering hope and developing a sense of identity are among the more individualist. We believe that values such as rationalism can include a collectivist sense of democratic participation; objectivism can include promoting argumentation processes and democratic debate. Awareness of the control of knowledge in society and the ecosystem is not necessarily included in these social-emotional skills, even while "understanding the feelings of others" is a collective value, relating to openness and ethics.

The new financial literacy strand also reflects various values. The overview of the strand emphasises developing "the ability to make informed decisions as consumers and citizens while taking into account the ethical, societal, environmental, and

Table 4.2 Skills and purposes of social-emotional learning (OME, 2020, p. 36)

Students will learn skills to:	So that they can:
Identify and manage emotions	Express their feelings and understand the feelings of others
Recognize sources of stress and cope with challenges	Develop personal resilience
Maintain positive motivation and perseverance	Foster a sense of optimism and hope
Build relationships and communicate effectively	Support healthy relationships and respect diversity
Develop self-awareness and self confidence	Develop a sense of identity and belonging
Think critically and creatively	Make informed decisions and solve problems

personal aspects of those decisions" (p. 89). Here we see mathematics in the service of consumption and a focus on "personal aspects". At the same time, there is scope for ethical, societal and environmental considerations to be incorporated and these aspects could involve a more collectivist orientation.

While we can see spaces for a range of values on the OMC, in general, we noticed that values promoted in the OMC are more individualist in nature.

4.6 Discussion

In this chapter, we have argued that in addition to examining values in relation to mathematics education, it is important to consider the individualist or collectivist orientation of these values. Drawing on critical mathematics education and post-normal science, we argue that a mathematics education that seeks to prepare learners to participate in democratic processes and contribute to tackling difficult complex problems like climate change, destruction of ecosystems, pollution or energy production needs to incorporate collectivist values. Using Bishop's six claims in relation to values, we developed a conceptualisation of the values in terms of individualist and collectivist orientations. The weighting of these orientations forms a series of continua.

To explore these ideas in more depth, we examined two mathematics curricula, one from Ontario, Canada, and the other from Norway to explore how different curricula not only reflect different types of values in relation to mathematics, but how these values may tend to be more individualist or more collectivist. We found that the values and practices that are endorsed in the two curricula are fundamentally different. In particular, the Ontario curriculum appears to focus on confidence and coping with challenges, which are about care for *individual* students, while the focus on acting ethically and environmentally in the revised Norwegian curriculum is about *collectivist* values. A comparison of the principles underpinning each curriculum is instructive. A direct comparison highlights how the two curricula include similar ideas but treats them in very different ways. For example, the Ontario curriculum emphasises motivation, while the Norway curriculum prefers joy, engagement and the urge to explore. Arguably the emphasis on motivation in Ontario is related to an individualist outcomes-based perspective, whereas the values in the Norway curriculum are more concerned with the experience of doing mathematics. For some of the principles, there is a clear contrast. Notably, the Ontario curriculum emphasises foundational concepts and skills, whereas the Norway curriculum talks about critical thinking and ethical awareness. Of course, both may be incorporated into mathematics teaching and may feature in the more detailed learning outcomes and examples. Nevertheless, our examination highlights these differences in the high-level framing of the curriculum in the two jurisdictions.

This difference suggests that mathematics curricula incorporate values to respond to different contemporary challenges, such as problems with sustaining democratic values in society or the mental health of children and youth, and also the more general point that the values proposed in mathematics curricula reflect broader societal values. In relation to the six values proposed by Bishop, we noted that the values of rationalism, objectivism and control are more frequently and broadly used in both curricula. The values of openness, progress and mystery were less apparent. Although the curricula highlight different values, less is known about how these values are operationalised in practice. For example, a study on feedback from mathematics teachers in Norway shows a focus on students' procedural skills rather than on engaging the pupils in mathematical practices (Stovner & Klette, 2022; Stovner et al., 2021). This suggests that individualist values may receive considerably more attention than collectivist values.

The examination of the two curricula highlighted for us once more that particularly in the context of mathematics classrooms and the teaching and learning of mathematics, "value" is a confusing and complex concept. In separating individualist and collectivist values, this chapter is an invitation to pay closer attention to collectivist values. Because we struggled somewhat with defining and analysing individualist and collectivist values using Bishop's framework, we welcome further development to analyse and understand values in mathematics education along the same lines. An awareness of what values are promoted in education is important as we believe that taking into account collectivist values is crucial in order for students to understand and cope with social and ecological challenges.

References

Abtahi, Y. (2022). What if I was harmful? Reflecting on the ethical tensions associated with teaching the dominant mathematics. *Educational Studies in Mathematics, 110*(1), 149–165.

Andersson, A., & Barwell, R. (2021). Applying critical mathematics education: An introduction. In A. Andersson & R. Barwell (Eds.), *Applying critical mathematics education* (pp. 1–23). Brill.

Biggeri, A., & Saltelli, A. (2021). The strange numbers of COVID-19. *Argumenta, 7*(1), 97–107.

Bishop, A. (1988). *Mathematical enculturation: A cultural perspective on mathematics education*. Kluwer.

D'Ambrosio, U. (1998). Mathematics and peace: Our responsibilities. *ZDM, 30*, 67–73.
D'Ambrosio, U., & Rosa, M. (2017). Ethnomathematics and its pedagogical action in mathematics education. In M. Rosa, L. Shirley, M. E. Gavarrete, & W. V. Alangui (Eds.), *Ethnomathematics and its diverse approaches for mathematics education* (pp. 285–305). Springer.
Funtowicz, S. O., & Ravetz, J. R. (1993). Science for the post-normal age. *Futures, 25*(7), 739–755.
Hauge, K. H., & Barwell, R. (2017). Post-normal science and mathematics education in uncertain times: Educating future citizens for extended peer communities. *Futures, 91*, 25–34.
Kuhn, T. S. (1962). *The structure of scientific revolutions*. University of Chicago Press.
Ministry of Education and Research. (2017). *Core curriculum—values and principles for primary and secondary education*. Available from: https://www.regjeringen.no/contentassets/53d21ea2bc3a4202b86b83cfe82da93e/core-curriculum.pdf
Ministry of Education and Research. (2019). *Curriculum for Mathematics year 1–10*. Available from: https://www.udir.no/lk20/mat01-05?lang=eng
Moses, R. P., & Cobb Jr, C. (2001). Organizing algebra: The need to voice a demand. *Social Policy, 31*(4), 4–4.
Ontario Ministry of Education (OME) (2020). *The Ontario curriculum Grades 1–8: Mathematics 2020*. Queen's Printer for Ontario.
Skovsmose, O. (2023). *Critical mathematics education*. Springer.
Sobe, N. (2021). Reworking four pillars of education to sustain the commons. Available from: https://www.unesco.org/en/articles/reworking-four-pillars-education-sustain-commons
Stovner, R. B., & Klette, K. (2022). Teacher feedback on procedural skills, conceptual understanding, and mathematical practices: A video study in lower secondary mathematics classrooms. *Teaching and Teacher Education, 110*, 1–12.
Stovner, R. B., Klette, K., & Nortvedt, G. A. (2021). The instructional situations in which mathematics teachers provide substantive feedback. *Educational Studies in Mathematics, 108*, 533–551.
Valero, P., García, G., Camelo, F., Mancera, G., & Romero, J. (2012). Mathematics education and the dignity of being. *Pythagoras, 33*(2), 1–9.

Chapter 5
Values into Pedagogical Practices in Mathematics: Promoting Prospective Teachers' Ethical Responsibility for Making Mathematics Meaningful

Huk Yuen LAW

5.1 Introduction: Making Meaning as Ethical Significance in Mathematics Education

Apparently, our technologically advancing society has granted the saying that mathematics is an essential discipline for enhancing human capabilities in the 21st Century. And it follows that mathematics would have to be taken for granted as a compulsory school subject for all to learn in our elementary education as a matter of course. The value that justifies the vital role of mathematics in our society is nonetheless not ready to be translated into the valuing of its learning by schoolchildren at large. It would be extremely difficult if not entirely impossible for the learners to see the significance of learning a subject called mathematics without a meaningful experience of learning it. The notion of "value" has become a critical concept in mathematics education in the past decade ever since Alan Bishop highlighted it from the cultural perspective of mathematics education in the 80 s of last century. The "WIFI (What I Find Important)" project has highlighted further the importance of what matters to learners in their learning of mathematics. In other words, what students value in mathematics cannot go unheeded if teachers want to deliver more effectively in the classroom of not just what the learners need to know but what they want to know. This chapter frames the ethical significance as a critical response to the ongoing expansion of valuing process of meaning-making in teaching and learning mathematics from the perspective of semioethics. Mathematics teaching and learning entails complex semiosis as constituted in the interpretative process of enhancing the functions of making meaning as co-constructed and co-shared by those who are involved in it. I cannot agree more of what Godino and Batanero's (2003, pp.12–13) claim that "the

H. Y. LAW (✉)
Faculty of Education, The Chinese University of Hong Kong, Hong Kong, China
e-mail: hylaw@cuhk.edu.hk

Y. Dede et al. (eds.), *Values and Valuing in Mathematics Education*,
https://doi.org/10.1007/978-981-99-9454-0_5

notion of meaning, in spite of its extraordinary complexity, may still play an essential role as a basis for research into the didactic of mathematics". The claim is grounded on the understanding that the learners' personal meanings are to be built not just on "cognitive factors, but rather on the semiotic-anthropological complex".

An impressive remark as valued by the learners of mathematics in the exploratory study of "Values in Mathematics Education" project that I took part in was to learn mathematics "with fun and having something to learn" (Law et al., 2011). Such a remark entails a complex fusion of cognitive endeavour and emotional anchoring of the learners' engagement in task activities in mathematics lessons. The subsequent analysis of interview data drawn from both the secondary mathematics teachers and their students unfolded three major themes of what they valued in the learning of mathematics, including "*meaningfulness*, *autonomy*, and *positive attitude*" (Law et al., 2012, p.49). In developing learner autonomy, teachers would have a critical role to play in supporting students to construct *the meaning of learning* itself as a kind of metacognitive strategy towards fostering students' positive attitude (see Thanasoulas, 2000).

Drawn from the semiotic perspective, the mathematical engagement in the form of semiotic actions involves a dynamic unity of thoughts and emotion (Raford, Schubring & Seeger, 2011). The demand for valuing of "with fun and having something to learn", as a way of developing students' own personal meanings, is no simple task for teachers to fulfil. I would argue that it requires teachers to see such demand as ethical value for making ethical choice in supporting students to learn in a meaningful way. In other words, teachers need to take up their ethical responsibilities of creating learning opportunities for endorsing students' personal meaning. Nonetheless, it would be a great challenge for teachers to exercise the kind of ethical agency in meeting both professional and institutional obligations in the schools they are practicing (see Ernest, 2019).

As a mathematics educator, I see "making mathematics meaningful" as ethical value that the teachers should be critically aware of in practicing their teaching. Last summer, I conceptualised an action research project designated as "WISE" ("What I See Ethical" in mathematics teaching) for the final year B.Ed. students. Prior to the Teaching Practicum in November, they needed to take the course "Mathematical Literacy and Problem Solving" that I taught in the first semester. Approaching the end of the course, I invited the class to join in the project on a voluntary basis. The project aims to answer the question, "How values in mathematics education can be put into pedagogical practices in mathematics classrooms through teachers' perception of their ethical responsibility for making mathematics meaningful?". The empirical material was drawn from the case stories of these prospective teachers who would value student's making of meaning in their learning of mathematics. The written materials would be drawn from their self-reflective journals and photo images to capture critical moments during the Teaching Practice (TP) period. By the end of the course, three students approached me with their signed consent form acknowledging their intentions of joining in the project during the TP period. I would delineate below the methodological orientation for conceptualizing the ethical responsibility from the perspective of global semiotics or semioethics grounded on the principles of action research. Through the three case stories, I would unfold the *art* of promoting teachers' ethical responsibility in practicing teaching of mathematics for making mathematics meaningful through the design of mathematical tasks.

5.2 Methodological Orientation of Values into Pedagogical Practices

The last decade witnessed a shift of focus from assessing what teachers and students value to how values can be put into pedagogical practices in mathematics education (see Seah, 2019). This is indeed a significant move in contributing to the benefits of promoting learners' well-beings in their life journeys of learning mathematics through research endeavours. Nonetheless, such a move invites us to think critically of more questions rather than the answers to what we mean exactly and in what ways we can do to 'apply' our knowledge of what teachers and students value as we learn from the previous research findings. Recent research (see Kalogeropoulos et al., 2021) have outlined the issue of value alignment that reflects the potential gap between what students and teachers value in practicing the learning and teaching of mathematics. Such kind of gap unfolds the complex and intriguing nature of values in mathematics education itself. Even if someone remains sceptical whether mathematics should be considered as a knowledge entity of context-free value (see Ernest, 2009), we should have no doubt of accepting the claim that putting values into practices is value-laden. It remains perplexing how to find out the absent piece of a jigsaw puzzle in terms of conative variables of the mathematics pedagogy in facilitating student learning of mathematics (Seah, 2019, p.101). Indeed, it would remain puzzling as ever if we consider seriously of how such a conative factor works with the cognitive and affective domains in the mind.

Conation itself is indeed a complex and hard-to-define notion entailing self-understanding of a person who strives with purposeful actions for fulfilling some targets or goals he or she wants to achieve. The recent literature review on the shifting framing of how we interpret the concept revealed five different characterisations of conation, namely, basic drives, values, short-term motivators, individual attributes, and action (Militello et al., 2006). Among these five characteristics, I find it particularly interesting to me is how values can be fed back into teacher actions to arouse students' basic drives to learn with human inherent curiosity and purpose. To me, it is not just a research issue but also an ethical concern ever since I turned myself into a university educator after practicing teaching school mathematics for more than two decades. Mathematics has long been recognised as a hard-to-learn school discipline owing to its inherent nature of abstraction in terms of numbers and symbols. The teaching and learning activities in the mathematics classroom are like a game of coding and decoding signs without authentic dialogues of making meaning of *discourses* in the forms of something (*text*) being spoken or written in the lessons. Drawn from the long-serving experiences that I had in the school contexts, I have endorsed "learning with *meaning*" as the core value of engaging in pedagogical practices.

Meaning-making, like conation, is a complex notion. It is hard for us to tell how individuals would make what kinds of meaning in what contexts. I understand that valuing and meaning-making cannot be equated in a simple way. In other words, students who value the learning of mathematics may not necessarily imply that they want to have meaning to be made in the first place whilst learning it. Nonetheless, I argue by drawing on the theoretical frameworks of the semiotics and semioethics that there would have an intriguing linkage of valuing with meaning-making. Before that, I would depict how value and meaning are related in terms of the teaching and learning of mathematics.

5.2.1 Value and Meaning: Can We See Something Significant Without Knowing What It is Meant to Be?

To be a learner who wants to learn mathematics meaningfully they would need to have the interpretation of mathematical facts in terms of meaning, of purpose, or of value, whereas the teacher would undergo a meaningful act of inquiring into what these facts are meant to the learner as such. As argued by Jared Moore (1914) more than a century ago, the notions of "value" and "meaning" are closely related to "purpose" as any purposeful actions entail values to be defined by their "affective-conative meanings". Learning mathematics, if it is regarded as a purposeful action relating to the value of learning, it requires learners to determine whether it is valuable to them by constructing meanings of their own. The way that the teacher expresses mathematics serves as a "sign" to her students to make sense of what is being taught through interpretation. What the students interpreted would have two kinds of "meaning": the semantic and the axiological. Such a distinction hints that the meaning entails "that which something signifies and the value or significance of what it is signified" (Morris, 1964, p.vii). In other words, asking whether the learning of mathematics is meaningful to a student is questioning the signification of "mathematics learning" that has been undergoing or about the value or significance of learning such a thing called mathematics as judged to have, or even both.

The purpose of teaching mathematics is, hopefully, to facilitate the learners to experience their learning with the two kinds of meaning not hardly distinguishable from each other during the process of learning in the classroom. To fulfil such a purpose, it requires teachers to exercise their ethical responsibilities to create the space for their students to construct meaning of both kinds if they want to put valuing of "meaningful learning" into their pedagogical practices. Teaching mathematics ethically invites teachers to reflect on signs and communication with an ethical concern of reducing the risk of incommunicability. In what follows is to discuss what can be done for a mathematics teacher to practice value judgement in terms of ethical responsibility for supporting students to make learning itself meaningful as drawn from the perspectives of semiotics and semioethics.

5.2.2 From Semiotics to Semioethics: Ethical Value for Making Mathematics Meaningful

The gap existing between value as perceived and value as enacted unfolds the complex and intriguing nature of values in mathematics education. To narrow if not able to close the gap, teachers need to be aware of their own unique role of promoting learners' well-being in going through the growth journey of learning mathematics. Drawn from Thomas Sebeok's "global semiotics" (1994), the gap entails the "translation" of signs in terms of mathematical communication between the teacher and the student. As "semiotic animals", both parties, as unique human beings of their own would value the meaning as co-constructed during the process of teaching and learning in the place called classroom. As such, it is a kind of semioethical responsibility with the commitment into the making of meanings to be shared by both teacher and the students during the classroom inter-actions (in the form of dialogue) as a process of "metasemiosis" (Urban & Brown, 2006). The value alignment phenomenon in terms of aligning the valuing making of meaning (semiosis) with the human agency for making meaning overshadows the traditional modality of classroom teaching in giving sheer explanations for supporting the individual learners to make sense of what is being taught.

Drawn from Thomas Sebeok's "global semiotics", Susan Petrilli and Augusto Ponzio developed the theory of 'semioethics' incorporating ethics with semiotics for the study of the relation "between signs and values", as well as "meaning and significance" (Petrilli, 2016, p.248). The genesis of the theoretical development of semioethics constitutes a trio of thinkers, including Charles Peirce, Vitoria Welby, and Thomas Sebeok. These great figures offer us a new perspective of how humans communicate through signs. The emergence of meaning and value in the chain of translative signifying processes serves as an ethical reminder of human responsibility for endeavouring to develop well-being globally from a biosemiotics perspective (see Petrilli & Ponzio, 2008; Petrilli, 2015). The genesis can be sketched with a beginning from Peirce's triadic semiotic relation between "objects, signs, and interpretants". And then, it goes with Welby's focuses of "significs" on the axiological implications in the relation between sign and meaning, as well as the relation between sign and value. Furthering on, it is Sebeok's "global semiotics" extended our understanding of human communication in developing meaning by taking up the responsibility for ensuring life to flourish. In relating "semioethics" to education, Petrilli (2016, p.270) argues that semiotic inquiry evidences ethical significance of developing "the unique individual's abilities together with a sense of responsibility for the other" rather than promoting "power, competition, and acquisitive success in preparation for a future career".

In mathematics classroom, the lesson would be initiated by the teacher with the objective of imparting students with the knowledge domain as specified in the curriculum. The teaching and learning process itself is a complex semiosis with intriguing flow of signs to signs waiting for individual modelling of meanings through the process of signification. Such kind of meaning modelling requires students to have acquired the ability to *attend* in order to *notice* what are regarded as something worthy of learning. To help students make meaning of what they are expected to learn, the teacher needs to have awareness and sensibility to any new responses as elicited as an art of practicing *mathematical noticing* (Mason, 2002). As teacher educators, we would have a crucial question to ask, "How can teachers develop their own awareness and sensibility to students' responses in complex classroom situations through such kind of noticing?". In making noticing productive, teachers may overlook the trick of how to help students "making sense" of what is being attended to in deciding in what ways new responses can be and should be made (Choy, 2013; Jacobs et al., 2011). I argue that it would be teachers who should develop their ethical value for making mathematics meaningful if they want their students *co-value* with them a discipline called mathematics. This kind of ethical value demands teachers see for themselves at enhancing students' meaning-making as an ethical responsibility of its own accord whilst practicing teaching mathematics in the classroom environment. Below, I would describe the methodological considerations behind the conceptualisation of WISE Project that I will detail later.

5.3 Methodological Considerations

As an action researcher, I have been critically aware of the ethical concern of the researcher's own ontological and epistemological assumption behind my inquiry into the comprehensible realities through the interpretations of the others' storied experiences through time. This kind of assumption is well understood not to be able to eliminate but to adapt or refine through the unfolding of the assumption being made in contributing to a shared understanding of worldly phenomena that we can possibly imagine. In the present study, I adopt the combined methodology approach of "hermeneutic phenomenological narrative enquiry" as proposed by Nashid Nigar (2020) to unmask the complex social phenomena of how the prospective teachers "see ethical" in practicing mathematics teaching in the classrooms during the Practicum. The individual enquiry approach (hermeneutic/ phenomenological/ narrative) is well established on its own and known to me not just as a tool for inquiry but also as a backup of philosophical reflection upon the action research work that I have undertaken. From a hermeneutic perspective, dialogue as I would enter with the student teachers enables us to understand the values in terms of meaning-making as endorsed in practicing their teaching as I as a researcher need to prepare myself to listen to the stories they contribute (Widdershoven, 2001). Adopting the phenomenological approach is to explore the lived experience of prospective teachers as they might encounter the value conflict with the mentors in the schools where they did their practicum (Tomkins & Eatough, 2013). And the narrative as adopted is to make sense of the student teachers' knowing of how meaning would be made through storying life events (Sandelowski, 1991). The rationale behind the choice of student teachers is to unfold the critical issue in teacher education for promoting teachers' ethical responsibility for supporting the learners to make meaning of what have been taught.

The need to adopt a combination of trio approach for the inquiry is to capture individual participants' unique interpretations of the struggle. It can be a struggle itself for them as experienced in undergoing the complex semiotic modelling of seeing ethical significance out of signifying process of practicing the valuing of supporting schoolchildren learn mathematics with meaning. The trio approach enables me to get in touch with the kind of ethical awareness that the participants might have developed in putting what they valued into classroom practicing of their teaching through the story (narrative) that they tell, the dialogue (hermeneutic) that unfolds the hidden insights, and the experience (phenomenological) that they reflect from what have been done through action taken for seeing what it is meant to attain some kind of ethical significance in teaching mathematics. To help further unfolding the 'capturing moment' of seeing ethical, I also adopted the 'photo-taking' as a reflective tool for visual narrative (Lemon, 2007) with which the participants would have deeper reflection through the story as told.

It would be a difficult task to interpret what valuing the goal of making mathematics meaningful to the students meant to the participants. Thus, I would further adopt the Greimas's (1966) "semiotic square" as a semiotic method of analysis (to be described later in the Section of Data Analysis) for interpreting of what they saw ethical as they were preparing themselves for enhancing students' learning in a meaningful way.

In what follows, I would depict the WISE Project that I developed to enhance the ethical awareness of mathematics education students for making mathematics meaningful in practicing teaching during practicum. I theorise the project grounded on action research which serves as narrative of self-interpretation for the participants to do the telling of their own stories (Law, 2023).

5.4 The WISE Project: Seeing Ethics in Teaching Mathematics

Prior to the Practicum in the first semester of final year of studies, the B.Ed. students would have to take the course of "Mathematical Literacy and Problem Solving", aiming at equipping the prospective teachers with not just the competence of teaching mathematical problem solving but also an understanding of what for in helping schoolchildren of becoming mathematically literate. The course comprises of 39 hours of lectures and seminars with two contact sessions conducted in a week. In preparing the teaching of the course, I conceptualised the WISE (What I See Ethical in mathematics teaching) and invited the class to join in on voluntary basis when the course came near to its end. The class size was 24, comprising 21 B.Ed. students and 3 students from Mathematics Department. I adopted a 'Thoughtful Dialogue' seminar approach for creating dialogic space in sharing with the class the ideas of how mathematics teaching would be put into classroom practice in an ethical way. The thoughtful dialogue is an open, cooperative, and critically examining interlocution. The students are expected to read the text as assigned and take part in the dialogic seminar by working cooperatively as a group. And they are also required to take notes from each seminar with reflection in the form of meta-dialogue.

One of the seminar themes was about "mathematics for all". The class had a very thoughtful and heated discussion of what it was meant to make it 'fair' for supporting learners to be able to have 'accessibility' to the knowing of mathematics itself. Their presentations clearly pointed out the value of facilitating the learners to engage in their learning with voice to be heard of the unique interpretations that each learner would have. Nonetheless, I could feel the worries of some of their concerns of how such a 'theoretical' thought could be practiced in the real classroom settings within the tight designated curriculum framework.

As the course was approaching its end by the end of October, I received from the class *three* signed consents of joining in the WISE Project with the aims to develop an ethical awareness of their own in practicing mathematics teaching during the Practicum. I invited them to have a briefing tea gathering in the campus with free and open conversations in the group meeting for about two hours on some issues of interest such as "Why do you want to join in the project?", "How would you compare the school culture or settings in the practicum schools with your own secondary schools that you had gone through as a student?", and "What would you do during the Practicum if you would see mathematics teaching as an ethical activity?". By the end of the meeting, we came up with the agreement that they would prepare the materials during the Practicum for the post-TP meeting. These include a self-reflective log, some sample student works to be collected, and some photo images (at least one) to illustrate how they "see ethical" at some of the critical moments during the Practicum. In response to their concern of keeping up students' habit-of-mind to do mathematics despite of the low motivation that the students would have in the practicum schools, I proposed to them that they could create a 'TAD' (Task-A-Day) learning journal for their students. The TAD features constitute a free and open invitation for the students to take part without the burdens of having any consequences for failing to submit the work. Submitting homework as a cultural value in most Asian regions, including Hong Kong, is regarded as a school duty that students need to comply with (Chen & Stevenson, 1989). The design of the TAD journal serves as a tool for the project participants to reflect on the ethical struggles that they would have gone through whilst making attempts of making mathematics meaningful in practicing their teaching in the classrooms.

5.4.1 Data Collection: The Participants

The participants—*Jenny*, *Kelly*, and *Wilson* had kindly provided me with rich data from the materials of their collections and recollections. Drawn from the materials, I could have a free and open dialogue with them in unmasking the unique experience that each of them had. The names (Jenny, Kelly and Wilson) as appeared in this article were in agreement with their suggestions in keeping the confidentiality of their identities. The participants, all of them were B.Ed. students, had similar educational backgrounds. They grew up in the 'ordinary' families with hard-working parents to provide them with good educational opportunity as far as the family finance could bear with. They were not particularly impressed with the way mathematics was taught in the 'ordinary' schools (not the 'elite' schools by tradition) where they were studying but were determined to becoming a mathematics teacher by earning excellent scores in the public examination.

The major source of data came from the materials as collected during the seven-week TP period. These include teacher self-reflective log, sampled student works, the photo images, and the TAD task sheets. To keep the intervention effect minimum, I would not collect the materials until the end of the Practicum and the participants would share with me all of the materials during the post-TP meeting. The free and open dialogue (as individual inter-viewing for about two hours) in the post-TP meeting served as another source of data with which I would make attempt of making sense of what being told. During the meeting, I asked the participants to share with me at least one photo (among the others) that they had taken during the Practicum. I asked them to do the finger-pointing to where they could possibly visualise as an ethical consequence of the task activities they had created for them to engage.

By the end of the post-TP meeting, I invited the participants to share with me through e-mail the experiences that they would have by writing up some reflections drawn from the dialogue as created in the meeting.

5.4.2 Data Analysis

The adoption of trio enquiry approach benefits me from using three kinds of data analysis (see Nigar, 2020) in making sense of the rich data as collected—the narrative/performative analysis for the *story* ("What was told?"), the interactional/dialogic analysis for the *dialogue* ("How the story was telling?"), and the thematic/ structural analysis for the *experience* ("In what way the story was told?"). I am going to elaborate below the 'findings' of the study in the form of frame creation—"the story" with "the experience" unfolded through "the dialogue". Through these frames, we hopefully can look into the participants' viewpoints of what and how they saw ethical in teaching mathematics in a meaningful way during the Practicum. To make sense of the frame creation in abductive reasoning (see Huh, 2016), I would adapt Dorst's (2011) equation of "What + How = Value" for looking into "how" (expression) the student teachers were practicing the aspired ethical "value" of making mathematics meaningful as implied in workings of "what" (content) they would have picked in fulfilling the goal.

Incorporating the Dorst's equation, the Greimas's (1966) "semiotic square" would be further adopted as a semiotic method of analysis for interpreting the student teachers' attempts at accomplishing valuing of the task goal of making mathematics meaningful to the students they taught. Such an analysis was aimed to help us to understand more of how the participants would become aware of ethical responsibility in practicing teaching of mathematics through the design of mathematical tasks. The semiotic square helps us unfold the meaning of "meaningful" in the relation of difference (Greimas & Rastier, 1968) through two operations of the "contradictory" and "contrary" relations (Greimas, 1987). Through the use of semiotic square, we can visualize four elements instead of the binary—(1) "meaningful" [S1], (2) "absent of meaning" [S2], (3) "non-meaningful and non-absent of meaning" (meaningful meaninglessness as absurdity of the first kind) [-S1 and -S2], and (4) "meaningful & non-meaningful" or "absent of meaning & non-absent of meaning" (absurdity of the second kind) [S1& -S1 or S2 & -S2]. I would clarify the classification of these two kinds of absurdity later on after the telling of the three student teachers' stories.

5.5 The Story

The TP period for the participants' final-year practicum lasted for seven weeks from October 31 to December 16. In telling each of their stories below (drawn from the self-reflective logs), the first-person "I" would be adopted for the stories as told to highlight a sense of 'self' in telling their stories as experienced.

5.5.1 The Story of Jenny

I was assigned to teach two classes—one S1 (Grade 7) and one S4 (Grade 10) in a secondary school quite far away from my home. On the first few days of the practicum, I was arranged to 'co-teach' with the school mentor who took up the key role of teaching the lesson. What I did was to walk around the class to give extra supports to the students in case they would come across any difficulty in doing the problems. Honestly, the teacher-centred lessons were quite ***boring with no responses*** from the students. The students were shy of approaching me even when they got mistakes in doing the problems.

Ten days later, I finally had the first lesson to teach entirely on my own. I could not wait to use my final-year project work "interactive e-book-based flipped classroom" to help students engage in the lesson. In the class, the students were grouped in three to discuss the assigned maths problems. I adopted the 'traffic lights' as tool for students' self-assessing of their own understanding—'red' for "I don't get it! I need some help!"; 'yellow' for "I think I understand but I need a little support"; 'green' for "I understand and can try this on my own". The lesson ended with students' impressive responses. To help students make meaning of what they learnt, I used ***story-telling*** in teaching the techniques of solving linear equations in the S1 class and vending machine in S4 class for introducing the concept of 'Function'.

Monday was always the most challenging day to keep students awake in the first lesson in the morning. It might let me feel miserable of being ***not able to keep them engaging*** in the lesson activities. Despite of this, I would make attempt of revising the lesson plan and hopefully would make them more engaged in doing the tasks.

Two weeks after the practicum began, I started using TAD in the form of student learning journals for students to do with the tasks that I designed for them. As it was not compulsory for students to complete the work, no one returned their work two days later. After another one week, I adopted a new strategy for encouraging them to do the TAD task. Instead of just distributed the task sheet to the class and asked them to take it back home, I spared some time before the end of the lesson and introduced the task in class so as to let them ***know more of what the task was about***. It was a task about a visit to Disneyland with questions set related to the topic that they were learning. I could see that some of the students were excited of the task and gave me positive responses to doing it. Some of them wanted me to take them to Disneyland if they could finish doing the task. Next day, at least I got *one* student handed in the work.

I noticed that students wanted to know more of ***why*** **they needed to learn the things taught** in the lesson. Like what happened today in the S1 class, one student asked, "why not I find the answer directly with the operations instead of setting up the equation for it?". I convinced her that the use of equations would be of great help in solving complex problems. To add up with more meaning for solving equations with unknowns, I told them, "Our ***lives are filled up with unknowns***! Using equations to solve for unknowns is a useful technique." It seemed that worked to let them get some sorts of meaning in learning it as I noticed that they nodded their heads though without saying anything.

I discovered some tricks that would be of help to make mathematics getting more meaningful to the students. One was to ***connect the tasks with students' life experiences*** and one was to have some ***hands-on activities*** (see Fig. 5.1) whether using some technological tools such as GeoGebra or some board games.

5.5.2 The Story of Kelly

The EMI (English as Medium of Instruction) school where I did the practicum was the one that I had been there for six years during my secondary education. In the first week of the practicum, I was arranged to have class observations of the school maths teachers in the classes that I would take over later. The teaching strategies they adopted were targeted for training students to get familiar with the skills for doing the maths problems with ***emphasis on attaining of procedural knowledge rather than***

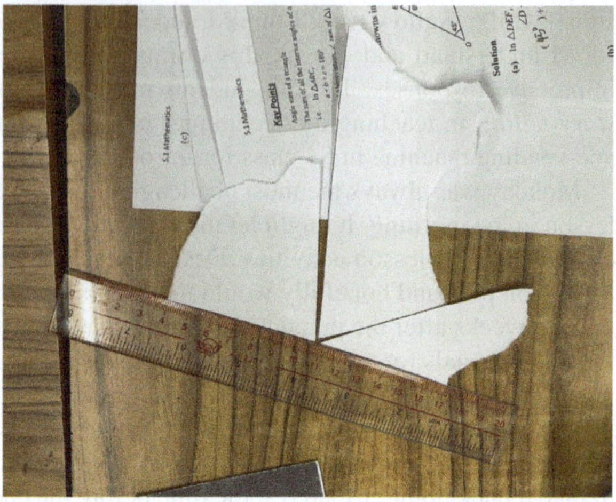

Fig. 5.1 Class activity—group work for angle sum of triangles

conceptual one. In one of the lessons that I observed, I was particularly impressed when the teacher ***used daily-life scenario*** to set up the task for students to make sense of "percentage error". The task enabled students to compare the measured weights of the baby and her mother based on same "maximum absolute error".

I started teaching on my own in the second week the topic of "rate and ratio" for S2 class. To make the examples more illustrating to the students, I searched in the web the prices of potato chips in two different shops online and designed the task for students to get touch with real life in learning the concept of "rate". For the concept of "ratio", I used the example of 1:99 diluted bleach for them to get touch with real-life application of the idea. This was what I thought to make them get touch with a sense of 'literacy' in learning mathematics. I found that it was not always easy to connect some of the topics, such as "law of indices" and "polynomials", to real life scenario. Then, I tried to *design some competition game activities and group tasks* for engaging students with the lessons. Yet, the students seemed not very used to taking part in the group discussions. The students were shy to ask questions even when they got difficulty of doing the tasks. It was not easy to draw students' attention. I knew not of what they knew from the explanations that I gave them.

In the fourth and the fifth weeks, I was asked to take up the classes of some other teachers who had to go for quarantine as they got infected with COVID. As a substitute teacher for the classes, I could have more freedom and space to try out some of ideas for making meaning in students' learning. I found that ***using daily-life objects*** such as different packing boxes to teach the concepts of "total surface area" was of great attraction to them in learning the ideas.

5.5.3 The Story of Wilson

Like Kelly, the school that I did the practicum was the one that I completed my secondary education. I started classroom teaching for S1 on the second day and S4 on the fourth day of the practicum. In S1 class, I tried to build up a closer relationship with the students by posing questions related to their ***daily-life experiences***, such as using some mathematics to do with travelling by bus. It seemed that worked! I got at least some ***responses*** **from the class**.

For S4 class, I adopted different teaching approach some challenging task sheet for guiding students learn as the school mentor was observing my class. I was feeling frustrated as the responses from students were not what I expected. It seemed to me that there were only two types of students in class—one was doing his own work and one was just staring at the teacher. Both types got one thing in common—no responses whatsoever!

I was feeling confused as the ***mentor's advices were of conflicting value with me in doing my teaching***. I was criticised ***not giving enough exercises for students' drilling*** of what they learnt on one hand and not teaching fast enough with a pace that could fit in the designated schedule on another hand. What made me feel particularly uncomfortable was not caring for whether students could make meaning from the learning but ***got job done by telling students what to do with the exercises***.

I was particularly impressed with a boy in S4 class. He was learning on his own some university-level mathematics, such as "number theory", "topology", "ordinary differential equations". I realised why the school teacher would ***not bother of his presence in class and left him alone of doing what he wanted to do***. He demonstrated not just the ability of doing maths far beyond the level of his learning in class but the competence that he demonstrated in sharing his notes of what he learnt in a highly organised way.

Back to S1 class, I felt irritated by the school's BYOD (Bring Your Own Device) policy that allowed students to bring in their iPad to class for learning. I found that some of the ***students using their iPads for watching YouTube and playing game***.

Back to the designs of the TAD tasks (see Fig. 5.2) I did for five weeks, I felt delighted of getting some responses from the students. I summarised below the result of TAD response rate in the table (see Table 5.1).

5.6 The Dialogue

The dialogue was conducted separately on individual basis after the TP period had ended. The meetings were arranged at the canteen or dining place with a setting of relaxing environment. The participants would bring their own computer notebook for showing the materials with me for the dialogue and would e-mail me the materials for record purpose. I did not do any video or audio recording of the meetings, but

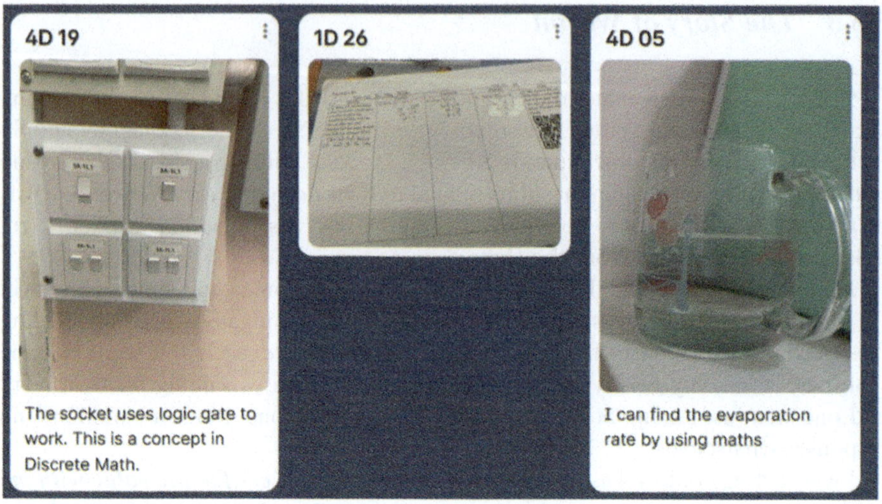

Fig. 5.2 Student's works on 'Task-A-Day' (TAD) with the task theme on 'Maths in Daily Life'.

Table 5.1 TAD response rate

	S1 (number of students with % * of submission)		S4 (number of students with %* of submission)	
Week 1	0	0	7	20%
Week 2	0	0	4	11%
Week 3	2	6%	1	3%
Week 4	0	0	3	9%
Week 5	0	0	1	3%

(* The percentages as shown in the table are round numbers)

instead I would jot down some notes in my sketchbook of what we were talking about during the meetings.

5.6.1 Dialogue with Jenny

I met Jenny on December 28 at 10am. She opened the computer notebook and began our conversation.

"What did you see any changes in your students' learning?" I asked.

"Changes? Students were getting more and more ***eager to ask the why-questions***. I think that I have ***created chances of encouraging them engage in the task activities in class***. By engaging in doing the tasks, they were supported to think in a mathematical way", Jenny answered without much hesitation.

During our conversations, Jenny mentioned four students that she had particular impressions of them—May, Haily, Joseph and Chi. The telling of the four students' stories enabled me to see of what she saw ethical in her teaching by ***seeing through the students' eyes***. Owing to lacking of space, I would adopt here only the story of Chi for illustration purpose.

5.6.1.1 Chi's Story

He was as bad as Haily (his classmate) in maths. He didn't concentrate in maths class but invited his groupmate to play instead. He was very shy in front of the teacher. When I approached him in class, he remained quiet and declined to answer my questions. Even when I asked him questions in private, he would only nod his head or respond by pointing with a finger. As I did not have much time to help him, I invited his group members to help him instead. His group member, Yuki, often stood next to him and would teach him after completing her own work. And yet, he rarely responded to Yuki.

During the 'numbers and angles' activity, I found that he was engaged enthusiastically in the activity. He used the protractor to find the angles actively (Fig. 5.3) and discussed with his group members of what he got. After this activity, I noticed that he began showing willingness of answering my questions. Although the questions that I raised were simple, such as "What is this line (radius)?", I just thought that it was already a big change in the way he did. I couldn't stop me from giving him sweets as a reward. Thereafter, he continued to take the initiative to answer the questions that I raised in class. In the activity lesson, he did not ask for any help from me or from the group members in doing the task. And then on, he worked hard to complete his homework on his own and seldom played with his classmates in class.

Jenny made use of the "Number and Angle" photo (Fig. 5.3) that she took to highlight how the activity had helped the student, Chi, to make meaning of what he was learning *by* ***pointing*** **to his hand working on the task**.

5.6.2 Dialogue with Kelly

I met Kelly in the afternoon on the same day of meeting Jenny. After we settled comfortably at the campus canteen with a nice view of clear sky through the opening of the window, she opened the computer notebook and began our conversation.

"What did you see any changes in your students' learning?" I asked with a similar opening question.

"I can see that there has been a ***change in the relationship with the students***. The change in relationship may come from the way of what I have made in teaching them. As I reflected on the design of task activities, I could see the importance of allowing students to have ***hands-on and minds-on experiences*** in their learning. The design process of supporting students learn maths meaningfully has opened up ***a space for growth not just for the students but for myself as a teacher*** as well.", Kelly replied with a reflective tone.

Fig. 5.3 Jenny's "pointing": The "Number and Angle" task activity

Kelly elaborated further some of her experiences in adopting different ways of doing classroom teaching. In the beginning of the practicum, she attempted to help students make sense of their learning through the uses of their ***daily-life experiences***. Nonetheless, she came across of the difficulty in using the same strategy in teaching some of the topics, such as "polynomial". She found that students responded actively when she used diluted bleach as a daily-life application of learning "ratio". In the Weeks 4 and 5, she got infected with COVID and took leave for quarantine. It might be a time for doing a deeper reflection for Kelly during her absence from the class teaching. She reflected from her observation in class and conversations with students and the school teacher as well—students would like to have challenging tasks to do as they (especially those of higher ability) put much emphasis on ***getting good examination results*** whilst they would expect to have something interesting in doing the tasks in the lessons.

Kelly made use of the three photos (Fig. 5.4) that she took for doing the "pointing" to highlight what she intended to help students learn in a meaningful way. The pointing showed that she felt sorry for the students with the difficulty of learning mathematics in an EMI school as they would have ***difficulty of reading maths problem using English to read and express themselves***. The last photo was a 'rewarding moment' for her improved relationship with the students by the end of the practicum.

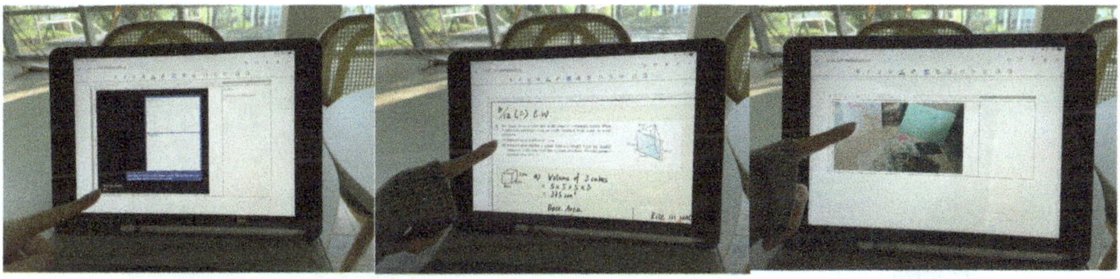

Fig. 5.4 Kelly's "pointing": (from left to right) The language problem in using English to learn maths, Reading problem for students, and Gifts from students as a 'reward' for the improved relationship with the students

5.6.3 Dialogue with Wilson

I had changed the meeting schedule with Wilson due his infection one day prior to our scheduled meeting. We at last met in the afternoon of January 13 at the canteen where I met Kelly. With the readiness of reading the screen of the computer notebook, we began our conversation.

> "What did you see any changes in your students' learning?" I asked Wilson with a similar opening question.
>
> "Students would have changed the way they learnt only if they could have noticed the extra efforts that teachers did have made. I could not agree more for ***making extra efforts at promoting learning of maths accessible to the students*** with fun and having something to learn in the lessons." Wilson said with an unwavering voice.

Wilson could not stop from sharing with me the experience of having ***value conflict*** with his school mentors in how maths should be taught. He did not agree with the school teacher who kept on asking him to assign more exercises for students to do so as to ensure them to have acquired skills for fluency. Wilson felt particularly confusing with the idea when he did the teaching for ***the remedial class with students who knew too little to do with the exercises***. He told me that he was not surprised with the nearly zero return rate of TAD task sheets in S1 class when he realised that only about 60% of the class would hand in their home assignments. Even in S4 class, the situation was not too much better in doing the TAD tasks though the class would have 100% success rate in handing in their homework. Another thing that annoyed him was about the BYOD policy for allowing students to use their own iPads in class. He found great difficulty of drawing students' attention without intention of using the tool for learning. For the "pointing photo" (see Fig. 5.5), Wilson highlighted the iPad that the 'super' student who used it in class to learn the university-level mathematics on his own.

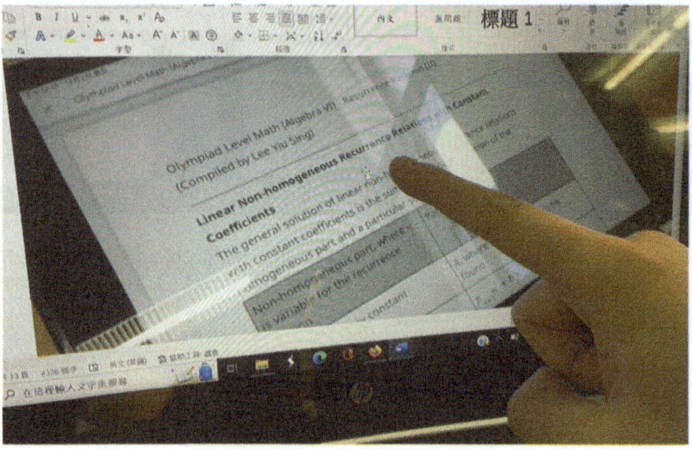

Fig. 5.5 Wilson's "pointing": The 'super' student using iPad for his own learning in class

5.6.4 The Experience

After the post-TP meetings, the participants e-mailed me their reflections on consolidating the experiences that they had in what they "saw ethical" in mathematics teaching.

5.6.5 Jenny's Experience

What impressed me in particular of Jenny's follow-up e-mail was the writing in much more details of the four students' stories (see Jenny's story) that she mentioned during the dialogue meeting. The detailing of student story reflects a kind of ***meta-reflection on seeing ethical in her teaching through the eyes of her students.*** In other words, the 'meta-telling' acts as mirror for Jenny to reflect on how she has put the ethical value into pedagogical practice from the beholders of the learning space. The ***student voice*** of such shows how a teacher's genuine concern of learners' (including the silent students') needs can help them make meaning of their learning despite of their low academic backgrounds. Jenny remarked that she would ***use story or even drama to attract students to engage*** more in the task activities if she could do the practicum once again. It would have been a meaningful learning experience to Jenny as she has prepared to commit herself to becoming a caring teacher in the professional field she has opted for.

5.6.6 *Kelly's Experience*

Kelly described her experience from three aspects, namely, (1) it matters for establishing good student–teacher relationship; (2) caring for student's emotional needs rather than for sheer academic achievement; (3) good teachers are life-long learners.

(1) **Student–teacher Relationship**

"Student-teacher relationship is something I value the most" as "a ***positive student-teacher relationship*** would lead to a positive learning experience. It is no doubt that knowledge is important but in my opinion, accompanying students in their growth is more vital. A professional teacher with an ethic of ***care is a gift to students***, whereas seeing the growth of students is a gift for a teacher".

(2) **Academic Achievement vs Emotional Need**

"Though it would take more time for students to complete the tasks, the engagement in the ***hands-on activities*** could bring positive emotions to students. This kind of learning experience is definitely more impressive to students who want to learn in a meaningful way. Students would think that learning mathematics is not so boring when participating in the activities. They may show more interest and engagement in the subject afterward".

(3) **Teachers as Learners**

"A good educator should be a lifelong learner. Each school year brings new students of unique selves of their own into your life. This leads to much of uncertainties in doing classroom teaching. Thus, a teacher should be ***open-minded to use different ways to teach and communicate with different students***. Besides, society is changing rapidly all the time, especially with the development of technology. ...in the past few years, teachers during the pandemic need to keep using a variety of teaching modes, such as the face-to-face, the online, and the hybrid. Hence, a teacher should keep learning new things to cope with the changes in the education field and in society".

5.6.7 *Wilson's Experience*

It is a regret for Wilson not showing good enough caring for the two particularly talented students by just leaving them alone in studying on their own. He felt sorry for them as he ***should guide every individual student to learn meaningfully***, including these two special learners.

"I should have given more care to students for coping with their individual difference. For the students who love mathematics, I can let them do more stimulating questions to attempt. For the low-ability students, I can let them try doing some guided questions with optimal hints. This would benefit students more. However, I think that I would need to bear with much greater teaching load for practicing this way of doing".

"The most interesting part of our discussion (during the post-TP meeting) is the *renaming* of remedial class". I have proposed to Wilson that we may redefine the "remedial lesson" as a special **tut**or**i**al – TTR (Time To Reflect) (Law, 2020). "In order to understand what students don't know, we should conduct the remedial class in form of interactive tutorial that would provide students with opportunity to express which part they are not familiar to. This idea is very good because I can know what they don't know". The renaming can help avoiding the labelling effect on those who need to attend the remedial class.

5.7 Interpreting Interpretive Action for Making Mathematics Meaningful

Peirce distinguished three kinds of phenomenological categories, designated as "Firstness" (sensation/ feeling), "Secondness" (perception/ reaction), and "Thirdness" (thought/ meaning) (Campbell, 2011, p.13). The "Story" as the Firstness unfolded the kind of *feeling* that the three participants had experienced the *struggles* that they shared with me through the "Dialogue" (the Secondness). The "Experience" as the Thirdness disclosed the thinking behind their representations of what "meaning" was meant to be. They had learnt from their TP experiences of how they define and redefine the meaning of "meaningfulness" as a notion of what they had undertaken in designing the task activities (including the TAD tasks) for their students. The kind of "meaning" they conceived subconsciously in doing classroom teaching might not be supported by experience as shared by the students. Nonetheless, they would benefit from developing the self-awareness in terms of conscious thinking to help them see how they are going to do with hopes in future practices of promoting meaningful learning for their students.

Within the limited scope of the present study, it is not intended to use the semiotic square for a 'full analysis' of the three student teachers' stories as design cases for practicing meaningful mathematics as an ethical value. And yet, its use helps us widen the perspective of interpreting their interpretive actions of striving to anchor classroom teaching practice with the value as aspired. The journey of the Practicum for the participants of the WISE Project was a process of *redefinition* of "meaningful learning" as drawn on their interpretations of what kinds of "meaningfulness" would be meant to the students. It was a kind of "abductive reasoning" for them to keep up with the awareness of making attempts at making mathematics meaningful as their aspired *value* in practicing their teaching. Throughout the TP period, they e-valuated and re-evaluated of "*how*" they were teaching from the daily observed outcomes as seen in the "problematic situation" of the classroom. And then, they reflected from a different viewpoint on "*what*" they could possibly re-invent their teaching in a meaningful way. Yet, they came up with the kind of puzzlement of what exactly "meaningful" was meant to be in the eyes of the students and the school mentors as beholders of how they taught behind what they saw as ethical in practicing their teaching.

As evidenced in the story of Jenny, what she learnt from observing and reflecting on the problem situation of "*boring with no responses*" from the students gave her the hints of adopting various ways of supporting students to make meaning of what they were learning through *story-telling*. When Jenny noticed that she was "*not able to keep them engaging*", she became aware of students' need to let them "*know more of what the task was about*". In teaching students for the use of equation, she made use of the "metaphorical explanation" of "*lives are filled up with unknowns*" to address their queries of "*why they needed to learn the things taught*". During the Practicum, Jenny strived to "*connect the tasks with students' life experiences*" and to design more "*hand-on activities*" for helping students to make sense of what they learnt. The dialogue with Jenny enable me to see what she saw ethical in making mathematics meaningful—Don't teach blindly without "*seeing through the students' eyes*". In sharing with me the photo she took, her "*pointing to his hand working on the task*" unfolded the 'capturing moment' of what she was seeing ethical.

In Kelly's story, she did not see meaningful learning with the "*emphasis on attaining of procedural knowledge rather than conceptual one*" by reflecting on what she observed in her mentor's class until the moment of how the teacher "*used daily-life scenario*" to help students to make sense of what was taught. Such an experience led her to "*design some competition game activities and group tasks*" as well as "*using daily-life objects*" for engaging students with the lessons. The dialogue with her unfolded a "*change in the relationship with the students*" as she saw classroom as "*a space for growth not just for the students but for* herself (*myself*) *as a teacher*". To her, what created meaning in students' learning came from their "*hands-on and minds-on experiences*" rather than from "*getting good examination results*". By the end of the Practicum, she was still baffled about what she could do to help students make their learning meaningful if they would still be trapped in the "language game" of having *"difficulty of reading maths problem using English to read and express themselves*". It remains a puzzlement as it appears to Kelly how she could learn from taking teaching as a serious "semiotic game" for making mathematics meaningful to students. It would not be too simple a task for each individual student to identify a particular classroom activity as either "meaningful" or "absent of meaning" in a binary way. Instead, the students may have a contradictory view of seeing mathematics as something absurd.

The feeling of absurdity, as I understand it in the context of learning mathematics, can come up with a further classification of two kinds. The first kind is a kind of absurdity in the form of "meaningful meaninglessness"—a contrary ("non-meaningful and non-absent of meaning") view on interpreting what "meaningful learning" is meant to be. And the second kind comes from the contradictory interpretation of "meaningful & non-meaningful" or "absent of meaning & non-absent of meaning". The feeling of absurdity comes from the experience of learning mathematics itself—learning of something that they cannot make meaning now-here and know no-where why that very something is really valuable in their future life. The classroom has turned itself into a restricted confine of "absurd walls" (Camus, 1991) within which they come to learn to reason with unreasonable silence of the world they are living in.

In Wilson's story, what he told us of his regret came from his ignoring act on the "super student" who was left to study alone in the lesson. Existence as a lonely learner in the classroom was "strange and beautiful—Every day, every moment, everything wavers between sameness and change, …, unity and fragmentation" (Genovese,

2010, p.12). It can be a kind of absurdity of the first kind in the form of "meaningful meaninglessness"—a contrary ("non-meaningful and non-absent of meaning") view on interpreting what "meaningful learning" is meant to be. Wilson further revealed that it could be "meaningful" to the students in terms of "*getting good examination results*" and yet "non-meaningful" in "*doing exercises for drilling*" purpose, like what we are told of how Sisyphus did to struggle for existence. In "*the remedial class with students who knew too little to do with the exercises*", it might be too harsh for us to ask the students to distinguish "absent of meaning" from "non-absent of meaning". The learning itself can be a struggle to create meaning out of meaninglessness. This can be a feeling of absurdity of second kind as I de-fine it.

5.8 Reprise: Seeing What Ethical in Teaching Mathematics is

From the semiotics perspective, the teaching and learning of mathematics as human activity involve signs subject to unique individual interpretations. And the further development of semioethics highlights the question of ethical responsibility of teachers who define commitments and values as grounded on their interpretations of the sign behaviour of humanity (Petrilli & Ponzio, 2005, p.547). In the study of signs, meaning, and interpretation, Vitoria Welby highlighted the trio relationship between sense, meaning, and significance, with its focus drawn on "the sign's ultimate value and significance beyond semantic meaning" (ibid, p.15). In conceptualising of the WISE project, I interpreted the ethical value of making mathematics meaningful as a sign waiting for the student teachers to make interpretations of their own. Such interpretations would afford them to take actions for creating task activities within and without the confine of the classrooms. Putting such kind of value into pedagogical practice is never simple but too complex to say in words of what exactly it is meant to be, despite of teacher's intention of valuing the practice. With ethical awareness of doing as such, they would have collaged "value" and "meaningfulness" into an open and unfinished design of pedagogical practice for promoting learning with meaning as conceivable by individual learners.

Recent years saw a growth of interest in adopting semiotics in the field of education as a new field of inquiry designated as "edusemiotics" (Deely & Semetsky, 2017). It is a move towards re-thinking and redefining of what exactly it is *meant* to do in education. If as humans we want to ask for the significance and meaning in life, we as educators need to ask why it is so hard to afford the learners of younger generation in making meaning of their learning especially in the process of the growth journey. Inna Semetsky (2015) in the conclusion of *Edusemiotics: The Tao of Education* asked critically, "why is edusemiotics so far absent in the departments of education, the theoretical foundation of this discipline thus remaining unchallenged and often counterproductive to the growth of students (and teachers for that matter)?". We may further ask, "why is it so hard to change teachers' mindsets of transforming their beliefs into making learning itself meaningful?". Semetsky (2015, p.141) would offer us an answer from the edusemiotical perspective—it is simply because "old habits are resilient and die hard!". What I did in the WISE Project was intended to make a humble contribution to raising teachers' awareness of seeing ethical in practicing their teaching. The use of Greimas's semiotic square was to help to broaden of our perceptions whilst making interpretations of what constitutes meaning in both teaching and learning, especially in the subject called mathematics. No-where might the project lead to perhaps, yet action is needed now-here to make a change in thinking seriously of what we can do in putting ethical value into pedagogical practice.

As I interpret interpretations of the three stories as told, I 'see' what they 'see' through the *story*, the *dialogue*, and the *experience*. I was particularly impressed by Jenny's detailed telling of the four students' stories, which signifies the significance of how she experienced the struggle behind making attempts at making mathematics meaningful to the learners. It appears to me that Jenny's telling is a kind of two-layered telling of a story in the form of 'meta-telling". I believe that such kind of meta-telling would be of benefit for developing a teacher's self-awareness in terms of uninterrupted movement of *redefinition* of what it is meant to teach for meaningful learning. Such unconscious thinking entails "paradox and misunderstanding, but also our most "authentic" possibility of *meaningfulness* (for ourselves and others)" (Possamai, 2022, p.142).

Seeing what ethical is in practicing teaching itself is an *art of seeing* how we value the making of meaning for the learners of mathematics. As I read and re-read the first 11 pages of John Berger's *Ways of Seeing* (1972), I could not stop myself from adopting the words he used for seeing the art of seeing the artworks in seeing what to see ethical in teaching mathematics.

> Seeing comes before words. The teacher looks and recognizes what ethical is before she can speak. It is seeing which establishes our classroom as our place in the surrounding world; we explain that world with words, but words can never undo the fact that we are surrounded by it. The relation between what we see and what we know is never settled. The way we see things is affected by what we know or what we believe. We only see what we look at. To look at is an act of choice. As a result of this act, what we see is brought within our reach. We never look at just one thing; we always looking at the relation between things and ourselves. Our vision is continually active, continually moving, continually holding things in circle around itself, constituting what is present to us as we are. And often dialogue is an attempt to verbalize this – an attempt to explain how, either metaphorically or literally, 'you see things, and an attempt to discover how 'she sees things'. When we 'see' a landscape, we situate ourselves in it.
>
> (Extracts in *italics* from John Berger's *Ways of Seeing*, pp.1-11)

As an educator, I see that making meaning for schoolchildren's learning itself is an ethical act. Nonetheless, it is always a great challenge for teachers to fulfil the ethical responsibility of making learning meaningful in general and in the study of mathematics in particular. Despite of the limitations of the present study (in terms of the small number of cases adopted and the short span of project duration), it is hoped that it serves twofold purposes—(1) to shed some light on highlighting the critical issue in teacher education for promoting teachers' ethical responsibility for making mathematics meaningful to the learners, and (2) to serve as an open invitation for all who are concerned with how we can put ethical value into pedagogical practice as a vitally important research agenda in mathematics education.

References

Berger, J. (1973). *Ways of seeing*. British Broadcasting Corporation and Penguin Books.

Campbell, P. L. (2011). *Peirce, pragmatism, and the right way of thinking* (No. SAND2011-5583). Sandia National Laboratories (SNL), Albuquerque, NM, and Livermore, CA (United States).

Camus, A. (1991). From 'The Myth of Sisyphus': An absurd reasoning. *The Myth of Sisyphus and Other Essays*, trans. Justin O'Brien, (pp. 1–24).

Chen, C., & Stevenson, H. W. (1989). Homework: A cross-cultural examination. *Child Development*, 551–561.

Choy, B. H. (2013). Productive mathematical noticing: What it is and why it matters. In V. Steinle, L. Ball & C. Bardini (Eds.), *Mathematics education: Yesterday, today and tomorrow*. Mathematics Education Research Group of Australasia. MERGA.

Corso, J. J. (2014). What does Greimas's semiotic square really do?. *Mosaic: A Journal for the Interdisciplinary Study of Literature*, 69–89.

Deely, J., & Semetsky, I. (2017). Semiotics, edusemiotics and the culture of education. *Educational Philosophy and Theory, 49*(3), 207–219.

Dorst, K. (2011). The core of 'design thinking' and its application. *Design Studies, 32*(6), 521–523.

Ernest, P. (2009). Values and the social responsibility of mathematics. *Critical Issues in Mathematics Education, 6*, 207.

Ernest, P. (2019). The ethical obligations of the mathematics teacher. *Journal of Pedagogical Research, 3*(1), 80–91.

Genovese, M. K. (2010). Meaningful meaninglessness: Albert Camu's presentation of absurdism as a foundation for goodness. *Pell Scholars and Senior Theses, 60*, 1–14.

Godino, J. D., & Batanero, C. (2003). Semiotic functions in teaching and learning mathematics. In M. Anderson, A. Sáenz-Ludlow, S. Zellweger, & V. V. Cifarelli (Eds.), *Educational perspectives on mathematics as semiosis: From thinking to interpreting to knowing* (pp. 149–167). LEGAS.

Greimas, A. J. (1966). *Sémantique structurale: Recherche de méthode*. Larousse.

Greimas, A. J., & Rastier, F. (1968). The interaction of semiotic constraints. *Yale French Studies, 4*, 86–105.

Greimas, A. J. (1987). *On meaning: Selected writings in semiotic theory*. (Theory and History of Literature 38), trans, by P. J. Perron & F. H. Collins (Ed.), The University of Minnesota Press.

Huh, J. (2016). Visual anatomy of frame creation in abductive reasoning by way of the semiotic square. *Archives of Design Research, 29*(2), 47–63.

Jacobs, V. R., Lamb, L. L., Philipp, R. A., & Schappelle, B. P. (2011). Deciding how to respond on the basis of children's understandings. In M. G. Sherin, V. R. Jacobs, & R. A. Philip (Eds.), *Mathematics teacher noticing: Seeing through teacher's eyes* (pp. 97–116). Routledge.

Kalogeropoulos, P., Russo, J. A., & Clarkson, P. (2021). Exploring educator values alignment strategies in an intervention context: The emergence of the Beacon strategy. *ECNU Review of Education, 4*(2), 327–348.

Law, H. Y., Wong, N. Y., & Lee, N. Y. (2011). The Third Wave studies of values in effective mathematics education: Developing student's mathematical autonomy in classroom learning. *The Mathematics Educator, 13*(1), 72–86.

Law, H. Y., Wong, N. Y., & Lee, N. Y. L. (2012). A study of espoused values in Hong Kong's mathematics classrooms. *ZDM, 44*(1), 45–57.

Law, H. Y. (2020). Theorising mentoring for the 21st Century teaching and learning: Making invisible professional growth visible through action research. In T. Barkatsas & T. MacLaughlin

(Eds.), *Championing cutting-edge 21st century mentoring and learning models and approaches* (pp. 135–156). Brill Sense Publisher.

Law, H. Y. (2023). Promoting learner agency for gifted, talented, creative and dissimilar learners: Developing STEM literacy for learners of all kinds in Primary mathematics classrooms. In T. Barkatsas (Ed). *Educating gifted, talented, creative and dissimilar learners: Imagining the future* (pp. 250–270). Brill Sense Publisher.

Lemon, N. (2007). Take a photograph: Teacher reflection through narrative. *Reflective Practice, 8*(2), 177–191.

Mason, J. (2002). *Researching your own practice: The discipline of noticing*. Routledge.

Militello, L. G., Gentner, F. C., Swindler, S. D., & Beisner, G. I. (2006, May). Conation: Its historical roots and implications for future research. In *International symposium on collaborative technologies and systems* (CTS'06) (pp. 240–247). IEEE.

Moore, J. S. (1914). Value in its relation to meaning and purpose. *The Journal of Philosophy, Psychology and Scientific Methods, 11*(7), 184–186.

Morris, C. (1964). *Signification and significance: A Study of the Relations of Signs and Values. Studies in Communication*. M. I. T. Press.

Nigar, N. (2020). Hermeneutic phenomenological narrative enquiry: A qualitative study design. *Theory and Practice in Language Studies, 10*(1), 10–18.

Pelkey, J. (2017). Greimas embodied: How kinesthetic opposition grounds the semiotic square. *Semiotica, 2017*(214), 277–305.

Petrilli, S. (2015). Sign, meaning, and understanding in Victoria Welby and Charles S Peirce. *Signs and Society, 3*(1), 71–102.

Petrilli, S. (2016). Semiotics and education, semioethic perspectives. *Semiotica, 2016*(213), 247–279.

Petrilli, S., & Ponzio, A. (2005). *Semiotics unbounded: Interpretive routes through the open network of signs*. University of Toronto Press.

Petrilli, S., & Ponzio, A. (2008). A tribute to Thomas A Sebeok. *Biosemiotics, 1*, 25–39.

Possamai, T. (2022). *Where thought hesitates: Gregory Bateson and the double bind*. Mimesis International.

Radford, L., Schubring, G., & Seeger, F. (2011). Signifying and meaning-making in mathematical thinking, teaching, and learning. *Educational Studies in Mathematics, 77*, 149–156.

Rosa, A., & Pievi, N. (2013). Semiotic methodology for the analysis of the cultural and individual dynamics of social representations: A view from cultural psychology. *Papers on Social Representations, 22*(2), 19–21.

Sandelowski, M. (1991). Telling stories: Narrative approaches in qualitative research. *Image: The Journal of Nursing Scholarship, 23*(3), 161–166.

Seah, W. T. (2019). Values in mathematics education: Its conative nature, and how it can be developed. *Research in Mathematical Education, 22*(2), 99–121.

Sebeok, T. A. (1994). *Global semiotics*. In Semiotics around the world: Synthesis in diversity (Proceedings of the fifth congress of the international association for semiotic studies, Berkeley 1994) (Vol. 1, pp. 105–30).

Semetsky, I. (2015). Edusemiotics: The tao of education. *Language and Semiotic Studies, 1*(1), 130–143.

Thanasoulas, D. (2000). What is learner autonomy and how can it be fostered? *Internet TESL Journal, 6*(11), November 2000 html. Retrieved on September 10, 2015 from (http://iteslj.org/Articles/Thanasoulas-Autonomy.html)

Tomkins, L., & Eatough, V. (2013). The feel of experience: Phenomenological ideas for organizational research. *Qualitative Research in Organizations and Management: An International Journal, 8*(3), 258–275.

Urban, G., & Brown, K. (2006). Metasemiosis and metapragmatics. *Linguistic Inquiry, 2*, 401–407.

Widdershoven, G. A. (2001). Dialogue in evaluation: A hermeneutic perspective. *Evaluation, 7*(2), 253–263.

Part II
Values Alignment and Classroom Practices

Chapter 6
Designing a Professional Development Model for Values Alignment Strategies in Inclusive Mathematics Instruction

Fatma Nur Aktaş

6.1 Introduction

Special education involves practices designed by considering the differences among students with mental, physical, social, or sensory impairments or talents. Therefore, educational opportunities are offered to disabled and gifted individuals in special education schools or inclusive/integration schools, or private educational institutions in line with the type of disabilities, impairments, or talents (Taller-Azulay et al., 2022). Due to the requirements in inclusive schools, various classroom cultures are created in the courses taught by different mathematics teachers in resource rooms and inclusive classrooms. Likewise, private educational institutions and support educational institutions, such as science and art centers (SACs), develop specific cultures (see Aktaş & Dede, 2023). Such diversity causes inclusion students to be exposed to different institutional and classroom cultures. Moreover, the values of mathematics teachers both in inclusive classrooms and resource rooms are apparent in inclusion practices (see Aktaş & Argün, 2018; Dede, 2015; Seah, 2019), and the alignment or conflict of these values is inevitable as students and teachers bring their values into inclusive education. Therefore, aligning the students' and teachers' values is important so that inclusive students can benefit from instructional practices as much as possible (see Kalogeropoulos et al., 2021). In this respect, mathematics teachers' values alignment strategies and developing their skills in choosing strategies are critical for the efficiency of inclusion practices. Thus, this chapter aims to design a professional development model for values alignment strategies used by mathematics teachers teaching inclusive students.

F. N. Aktaş (✉)
Kahramanmaras Sutcu Imam University, Onikişubat, Turkey
e-mail: fatmanuraktas@ksu.edu.tr

Y. Dede et al. (eds.), *Values and Valuing in Mathematics Education*,
https://doi.org/10.1007/978-981-99-9454-0_6

6.2 Theoretical Background

The theoretical background is discussed under three subheadings.

6.2.1 Values and Values Alignment in Mathematics Education

The term value is defined as "an individual's internalization, 'cognitisation' and decontextualization of affective constructs (such as beliefs and attitudes) in his/her socio-cultural context" (Seah, 2003, p. 2). Adopting a socio-cultural lens, Bishop (2001) has the following to say about values:

> Values exist on all levels of human relationships. On the individual level, learners have their own preferences and abilities that predispose them to certain activities more than others. In the classroom, values are inherent in the negotiation of meanings between teacher and students and among the students themselves. […] The larger political scene is at the societal level, where powerful institutions determine national and state priorities for mathematics curricula, teacher-preparation requirements, and other issues. Finally at the cultural level, there are sources of knowledge and beliefs, and the language influences our values in mathematics education (p. 347).

From this perspective, two critical elements are (1) teachers and students, who shape the classroom culture, and (2) the values they convey to mathematics classrooms. Bishop (1996) classified the values reflected in mathematics classes into three categories: general educational values, mathematical values, and mathematics educational values. Educational values carry the traces of educational vision, and mathematical values reflect the nature of mathematics. Mathematics educational values, on the other hand, are discussed concerning teachers, students, and classroom practices. For teachers and students, these values mean how to teach and learn mathematics effectively. Thus, there are some differences or similarities between the values of the teachers and those of the students in the mathematics classroom. Values alignment is the coexistence and prioritization of these values during conflict or adaptation processes. However, it does not aim to ensure "that the students' values are the same as those of their teachers" (Seah & Andersson, 2015a, p. 3124). Also, this process is not intended to force one side of the teacher and students to accept the value of the other.

6.2.2 Mathematics Teachers' Values Alignment Strategies and Professional Development

Teacher and student values should be harmonious for efficient classroom practices facilitating learning. Seah and Andersson (2015a) emphasised that "for a teacher, being able to facilitate values alignment between what s/he values and what his/her students value promises to strengthen the relationships and is one of the keys to nourishing teaching and learning practices" (p. 177). Likewise, Seah (2019) stressed that "the ability of effective teachers to align their values with those of the students is instrumental in facilitating mathematics learning" (p. 110). The teacher's values alignment strategies consist of choosing between her/his own values and those of the students and aligning these two groups (Kalogeropoulos & Bishop, 2017). Various classifications have been offered for these strategies (Kalogeropoulos & Bishop, 2017; Kalogeropoulos & Clarkson, 2019; Kalogeropoulos et al., 2021; Seah & Andersson, 2015a). In this sense, Seah and Andersson (2015a) designed their categories of strategies by considering student thinking: *redefining*, *reprioritising*, and *complementing*. *Redefining* refers to the teacher's revising the lesson as s/he recognises that student values are compatible with her/his own. *Reprioritising* involves the process of re-deciding between the teacher's own values and those of the students. However, the teacher delivers the lesson by giving importance to the students values without giving up his own values. *Complementing* refers to the teacher's incorporating different values that complement or emphasize each other into the lesson. On the other hand, Kalogeropoulos and Bishop (2017) suggested the main strategies of scaffolding, equilibrium, intervention, and refuge. *Scaffolding* is based on adhering to the teacher's designed lesson plan and involves allowing for small changes, such as peer support in practices in which students have difficulty. *Equilibrium* is the teacher's adaptation of the lesson plan by interpreting the students' values according to his/her own values in unexpected situations. This strategy was later called the 'balancing strategy' by Kalogeropoulos and Clarkson (2019) as it involves establishing a balance between student values and teacher values. The *intervention* strategy involves suspending the values at the beginning of the implementation and emphasizing human values (Kalogeropoulos & Clarkson, 2019) or focusing on the thoughts of a student or a group of students (Kalogeropoulos et al., 2021). The *refuge* strategy involves ignoring the teacher's values and those of the lesson plan and conducting the lesson based on the students' values during the instructional activities. On the other hand, in *beacon*, a strategy introduced by Kalogeropoulos et al. (2021), the teacher strictly adheres to the application s/he has designed based on her/his opinion and experience so that the student will learn better to maximise student learning according to the educator's expertise. When aligning teacher and student values, strategies can be ordered in descending order based on the level of commitment to teacher values: *beacon*, *scaffolding*, *balancing*, *intervention*, and *refuge* (Kalogeropoulos et al., 2021). The main difference between these classifications is that Seah and Andersson (2015a) focus on student thinking, while Kalogeropoulos and Bishop (2017) focus on teacher thinking and lesson design considering students' values. However, Kalogeropoulos et al. (2021) focused on the idea of scaling strategies between the values of these two actors.

Recognizing the values emerging during classroom practices and maintaining the adaptation process are key instructional skills for teachers (see Aktaş & Argün, 2018; Seah & Andersson, 2015a). Similarly, it is a professional skill for them to decide on values alignment strategies and use them in their practice (Seah, 2019). Helping teachers develop their ability to choose and apply values alignment strategies based on the emerging values alignment literature is a remarkable contribution to the field, as well as to the professional development literature.

Professional development is a process in which a teacher's professional identity is formed, and it is a process of personal and professional empowerment within the realm of the teacher's expertise. Also, it aims to promote and improve her/his knowledge, skills, and values (Fraser et al., 2007). Therefore, teachers can re-evaluate their beliefs, skills, and values through professional development models by considering the changes in the classroom practices of in-service teachers and the developments in students' thinking and attention skills (Clarke & Hollingsworth, 2002). Also, video use helps the teacher notice the salient aspects of instruction, including student thinking and interactions between the teacher and students (Borko et al., 2008; Büscher & Prediger, 2022; van Es & Sherin, 2010). Indeed, the self-video analysis provides objective data for identifying teachers' values in decision-making processes (see Aktaş et al., 2019). Thus, teacher professional development has its place in mathematics education research to improve or influence mathematics content and pedagogical knowledge, belief, and practice based on cognitive and sociocultural perspectives (Lin & Rowland, 2016). Teaching mathematics to students who need special education also necessitates professional development considering students' knowledge. Therefore, mathematics teachers need professional development processes to improve students' opportunities to learn advanced mathematics and provide enrichment opportunities for gifted students in inclusive classrooms (Even et al., 2009).

6.2.3 Inclusive Education for Gifted Students and Their Education in Türkiye

Inclusive education is defined differently by different countries, schools, and laws. The definitions refer not only to the principle of equal opportunity in education but also to educational opportunities in mainstream classrooms. Waitoller and Artiles (2013) examined the definitions and identified three key aspects: (1) the issues related only to ability differences, (2) those concerned with revising the curriculum based on gender and cultural differences, and (3) the process of overcoming barriers to participation and learning for all students. In the present chapter, inclusive education is defined as an ongoing and systemic process of changing the school culture to inform the practices that facilitate access, participation, and learning among students with diverse abilities. It could be inferred from this definition that differentiated and enriched educational practices are essential not only for disabled students but also for talented students. This is because a gifted student "is an individual who learns faster than her/his peers, is ahead in creativity, art, and leadership, has special academic abilities, can understand abstract ideas, likes to act independently in her/his interests and performs at a high level" (Ministry of National Education SAC Directive, 2016,

p. 2). Leikin (2019) describes the characteristics of gifted students: "A student is mathematically gifted if s/he exhibits a high level of mathematical performance within the reference group and can create mathematical ideas which are new with respect to his/her educational history" (p. 3). This implies that gifted students should be educated through effective classroom practices to become advanced individuals who can contribute to society and humanity. Focusing on gifted students' thinking skills is important to maintain harmony between student and teacher values (Aktaş & Dede, 2023).

Gifted students are identified using data from intelligence tests, along with information about students' characteristics and mathematical abilities. In Turkey, they are identified using general tests and individual interviews in the 2nd to the 4th grade of primary school. After that, special education opportunities are offered to them at the SACs based on specific types of intelligence and abilities in visual arts and music. Based on the tests of annual evaluations in the fifth to eighth grades at elementary school, gifted students receive special education in mathematics, literature, science, arts, and technology in SACs in addition to general education in inclusive classrooms. At the high school level, gifted students carry out projects based on their interests (MEB SAC Directive, 2016). They also receive K12 education in inclusive schools in Turkey. In inclusive schools, they receive education not only with their peers in inclusive classes but also retain the right to work with their individual teachers in resource rooms. It is optional for the gifted student to receive training in resource rooms. Therefore, gifted students are exposed to diverse classroom cultures and teacher values. For efficient and healthy instructional practices, teachers should have the ability to manage their choices of values alignment strategies. However, there is a paucity of systematic research in the literature about professional development in inclusion classes (Waitoller & Artiles, 2013). Therefore, the main purpose of this chapter is to design a professional development model for values alignment strategies used by mathematics teachers teaching inclusive students. As a result of the intervention phases followed in the present study, a model was presented to develop the skills of mathematics teachers to select values alignment strategies. For this purpose, the present study sought an answer to the following research question:

What characteristics should a professional development model possess to improve mathematics teachers' skills to employ values alignment strategies in inclusion classes with gifted students?

6.3 Method

This chapter aimed to develop the skills of middle school mathematics teachers to choose values alignment strategies. It considers the values of the inclusion students, who not only have their own values but also have been influenced by the educational practices fed by the values of various mathematics teachers. This chapter reports the results of an interpretive case study designed to examine the effectiveness of the intervention designed to achieve this. Indeed, "a case study can test theory as well as build theory, and use data gathering and data analysis techniques common to traditional forms of research." (Merriam, 1985, p. 206). So, the case study is adequate for exploring, understanding, or explaining an event, process, decision, or intervention that is still unknown and not well-researched. Therefore, Merriam (1998) defined the interpretive case study design, using the rich and think description obtained to construct and test theory. The applications made to create development and awareness constitute an intervention. Indeed, video analyses and interpretation and video clubs are tools that influence the thoughts and decisions of individuals (Borko et al., 2008; van Es & Sherin, 2010). Therefore, each model phase indicates an intervention process.

The participants, whose names are anonymous, were Eda, a gifted 7th-grade student, and her mathematics teachers in the middle school and the SAC. Eda was diagnosed as a gifted student in the second grade of primary school by the Guidance Research Center and has received supportive training at the SAC since then. She received an education that aimed to identify her general abilities until the 7th grade, and this year, she is taking only mathematics, chemistry, and philosophy courses. Eda was receiving support training at SAC three days a week. At any time, she could discuss and reflect with his teachers at SAC and follow their lectures. Ilgaz is Eda's supervising mathematics teacher at the SAC. He has been teaching mathematics with two of his peers for a year. Ilgaz has been a supervising mathematics teacher for gifted students for 16 years and has taught mathematical reasoning courses, carried out related projects, and organized competitions for gifted students. Mert was Eda's mathematics teacher at the inclusion school and had taught in the same class for two years. At the time of the study, Mert had taught mathematics in inclusive classrooms with gifted students for several years. Ilgaz and Mert are male teachers with 18 and 16 years of professional experience, respectively. The SACs do not offer a specific mathematics curriculum, so teachers can use whatever content they want. However, inclusion classes in public schools follow the same curriculum as others in Turkey.

Teacher professional development is built on the teacher, student, content, and affective factors (see Lin & Rowland, 2016). Likewise, values alignment situations have two main classroom actors: teacher and student. Therefore, according to the purpose of this chapter, students and their teachers are essential in determining the model phases. The student feedback phase is included to evaluate the development process from the student's perspective. Video-based interviews were discussed to reveal the interaction between these actors and also the values alignment strategies (Aktaş et al., 2019; Borko et al., 2008). Self-, peer-, and group analysis phases are among the effective strategies for professional development and video-based attending processes (see Aktaş et al., 2019; Borko et al., 2008; Büscher & Prediger, 2022; van Es & Sherin, 2010). Therefore, the development model in the current chapter is limited to self-video analysis, peer-video analysis, student feedback, and group-video analysis phases.

The data collection tools were the video recordings of the lessons taught to Eda in the SAC and the inclusive classroom and the interviews in which Eda and her mathematics teachers analysed these recordings. The phases of the teacher's skill development model were self-analysis, peer-analysis, student feedback, and group analysis. Classroom videos were recorded for each phase, and the interviews were conducted every week (see Fig. 6.1). The critical moments with values conflicts and alignment in the videos were considered critical situations. However, it was expected that the teachers could first notice the critical situations in the interviews. Then, the interviews were conducted to discuss the critical situations the teachers failed to notice. Here, noticing includes teachers' attending to and interpreting values alignment and/or conflict processes (van Es & Sherin, 2010).

While self-analysis refers to the teachers' reviewing her/his own lecture videos, peer analysis involves the review of the lecture videos of other participating teachers. Group analysis occurs when the teachers review their videos together. In these interviews, the teachers were asked various questions that revealed their experience of values alignment strategies, such as "What is the rationale for this decision?", "Why did you give this feedback to Eda?", "What would you do in this situation?" and "How can we improve this moment?" These questions aimed to help the teachers develop more diverse strategies through exchanging opinions and instructional experience. Weekly video recordings lasted two class hours in both instructional environments, and the interviews lasted approximately three hours for each teacher.

Student feedback refers to the practice of eliciting reasons for Eda's answers, reactions, or decisions in both classes and communicating them to the participating teachers. For this, the video recordings obtained for the three weeks determined the critical situations in which Eda existed. These critical situations particularly refer to moments when the teachers failed to interpret what Eda thought when value conflicts and alignments were observed or when the teachers advocated different strategies during the peer-video analysis process. Eda was asked various questions, such as "Why were you raising your hand reluctantly here?", "Why were you insistent here?", "What were you thinking here?", "What would you say about this teacher's feedback?" These questions helped identify Eda's preferences for value alignment strategies in two-session interviews that lasted approximately six hours. Afterward, the participant teachers were asked to watch the videos of the critical situations in their classes again, Eda's feedback was given to them, and their opinions about preferences of values alignment strategies were sought again.

The phases of the model were formatively evaluated using ongoing and retrospective analyses. The ongoing analysis refers to the practice of identifying critical situations in weekly video recordings and analysing interviews for peer and group analyses in the following weeks. Thus, the effect of the weekly intervention phases on the strategy change was also investigated. After the research was completed, a retrospective analysis was carried out to identify the differences among the participating teachers and to compare the model phases by analysing the changes in strategy choices. The teachers' values alignment strategies were examined using both video analysis and the analysis of the interviews. The analysis of the data is exemplified in Fig. 6.2.

The categories that emerged in this chapter were compared with various categories of values alignment strategies used in different contexts in the literature, such as focusing on students or teachers (Kalogeropoulos & Bishop, 2017; Kalogeropoulos & Clarkson, 2019; Kalogeropoulos et al., 2021; Seah & Andersson, 2015a), to ensure

Fig. 6.1 The phases involved in the professional development model and research process

Critical Situation

Eda asked a question about a different content from the problem that another student had solved on the board -due to the subtraction in the solution-. Eda visited Ilgaz's desk to ask a different question.

Eda: I do subtraction differently. When subtracting 9 from 11, I subtract 9 from 10 and add 1.

Ilgaz: OK. That is also correct.

Self-analysis in the Interview

Ilgaz: Students often think of one thing but find something else. Whenever I give up listening, they stop thinking. I listen to them even if what they say is not relevant. Maybe they will find something relevant.

Values Alignment Strategies

Suspending the lesson plan and focusing on student values by prioritizing student thinking

Refuge Strategy

Fig. 6.2 Data analysis

the theoretical triangulation during categorisation. The data collection also involved triangulation through self, peer, and group analysis. The member check and peer review ensured the research data's reliability (Merriam, 1985). A colleague checked the themes and sub-categories created by the researcher with a Ph.D. in mathematics education. The sub-categories were revised based on expert opinion. For example, it was agreed that the refuge and reprioritizing strategies differ regarding the teacher's greater consideration of student opinions. Considering the definitions of these strategies (Kalogeropoulos et al., 2021; Seah & Andersson, 2015a), analyses focused on teachers' prioritizing their own and students' values.

6.4 Results

The strategy preferences of the teachers determined through the weekly self-video analysis are presented under the themes specifying the model's phases. This helps observe the reflection of each type of intervention. Also, the results were presented using the same critical situations to highlight the reflections on the student feedback and group-video analyses. In this chapter, the critical situations were numbered in the order in which they were presented to compare the developmental process globally.

6.4.1 Values Alignment Strategies

6.4.1.1 Initial Values Alignment Strategies

Mert communicated with Eda only twice during the two class hours.

Critical Situation I

In the problem, they were asked to determine the rational numbers represented by the points A, B, and C marked on the number line from among the options. A student solved it by equating the denominators of the rational numbers in the options.

Mert: Is there a more practical way to do it?

Eda and another student were raising their hands. Mert called on Eda.

Mert: Your friend equated the denominators; another student could have equated the numerators. Eda is also comparing more or less than half (He interrupted Eda's solution and began to offer explanations).

Eda marked the correct option and took her seat. Mert asked Eda to explain the correct option again.

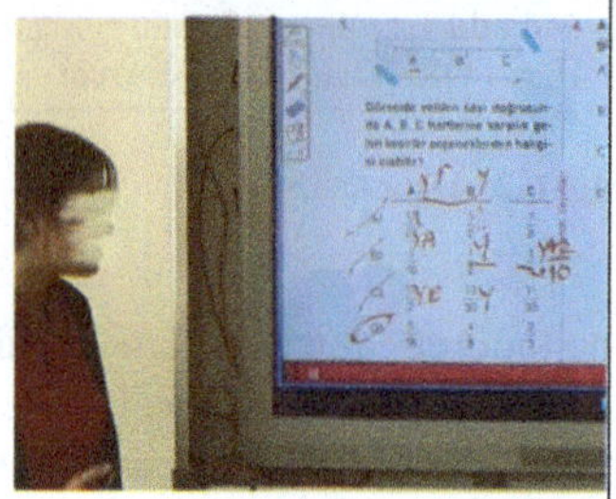

Mert: Can you go on with the half and quarter comparison here?

Eda: How? Do you need to equate the denominators?

Mert: For example, is 5/18 less than half or more?

Eda: Less.

Mert: How about 4/9?

Eda: Less.

Mert: And 2/3?

Eda: More.

Mert: Then you need to equate the denominators with being able to compare the first two.

Eda completes the operations.

Fig. 6.3 Critical situation I

In the interview about Critical Situation I (see Fig. 6.3), it was found that Mert invited Eda to the board because he thought she would continue the lesson as he had planned in line with his own values:

Mert: She showed that the options she had eliminated were incorrect but did not question option D. The same solution cannot be used here. She gets confused when I question her. It was this strategy for which I chose Eda.

Eda's intervention and guidance in the problem-solving strategy also support the results above, pointing to the *scaffolding* strategy.

Mert: Because she has progressed fast, she has solved another problem. However, she could not remember something she had done. I respond to everyone. It is not just for Eda. […] She is explaining it to her friends. In fact, it should not be done in class. However, you cannot stop Eda. Moreover, blocking her is not right. She finds it boring to listen to a solution she has already found. This is actually peer education.

Mert's words in the interview for Critical Situation II (see Fig. 6.4) revealed that he employed the *intervention* strategy by focusing on the wishes of individual students at times in his lessons. However, not interfering with Eda's helping her peers in problem-solving and prioritizing her values are reflections of the teacher's applying the *reprioritising* strategy.

It is clear that Ilgaz does not focus on Eda's thoughts. In Critical situation III (see Fig. 6.5), Ilgaz employed the *scaffolding* strategy as he focused on his own values

Critical Situation II	Mert walked around the classroom, waiting for students to solve a problem. Eda came to where Mert was to seek explanations about another problem she could not solve. They examined the problem together. Then Eda began to explain the solution to her friends. Meanwhile, Mert explained the answer to a question he had asked the class earlier.	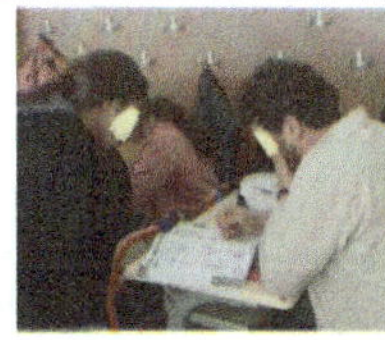

Fig. 6.4 Critical situation II

Critical Situation III	Ilgaz had just introduced the patterns concept. He stated that patterns could be created using letters, numbers, or shapes. Ilgaz: Let us get away from mathematics first. Let us start with letters. He writes the following letters and question marks on the board: a, e, i, i, o, ?. Let this string of letters be a pattern. What do you think will replace the question mark? Student A: The first letter. Ilgaz: It means this sequence will repeat. Student B: ö (The letter that comes after the letter o in the Turkish alphabet). Eda: A number must be between a and e, and another between e and i. Ilgaz: This part of the lesson has nothing to do with mathematics. Student B: It could be ö. Ilgaz: Yeah. That is the reason that I had in mind. Why? […]

Fig. 6.5 Critical situation III

and conducted the lesson based on the lesson plan by considering the opinions of other students.

Ilgaz: I enjoy hearing my students answer incorrectly, as we can hardly find the correct solution. However, Eda is struggling to thrust herself on me. She unnecessarily inserts numbers.

In the interview for Critical situation IV (see Fig. 6.6), Ilgaz stressed his prejudice against Eda. Moreover, while working with gifted students, he continued using the scaffolding strategy by not focusing on student values as he thought it prompted them to think. Although presenting daily life examples gives the impression that Ilgaz focuses on student values, he did not consider student thinking in the interview.

Ilgaz: One of the first topics they studied at school this year was negative numbers; she wanted to show that two minuses make a plus. I think she is trying to show that she knows it […] We are concerned about giving real-life examples in mathematics instruction. However, I decided I was wrong because these children live in the same world with us, but they hardly experience the same things. For example, this bank account example is not good enough for a child. I did not provide the whole pattern for a few terms here so they could deal with these patterns themselves.

Critical Situation IV	Student A tried to find the general term for the pattern "27, 24, 21, 18, …". He determined the common difference and was trying to determine the constant term by using 27: Student A: -3n minus -30. Ilgaz: So +30. Student A deleted the two minus symbols and wrote +30. Eda: Sir, can we retain both minuses? Ilgaz: It would be unnecessary. Two negatives make a positive.  Student A wrote the general term. Meanwhile, Ilgaz was writing a new pattern. Eda: Sir, shouldn't the result be +15 (for the fifteenth term)? Ilgaz: Well, think like this. You have a bank account. You regularly withdraw money. When you run out of money in your account, the bank continues to give you a loan. It means your balance is minus.

Fig. 6.6 Critical situation IV

6.4.1.2 Values Alignment Strategies Following the First Self-video Analysis

It was found that Mert allowed Eda to speak more frequently in the self-video analysis interview of the second week's lessons. In these discussions and problem-solving processes, it was found that Mert employed the *scaffolding* strategy and remained loyal to the lesson plan. However, in Critical Situation V (see Fig. 6.7), he focused on Eda's natural abilities and used the *reprioritising* strategy.

Mert: She made progress because she was a little faster. Usually, I do not like moving on to the next topic in the textbook. She was bored in class and was on the right track. I let her.

As in Critical Situation VI (see Fig. 6.8), Ilgaz frequently employed the *scaffolding* strategy, focusing on problem-solving processes according to his designed plan. However, giving Eda the opportunity to focus and express herself indicates the *intervention* strategy.

Ilgaz: It is less practical for them to find the general term and find the difference between the terms. As a shortcut, I tried to guide the students to see if it related to the common difference. There was none other than Student A. I wanted to see a long way first. Eda still hesitated when writing the general term. I chose Eda to help

Critical Situation V	Mert expected the students to solve problems related to the addition and subtraction of fractions one by one. Then, they began to solve problems with the students on the board. Mert carefully looked at Eda's textbook while walking around the classroom. Eda: I got bored, so I passed on to multiplication. Mert: (He checked the pages of the book backwards) Good!

Fig. 6.7 Critical situation V

Critical Situation VI	Ilgaz asked the students to find the difference between the 15th and 20th terms in the pattern "11, 15, 19, 23, …". Eda solved the problem on the board by conducting several operations. Ilgaz strongly suggested simplification while Eda was doing the operations. However, Eda was hardly willing to do it. Coming to the board later, Student A solved it more practically using the terms' differences.

Fig. 6.8 Critical situation VI

her overcome her hesitations. Then we listened to Student A, so that others saw the different solution. […] I force the students to get into the habit of simplification. This is because students often think that they can find the result through multiplication.

6.4.1.3 Values Alignment Strategies in the Peer-Video Analysis Phase

In the interviews conducted for the peer observation of Mert and Ilgaz and the observation of Eda's thinking in a different classroom environment, they were asked to specify which strategies they would prefer if they were in the place of their peers.

Critical situation III was discussed with Mert, and Ilgaz's feedback to Eda was examined:

Mert: If I were his, I would ask why there should be a number between a and e. She may have decided on the order of the letters in the alphabet.

Eda's desire to include both minuses in Critical situation IV attracted Mert's attention.

Mert: She remembers the four operations in integers, which we have just studied at school. Minus multiplied by a minus. The thing we use the most these days […] I think it is more comprehensible to write $30 - 3n$ than to write $-3n + 30$. The student could more easily understand that it is 27 when n is replaced with 1. However, the student profile is different. We always favour simplification […] If the children owned a bank account themselves, the example of a bank would be useful. Although they are gifted, they may have difficulty understanding real-life terms.

As understood from Mert's words, he emphasizes *reprioritising* strategy by considering Eda's values or the *intervention* strategy as he draws attention to the change in the instruction.

In Critical Situation I, Mert's allowing Eda to speak was examined with Ilgaz.

Ilgaz: The gifted student will certainly find something different. I cannot say the same thing for the other student. However, I would not interfere too much if I were you because Eda's thinking could be distracted. Eda has sat down; she might have even forgotten how she solved the problem.

Eda's individual questioning in Critical situation II has grabbed Ilgaz's attention:

Ilgaz: Mert is right; answering the question immediately while the student is still focused will make it easier for her/him to understand it. Things are easier as it is a very enthusiastic class. […] It is nice of him to offer his friends explanations. A person can only explain something that s/he understands. It is Eda who completes the process, not Mert. It has been an effective learning environment.

Ilgaz evaluated Mert's lessons by prioritizing Eda's values more than he did in his own classes and preferred the intervention strategy. In addition to expressing similar ideas with Mert about learning environments and preferences, he drew attention to Eda's thinking processes.

6.4.1.4 Values Alignment Strategies After Peer-Video Analyses

As a result of the self-analyses of the third week's videos, Mert employed the *refuge* strategy to allow each student to express their opinions, as in Critical Situation VII (see Fig. 6.9). However, he emphasized that he applied the *intervention* strategy by focusing on Eda, a gifted student, by considering the steps in the lesson plan.

Mert: As I had more confidence in Eda, I initially wanted to see what the other student would do. I checked everyone's notebook because if someone provided the solution, I did not want others to skip the question. However, I would have given Eda the floor first if it were a higher-level question. She gives ideas to her peers.

Ilgaz decided to use the opportunity to see the values of the plan he designed and those of Eda together, thereby employing the *balancing* strategy. This approach helped him integrate the refuge strategy by allowing other students to express their thoughts (see Fig. 6.10).

Ilgaz: In the model, it was unclear which one was the adult and the infant. I did not want her to try because I knew the process would be problematic. Then, I wanted her to do it anyway and see if the answer was wrong. [...] It is helpful for her to be self-assertive. I let her defend her ideas. [...] The lesson was not under my control.

Critical Situation VII	Mert wanted to explain multiplication in fractions with the help of a model. Eda and Student A were raising their hands. Mert first gave the floor to Student A. Other students then raised their hands. Mert walked around the classroom, checking each student's notebook.

Fig. 6.9 Critical situation VII

Critical Situation VIII	In a pattern problem based on rabbits' breeding, Eda wanted to model the pattern on the board. Ilgaz guided her to illustrate the problem using a few lines. Eda insisted on increasing the number of terms. However, Ilgaz helped Eda solve the problem as she confused the representations in her model. Then, because Student A was curious about the concept of the golden mean, they watched a video on the Internet.

Fig. 6.10 Critical situation VIII

The lesson did not progress as I had planned. The children's questions were very good; I would have missed a lot if I had blocked them.

6.4.1.5 Self-video Analyses in the Student Feedback Phase

Based on Eda's feedback, the teachers analysed their own videos in the first three weeks. In this process, the rationale for the strategy choices was determined. In addition, the teachers' awareness of Eda's values increased. In particular, the thoughts that Eda could not express about Ilgaz's lessons were identified. For example, for Critical Situation IV, Eda still defended her point of view:

Eda: I will write both minuses. I use something a couple of times if it confuses me. I also asked my teacher (Mert) about this previously; he told us to be careful. Maybe I will change it later. I wish the teacher (Ilgaz) had let me go to the board. I would have expressed myself better. Let me first explain what I think, and then he could give me feedback […] After the bank example, it made sense. However, it would have been better to write a few more terms.

Ilgaz: Since SAC students can do this very easily, I never thought they could make a mistake. It was because of me. I get to know her better now. However, if Eda is going in the wrong direction while on the board, the class might think she is right. I have to intervene. Eda was dumbfounded after I intervened, as I disturbed her thinking. It is better to contemplate it using a pen and paper first and to illustrate it on the board best […] When the other two children do it but Eda cannot, I directly skip the problem. I feel that the other two could lose interest when I provide additional explanations so that Eda can understand. I am trying to find another example that Eda could understand.

It is clear that Ilgaz focuses on the experiences of gifted students and the entire class rather than focusing on individual student values. Therefore, he explained the reason for his more frequent use of the *scaffolding* strategy.

6.4.1.6 Values Alignment Strategies Following Student Feedback

Mert tried to reveal not only the thoughts of Eda but also those of every student in the class after Eda's feedback "I get excited when my teacher (Mert) answers my questions". As in Critical situation IX, Eda reflected the *refuge* strategy by focusing on individual student values throughout the course (see Fig. 6.11).

Due to Eda's feedback about the freedom of expression, Ilgaz listened to her thoughts individually and allowed her to express her emerging ideas on the board. Ilgaz employed the *intervention* and *refuge* strategies in his class by going through processes similar to those in Critical Situation X (see Fig. 6.12).

Critical Situation IX	Mert examined the cause of the error on the board with Student B, who had supplied an incorrect answer regarding the order of operations in rational numbers. On the other hand, Eda progressed and focused on a problem related to complex operations in rational numbers. Eda asked Mert a question about a problem, and other students also wanted to ask their questions. Mert first listened to the student on the board. Meanwhile, a student was on the board to solve the following problem while Mert answered Eda's and the other students' questions.

Fig. 6.11 Critical situation IX

Critical Situation X	Eda tried to count the dots to find the 64th term in the figure pattern. Ilgaz came to Eda's desk and began to ask her questions. He helped Eda review a few terms. After listening to other students, she invited Eda to the board to draw the different figure pattern she had discovered for the general term.	

Fig. 6.12 Critical situation X

6.4.1.7 Values Alignment Strategies in the Group-Video Analysis Phase

The recordings of the lessons taught by Mert and Ilgaz in the first three weeks were watched and discussed together. Thus, the teachers had the chance to compare Eda's thoughts in two different classes and obtained comments on how they made decisions during classroom practices.

They discussed Critical Situation I and sought solutions for Eda's similar behaviour in the two classrooms and discussed the indicators of the *intervention* strategy:

Ilgaz: When you intervene while Eda explains her ideas, she gets distracted. It often happens in my class, too.

Mert: It is a problem. She was about to walk to the board after solving one of the problems. She said, "I have just done it, but now I cannot do it again".

Ilgaz: If she has finished it in her head, she has to re-read and re-evaluate the task, and she fails to do it again.

Researcher: So what should we do?

Mert: Maybe it would have been better if I had asked her to continue her explanations before she sat down. However, when we leave it totally to the student, the other students get distracted. Students can be dependent on the teacher.

Ilgaz: Our main goal here is to ensure comprehension in the whole class rather than just Eda's understanding.

Ilgaz's attitude is discussed in Critical situation III. Eda's attention problem is focused on the class as a whole while sticking to her lesson plan, with an emphasis on the *scaffolding* strategy:

Ilgaz: I emphasized that patterns can also be formed using letters. I wanted to remind her of this when she provided a numerical answer.

Mert: She thought of a number as this is a maths class.

Researcher: In our interview with Eda, she stated that she first thought in numbers, then letters, and then when she needed to think numerically again, she could not focus on the task.

Ilgaz: It is normal for her to get distracted. However, she got more distracted than the other kids.

Mert: She often gets distracted quickly. She gets stuck while being asked questions.

Ilgaz: Upon realising it, she pulls back, and then she gets even more distracted. After all, there is the class order. The lesson has a flow in my mind.

Critical Situation IV was discussed, and it was concluded that both teachers made decisions based on their experiences. It was noted that the *intervention* strategy should be employed based on student thinking:

Mert: We discussed that topic that week. I asked them to write -(-3) as they made mistakes on it.

Ilgaz: It was not about operations. I thought she stressed that she knew it.

Mert: If I were you, I would write the generalization as 30 - 3n. Of course, the student's profile is a key factor. My students confuse it because the minus is at the beginning.

Ilgaz: It was no problem for the students at the SAC. It never occurred to me to write it the other way.

6.4.1.8 Values Alignment Strategies After Group-Video Analyses Phase

In his lectures delivered this week, Mert employed the strategy of reprioritising and emphasized Eda's values. For example, in the group-video analyses in Critical Situation XI, in agreement with Ilgaz, they waited for Eda to solve the problem without intervening (see Fig. 6.13). He also employed the balancing strategy as he made this decision by considering the values of other students without sacrificing his own values.

Ilgaz tried to examine Eda's thoughts, as he noticed them in the group analyses in Critical Situation XII. He employed the *refuge* strategy by focusing on each student's thoughts (see Fig. 6.14).

Critical Situation XI	They were examining the end-of-unit problems, which were problematic for the students. From the beginning, Eda sought an explanation for the 22nd problem. While Mert continued to examine the problems, Eda solved the 22nd problem. Eda began to answer the questions in the next assessment test. For the 22nd item, Mert chose Eda, although a few students raised their hands. He did not interfere with Eda's work on the board. However, Eda could not solve the problem. Mert helped Eda solve the problem by eliciting questions such as "Are you trying to find half of 20 loaves of bread?" Later, Mert again provided the solution.

Fig. 6.13 Critical situation XI

Critical Situation XII	They were trying to find the general term for the figure pattern on the worksheet. Eda often visited Ilgaz's desk to question her reasoning. Ilgaz examined the solutions found by the students and provided feedback. He was trying a term for the general term Eda had identified and asked her to do it again if it was incorrect. He took his time so that each student in the class could find the correct answer and invited each student who came up with a different solution to the board to listen to her/his solution.

Fig. 6.14 Critical situation XII

6.4.2 *The Phases of the Professional Development Model for Values Alignment Strategies*

Mert and Ilgaz evaluated four different intervention phases, specifically referring to awareness, thoughts, and decisions that affect the choices of values alignment strategies. Mert stressed that he gained a different perspective through the peer-video analysis and group-video analysis phases:

Mert: I am worried I might make similar mistakes again during instruction as every student in the class is different, and heterogeneous classes are challenging to handle. I got the ideas of Ilgaz. I never knew about the instructional practices in the SAC.

Emphasizing the phases of self-video analysis, student feedback, and group-video analysis, Ilgaz suggested that the model should be used, particularly by teachers in the SAC.

Ilgaz: I have never watched myself while teaching. It was a different experience. When I started teaching at the SAC, I expected to follow a curriculum, but there was none. That is why I am still struggling. Expert and novice teachers at the SACs should come together and do peer observation. Every new teacher should observe an expert teacher's class. If I had done it earlier, I would have had less trouble. Our environments and objectives are different. Naturally, how we get to know our students is also different. In the interviews, I got unexpected results about Mert and Eda.

6.5 Discussion

While designing the professional development model, the purpose was to help teachers realize the ideas and values that the inclusive student has and carries from the other classroom culture and to incorporate the values alignment strategies that could support the gifted student's learning and thinking processes into their practices. Indeed, in the interviews carried out as a part of the phases in the model, it was found that the teachers were not aware of the thoughts of the inclusion student and the classroom practices in other institutions and that they made progress while putting the model into practice. Considering the researches at the intersection of professional development and teacher noticing theoretical frameworks, the development is an inevitable result regardless of its quality (see Clarke & Hollingsworth, 2002; van Es & Sherin, 2010). However, since inclusive mathematics education opportunities contain challenging demands for advanced and enrichment practices, it is expected that there should be differences among teachers due to institutional purposes (Büscher & Prediger, 2022; Even et al., 2009). Indeed, the results reveal differences between the initial strategies and those used during the process by the mathematics teachers in the inclusive classroom and the SAC (see Fig. 6.15). Of course, as noted by the participants, this difference is caused by institutional and

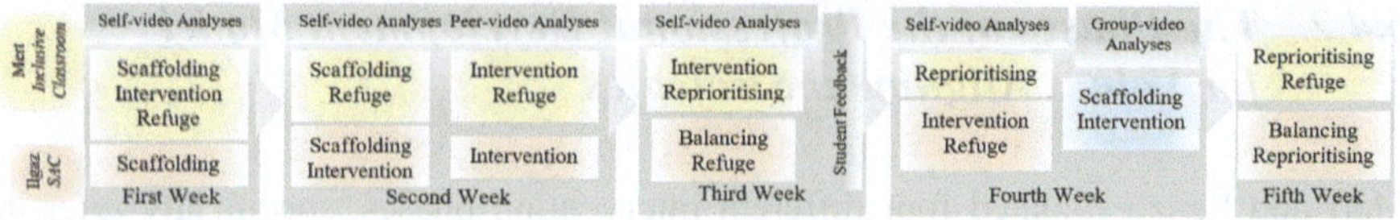

Fig. 6.15 The results of the professional development model

cultural (Waitoller & Artiles, 2013) factors, which are shaped by the objectives, curriculum, and heterogeneous student profiles. Büscher and Prediger (2022) emphasize that professional development in inclusive education is challenging because of curricular and institutional diversity and limitations. Also, teacher values shape how they design instructional practices (Seah, 2019), what decisions they make in the implementation process (Aktaş et al., 2019), and the extent to which student values are reflected during the instruction (Kalogeropoulos & Bishop, 2017). Ilgaz began to pay attention to student thinking after the peer-video analyses and student feedback sessions, which eliminated Ilgaz's judgments about gifted students and the practices in the inclusive classroom. This is an indication of teacher values.

In Fig. 6.15, the mathematics teacher in the inclusive classroom focuses on teacher values, the values of a particular group or the values of the inclusive student, and those of each student throughout the process. However, after the peer-video analysis phase, there was an increase in the focus on student thinking and student values. This indicates that the values alignment strategies that prioritize student values were employed. While the initial strategies of the SAC teacher centred on teacher values, the strategy choices gradually shifted their focus on the values of the gifted student after the self- and peer-video analysis phases. The effort to balance teacher and student values after the peer-video analysis phase and the increased attention to student thinking is critical here. Unsurprisingly, self- and peer-video analyses cause remarkable changes. Because research indicates that video analysis helps teachers recognize their students' thinking skills and focus on student–teacher interaction (Borko et al., 2008; van Es & Sherin, 2010). Therefore, the prominence of *intervention, refuge,* and *reprioritising* strategies, in which there is an increased focus on student values, is a possible outcome of using the professional development model. This result does not imply that student values are more important for effective learning in classroom practices. This is because no strategy is superior to others (Kalogeropoulos et al., 2021). Merely, it was aimed to draw attention to inclusive student values in classroom environments where different cultures and values prevail and to raise teacher's awareness of the revisions that learning environments need. Furthermore, in this chapter, a subtle nuance between the *refuge* and *reprioritising* strategies was explained to distinguish between them, and the results were refined based on this. The *refuge* is a strategy in which the teacher interprets student values by disregarding his own values and finds new values and uses them during the instruction (Kalogeropoulos et al., 2021). *Reprioritising*, on the other hand, is a strategy in which the teacher prioritizes student values or mediating values without giving up their own values (Seah & Andersson, 2015a).

The student feedback phase was the breaking point concerning the teachers' preferences for values alignment strategies (see Fig. 6.15). In the fourth week, classroom practices prioritized not only the values of the inclusion student but also those of their peers in both the inclusion class and the SAC. The inclusion teacher was impressed by the discussion processes at the SAC and the desire to find answers to the questions of the inclusion student. Thus, discussion, peer teaching, reasoning, and exploration strategies have found more place in practice. Also, especially gifted students have more opportunities to learn about their individual values in practices in the inclusive classroom. Therefore, as the model phase progress, it is possible to encounter practices that consider gifted student's values based on reasoning and flexible thinking (see Aktaş & Dede, 2023). On the other hand, the teacher in the SAC was able to interpret the thoughts of the inclusion student. These results support Seah's (2019) assumption that "the ability of effective teachers to align their values with those of the students is instrumental in facilitating mathematics learning" (p. 110). To consciously and subconsciously identify or perceive students' feedback during the values alignment process is the main way for teachers not only to interpret their students' thoughts about the values reflected in practice but also to understand whether learning has taken place (Seah & Andersson, 2015). However, it is noteworthy that the teachers emphasised teacher values in the group-video analysis phase. The first factor here is that teachers focus on the strategies in their videos in the first two weeks. Therefore, the teachers tried to explain the rationale for their strategy preferences. Thus, the results indicate that the professional development model enabled the teachers to attach more importance to student thinking and values. The observations made to evaluate the model's functionality in the fifth week also supported this finding. However, the group discussions, the teachers' evaluation of the model's phases, and the final week's strategy preferences indicate that even though there is a change in the teachers' strategy preferences, they may revert to their original strategies. The two main ideas for which van Es and Sherin (2010) called for further research are also consistent with the results of the current study. First, video analyses led to fundamental changes in the teachers' affective perceptions. Second, the professional development model should be implemented over a wider period. The repetition of the phases in the model allows for developing and changing the strategies in the fifth week. Moreover, Büscher and Prediger (2022) suggest that the process of professional development in inclusive mathematics classrooms should systematically continue with the analysis of classroom videos in the long run.

6.6 Moving On

One of the limitations of the present study is the lack of a second cycle that could be used to test the effectiveness of the suggested professional development model and to revise it based on it. However, this chapter explores what procedures mathematics teachers could follow within a professional development model to address the multicultural nature of inclusive education. Therefore, further research could be carried out to replicate the model's phases or develop new interventions. In the replication or further development of the present model, student feedback can be presented simultaneously with the self- and peer-analyses processes at each phase. This is because the results revealed that student feedback is critical in strategy choices. In addition, the feedback of not only the inclusion students but also their peers could be shared with the teachers for the practices in the inclusion classrooms. Thus, teachers will likely gain new values and value alignment strategies for enriching, redesigning, and improving practice. Moreover, teachers working in the same institution can participate in group discussions. For example, classroom teachers in mainstream schools and mathematics teachers in the resource room could be involved in more critical professional development practices. This is because the results indicate that institutional values and goals are reflected in teachers' preferences of values alignment strategies, and therefore, interaction gets more complicated in group discussions.

Another limitation is that only gifted students are selected for inclusive practices. For the purpose of this chapter, gifted students were chosen to identify student thinking more clearly. Of course, it is possible to develop a professional development model for teachers teaching different inclusion students, such as students with learning disabilities or those with visual or hearing impairments. Diverse classroom cultures, ways of student thinking, and values should be considered in further studies. Teachers' strategy preferences and the effectiveness of model phases will vary in such studies. Considering not only the educational practices based on the inclusion of students and their qualifications but also the institutional and cultural values, the results of the current study may pioneer further studies on values alignment and special education in mathematics education in many different cultures.

References

Aktaş, F. N., & Argün, Z. (2018). Examination of mathematical values in classroom practices: A case study of secondary mathematics teachers. *Education and Science, 43*(193), 121–141. https://doi.org/10.15390/EB.2018.7177

Aktaş, F. N., & Dede, Y. (2023). Gifted students' mathematics educational and socio-cultural values in the decision-making process through STEM tasks. In T. Barkatsas (Ed.), *Educating gifted, talented, creative and dissimilar learners: Imagining the future* (pp. 36–63). Brill. https://doi.org/10.1163/9789004532571_003

Aktaş, F. N., Yakıcı-Topbaş, E. S., & Dede, Y. (2019). The elementary mathematics teachers' values underlying teacher noticing: The context of polygons. In P. Clarkson, W. T. Seah, & J. Pank (Eds.), *Values and valuing in mathematics education* (pp. 209–222). Springer International Publishing. https://doi.org/10.1007/978-3-030-16892-6_15

Bishop, A. J. (1996, June). *How should mathematics teaching in modern societies relate to cultural values some preliminary questions* [Paper presentation]. *The Seventh Southeast Asian Conference on Mathematics Education*, Hanoi, Vietnam.

Bishop, A. J. (2001). Educating student teachers about values in mathematics education. In F. L. Lin & T. Cooney (Eds.), *Making sense of mathematics teacher education* (pp. 233–246). Kluwer Academic Publishers. https://doi.org/10.1007/978-94-010-0828-0_11

Borko, H., Jacobs, J., Eiteljorg, E., & Pittman, M. E. (2008). Video as a tool for fostering productive discussions in mathematics professional development. *Teaching and Teacher Education, 24*(2), 417–436. https://doi.org/10.1016/j.tate.2006.11.012

Büscher, C., & Prediger, S. (2022). Teachers' practices of integrating challenging demands of inclusive mathematics education in a professional development program. *Journal of Mathematics Teacher Education*. Advance Online Publication. https://doi.org/10.1007/s10857-022-09560-5

Clarke, D., & Hollingsworth, H. (2002). Elaborating a model of teacher professional growth. *Teaching and Teacher Education, 18*(8), 947–967. https://doi.org/10.1016/S0742-051X(02)00053-7

Dede, Y. (2015). Comparing primary and secondary mathematics teachers' preferences regarding values about mathematics teaching in Turkey and Germany. *International Journal of Science and Mathematics Education, 13*(1), 227–255. https://doi.org/10.1007/s10763-013-9483-z

Even, R., Karsenty, R., & Friedlander, A. (2009). Mathematical creativity and giftedness in the professional development of teachers. In R. Leikin, A. Berman, & B. Koichu (Eds.), *Creativity in mathematics and the education of gifted students* (pp. 309–324). Sense Publisher. https://doi.org/10.1163/9789087909352_020

Fraser, C., Kennedy, A., Reid, L., & Mckinney, S. (2007). Teachers' continuing professional development: Contested concepts, understandings and models. *Journal of in-Service Education, 33*(2), 153–169. https://doi.org/10.1080/13674580701292913

Kalogeropoulos, P., & Bishop, A. J. (2017). What is the role of value alignment in engaging mathematics learners? In A. Chronaki (Ed.), *Mathematics education and life at times of crisis: Proceedings of the 9th International Mathematics Education and Society Conference* (Vol. 2, pp. 603–610). University of Thessaly Press.

Kalogeropoulos, P., & Clarkson, P. (2019). The role of value alignment in levels of engagement of mathematics learning. In P. Clarkson, W. T. Seah, & J. S. Pang (Eds.), *Values and valuing in mathematics education* (pp. 115–127). Springer. https://doi.org/10.1007/978-3-030-16892-6_8

Kalogeropoulos, P., Russo, J. A., & Clarkson, P. (2021). Exploring educator values alignment strategies in an intervention context: The emergence of the beacon strategy. *ECNU Review of Education, 4*(2), 327–348. https://doi.org/10.1177/2096531120923127

Leikin, R. (2019). Giftedness and high ability in mathematics. In S. Lerman (Ed.), *Encyclopedia of mathematics education* (10-page entry). Springer. https://doi.org/10.1007/978-3-319-77487-9_65-4

Lin, F. L., & Rowland, T. (2016). Pre-service and in-service mathematics teachers' knowledge and professional development. In Á. Gutiérrez, G. C. Leder, & P. Boero (Eds.), *The second handbook of research on the psychology of mathematics education: The journey continues* (pp. 482–520). Brill.

Merriam, S. B. (1985). The case study in educational research: A review of selected literature. *The Journal of Educational Thought (JET), 19*(3), 204–217.

Merriam, S. B. (1998). *Qualitative research and case study applications in education*. Jossey-Bass.

Ministry of National Education Science and Art Centre Directive. (2016). *Author*. https://orgm.meb.gov.tr/meb_iys_dosyalar/2016_10/07031350_bilsem_yonergesi.pdf

Seah, W. T. (2003, November). *The professional socialisation of teachers in transition: A values perspective* [Paper presentation]. *The International Education Research Conference AARE-NZARE*, Auckland, New Zealand.

Seah, W. T. (2019). Values in mathematics education: Its conative nature, and how it can be developed. *Research in Mathematics Education, 22*(2), 99–121. https://doi.org/10.7468/jksmed.2019.22.2.99

Seah, W. T., & Andersson, A. (2015a). Valuing diversity in mathematics pedagogy through the volitional nature and alignment of values. In A. J. Bishop, H. Tan, & T. Barkatsas (Eds.), *Diversity in mathematics education: Towards inclusive practices* (pp. 167–183). Springer. https://doi.org/10.1007/978-3-319-05978-5_10

Seah, W. T., & Andersson, A. (2015b, February). *Teacher alignment of values in mathematics classrooms* [Paper presentation]. *The Ninth Congress of the European Society for Research in Mathematics Education*, Prague, Czech Republic. https://hal.archives-ouvertes.fr/hal-01289811/

Taller-Azulay, G., Heiman, T., & Olenik-Shemesh, D. (2022). The role of the educational counsellor in the inclusion and integration processes. *European Journal of Special Education Research, 8*(2), 60–80. https://doi.org/10.46827/ejse.v8i2.4249

van Es, E. A., & Sherin, M. G. (2010). The influence of video clubs on teachers' thinking and practices. *Journal of Mathematics Teacher Education, 13*(2), 155–176. https://doi.org/10.1007/s10857-009-9130-3

Waitoller, F. R., & Artiles, A. J. (2013). A decade of professional development research for inclusive education: A critical review and notes for a research program. *Review of Educational Research, 83*(3), 319–356. https://doi.org/10.3102/003465431348390

Chapter 7
The Text Mining Approach to Identifying What Students Value in Mathematics Learning

JeongSuk Pang, Wee Tiong Seah, Leena Kim, and SeungMin Kim

7.1 Introduction

It is well known that students' values in learning mathematics is key to understanding and facilitating their learning (Bishop, 1988). If what the teacher regards as important in teaching mathematics are aligned with what students regard as important equally, this would lay the foundation for effective teaching. In this respect, the significance of considering what students value when learning mathematics cannot be overemphasized.

In the context of mathematics education, values are regarded as any attribute of mathematics or of mathematics teaching/learning which is considered personally important (Seah, 2019). Examples include *rationalism, practice*, and *creativity*. From this perspective, several international studies identifying students' values of mathematics and its learning have been conducted, such as the "What I Find Important [WIFI] (in my mathematics learning)" study (Seah et al., 2017). Korea's participation in the WIFI study identified Korean students' dominant valuing of *understanding* and *connections* in mathematics learning (Pang & Seah, 2021). Here the

J. Pang (✉)
Korea National University of Education, Cheongju-si, South Korea
e-mail: jeongsuk@knue.ac.kr

W. T. Seah
Faculty of Education, The University of Melbourne, Melbourne, Australia
e-mail: wt.seah@unimelb.edu.au

L. Kim
Graduate School of KNUE, Cheongju-si, South Korea
e-mail: kimleena@hanmail.net

S. Kim
Seoul Jungbu District Office of Education, Seoul-si, South Korea
e-mail: sktsemm0@gmail.com

Y. Dede et al. (eds.), *Values and Valuing in Mathematics Education*,
https://doi.org/10.1007/978-981-99-9454-0_7

valuing of *understanding* was mainly associated with the students placing importance on understanding mathematical concepts and processes. On the other hand, students' valuing of *connections* was associated with relating mathematics to external contexts including other school subjects or real life, and/or to internal connections within mathematics. These values were identified by researchers based on the extent to which students found particular learning activities in the questionnaire important. While this chapter still focuses on investigating students' values when learning mathematics, it extends the prior quantitative study in the following three aspects.

Firstly, we asked students to write three attributes that they thought were important when learning mathematics, along with the reason why the three attributes would be important. We then analysed the characteristics of student responses based on the connectivity between words and the centrality of specific words. It is expected to provide an innovative method of analysing students' mathematics educational values while utilizing their open-ended descriptive responses as much as possible.

Secondly, we differentiated what would be important for anyone to learn mathematics from what would be important for students themselves to learn mathematics after recalling their own learning experiences. We regard the former as reflecting students' general values of learning mathematics and the latter as reflecting their personal values of learning mathematics. Regarding the former, students may refer to more desirable or ideal values related to mathematics learning, which may not reveal exactly what they think is important for themselves to learn mathematics. This led us to include two questions: One concerning students' general values of learning mathematics and the other for their personal values of learning mathematics. It is expected to identify any similarities and/or differences of students' valuing of learning mathematics.

Thirdly, we compared what students thought was important when learning mathematics with what they thought was important to their mathematics teachers in teaching mathematics. Note that we were not analysing teachers' own values of teaching mathematics but students' perception of their teachers' values. The intention here is to understand how students perceived their teachers' values of teaching mathematics and to analyse to what extent students' own values of learning mathematics would match their teachers' values as perceived by them.

This chapter is significant because it identifies the mathematics educational values embraced by Korean students, which have not been researched much before as demonstrated through our extensive literature search. Another significance is the novel application of text mining to analyse students' values. The overarching research question of this study is, what do Korean students value in their mathematics learning? Specifically, what are their general and personal values of learning mathematics? What are their perceptions of teacher values of teaching mathematics? How similar are these categories of values?

7.2 Background to the Study

7.2.1 Values in Effective Mathematics Pedagogy

In the context of mathematics education,

> valuing is defined as an individual's embracing of convictions in mathematics pedagogy which are of importance and worth personally. It shapes the individual's willpower to embody the convictions in the choice of actions, …. [and] also regulates the individual's activation of cognitive skills and affective dispositions in complementary ways. (Seah, 2019, p. 107)

That teaching and learning mathematics is a social process means that in any (mathematics) teaching and/or learning situation, what the different individuals participating in the didactic interaction value can often be different from one another. For example, some students may value *rationalism* and embrace algebraic proofs, some others in their class might value *visualisation* instead. If their mathematics teacher is planning to show how angles on a straight line add up to 180 degrees, they will need to take into consideration these conflicting student values, given constraints such as time limits. How might the mathematics teacher's own values guide their decisions and actions?

The point is, what the teacher considers important in their teaching, and what each student considers important in their learning, mediate the quality of pedagogical interactions between teacher and students, and amongst students themselves. As another example, a teacher may value *collaboration* but their students may value *independent work* or *competition* instead. The assumption that is being made here is that teacher expertise is very much a function of a teacher's capability to recognize and respond to any difference between teacher and students' valuing.

It is reasonable to assume that effective teachers are aware of what their students value in mathematics learning and are proactive in resolving any (potential) value difference between their own valuing and students' valuing, or amongst the students. What is important is that as a starting point, teacher expertise in identifying students' values pertinent to mathematics learning is crucial.

Teachers' values in mathematics teaching and learning can affect students through mathematics lessons. Pang and Yim (2019) showed that Grade 5 students tended to interpret in various ways their master teacher's values of mathematics learning rather than accepting them as they were. The students also perceived the teacher's values through not only what the teacher said in mathematics lessons, but also how the teacher acted in classroom interactions. In this respect, it is worthwhile to explore the teacher values of mathematics teaching and learning from the lens of students.

7.2.2 Korean Students' Values of Mathematics and Its Learning

So, what do Korean students generally value in their mathematics learning? It has been well-known that Korean students demonstrate superior mathematics achievement in international comparative studies. For instance, both fourth graders and eighth graders came in third in Trends in International Mathematics and Science Study [TIMSS] 2019 (Mullis et al., 2020). In the latest Programme for International Student Assessment [PISA] 2018 (Organization for Economic Co-operation and Development [OECD], 2019), as much as 21% of the 15-year-old students in Korea performed at Level 5 or higher in mathematics (OECD average being 11%) and 85% of the students attained at least Level 2 proficiency in mathematics (OECD average being 76%). In contrast, Korean students' negative attitudes toward mathematics is also notorious (e.g., feeling less confident in mathematics or disliking learning mathematics). Korea remains at the bottom of the affective rankings in TIMSS and PISA (Mullis et al., 2020; OECD, 2019).

Given the mismatch between Korean students' excellent performance in mathematics and their unfavourable attitude towards mathematics along with the low valuing of mathematics in the international context, Pang and Seah (2021) provided an explanation from the perspective of values and valuing. Using the 64 Likert-type scale items of the WIFI questionnaire, they investigated which attributes of mathematics pedagogy were valued by Korean students. For each item of specific learning activities (e.g., problem-solving), students were asked to respond to how important it was for them in learning mathematics. Pang and Seah (2021) identified *understanding* and *connections* as the top two values, followed by *fun, accuracy*, and *efficiency*. In particular, the valuing of *understanding* was associated with as many as 26 items from the remaining 49 items after the principal component analysis (PCA). Such items include understanding concepts and processes, using concrete materials, practicing with lots of mathematics questions, writing out the solutions in the process, learning through mistakes, for examples. The valuing of *connections* was associated with 13 items, such as relating mathematics to other subjects in school, connecting mathematics to real life, looking out for mathematics in real life, and appreciating the beauty of mathematics. These associations are significant and meaningful, as they reveal the nature of what understanding and connections mean in the Korean mathematics education culture. On the one hand, Korean students' dominant valuing of *understanding* and *connections* may help us to understand to some extent their exceptionally high performance in a series of the international mathematics achievement despite a generally low affective mode toward mathematics, given that deep understanding based on connections may lead to enhanced performance. On the other hand, the learning activities grouped together under the same PCA component (here labelled as *understanding* or *connections*) are pretty much broad-based and sporadic. We need to further investigate what it really means for students to place importance on *understanding* or *connections* when they learn mathematics.

Pang et al. (2016) also analysed which mathematical values and mathematics educational values were deemed as being important by Korean Grade 9 students using the ten slider scale items of the WIFI questionnaire. For each pair of phrases representing apparently opposite values, students were asked to indicate how more important one phrase would be for them in mathematics learning than the other phrase (e.g., "*How* the answer to a problem is obtained" versus "*What* the answer to a problem is"). Among the mathematical values (Bishop, 1988), Pang et al. (2016) identified that Korean students valued more *rationalism* than *objectism*, more *control* than *progress*, and more *openness* than *mystery*. Among the mathematics educational values (Seah, 2005), Korean students were regarded as placing more importance on *effort* than *ability*, more *well-being* than *hardship*, more *process* than *product*, more *application* than *computation*, more *ideas and practice* than *facts and theories*, more *exposition* than *exploration*, and more *recalling* than *creating*. On the one hand, these results help us better understand to what extent Korean students chose one value over the other competing or complementary value. On the other hand, the pair of phrases in the questionnaire does not necessarily reflect the opposing values. For instance, both *application* and *computation* may be equally important for students. In fact, except for the pair of *process* and *product*, Korean students' overall responses in Pang et al. (2016) tended to mark roughly around the middle between the two competing phrases. In addition, the same word or even phrase can be interpreted differently by students according to their social, cultural, or educational background. According to Seah (2005), for instance, the word 'computation' was deemed to include cognitive aspects such as the reconstruction of mental computation or basic operation for Chinese students, whereas it was regarded only as simple computation for Australian students. Given these, we need to consider the context in which a specific word or phrase was used to explore what is considered important and worthwhile by students.

7.2.3 Identifying Students' Values

Even though the values and valuing perspective has received consistent attention in mathematics education (Clarkson et al., 2019), the issues of how to identify the hidden nature of values have been challenging (Chan & Wong, 2019). On the one hand, employing lesson observations or interviews to investigate the participants' values might have the advantage of identifying the values as being emergent through lessons or interviews, the small sample sizes involved often mean that generalizability cannot be made. On the other hand, the questionnaire provides a quick and efficient method to survey large research populations, thus facilitating generalizability. In fact, the questionnaire method was employed in the WIFI study. Specifically, the WIFI questionnaire consisted of 64 Likert-type scale items, 10 slider scale items, and one open-ended question (see Seah et al., 2017 for details). While the first two groups of items have been analysed among participant countries, large-scale analysis of the open-ended item has been problematic, probably because of the difficulty of identifying values from free-form prose (e.g., Seah et al., 2017; Zhang et al., 2016).

In addition, the experience of the WIFI study had highlighted that the same learning activity might reflect different values across different education systems. For instance, "knowing which formula to use" may reflect *fluency, efficiency*, or *product*, depending on the sociocultural context within which the classroom is located (Pang & Seah, 2021; Seah et al., 2017). This adds another layer of uncertainty to confirming the values underlying learning activities as these were written in the open-ended responses.

These research experiences have led us to seek an alternative efficient method to identify students' values. Here we propose a text mining approach which enables us to extract meaningful patterns and relationships, while keeping up with the large volume of students' descriptive responses.

7.3 Methods

7.3.1 Participants

The data for this chapter were students' written responses to the questionnaire of a recent values-based study, Values Alignment Study [VAS].[1] This study investigates how teachers respond to perceived value differences in mathematics lessons.

20 middle schools were selected across Korea through stratified probability sampling based on school location, resulting in four schools in Seoul, seven in metropolitan cities, four in small cities, and five in rural areas. The targeted participants were Grade 9 middle school students (who were mostly 14 years old) and their mathematics teachers. A total of 832 students participated, completing the VAS questionnaire.

7.3.2 Questionnaire

The VAS has a questionnaire for student participants and another similar one for their teachers.[2] The student questionnaire is made up of two sections. Section A surveys student respondents' demographic and personal information along with their views on the usefulness of mathematics and the nature of family support in mathematics learning. Section B of the student questionnaire examines, amongst other things, what is important for students when engaging with mathematics learning. Specifically, student responses to three items were analysed and reported in this chapter. The first item seeks to identify students' general values associated with mathematics learning

[1] Details of the study are available from https://thirdwavelab.education.unimelb.edu.au/study-5-values-alignment-study-vas/

[2] The Korean version of the student questionnaire may be accessed at: https://thirdwavelab.education.unimelb.edu.au/files/2023/02/WIFItoo_KOR_student.pdf

mathematics, with the question being "What do you think are important when anyone learns mathematics?" The second item covers students' personal values of learning mathematics: "Think about your own experience of learning mathematics. What do you think are important when you learn mathematics?" The third item asks for students' perception of their respective teachers' values in teaching mathematics. Here, student respondents were asked to "Think about your mathematics teacher this year. What do you think are important to him/her in mathematics teaching?" In each of these three items, students were asked to list and explain up to three points corresponding to what they think were important.

7.3.3 Data Collection and Analysis

The student questionnaire used for our study had been translated into Korean from the original version which is in English. The language validity of the questionnaire was achieved through back translations. A pilot test with two classes of students in a middle school was also conducted to check the appropriateness of the student questionnaire to the target research population.

Basic information about the study, including its purpose and facilitator's guide for data collection, was distributed to the 20 mathematics teacher participants. They were also informed of the website of the questionnaire. The data were collected using the SurveyMonkey platform to facilitate online administration and more accurate data entry.

That the students' descriptive responses to the Section B items constituted unstructured data has provided us with the opportunity to utilise text mining to analyse these data. After all, text mining involves structuring unstructured data, deriving patterns within the data, and evaluating and interpreting the results (Feldman & Sanger, 2007). Thus, text mining allowed us to extract meaningful patterns or relationships from the data. In doing so, we decided against performing thematic analysis because doing so would not capture the richness of the data that were the students' writings.

Data analysis began with refining students' descriptive responses using Korean Natural Language Processing (KoNLP).[3] The data were then analysed using the R version 4.1.2 software, involving frequency analysis, Term Frequency-Inverse Document Frequency (TF-IDF) analysis, and co-occurrence network analysis. Here frequency analysis helped to identify the overall trend of student responses. Considering that the high frequency of occurrence of a word does not necessarily indicate

---

[3] Specifically, the spelling and spacing of students' responses were unified so that the data could be easily analysed. Nouns were extracted after removing stopwords such as meaningless punctuation marks or special characters. The data were further refined through keyword cleansing, the process of deleting common words and meaningless words from student responses. For instance, mathematics (in Korean) was deleted because it was the common word across almost all the answers. In addition, the words related to *important* were also deleted because they were presented in the questionnaire items. Word preprocessing was performed to combine synonyms. For example, *ability* to concentrate and *concentration* were regarded as synonyms so that they were unified into *concentration*.

importance, however, both TF-IDF analysis and co-occurrence network analysis were additionally performed. Since TF-IDF analysis extracts statistical values indicating how important a particular word is in a specific document among a group of multiple documents (Berry, 2004), words with high TF-IDF values can be considered to characterize the responses of the corresponding questionnaire items.

Co-occurrence network analysis facilitates the representation of relationships between words, using co-occurrence frequencies. Using the network graphs to show these relationships, the words will be represented by numerous nodes. However, it is difficult to determine which word should be interpreted as the center of a network. This resulted in the subsequent centrality analysis, which reveals the topological structure of words. Specifically, Degree Centrality (DC) represents how many words are adjacent to one word. Closeness Centrality (CC) indicates the proximity between words. Eigenvector Centrality (EC) indicates the degree to which a word plays a central role in the network. Betweenness Centrality (BC) represents the extent to which a word plays an intermediate role between words, and when the relationships between words are frequently connected, a community (or group) is formed (Newman, 2008). We drew the network graphs through the R software based on the centrality analysis. As the label of each node in the graphs was in Korean, we also added the English translation to each node for the purpose of this chapter (see Sect. 7.4 for graphs).

After centrality analysis and the subsequent construction of network graphs, representative student responses were selected to better understand the group of words extracted from the text mining process. In other words, the meaning of the words constituting a specific group was checked against the students' sampled responses (see Sect. 7.4).

Since these student-generated words were in response to questions with the stem 'what do you think are important …', and were in the form of singular words or terms, we can consider them as representing values (see Seah, 2019). Yet, just like a valuing of *technology* is a reflection of deeper core values such as *efficiency* or *understanding*, we differentiate between instrumental values (such as *technology*) and ultimate values (such as *efficiency* and *understanding*) (e.g., Tiberius, 2018). That is, ultimate values are the core values served by different instrumental values. The strength of our methodology here is the construction of network graphs to cluster the student-generated words (which are either ultimate or instrumental values) into groups, essentially providing a visual categorisation and identification of ultimate values. This is, however, beyond the scope of this chapter.

Table 7.1 Frequency analysis, TF-IDF analysis, centrality analysis, and the grouping related to students' general values of learning mathematics

Value	N	TF-IDF	DC	CC	EC	BC	Group
Understanding	305	55.66	22	0.08	1.00	26	1
Problem	330	42.96	22	0.08	0.99	28	2
Review	139	52.00	6	0.05	0.44	1	1
Preparation	95	44.78	4	0.04	0.28	0	1
Persistence	133	40.26	6	0.05	0.48	0	3
Thinking	139	34.26	6	0.05	0.48	0	3
Formula	69	24.44	4	0.05	0.39	0	1
Concept	76	30.31	4	0.05	0.39	0	2
Concentration	108	43.62	4	0.05	0.39	0	2
Computation	94	31.45	6	0.05	0.48	0	4
Ability	67	19.72	6	0.05	0.48	0	4
Creativity	46	22.05	2	0.04	0.19	0	2
Basics	60	20.46	4	0.05	0.39	0	1

7.4 Results

7.4.1 Students' General Values of Learning Mathematics

A total of 922 words were extracted to the item, "what do you think are important when anyone learns mathematics?" Table 7.1 shows the results of frequency analysis, TF-IDF analysis, centrality analysis, and the grouping of the words into value categories. Considering the frequency analysis, both *problem* (n = 330) and *understanding* (n = 305) were the most important, followed by *review* (n = 139), *thinking* (n = 139), *persistence* (n = 133), and *concentration* (n = 108). The results of TF-IDF analysis[4] indicated that *understanding* (55.66), *review* (52.00), *preparation* (44.78), *concentration* (43.62), *problem* (42.96), and *persistence* (40.26) had important meanings to the item, regardless of whether each of them is an instrumental or ultimate value. In addition, the centrality analysis of word pairs with more than 19-word connections showed that the centrality of *understanding* and *problem* was noticeably high. This is in line with the results of the frequency analysis. In other words, Korean middle school students identified *understanding* and *problem* as the top two values for learning mathematics in general.

As shown in Table 7.1, the centrality analysis resulted in four groups of values. Group 1 consisted of *understanding, review, preparation, formula,* and *basics.* Group 2 included *problem, concept, concentration,* and *creativity.* Group 3 consisted of *persistence* and *thinking*, while Group 4 included *computation* and *ability*. Figure 7.1

[4] The TF-IDF value means term-weighting, which is a numerical representation of how important a word is in the corpus.

Fig. 7.1 Network graph related to students' general values of learning mathematics

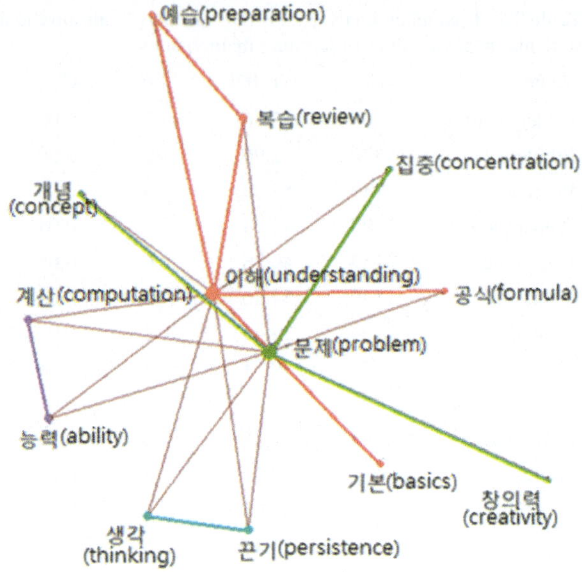

is a visual arrangement of these four groups of values while reflecting the centrality analysis.[5] As such, both *understanding* and *problem* occupy the center of the graph, with the biggest node size indicating their high centrality. Within each group, the stronger the connection between values, the closer the distance between them. For instance, the value *creativity* was grouped together with the *problem*, but the DC of *creativity* is low so that the value is far from the center and far away from *problem*. The values *thinking* and *persistence* are far from the center because their DCs are low, but the values are close to each other in the network, resulting in a separate group.

In Group 1, *review* and *preparation* (or pre-study) mainly appeared, based on *understanding*. The frequency of simultaneous appearance of value pairs was 71 for *review* and *preparation*, 35 for *understanding* and *review*, 24 for *understanding* and *formula*, 22 for *understanding* and *preparation*, and 19 for *understanding* and basics, respectively. As evidenced by the sampled student responses in Table 7.2, students described the importance of understanding in reviewing what someone has learned, studying in advance what he or she will learn, learning a formula, etc. In addition, students recognized the importance of *preparation* and *review* in learning mathematics.

---

[5] Note that a network graph uses the same color for each group of words. The value of DC determines the node's size in each graph, indicating that the big size of a specific word means more words were adjacent to it. A rink (or line) connecting a node (or word) to another node in the graph represents a relationship between the nodes. The length of a rink indicates the distance between nodes, while the thickness of a rink is used only to distinguish words grouped together. The node with a high connection centrality is at the center of a network graph.

Table 7.2 Sampled student responses related to their general values of learning mathematics by the groups of values

Group	Value	Sampled student response
1	Understanding, review, preparation, formula, basics	• Reviewing what you have learned promotes your understanding and preparation makes you do well by practicing what you will learn later • Understanding should be the basis in learning a formula • If you know the basics, you can better understand what you learn
2	Problem, concept, concentration, creativity	• You will not forget a certain concept or principle by steadily solving problems. You can understand a concept deeper and wider while solving various types of problems • It is easy to solve problems only when you have concentration • You need creativity when you face with a creative problem
3	Persistence, thinking	• This is because when solving a problem, you must think about a formula and how to use it with persistence • To do mathematics, you need persistence and continuous thinking
4	Computation, ability	• Computational ability is fundamental to mathematics • Computational ability is important, and if you can compute quickly, you can do mathematics quickly

In Group 2, the three values, *concept*, *concentration*, and *creativity* were combined to *problem*. The frequency of simultaneous appearance of value pairs was 33 for *problem* and *concept*, 31 for *problem* and *concentration*, and 22 for *problem* and *creativity*. As illustrated in Table 7.2, students stated that someone needs to know the concept and to concentrate to solve a problem. They also described that creativity would also be necessary to solve a creative problem.

In Group 3, the frequency of simultaneous appearance of value pairs was 22 for *persistence* and *thinking*. The representative response was that someone needs to think persistently to do mathematics.

Finally, in Group 4, the frequency of simultaneous appearance of value pairs was 22 for *computation* and *ability*. As shown in Table 7.2, students stated that computation is a fundamental ability to do mathematics.

In summary, students regarded *understanding* and *problem* as the most important variables for anyone to learn mathematics. Specifically, both *preparation* and *review* were worthwhile to understand formulas and basics. One also must know a concept to solve a problem with concentration and creativity. In addition, persistent thinking and computational ability were also considered important and valued.

7.4.2 Students' Personal Values of Learning Mathematics

A total of 879 words were extracted to the item, "Think about your own experience of learning mathematics. What do you think are important when you learn mathematics?" Table 7.3 shows the results of frequency analysis, TF-IDF analysis, centrality analysis, and the grouping of the words into value categories. Considering the frequency analysis, the top three values were *problem* (n = 404), *understanding* (n = 278), and *review* (n = 216), followed by *concentration* (n = 172), *preparation* (n = 133), and *persistence* (n = 101). Note that the top three values by frequency are the same between students' general values and personal values of learning mathematics.

The results of TF-IDF analysis indicated that *review* (60.55), *concentration* (55.55), *preparation* (55.24), *understanding* (54.72), *problem* (49.92), and *persistence* (40.46) had important meanings to this specific item. In other words, students regarded *review*, concentration, and *preparation* as more important than *understanding* in their own mathematics learning. In addition, the centrality analysis of value pairs with more than 21-word connections showed that the centrality of *problem* was the highest. This is different from the students' general values of learning mathematics in which the centrality of *understanding* was prominent.

As shown in Table 7.3, the centrality analysis pointed to three groups of values. Group 1 consisted of *problem, formula, concept, thinking, solution, ability, computation, basics,* and *types*. Group 2 included *understanding, concentration*, and *persistence.* Group 3 consisted of *preparation, review,* and *content*. Figure 7.2 is a visual

Table 7.3 Frequency analysis, centrality analysis, and the grouping related to students' personal values of learning mathematics

Value	N	TF-IDF	DC	CC	EC	BC	Group
Problem	404	49.92	26	0.07	1.00	70.83	1
Understanding	278	54.72	14	0.05	0.78	7.83	2
Preparation	133	55.24	6	0.04	0.50	0.00	3
Review	216	60.55	10	0.04	0.61	13.33	3
Concentration	172	55.55	6	0.04	0.50	0.00	2
Formula	73	24.31	4	0.04	0.37	0.00	1
Persistence	101	40.46	4	0.04	0.37	0.00	2
Concept	82	37.23	2	0.04	0.21	0.00	1
Thinking	96	24.81	4	0.04	0.37	0.00	1
Solution	42	14.47	2	0.04	0.21	0.00	1
Ability	56	16.32	2	0.04	0.21	0.00	1
Computation	66	28.69	2	0.04	0.21	0.00	1
Content	34	11.81	2	0.03	0.13	0.00	3
Basics	47	18.22	2	0.04	0.21	0.00	1
Type	35	13.13	2	0.04	0.21	0.00	1

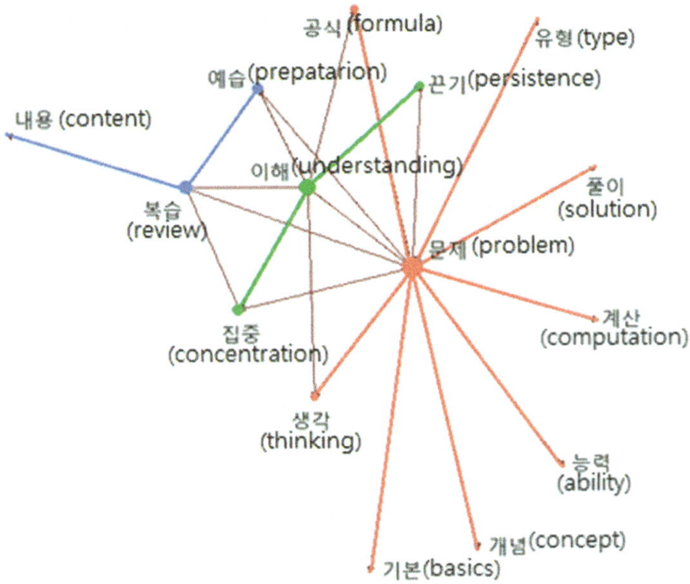

Fig. 7.2 Network graph related to students' personal values of learning mathematics

arrangement of these groups, indicating the overall centrality of each value and the connections between values. For instance, the value *problem*, being the most centralized, is at the center of the graph. Eight other values were connected to the value *problem,* but *thinking* is much closer to *problem* than other values such as *basics* or *concept*.

In Group 1, all the values are connected to *problem*. Specifically, the frequency of simultaneous appearance of value pairs was 39 for *formula*, 33 for *concept*, 31 for *thinking*, 26 for *solution*, 25 for *ability*, 24 for *computation*, 21 for *basics*, and 21 for *types*. Note that some values that had been grouped together with different values in the results pertinent to students' general values of learning mathematics (see Table 7.2) were connected to the value *problem*. For instance, the values *formula* and *basics* were connected to *understanding* regarding students' general values of learning mathematics, but these two values were connected to *problem* in their personal values of learning mathematics. Additionally, new values were connected to *problem*. Specifically, the values *solution* and *types* did not appear in students' general values of learning mathematics (see Table 7.2), but these two values were connected to *problem* in their personal values of learning mathematics. As evidenced by the sampled student responses in Table 7.4, students described the importance of *problem* along with its related words. For instance, students responded that in order to solve a problem, they need to know a formula, employ a concept, and think a lot. They also described of solving basic problems and various types of problems, based on problem-solving ability and computation ability.

Table 7.4 Sampled student responses related to their personal values of learning mathematics by the groups of values

Group	Value	Sampled student response
1	Problem, formula, concept, thinking, solution, ability, computation, basics, types	• I need to solve difficult problems several times after memorizing the formula well • I must have the ability to solve problems using concepts • The ability to think is important, which means I must think in solving a problem • My own solution method is important. I look for my style while solving problems • Reducing mistakes in computation is important for me to solve problems well • If you don't understand the concept, you can't solve even the basic problem • You can broaden your thinking by solving various types of problems
2	Understanding, concentration, persistence	• I must concentrate so that I can learn by understanding without missing anything in the middle • Understanding is important in learning mathematics, so I need my own persistent efforts
3	Preparation, review, content	• It is important to prepare and review. Namely, I can master the content only when I study in advance or review it • I think I can learn by reviewing the content I have learned at school • The more difficult content I learn, the more I forget, so it is important to review the content

In Group 2, the two values, *concentration* and persistence were connected to *understanding*. Specifically, the frequency of simultaneous appearance of value pairs was 36 for *understanding* and concentration, and 21 for *understanding* and *persistence*. As illustrated in Table 7.4, students stated the importance of *understanding* through *concentration* and *persistence*.

In Group 3, the two values, *preparation* and *content*, were connected to *review*. Specifically, the frequency of simultaneous appearance of value pairs was 91 for *preparation* and *review*, and 21 for *content* and *review*. Students recognized the importance of studying what they would learn in advance and reviewing the content they had learned.

In summary, students described *problem* as the most important variable for their own learning of mathematics. Specifically, they mentioned that formula, concept, thinking, solution, ability, and computation would be needed to solve a problem, while recognizing the importance of solving various types of problems including basic problems. Students also mentioned *concentration* and *persistence* for understanding. In addition, they underscored the importance of preparation and review of the content they have learned.

7.4.3 Students' Perceived Teacher Values of Teaching Mathematics

A total of 921 words were extracted to the item, "Think about your mathematics teacher this year. What do you think are important to him/her in mathematics teaching?" Table 7.5 shows the results of frequency analysis, TF-IDF analysis, centrality analysis, and the grouping of the words into value categories. Considering the frequency analysis, the top three values were *problem* (n = 311), *understanding* (n = 256), and *review* (n = 164), followed by *oneself* (n = 148), *concept* (n = 144), and *concentration* (n = 110). Note that the top three values by frequency were the same between students' personal values of learning mathematics and their perceived teacher values of teaching mathematics. In other words, what the students think is important for their own learning of mathematics is aligned with what they perceived their teachers think is important in mathematics teaching.

The results of TF-IDF analysis indicated that *understanding* (62.08), *review* (58.26), *concept* (49.37), *problem* (46.49), and *concentration* (44.00) had important meanings to the item. Compared to the results of student values of learning mathematics, the value *concept* appeared as more important, whereas *preparation* appeared as less important. In addition, the centrality analysis of value pairs with more than 15-word connections showed that the centrality of *problem* and *understanding* was high. This is similar to the students' personal values of learning mathematics.

Table 7.5 Frequency analysis, TF-IDF analysis, centrality analysis, and the grouping related to students' perceived teacher values of teaching mathematics

Value	N	TF-IDF	DC	CC	EC	BC	Group
Problem	311	46.49	24	0.06	1.00	62	1
Understanding	256	62.08	16	0.05	0.82	22	2
Review	164	58.26	8	0.04	0.52	13	3
Oneself	148	31.86	6	0.04	0.46	1	4
Concept	144	49.37	6	0.04	0.50	0	1
Thinking	95	28.98	4	0.03	0.39	0	2
Solution	58	23.12	2	0.03	0.21	0	1
Concentration	110	44.00	4	0.04	0.39	0	2
Preparation	36	30.19	2	0.03	0.11	0	3
Lesson	84	28.26	4	0.04	0.31	0	4
Formula	40	21.17	2	0.03	0.21	0	1
Basics	48	22.34	4	0.04	0.39	0	1
Method	31	14.69	2	0.03	0.21	0	1
Explanation	45	21.59	2	0.03	0.18	0	2
Variety	19	7.68	2	0.03	0.21	0	1

Fig. 7.3 Network graph related to students' perceived teacher values of teaching mathematics

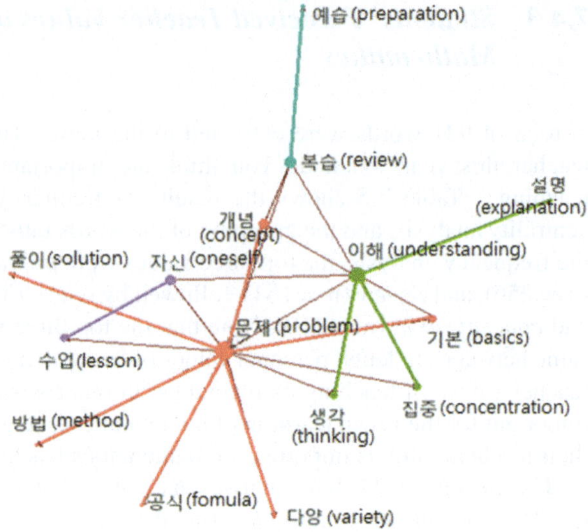

As shown in Table 7.5, the centrality analysis resulted in four groups of values. Group 1 consisted of *problem, concept, solution, formula, basics, method,* and *variety*. Group 2 included *understanding, thinking, concentration,* and *explanation.* Group 3 consisted of *review* and *preparation*, while Group 4 included *oneself* and *lesson*. Figure 7.3 is a visual representation of these groups.

In Group 1, all the values were connected to *problem*. Specifically, the frequency of simultaneous appearance of value pairs was 38 for *concept*, 28 for *solution*, 18 for *formula*, 17 for *basics*, 16 for *method*, and 15 for *variety*. As evidenced by the sampled student responses in Table 7.6, students felt that their teachers emphasized the concept, basics, solution, and method to solve a mathematics problem. Students also described that their teachers regarded it as being important to solve various problems while employing formulas. The word 'method' appeared new, indicating that the teachers valued various solution methods.

In Group 2, the three values, *thinking, concentration,* and *explanation* were connected to the value *understanding*. Specifically, the frequency of simultaneous appearance of value pairs was 26 for *understanding* and *concentration*, 18 for *thinking*, and 15 for *explanation*. As illustrated in Table 7.6, students stated that they were asked to concentrate and think for understanding by their teachers. They also described that their teachers provided explanations for all students' understanding.

In Group 3, the two values, *review* and *preparation*, were connected. Specifically, the frequency of simultaneous appearance of value pairs was 25 for *review* and *preparation*. This frequency was relatively low, considering that the value *review* was mentioned 164 times. The students described that their teachers emphasized reviewing what they had learned, for instance by saying "The first thing my teacher does every day at the beginning of a lesson is to review what we learned in the previous lesson. When we learn something new, my teacher reviews the most basic

Table 7.6 Sampled student responses related to their perceived teacher values of teaching mathematics by the groups of values

Group	Value	Sampled student response
1	Problem, concept, solution, formula, basics, method, variety	• Only when you know the basics can you access difficult problems • My teacher values problem-solutions, expecting me to solve many problems to better understand the types of such problems • My teacher values the method to solve a problem • The ability to solve a problem by applying a formula is important • My teacher encourages me to solve a variety of problems
2	Understanding, concentration, thinking, explanation	• When you concentrate, you can understand and solve mathematics • You must concentrate on understanding in learning mathematics • You can understand mathematics when you solve it by thinking • My teacher explains mathematics easily and slowly at a low level of difficulty for all students to understand it
3	Preparation, review	• Review is better than studying in advance or preparation • My teacher values thorough review and preparation
4	Oneself, lesson	• My teacher expects me to engage in her lesson myself • It is important to raise a question about what you don't know in the lesson

content related to it." When students mentioned both *review* and *preparation*, there were two cases: (1) *Review* is more effective than *preparation*, and (2) Thorough *review* and *preparation* are important in learning mathematics.

In Group 4, the frequency of simultaneous appearance of value pairs was 22 for *oneself* and *lesson*. As illustrated in Table 7.6, students described that their teachers emphasized the need for students to engage in the lesson for themselves.

In summary, students described that their teachers emphasized both *problem* and *understanding* in teaching mathematics. Specifically, they thought that basic concepts, various solution methods, and formulas to solve a problem were underscored by their mathematics teachers. Students also perceived that their teachers provided them with detailed explanations for better understanding of the mathematical content and they themselves need concentration and thinking for understanding. A review of what students had learned was perceived as being more important by their teachers than the preparation of what they would be learning. Students' participation in the mathematics lessons was also valued by teachers.

7.5 Discussion

7.5.1 Students' Top Three Values of Mathematics Learning and Teaching

The frequency analyses among students' general values of learning mathematics, their personal values of learning mathematics, and their perceived teacher values of teaching mathematics show a remarkably similar trend, given that the top three values, namely *problem, understanding*, and *review*, were the same. In addition, the grouping results by the centrality analysis demonstrate the consistently close connections between *review* and *preparation* as well as between *problem* and *concept*.

However, the results of the TF-IDF analysis and the centrality analysis reveal subtle but important differences in what was valued by students. Most of all, students regarded *understanding* as being the most important for mathematics learning in general, and for what students perceived their teachers to be valuing. It is desirable for students to value *understanding*, which is the essence of mathematics learning and teaching. Note that almost all Korean students agree that they need to do well in mathematics to get into the university of their choice (Mullis et al., 2020), which is the instrumental purpose of learning mathematics. Nevertheless, the students acknowledged the importance of *understanding* in learning and teaching mathematics. Specifically, values such as *review, preparation, formula,* and *basics* were connected to the value *understanding* as being important for anyone learning mathematics. Values such as *concentration, thinking*, and *explanation* were also connected to the value *understanding* insofar as teachers' valuing was concerned. Given that Pang and Seah's (2021) earlier quantitative study had found that Korean students value *understanding* in mathematics learning, the current study confirms this earlier claim through a TF-IDF analysis and a centrality analysis with students' own descriptive data.

Another noticeable difference is that the value *review* was the most characteristic in describing what would be important for students' own learning of mathematics, according to the TF-IDF analysis. In other words, reviewing the mathematical content was more valued than understanding it. Students tended to mention both *preparation* and *review* together, resulting in the two values grouped together, but it turned out that reviewing what they had learned was deemed to be much more important.

The centrality of *problem* is worthwhile to mention. Words associated with the valuing of *problem* constituted the highest frequency across all three descriptive items. Even though it did not capture the characteristics of student responses to each item, considering the TF-IDF analysis, the value *problem* was centralized because most values were adjacent to it in the three items. Specifically, values such as *concept, concentration*, and *creativity* were connected to the value *problem* as being important for mathematics learning in general. Values such as *concept, formula, thinking, solution, ability, computation, basics*, and *types* were connected to the value *problem* associated with mathematics learning personally. On the other hand, values such as *concept, formula, solution, methods, variety,* and *basics* were connected to *problem* in the context of perceived teacher valuing. To emphasize, the centrality of the value *problem* was revealed in common in the three items. However, different values were connected to *problem* across the items. In this respect, to better understand students' mathematics educational values, it is necessary to articulate in what context certain values would be connected with.

7.5.2 Comparing Students' Personal Values of Mathematics Learning and Their Perceived Teacher Values of Mathematics Teaching

The network graph of students' personal values and the graph of their perceived teacher values display a similar topological structure. Specifically, in both graphs, Group 1 included *problem, concept, formula, solution,* and *basics*; Group 2 included *understanding* and *concentration*; and Group 3 included *preparation* and *review* in common. In both graphs, the value *problem* was centralized as being the most important for their own learning of mathematics and for their teachers to teach mathematics. This is different from the network graph displaying values associated with mathematics learning in general, wherein not only was *problem* but also *understanding* were underscored, according to the DC analysis. For instance, the two values, *basics* and *formula,* were connected to the value *problem* in the contexts of students' personal values of learning mathematics and also of their perceived teacher values of teaching mathematics. However, the same values were connected to the value *understanding* for mathematics learning in general. Given this, we may infer an alignment between what the students thought was important for their learning of mathematics and what they thought was important for their mathematics teachers in teaching mathematics. In other words, there appears to be values alignment in Korean mathematics classrooms, as perceived by students. The affordances this implies for quality pedagogical interactions, communication and relationships between teacher and students in the Korean mathematics classroom are reflected in the excellent student performance, as would be expected.

There are three additional aspects that are noteworthy when comparing students' personal values of mathematics learning with their perceived teacher values of mathematics teaching. Firstly, more values (including *thinking, ability, computation,* and *type*) were adjacent to the value *problem* in describing students' own values of mathematics learning. This indicates how important the value *problem* is for oneself to learn mathematics. Secondly, a new group with specific keywords (i.e., *oneself* and *lesson*) was emergent only regarding what would be important for the teachers in mathematics teaching. In other words, students described that their teachers expect them to engage in a mathematics lesson for themselves. Given that students' active engagement in a mathematics lesson has been called for in Korea (MOE, 2022), it is desirable for students to perceive that their teachers value it. Lastly, according to the TF-IDF analysis of both students' personal values in mathematics learning and their perceived teacher values in mathematics teaching, the values *review, concentration, understanding,* and *problem* were commonly included. However, the value *preparation* was regarded as being important by students, whereas the value *concept* was considered mainly by their teachers.

On the one hand, the similarities between students' personal values of mathematics learning and their perceived teacher values of mathematics teaching suggests that students and teachers' values of mathematics education are closely related. It is encouraging that the students in this study were able to perceive their teachers placed importance on *engagement* in a mathematics lesson. Students' active engagement in mathematics activities, as closely related to their values (Pinto, 2021), is regarded as a common aspect of effective mathematics teaching (Cai et al., 2009). Teachers would be able to orchestrate more effective mathematics teaching if what they think is important in teaching mathematics would be aligned with what their students regard as being important in learning mathematics, or at least when both a teacher and students better understand each other regarding their values of mathematics teaching and learning.

On the other hand, the subtle differences between students' personal values of mathematics learning and their perceived teacher values of mathematics teaching may indicate practical implications on how to further improve mathematics instruction based on teacher knowledge of students' values. For instance, considering that the students in this study valued *preparation* for their mathematics learning, teachers may explore strategies that might enhance students' sense of feeling more prepared and ready, such as providing summaries of next lessons, and/or making adjustments to the pacing and reach of each lesson.

In this respect, teacher expertise should also include a capability to align the different values that are espoused by teachers and their students in classroom interactions, so that intended pedagogical goals may be achieved effectively. Such an alignment is often concerned with a middle-path solution in which all parties involved retain varying aspects of what they value, and take on board varying aspects of what others value (Kalogeropoulos et al., 2021), resulting in a harmonious relationship between teachers and their students. There are certainly implications here for pre-service teaching programs and also for in-service professional learning.

7.5.3 Methodological Considerations in Identifying Students' Values of Mathematics Learning

The research method used in this study may serve as an useful analytic tool while working with students' descriptive data. To emphasize, the TF-IDF analysis can indicate which word would characterize the student responses, tailored to a specific item. The centrality analysis, specifically DC analysis, can display which word would be the center in connecting other words with one another.

The earlier finding that the valuing of *understanding* in mathematics learning was dominant among Korean students (Pang & Seah, 2021) is confirmed in the current study through this analytic method. However, Korean students' valuing of *connections* in the earlier study was not found in the current study. Reminding ourselves of the different contexts in which data were collected in these two studies offers us a clue to explaining why there is this apparent discrepancy. In the previous study, students were given a list of specific learning activities to which they indicated the extent each was important to them. In this study, students responded to stimuli questions in an open manner, in which they could write and explain what they considered important to them during their mathematics learning. Then, given Korean students' commitment to achieve in mathematics (Mullis et al., 2020), it is understandable why the student participants were emphasizing words such as 'review', 'solution', and 'preparation' (Table 7.3). While Korean students might value *connections* when asked such as in Pang and Seah (2021), it is likely that such words reflecting more intrinsic valuing might be more elusive when indirectly probed. In this context, we emphasize the usefulness of eliciting participant responses in multiple ways when researching. For classroom teachers, this should reinforce for them what they already know, that is, the importance of gathering student information from various sources such as teacher observation and students' written work.

A feature of our questionnaire items has been the ability to tease out what students felt should be valued in mathematics learning from what they felt they themselves valued as learners. That is, as we have seen above, the results were different when students were asked: (1) What do you think is important when anyone learns mathematics? and (2) Think about your own experience of learning mathematics; What do you think are important when you learn mathematics? Both 'understanding' and 'problem' were the centralized words of students' general values of mathematics learning, whereas the word 'problem' only played such a role with relation to their personal values of mathematics learning. The grouping results of the centrality analysis were also different. In addition, the word 'creativity' was emergent only with relation to students' general values of mathematics learning. The valuing of *creativity* in mathematics learning seemed rather distinctive in that only Japanese students had been observed to be embracing it (Seah et al., 2017). In a similar vein, valuing of *creativity* has not been reported in previous studies of Korean students. This implies the need for the use of precise language when surveying what respondents value (in mathematics education). This is especially important given that valuing is not purely all or nothing, but a phenomenon that is defined along a value continuum (Hofstede et al., 2010).

Knowing what students value (in their mathematics learning) can be hard to discern, especially when the data is in free form. Despite the limitation of using only three items, the text-mining method that was deployed in our study represents an alternative analytic method of assessing students' values through open-ended descriptive responses. The co-occurrence network analysis and the network graphs which are created as a result also hold promise for an objective identification of ultimate values from amongst the instrumental values. Further research is needed to confirm the usefulness of this methodology with different data sets encoded in different languages in the wider values-in-mathematics-education research community.

Last but not least, in the context of Korean effective mathematics education, our study involving 20 middle schools across Korean urban, city and rural communities has documented values alignment between teachers and their students. Developing teachers' capacity and expertise to orchestrate an alignment between what they themselves and what their own students value in mathematics teaching and learning respectively would certainly complement – indeed, facilitate the application of – excellent pedagogical content knowledge.

References

Berry, M. W. (2004). Survey of text mining. *Computing Reviews, 45*(9), 548.

Bishop, A. J. (1988). *Mathematical enculturation: A cultural perspective on mathematics education.* Kluwer.

Cai, J., Wang, T., Wang, N., & Garber, T. (2009). Studying effective teaching from teachers' perspectives: The journey has just begun. In J. Cai, G. Kaiser, B. Perry, & N-Y. Wong (Eds.), *Effective mathematics teaching from teachers' perspectives: National and cross-national studies* (pp. 303–317). Sense.

Chan, Y.-C., & Wong, N.-Y. (2019). Methodological issues in the investigation of values in mathematics. In P. Clarkson, W. T. Seah, & J. Pang (Eds.), *Values and valuing in mathematics education: Scanning and scoping the territory* (pp. 197–208). Springer.

Clarkson, P., Seah, W. T., & Pang, J. (Eds.). (2019). *Values and valuing in mathematics education: Scanning and scoping the territory.* Springer.

Feldman, R., & Sanger, J. (2007). *The text mining handbook: Advanced approaches in analyzing unstructured data.* Cambridge university press.

Hofstede, G., Hofstede, G. J., & Minkov, M. (2010). *Cultures and organizations: Software of the mind* (3rd ed.). McGraw-Hill.

Kalogeropoulos, P., Russo, J. A., & Clarkson, P. (2021). Exploring educator values alignment strategies in an intervention context: The emergence of the Beacon strategy. *ECNU Review of Education, 4*(2), 327–348. https://doi.org/10.1177/2096531120923127

Ministry of Education. (2022). *The national mathematics curriculum* (in Korean).

Mullis, I. V. S., Martin, M. O., Foy, P., Kelly, D. L., & Fishbein, B. (2020). *TIMSS 2019 international results in mathematics and science.* Retrieved from Boston College, TIMSS & PIRLS International Study Center. https://timssandpirls.bc.edu/timss2019/international-results/

Newman, M. E. (2008). The mathematics of networks. *The New Palgrave Encyclopedia of Economics, 2*(2008), 1–12.

OECD. (2019). *Country note (Korea): PISA 2018 results.* http://www.oecd.org/pisa/publications/PISA2018_CN_KOR.pdf

Pang, J., Cho, S., & Seah, W. T. (2016). An investigation of what Korean students valued with regards to mathematics and mathematics learning: A study with sixth and ninth graders [in Korean with an English abstract]. *The Mathematical Education, 55*(4), 467–484.

Pang, J., & Seah, W. T. (2021). Excellent mathematical performance despite "negative" affect of students in Korea: The values perspective. *ECNU Review of Education, 4*(2), 285–306.

Pang, J., & Yim, M. J. (2019). Perception and change of the values on mathematics learning by fifth graders through the teaching of a master teacher [in Korean with an English abstract]. *Journal of Elementary Mathematics Education in Korea, 23*(4), 405–435.

Pinto, M. M. F. (2021). Values and valuing across cultures and time zones. *ECNU Review of Education, 4*(2), 376–387.

Seah, W. T. (2005). *The negotiation of perceived value differences by immigrant teachers of mathematics in Australia* (Unpublished Ph.D. dissertation). Monash University, Victoria, Australia.

Seah, W. T. (2019). Values in mathematics education: Its conative nature, and how it can be developed. *Research in Mathematical Education, 22*(2), 99–121.

Seah, W. T., Baba, T., & Zhang, Q. (2017). The WIFI study: Students' valuing of mathematics learning in Hong Kong and Japan. In J. W. Son, T. Watanabe, & J. J. Lo (Eds.), *What matters? Research trends in international comparative studies in mathematics education* (pp. 333–354). Springer.

The Third Wave Lab. (2023). *Study 4: What I Find Important (in my maths learning) (WIFI).* https://thirdwavelab.education.unimelb.edu.au/study-4-what-i-find-important-in-my-maths-learning-wifi/

Tiberius, V. (2018). *Well-being as value fulfillment: How we can help each other to live well.* Oxford University Press.

Zhang, Q., Barkatsas, T., Law, H. Y., Leu, Y. C., Seah, W. T., & Wong, N. Y. (2016). What primary students in the Chinese mainland, Hong Kong and Taiwan value in mathematics learning: A comparative analysis. *International Journal of Science and Mathematics Education, 14*(5), 907–924.

Chapter 8
Values Alignment as Teacher Craft for Effective Mathematics Teaching and Learning

Penelope Kalogeropoulos, James Anthony Russo, and Angela Liyanage

8.1 Values and Valuing in Mathematics Education

The things that we believe in shape the way we view our world and the way we act within the world through the decisions we make, indicating our values (Kalogeropoulos & Clarkson, 2019). In a mathematics classroom, the decisions and actions relating to the teaching and learning of mathematics education reflect directly what teachers and students' value (Seah & Andersson, 2015).

> Values are the convictions which an individual has internalised as being the things of importance and worth. What an individual values defines for her/him a window through which s/he views the world around her/him. Valuing provides the individual with the will and determination to maintain any course of action chosen in the learning and teaching of mathematics. They regulate the ways in which a learner's/teacher's cognitive skills and emotional dispositions are aligned to learning/teaching in any given educational context. (Seah & Andersson, 2015 p. 169).

8.1.1 Values Alignment Strategies

Values and valuing are sociocultural in nature. Values are abstract qualities that we notice when we see them in action through decision making, reactions to critical incidents (Tripp, 1993), and engagement in any particular situation. What we value in an educational context reflect years of learning and is influenced by our experiences and social interactions (Seah, 2018). The role of a teacher has always involved the teaching of values, even though such teaching is often implicit. For example,

P. Kalogeropoulos (✉) · J. A. Russo · A. Liyanage
Monash University, Melbourne, Australia
e-mail: Penelope.kalogeropoulos@monash.edu

J. A. Russo
e-mail: james.russo@monash.edu

Y. Dede et al. (eds.), *Values and Valuing in Mathematics Education*,
https://doi.org/10.1007/978-981-99-9454-0_8

if a teacher communicates that they value doing arithmetic quickly through regularly incorporating "Timetables Races" into their mathematics lessons, students are likely to come to equate mathematical proficiency with being fast at arithmetic. This illustrates how pedagogical decisions of teachers have the power to shape students' understanding of the nature of mathematics and their feelings of what it means to be learners of mathematics. In addition to the teacher, parents are the other significant adults that have a substantial influence on the specific mathematical values a student develops, as well as the student's valuing of mathematics more generally (Harackiewicz et al., 2012; Hill & Seah, 2023; Zhang et al., 2016).

What a teacher considers important in their teaching, and what students consider important in their learning, mediate the quality of pedagogical interactions between the two parties (Seah, 2020). Teachers capable of aligning different valuing so that intended pedagogical goals may be achieved in a productive manner demonstrate a professional skill which is termed 'value alignment'.

In effective mathematics classrooms, values are aligned or negotiated by the different parties to maintain functioning activities in interaction (Kalogeropoulos & Clarkson, 2019) and to support student mathematics learning (Seah & Andersson, 2015). Hence, in sociocultural contexts, values are acquired over time but are also re-evaluated and re-prioritised depending on the situation at hand.

Classroom interactions between teachers and students and amongst students, represent sites of contestations and conflicts. Although some studies have revealed common values between teachers and students within particular culturally coherent communities (Dede et al., 2023), more generally, as students enter their classrooms, they have their own sets of values that may or may not be the same, similar or different to their teachers' values. The approach of aligning student and teacher values facilitates student engagement in mathematics classrooms in ways that optimise student learning of mathematics (Seah & Andersson, 2015). This is dependent on the pedagogical repertoires employed by the teachers and by the pedagogical relationship that is generated between the teacher and their students (Attard, 2011).

Strategies that educators use to resolve critical incidents are based on the extent to which an educator retains their values after negotiation has taken place (Kalogeropoulos et al., 2021). It is part of a teacher's craft to identify triggers that lead to student disengagement and use their resourcefulness and creativity to adopt a value alignment strategy to reengage students (Kalogeropoulos et al., 2021). Student values are being shaped in the mathematics education process with the teacher who is playing the role of value agent in mathematics teaching.

Five alignment strategies (see Fig. 8.1) have been suggested that educators call upon to facilitate the coexistence of different values: Beacon, Scaffolding, Balancing (or Equilibrium), Intervention and Refuge (Kalogeropoulos et al., 2021). These five strategies represent a continuum of gradual decrease in the number and/or valence of an educator's values that are retained after the values alignment process. The strategy that is deployed depends on the context, with the main focus being on maintaining student engagement. It is suggested that teacher effectiveness can depend on the extent to which teachers are able to negotiate these inevitable value differences (Seah, 2018).

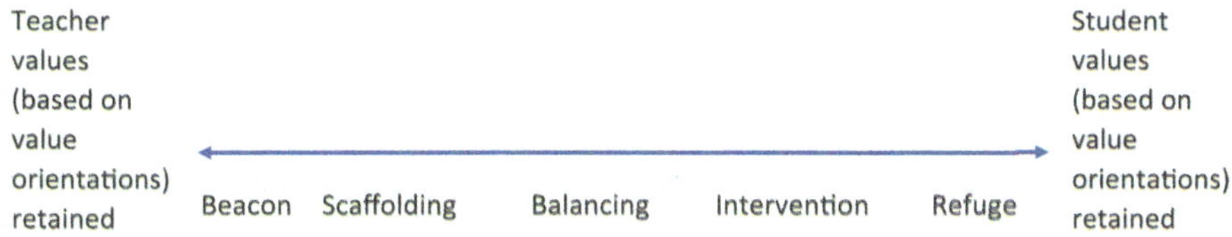

Fig. 8.1 Five value alignment strategies (Kalogeropoulos et al., 2021)

8.1.2 The Beacon Strategy

The Beacon strategy involves teachers reasserting their own values and attempting to persuade students involved as to the merits of their position. This strategy is most likely to be successful when the student trusts, values and invests in the relationship with the teacher (Kalogeropoulos et al., 2021). For example, a teacher communicating to their students that they must persist in a challenging task despite feelings of confusion and at times, frustration (Russo, 2019).

8.1.3 The Scaffolding Strategy

The Scaffolding strategy is adopted by a teacher who has come to the lesson prepared to scaffold the intended learning objective, including maintaining most of their initial values by having a system organised to avoid states of disengagement (Kalogeropoulos et al., 2021). For example, providing an enabling prompt when students are working on a challenging task (Sullivan et al., 2009), to ensure that all students can engage with the primary mathematical learning objective, albeit at different levels of depth (Russo et al., 2020).

8.1.4 The Balancing Strategy

The Balancing strategy is used by teachers when students unexpectedly refer to values that are not being catered for in the mathematics lesson (Kalogeropoulos et al., 2021). It involves the teacher listening to students' requests and responding to them but without sacrificing their own initial values in the learning situation. This strategy involves a teacher identifying the values that sit behind the student behaviour through gentle questioning, and generating an alternative action that effectively addresses both student and teacher values. For example, a decision by a teacher to allow a usually disruptive student to share their success in a mathematics lesson through 'spotlighting' their work (Hubbard et al., 2023), even though their work sample is, although relevant, not the one most closely aligned with the mathematical learning focus.

8.1.5 The Intervention Strategy

The Intervention strategy relies on strong interpersonal skills of the teacher who takes a pastoral role to address a situation, such as allowing a student to work with a friend if they are feeling overwhelmed with a task during independent learning time (Kalogeropoulos et al., 2021).

8.1.6 The Refuge Strategy

The Refuge strategy is a value alignment strategy used when the teacher puts most (if not all) of their values to one side and uses their authority in a manner that postpones their proposed learning and lesson objectives, and instead focusses on the value orientations of the students. In this situation, the teacher finds new values that align with their own and those of their students (Kalogeropoulos & Clarkson, 2019).

The identification of five alignment strategies on a continuum determined by whose initial values are retained (i.e., teachers or students), suggests potential overlap between two neighbouring value alignment strategies; and, therefore, some fluidity in classification. For example, during a mathematics intervention program, the tutor adopted the scaffolding strategy to support their students to develop a growth mindset (Boaler, 2022), and this pedagogical approach led to an increase in student confidence. This provided an opportunity for the balancing strategy to then be adopted by the tutor who now expected students to persist with challenging mathematical problems (Kalogeropoulos et al., 2021). It is suggested that the effective deployment of any value alignment strategy necessitates, to varying degrees, the assumption that an educator knows their students well (Kalogeropoulos et al., 2021).

8.2 The Current Study

The current study had three aims. The first aim is to deepen our understanding of student valuing of mathematics, as well as the factors students think are important when learning mathematics. The second is to explore the extent to which the teacher-participant's description of the teaching and learning of mathematics in their classroom can be mapped onto what students think is important about their mathematics learning. The third and final aim of the study is to examine the nature of the values alignment strategies adopted by the teacher-participant, and whether these strategies can be interpreted through the lens of the five value alignment strategies previously identified in the literature (see Kalogeropoulos et al., 2021).

8.3 Method

The current study can be described as a single case study of a classroom teacher and her students. A case study approach is a useful methodology focussed on understanding a phenomenon in more depth (Gerring, 2004), in this instance, student and teacher values and the process of achieving value alignment in a mathematics classroom. Regarding how we approached the research, we sought to identify a potential teacher-participant through our social networks with particular characteristics. Specifically, we were interested in identifying a mathematics teacher who had developed strong, high-trust relationships with her students, as described by her colleagues. We were of the view that such a case study would provide insights into how a teacher successfully manages to navigate value alignment in a classroom setting, and provide an opportunity to look for further evidence for the use of the five value alignment strategies previously identified in the literature (Kalogeropoulos et al., 2021).

8.3.1 *Participants*

Participants included two classes of upper secondary students at a Catholic boy's school in Metropolitan Melbourne (Australia), and their teacher. One of the classes were Year 10 students ($n = 25$) undertaking an advanced mathematics course, whilst the other class involved Year 12 students ($n = 16$) enrolled in an elementary mathematics course (i.e., General Mathematics). The teacher had five years of teaching experience at the time of the interview, and was an 'out-of-field' mathematics teacher, with a background in economics and commerce, rather than mathematics specifically. The school in scope for the study had an Index of Community Socio-Educational Advantage score of 1082, placing it in the 80th percentile for socio-educational advantage for the country as a whole.

8.3.2 *Procedure*

Student-participants completed part of the VAS (Seah, 2020) during a regular mathematics lesson. This involved students considering a series of statements that reflected different aspects of their valuing of mathematics, and responding to each of these on a 4-point scale (disagree a lot; disagree a little; agree a little; agree a lot). The statements included:

- I think learning mathematics will help me in daily life
- Mathematics will assist me with my learning of other school subjects
- I need to do well in mathematics to get into the college or university of my choice
- I need to do well in mathematics to get the job I want

- I would like a job that involves using mathematics
- It is important to learn about mathematics to get ahead in the world
- My parents think that it is important that I do well in mathematics
- My family discusses mathematics at home
- My parents give me support when completing my maths homework/ revision
- My parents expect me to do well in mathematics at school.

In addition, students responded to an open-ended item that asked them to note the most important things when learning mathematics (and to list up to three important things).

Following on from this, the teacher-participant (Donna) participated in a semi-structured interview with the first two authors that focussed on their approach to teaching mathematics and how it had shifted over time. The interview lasted approximately 50 min and was audio recorded and subsequently transcribed. Examples of prompting questions used during the semi-structured interview included:

- How did you help students transition from having an interest in mathematics to believing they could be good at mathematics?
- Could you please elaborate more on what you mean when you say "What would you like me to keep doing, start doing and stop doing"?
- How did the value alignment transitions occur?
- Do you believe that the reason that you have these positive relationships with the students is that you both have similar values or did you accommodate student values in your teaching?
- How long do you need to have a teacher that uses value alignment strategies to ingrain long-term effects for effective mathematics learning and subject appreciation?

8.3.3 Approach to Data Analysis

Student responses to the Likert-scale items are presented descriptively, whilst student responses to the open-ended item were analysed thematically and inductively, approximating the stages outlined by Braun and Clarke (2006) using the nVivo software package. Specifically, the data was first read holistically, and then read again with a view of generating a series of codes that captured the full diversity of student responses. These codes were then aggregated into themes, and the raw data revisited to ensure that relevant segments of student responses were categorised under one or more of the identified themes.

By contrast, the teacher interview data was analysed thematically and deductively to address the second and third research aims. The first iteration of deductive analysis involved using the themes that emerged from the qualitative analysis of the student data to attempt to capture the teacher participant's view about teaching and learning mathematics. The second iteration of deductive analysis involved using the five values alignment strategies identified by Kalogeropoulos et al. (2021) to identify how the teacher participant pursued value alignment in her mathematics classrooms.

8.4 Results: Student Participants

8.4.1 Quantitative Data: Likert Scales

The extent to which students agreed or disagreed with the various value statements is presented in Table 8.1. The statements are organised in order of most endorsed to least endorsed. Overall, with the exception of 'My family discusses mathematics at home', the majority of students agreed with each statement. The most endorsed statements related to parents valuing of a student's mathematics achievements, as well as parental expectations that a student did well in mathematics, with around 9 in 10 students agreeing with each of these statements. By contrast, less than two-thirds of students agreed that their parents supported them with their mathematical work at home, whilst less than half discussed mathematics at home.

Approximately 8 in 10 students agreed that they valued mathematics for instrumental reasons, specifically because it would: help them get into a preferred university course; support them to have success in other school subjects; eventually land them the job they want; and help them in their daily life more generally. By contrast, only around 6 in 10 students agreed that they would like a job that involved doing mathematics, suggesting that students valued the utility of mathematics as a discipline more than actually engaging in mathematical work out of interest or because such work was intrinsically satisfying.

When comparing mean scores across the statements for Year 10 and Year 12 students, there were a number of notable differences. First, Year 10 students were more likely to endorse each of the statements than Year 12 students on average. Second, the largest differences between Year 10 and Year 12 students (mean differences greater than 1) were noted for those items relating to instrumental reasons for valuing mathematics, including: that it was important for getting entry into a preferred university course; that it supported student work in other subjects; and that it put the student on a path to get a job they wanted. It is perhaps not surprising that these differences between Year 10 and Year 12 students existed, given that this group of Year 10 students were enrolled in a course designed to enable students to study advanced mathematics in Year 12 and beyond, whereas the Year 12 students were undertaking a General Mathematics course focussed more on equipping them with practical mathematical and numeracy skills to navigate adult life. This confounding factor of the two cohorts of students being notably different means we are unable to attribute any of the noted differences to age or schooling stage.

Table 8.1 Student endorsement of statements from the VAS

Items	Disagree a lot (1)	Disagree a little (2)	Agree a little (3)	Agree a lot (4)	Mean (SD) Overall	Mean (SD) Year 10	Mean (SD) Year 12
My parents think that it is important that I do well in mathematics	0 (0%)	5 (13%)	14 (35%)	21 (53%)	3.40 (0.71)	3.58 (0.58)	3.13 (0.81)
My parents expect me to do well in mathematics at school	2 (5%)	2 (5%)	16 (40%)	20 (50%)	3.35 (0.80)	3.54 (0.78)	3.06 (0.77)
I need to do well in mathematics to get into the college or university of my choice	3 (8%)	4 (10%)	11 (28%)	21 (54%)	3.28 (0.94)	3.74 (0.45)	2.63 (1.09)
Mathematics will assist me with my learning of other school subjects	2 (5%)	5 (12%)	18 (44%)	16 (39%)	3.17 (0.83)	3.60 (0.58)	2.50 (0.73)
I need to do well in mathematics to get the job I want	4 (11%)	4 (11%)	13 (34%)	17 (45%)	3.13 (0.99)	3.64 (0.49)	2.44 (1.09)
I think learning mathematics will help me in daily life	0 (0%)	8 (20%)	22 (54%)	11 (27%)	3.07 (0.69)	3.36 (0.64)	2.63 (0.50)
My parents give me support when completing my maths homework/ revision	10 (25%)	5 (13%)	11 28%)	14 (35%)	2.73 (1.20)	2.96 (1.16)	2.38 (1.20)
I would like a job that involves using mathematics	6 (15%)	9 (23%)	19 (48%)	6 (15%)	2.63 (0.93)	3.00 (0.72)	2.06 (0.93)
It is important to learn about mathematics to get ahead in the world	4 (10%)	13 (33%)	21 (53%)	2 (5%)	2.53 (0.75)	2.79 (0.59)	2.13 (0.81)
My family discusses mathematics at home	11 (28%)	12 (30%)	13 (33%)	4 (10%)	2.25 (0.98)	2.58 (0.93)	1.75 (0.86)

8.4.2 Qualitative Data: Thematic Analysis

Recall that for the open-ended item, students were asked to note what are the most important things when learning mathematics, and to record up to three things they thought were important. Results of the thematic analysis of this data is presented in Table 8.2. Six themes emerged that comprehensively described student responses to this item, whilst there were some notable differences between the Year 10 student cohort and the Year 12 student cohort. Each of these themes will now be considered in turn, from most prevalent to least prevalent, with illustrative student quotations presented.

One of the most prominent themes to emerge, *Application of knowledge*, referred to the learning and application of procedures, and using knowledge of procedures to engage in problem solving. For example:

> Being able to identify and replicate the processes used to calculate an answer. Year 10 student.

> Being able to use methods previously learnt and apply them to future questions. Year 10 student.

> Applying rules to wide variety (of) contexts. Year 12 student.

Application of knowledge also encompassed those responses that made reference to opportunities to learn about specific mathematical content areas, including measurement, algebra, patterns and particularly financial mathematics. In fact, nine students specifically mentioned learning about financial mathematics as of central importance, including six (38%) Year 12 students. An exemplar quote includes:

> Definitely understanding the fundamentals of finance and how to calculate interest etc. Year 12 student.

Another prominent theme was the importance of developing an *interconnected web of knowledge* to underpin one's mathematical learning. This theme encompassed the importance of developing strong conceptual understandings of relevant mathematical concepts, having sound foundational mathematical knowledge, and connecting new material to previously understood concepts. Student quotes that encapsulated the three aspects of this theme include:

> Developing an in-depth understanding of key concepts. Year 10 student.

> (You) need to understand the basics. Year 10 student.

> Constantly using concepts used in the past for current work. Year 10 student.

Students also mentioned the importance of particular *emotions and dispositions* when learning mathematics, highlighting various aspects such as: passion and enjoyment; independence; and resilience. Three quotes which captured these aspects (respectively) include:

Table 8.2 Summary of themes relating to what students think is most important when learning mathematics

Theme	Year 10 (n = 25)	Year 12 (n = 16)	Total (n = 41)
Application of knowledge	14 (56%)	10 (63%)	24 (59%)
Interconnected web of knowledge	15 (60%)	8 (50%)	23 (56%)
Emotions and dispositions	12 (48%)	5 (31%)	17 (41%)
Real-world connections	4 (16%)	7 (44%)	11 (27%)
Consolidation and practice	7 (28%)	4 (25%)	11 (27%)
Support	8 (32%)	1 (6%)	9 (22%)

> Being able to enjoy mathematics to learn it to its full extent. Year 10 student.

> Doing a problem without assistance. Year 10 student.

> Trial and error/keep trying if something doesn't work. Year 10 student.

In addition, students also noted: the importance of developing a growth mindset; collaborating effectively with peers; and curiosity. However, the most frequently emphasised emotion and disposition was conscientiousness, which was highlighted by eight students. For example:

> Staying focused and attentive at all times. Year 10 student.

> Working hard. Year 12 student.

Another theme related to learning mathematics through making *real-world connections*. Exemplary quotes included:

> To understand where/how it may be relevant in everyday life. Year 10 student.

> When you will use it in your everyday life and how it can be more useful for life. Year 12 student.

The final two themes related to the importance of consolidation and practice, and the need to proactively seek support to improve one's mathematical knowledge and skills. Respective student quotes relating to these themes included:

> The topics should be gone over more than once so that the knowledge and understanding of the topic is retained. Year 10 student.

> Don't be afraid to ask questions/for help. Year 10 student.

8.5 Results: Teacher-Participant

The results regarding the teacher-participant Donna are presented under two distinct subsections. The first subsection involves mapping Donna's perceptions about her practice onto what students think is important when learning mathematics. By contrast, the second subsection explores the value alignment strategies adopted by Donna as she endeavoured to navigate potential differences between her values and those of her students.

8.5.1 Mapping Teacher Perceptions onto What Students Think is Important When Learning Mathematics

Overall, deductive thematic analysis of the interview data revealed that the manner in which Donna described her teaching and learning of mathematics resonated with what students thought was important about learning mathematics. Specifically, four of the themes featured prominently in Donna's interview (interconnected web of knowledge; emotions and dispositions; real-world connections; support), whilst aspects of the two other themes (application of knowledge; consolidation and practice) were also mentioned by Donna. Table 8.2 presents the six themes that represent what Donna's two classes of students' valued most about learning mathematics, and endeavours to provide some illustrative quotations connecting comments made by Donna during the interview to each of these themes.

Despite these overall similarities, there were noteworthy differences between what students thought was important and the issues highlighted by Donna. Perhaps the key difference between Donna's interview and student questionnaire responses related to the first theme, *application of knowledge*. Although Donna referred to the importance of problem solving, at no point in her interview did she emphasise the importance of learning and applying procedural knowledge, nor did she highlight particular mathematics content areas, somewhat in contrast to her students' collective responses.

By contrast, Donna made multiple references to the importance of an *interconnected web of knowledge,* highlighting on several occasions the need to encourage students to draw on past learning to make sense of new concepts, and to develop a robust conceptual understanding of key mathematical ideas more generally. She

Table 8.3 Mapping teacher-participant interview data on to what students think is most important when learning mathematics

Theme	Illustrative teacher quotation
Application of knowledge	I tried to make it seem really random so they could build on those social skills, and they could build on the collaboration of problem solving, critical thinking
Interconnected web of knowledge	You need to understand how what we're learning has come to be. So that if you sit there in a test for example and you forget everything that you've learnt to that point how far can you go back before you can slowly start to build… With the unit circle for example reminding them that we can look at Pythagoras and an equilateral triangle and quickly build up from there, and suddenly you'll remember everything versus just memorising the table of exact values
Emotions and dispositions	So they found I think that there was a confidence coming by answering the questions in class, and even if they got it wrong they were like "Oh okay" and they'd self-reflect and just move on from that
Real-world connections	But I think for them what they meant was more of that discovery or the inquiry into the relevancy between the maths that they're learning and the real world
Consolidation and practice	So I think for those two (students) as well I do something like practice sheets at the start of every class and the past exam questions from various topics that we do, so the ideas is they are constantly keeping up with the skills
Support	And when they don't understand something, they'll usually go "No, I don't understand this, can you please explain this?" or they'll try and teach me what they know

also referred to student *emotions and dispositions*, including: their enjoyment and interest in learning mathematics; their confidence; and the importance of developing resilience as learners. Interestingly, Donna highlighted the importance of allowing students to work collaboratively and to communicate their mathematical thinking on half a dozen separate occasions during the interview, whereas the importance of collaboration, group work or peer learning was only mentioned by two of her 41 students.

Again, resonating with student data, Donna made multiple references to *real-world connections*, which were often discussed in the context of employing inquiry-based pedagogies, as well as the importance of students proactively seeking *support*. Finally, although the importance of *practice and consolidation* was mentioned by Donna when discussing two students who required additional support, it was emphasised less in her interview when contrasted with the collective responses of her students.

To summarise, despite the considerable overlap in what Donna and her students valued about mathematics learning, Donna placed relatively less emphasis on the more process-oriented aspects of mathematics learning, such as learning and applying procedures and consolidation and practice.

8.5.2 Exploring Values Alignment Strategies Adopted by the Teacher-Participant

This section outlines the value alignment strategies that Donna utilised in her attempts to reduce the values gap between herself as a mathematics teacher and her students. Detailed quotations from the interview with Donna have been included throughout this section, both to give voice to Donna to illustrate these ideas in her own words, as well as to provide transparency around our interpretation of the data. Although the interview data was analysed using deductive thematic analysis (as noted in the methods section), it has been reconstructed and presented here as a chronological narrative to capture how Donna's use of various value alignment strategies has evolved over the course of her teaching career.

When Donna first began in her role, there was evidence that she relied on the *refuge* strategy, the value alignment strategy that effectively involves a teacher suspending their own values and immediately prioritising the values of students. Donna essentially dedicated herself to trying to ascertain her students' values in relation to learning mathematics, and to present a program of work that resonated with their values. This approach was a consequence of Donna taking over a class part way through the year, combined with the fact that it was her first teaching role. Effectively Donna prioritised building relationships with students through identifying the student values of play and discussion, and consequently providing hands-on, playful tasks, and continually encouraging opportunities for discourse in relation to mathematical work.

> At the start, when I first started my career in teaching in the last five years, I inherited a class that obviously I didn't know what they were doing, what the previous teaching style was, and it was a lot of putting out little spot fires everywhere. But eventually you get talking to the kids and you build on – I started just by building on their interests and the kids would say they liked play, all right cool, so what tactile tasks can I do? They liked talking, okay well can I build in an element of discussion? And then I think I've built from that.

However, Donna was aware that by adopting the *refuge* value alignment strategy, she was building relationships and a broader reputation for being an empathetic and trustworthy teacher that would in turn empower her to shape her future work with these and other groups of students.

> I mean they always say when you start a new job the hardest part is building a reputation right. So once I'd had six months of building a reputation, it made it easier for me to be accepted when I moved over to the year 9 campus with those boys to suddenly go "Oh okay, she's a bit different, let's go with this" … I've entered into their territory by coming in halfway through the year. Their class is already established. They've got the dynamics down. I'm the intruder into the class. So in a very short amount of time I need to buy their trust… I had to meet them halfway and then like you said slowly go "Okay, ball's in my court now, let me do it my way".

Having first identified her students' values and having built a reputation as a trustworthy teacher, Donna was then able to turn her attention to trying to ensure her values were enacted in her mathematics classrooms. Donna clearly valued teaching mathematics for conceptual understanding, and ensuring that the mathematics was comprehendible, relatable and useful for her students. These values likely reflect Donna's background as an out-of-field mathematics teacher with a specialisation in economics and commerce; a discipline where mathematics is applied both to make sense of, and predict, observed and anticipated real world phenomena.

> So my background is in commerce and economics… my teaching rounds were all maths but my methods, so to speak, were accounting and economics. So I've never actually taught accounting or economics… [Interviewer: That's why I think it's interesting because you see mathematics as an applied subject already, rather than seeing it as the theoretical subject]. Yeah… And I think that's what they seem to value as well. A lot of the time, and again, I'm – this is based off the conversations I am constantly privy to with the year 12s, is that they're constantly talking about how "it was so useful" or "it was just so fascinating" or "it was so interesting". I'm like good, this is what I want you guys to take away from it.

This emphasis on real-world connections is clearly a value Donna perceives she shares with some classes of students, particularly those enrolled in more rudimentary mathematics courses, however not necessarily others, particularly her students in more advanced courses. The two contrasting reflections below provide a sense of the differences between her classes of students:

> They [my General Mathematics class] have actually pushed me to make sure that they understand in that capacity or have the real world comparison or that real world link.

> They [my advanced Year 10 class] were challenging in and of themselves because they didn't value inquiry, they valued formulas, process, so very algorithmic, very close-minded thinking. Because I think it's that mentality of "I don't want to be seen as not being smart". So they didn't like to be challenged. So for them, my sole focus that year was really building

> those fundamental skills, like back to the framework, the collaboration of social skills because I thought without those you're not actually going to be able to articulate your thinking, critical thinking for problem solving.

Importantly however, inquiry-based learning provided a framework that allowed for Donna's valuing of mathematics as being both understandable to her students, and applicable to relevant real-world phenomena, to be married with her students' interest in specific topics and content areas. Donna perceived inquiry-based learning opportunities to facilitate both meaningful and memorable learning experiences for students in a manner that procedurally-focussed mathematics could not contend with.

> So that's where that inquiry mindset I think has played such an integral part in how I run my classes is because I'm so conscious of the fact that at the end of the day they're not going to remember necessarily the special triangles for trigonometry. But what they will take away is that unit of work that we did where we explored viruses for exponentials, or they might look at a graph one day and go "I remember the project that I did at year 10 on that". And I think that is something that they want to take away from them.

Donna viewed inquiry-based learning, including the opportunities to work on larger projects, as a means of connecting student subject matter interest in other areas, and pursuits beyond school, with the mathematics curriculum.

> That's where I took their interest into consideration when I actually developed the project in mind because I had so many kids that had told me they wanted to do – so this was around subject selection time when I wrote the project. So, a lot of them wanted to do General or they said they said they were interested in doing Carpentry or a number of them had, I think, interest in Biology. So with that project it was open enough for them to double in subject areas that they were interested in. So that's something I've tried to embed moving forward.

However, in addition to *refuge* and achieving value alignment through employing inquiry-based pedagogies, there was also evidence that Donna drew on the *beacon* value alignment strategy. Recall that the *beacon* strategy involves a teacher achieving value alignment through asserting deeply held values and expecting students to adjust their values accordingly. In Donna's case, having established that inquiry-based mathematics learning was central to her practice, she outlined her expectations that students will engage in mathematics classes and be accountable for their learning. This is consistent with the notion that a teacher having both insight into their own values, and the willingness and capacity to clearly communicate them to students, is central to effectively utilising the *beacon* strategy.

> Now that I'm teaching obviously year levels that I may have been in for the first time, I think I've managed to refine my practice in such a way they don't know me and they don't know what to expect when they walk in, right, so I make it a very clear point to go "Our classroom's going to be different. We're going to do more inquiry-based work. My expectation though is that you get involved". And I'm like "I will not stand away from that". So if I have to do high risk strategies like threaten them, an email gets sent home for not engaging, I will do that.

In addition, Donna also facilitated the use of the *beacon* strategy through leveraging off her strong relationships with students.

> I think it was the rapport that I built with them. So I've actually taught this year level from when they were year 8. I taught them in year 9, 10, 11 and then obviously 12. They are looking to master the subject too now because I think they feel comfortable asking for help, and suddenly because they feel like they want to do well or they're seeing the idea of making me happy as a goal for them, that they're trying to understand the units better.

One of the means in which Donna invested in her relationships with her students was to invite their input into the teaching and learning process. In addition to strengthening relationships and thus facilitating the *beacon* strategy, this also allowed Donna to uncover potentially hidden values held by students and adjust her practice accordingly; an action that resonates with the *balancing* value alignment strategy.

> I'll do a survey at the start, like when I first meet the kids and I'll go "What are three things you expect from me? What do you think -?" and I'll always go "What do you think three things are that I expect from you?". So before I even articulate my expectations I want to know what they think my expectations will be because it'll be quite interesting then to see. And nine times out of ten they always go "You want us to try out best". Excellent, first cab off the rank… But I'll [also] ask them "What would you like me to keep doing, start doing and stop doing?". And very rarely will they put down anything to stop doing, which is nice, but all the time they'll say, "Please keep doing what you're doing, we really appreciate it". Every now and then they might say, "Can we have more practice?" or "Can we have more class time to do practice questions?" or whatever…

This process also allowed Donna to publicly demonstrate and model being vulnerable and open to feedback, and what it means to be a resilient learner.

> I show them that I'm vulnerable to learning as well because I think if I don't model the behaviours, I want from them then I'm a – I'd feel like a hypocrite as well. If I want them to learn, I show them I'm vulnerable. I'll make a mistake. I'll forget a negative in front of the two and I definitely have made mistakes, and they go "Oh Miss you've messed up". "Cool, thanks for letting me know". It's not a big deal. But if I show them that it's not a big deal then suddenly the emphasis on it is not actually that bad.

To summarise, Donna initially adopted the *refuge* value alignment strategy as she looked to build rapport with students as an outsider in an unfamiliar situation. However, through her commitment to building strong relationships with students, Donna was able to pivot to using the *beacon* strategy, expecting students to engage in mathematics class on her terms. This was supported by her commitment to inquiry-oriented pedagogies, which served to align her values of ensuring that mathematics was comprehendible and relevant to students, to students' motivation in pursuing topics and subject areas beyond mathematics that were of inherent interest (and value) to them. This led to Donna and many of her students developing a shared value that good mathematics instruction meant that mathematics was taught in a manner that was connected meaningfully to the world around them. Finally, there was also evidence that Donna set up processes in her classroom that allowed students to have regular input into teaching and learning processes, which served to both deepen rapport and support the *balancing* value alignment strategy.

8.6 Discussion

The first aim of the current study was to establish what upper secondary students in a particular educational context valued about learning mathematics. Overall, there was clear evidence that students valued their mathematics learning, although largely for instrumental reasons—as a means to pursuing valued ends such as performing well in Year 12, or earning a good job—rather than because they were inherently interested in, or intrinsically motivated to learn about, mathematics. This finding is consistent with the idea that even as enjoyment associated with learning mathematics declines, the valuing of mathematics as a discipline remains relatively robust as students move through their schooling (Mullis et al., 2012). We also found some indirect evidence that higher performing students (in this case the Year 10 cohort undertaking the advanced mathematics course) valued mathematics more than lower performing students (that is, Year 12 students enrolled in a General Mathematics course); a finding that also resonates with past research (Mullis et al., 2012).

The other notable point regarding this first aim was that parental expectations around their student's mathematical performance, and parenting valuing of mathematics as a discipline, were amongst the most strongly endorsed items; despite many students spending little time at home discussing mathematics, nor getting support or assistance from parents with learning the mathematical content. This paradox of many parents being relatively disengaged from their child's mathematics learning in any specific sense and showing relatively little interest in discussing mathematics in the home environment on the one hand, whilst being highly invested in their child's mathematical performance, has been noted elsewhere in the literature. For example, Cao et al. (2007) found that, although Australian students' perceptions of parental encouragement, support and help declined notably between Year 5 and Year 9, parental achievement expectations remained relatively stable. It is a powerful reminder that although teachers are important in influencing a student's values with regards to mathematics, parents also have a pivotal role to play (Harackiewicz et al., 2012).

Our second aim was to explore how the teacher participant's (Donna) description of the teaching and learning of mathematics in their classroom resonated with what students think is important about learning mathematics. Essentially this involved attempting to map the values articulated by Donna onto the values articulated by her students. One way of interpreting the similarities and differences between what Donna and her students valued about learning mathematics is through the lens of the five mathematical proficiencies noted by Kilpatrick et al. (2001) and embedded in the Australian curriculum (Burrows et al., 2020). Whereas both Donna and her students clearly valued (conceptual) understanding, problem solving (strategic competence) and having a productive disposition towards mathematical work, Donna placed considerably more emphasis on (adaptive) reasoning compared with her students. This was evidenced by her frequent and relatively in-depth references to the importance of working with others and communicating mathematical thinking when describing the teaching and learning of mathematics in her classroom. By contrast,

Donna placed far less emphasis on (procedural) fluency than her students, neglecting to mention the importance of students learning and applying mathematical formulas, and making fewer and more ambivalent references to the importance of consolidation and practice. Perhaps Donna is more conscious of addressing skills and values that are required for the 21st Century, such as communication and collaboration.

Our third aim was to explore the value alignment strategies adopted by Donna to navigate value conflicts as they arose. Our overall findings were consistent with previous research in that they confirmed that a teacher will deploy value alignment strategies depending on their skills, knowledge, and experiences (Kalogeropoulos et al., 2021). As a novice teacher, Donna used more of the value alignment strategies on the right side of the continuum (e.g., refuge), which involves more compromise of her values. One could argue that this was purposeful on Donna's behalf in her attempt to recognise and acknowledge her students' values. This provided Donna with opportunities to know her students, earn their trust and show her investment in knowing how they prefer to learn mathematics. As time progressed and Donna became more experienced as a mathematics educator and cemented her positive relationships with her students, she was more confident to embrace her values for effective teaching and learning pedagogies. Even in new classes, where Donna has not yet established personal relationships with the students but has earned a reputation as an innovative teacher, she conveys her values at the beginning of the school year, communicating to students that they will be expected to be actively involved in inquiry-based learning. Hence, she more recently deployed the value alignment strategies to the left of the continuum, the beacon strategy in particular, which involves less compromise of her values.

The idea that value alignment strategies are employed by teachers dynamically, with the same teachers adopting different strategies in different circumstances, resonates with prior research in this area (Kalogeropoulos & Clarkson, 2019; Kalogeropoulos et al., 2021). However, the notion that the profile of value alignment strategies deployed may evolve in a predictable manner as one becomes more experienced as a teacher is a relatively novel proposition supported by data from the current study. Through considering the case of Donna, we have seen how an educator can pivot through the different value alignment strategies through her teaching career thus far. An area for future research is looking at the extent to which teachers are able to negotiate inevitable value differences in-the-moment and whether more experienced teachers face less conflict situations due to having more opportunities to understand the educational context and to develop their interpersonal and pedagogical strengths. Another related issue to consider is whether the five previously value alignment strategies need to emerge naturally through teaching experience, or whether they can be more explicitly and deliberately developed through pre-service or in-service teacher professional learning.

In addition, it is noteworthy that Donna is an 'out-of-field' mathematics teacher, an aspect of her background that was pivotal to her self-described narrative of her teaching career and that appeared to influence her values as an educator. Consequently, future research might also consider how teacher qualifications interact with a teacher's level of teaching experience to shape the types of value alignment strategies adopted by the teacher. This would seem a particularly important consideration in an Australian context, given that almost one-quarter of secondary mathematics students in Australia are taught be an out-of-field teacher (Mullis et al., 2016).

8.7 Conclusions and Limitations

Our study has limitations that need to be acknowledged. Most obviously, the fact that it was limited to a single teacher-participant and her students in a particular educational setting (upper secondary mathematics classes in a Catholic boy's school) limits the generalisability of our findings. In addition, both the teacher and student data collected relied on retrospective reflections and reconstructions of the teaching and learning experience, and no attempt was made to triangulate this perception-based data with classroom observations. Finally, a further limitation arose from our decision to treat the teacher-participant Donna as the case study unit, and her students as effectively a monolithic group. This analytical lens did not allow us to explore potentially conflicting values between students, and how Donna navigated this conflict. This remains an important direction for future research, given that conflicting student values would likely add substantial complexity to teacher efforts at navigating value alignment. Notwithstanding these limitations, the current study sheds further light on the utility of pursuing values-based research in mathematics education through two further corollaries that arise from the current findings.

First, teacher-student relationships can be nurtured through values with open and honest discussions that relate to students' thoughts on pedagogical tasks and activities, as this will bring to the fore what students' value similarly and differently to their educator (Seah, 2018). A confident teacher in mathematics will have insight into their own system of values and clearly communicate these to their students for transparency and to justify their preferred teaching methods. Allowing student voice provides opportunities for value negotiation (Seah, 2018) and ultimately more student engagement. "Thus, for a teacher, being able to facilitate values alignment between what they value and what their students value promises to strengthen the relationships, and is one of the keys to nourishing teaching and learning practices" (Seah & Andersson, 2015, p. 177).

Second, mathematics learning can be optimised through the harnessing of values (Seah & Andersson, 2015). In this case study, Donna's description of her teaching and learning of mathematics resonated with what students thought was important about learning mathematics. The mapping of student values provides teachers with information about their students and their preferences when learning. Values provide students "with the ability to focus and to maintain persistent effort" (Seah, 2019, p. 103); perhaps the most important qualities in mathematics learning. Although the content of mathematics may not change, the way information is delivered to become valuable knowledge may influence the proportion of students selecting advanced level mathematics subjects in upper secondary schooling.

References

Attard, C. (2011). Engagement with mathematics: The influence of teachers. *South East Asian Mathematics Education Journal, 1*(1), 31–39. https://doi.org/10.46517/seamej.v1i1.8

Boaler, J. (2022). *Mathematical mindsets: Unleashing students' potential through creative mathematics, inspiring messages and innovative teaching.* Wiley.

Braun, V., & Clarke, V. (2006). Using thematic analysis in psychology. *Qualitative Research in Psychology, 3*(2), 77–101. https://doi.org/10.1191/1478088706qp063oa

Burrows, P., Raymond, L., & Clarke, D. (2020). A powerful image of mathematical thinking, doing and being: The four proficiencies as verbs. *Australian Primary Mathematics Classroom, 25*(3), 3–5.

Cao, Z., Bishop, A., & Forgasz, H. (2007). Perceived parental influence on mathematics learning: A comparison among students in China and Australia. *Educational Studies in Mathematics, 64*(1), 85–106. https://doi.org/10.1007/s10649-006-9033-5

Dede, Y., Aktaş, F.N., & Akyıldız, P. (2023). Mathematics educational values of students and teachers in Imam-Hatip high schools. *International Journal of Mathematical Education in Science and Technology*, (in print) 1–17. https://doi.org/10.1080/0020739X.2023.2172621

Gerring, J. (2004). What is a case study and what is it good for? *American Political Science Review, 98*(2), 341–354.

Harackiewicz, J. M., Rozek, C. S., Hulleman, C. S., & Hyde, J. S. (2012). Helping parents to motivate adolescents in mathematics and science: An experimental test of a utility-value intervention. *Psychological Science, 23*(8), 899–906. https://doi.org/10.1177/0956797611435530

Hill, J. L., & Seah, W. T. (2023). Student values and wellbeing in mathematics education: Perspectives of Chinese primary students. *ZDM–Mathematics Education, 55*(2), 385–398. https://doi.org/10.1007/s11858-022-01418-7

Hubbard, J., Russo, J., & Sullivan, P. (2023). Spotlighting: Helping to scaffold the 'explore phase' of the lesson when teaching with challenging tasks. *Australian Primary Mathematics Classroom, 28*(1), 31–37.

Kalogeropoulos, P., Russo, J. A., & Clarkson, P. (2021). Exploring educator values alignment strategies in an intervention context: The emergence of the Beacon strategy. *ECNU Review of Education, 4*(2), 327–348. https://doi.org/10.1177/2096531120923127

Kalogeropoulos, P., & Clarkson, P. (2019). The role of value alignment in levels of engagement of mathematics learning. In P. C. Clarkson, W. T. Seah, & J. Pang (Eds.), *Values and valuing in mathematics education: Scanning and scoping the territory* (pp. 115–127). Springer. https://doi.org/10.1007/978-3-030-16892-6_8

Kilpatrick, J., Swafford, J., & Findell, B. (Eds.). (2001). *Adding it up: Helping children learn mathematics.* National Academy Press.

Mullis, I. V. S., Martin, M. O., Foy, P., & Arora, A. (2012). *TIMSS 2011 international results in mathematics.* TIMSS & PIRLS International Study Center.

Mullis, I. V. S., Martin, M. O., Foy, P., & Hooper, M. (2016). *TIMSS 2015 international results in mathematics.* Boston College, TIMSS & PIRLS International Study Center.

Russo, J. A. (2019). Walking the line between order and chaos: A teacher-researcher's reflection on teaching mathematics with challenging tasks in primary classrooms. *International Journal of Innovation in Science and Mathematics Education, 27*(3), 14–24. https://doi.org/10.30722/IJISME.27.03.002

Russo, J., Minas, M., Hewish, T., & McCosh, J. (2020). Using prompts to empower learners: Exploring primary students' attitudes towards enabling prompts when learning mathematics through problem solving. *Mathematics Teacher Education and Development, 22*(1), 48–67. https://mted.merga.net.au/index.php/mted/article/download/592/388

Seah, W. T., & Andersson, A. (2015). Valuing diversity in mathematics pedagogy through the volitional nature and alignment of values. In A. Bishop, H. Tan, & T. Barkatsas (Eds.), *Diversity in mathematics education: Towards inclusive practices* (pp. 167–183). Springer. https://doi.org/10.1007/978-3-319-05978-5_10

Seah, W. T. (2018). Improving mathematics pedagogy through student/teacher valuing: Lessons from five continents. In G. Kaiser, H. Forgasz, M. Graven, A. Kuzniak, E. Simmt, & B. Xu (Eds.), *Invited lecturers from the 13th international congress on mathematical education* (pp. 561–580). Springer. https://doi.org/10.1007/978-3-319-72170-5_31

Seah, W. T. (2019). Values in mathematics education: Its conative nature, and how it can be developed. *Journal of the Korean Society of Mathematics Education, 22*(2), 99–121. https://doi.org/10.7468/jksmed.2019.22.2.99

Seah, W. T. (2020). *Study 5 the values alignment study: Discussion paper*. Unpublished manuscript, Melbourne, Australia.

Sullivan, P., Mousley, J., & Jorgensen, R. (2009). Tasks and pedagogies that facilitate mathematical problem solving. In B. Kaur, Y. B. Har & M. Kapur (Eds.), *Mathematical problem solving: Yearbook 2009* (pp. 17–42). World Scientific Publishing Co. https://doi.org/10.1142/9789814277228_0002

Tripp, D. (1993). *Critical incidents in teaching: The development of professional judgement*. Routledge.

Zhang, Q., Barkatsas, T., Law, H. Y., Leu, Y. C., Seah, W. T., & Wong, N. Y. (2016). What primary students in the Chinese Mainland, Hong Kong and Taiwan value in mathematics learning: A comparative analysis. *International Journal of Science and Mathematics Education, 14*, 907–924. https://doi.org/10.1007/s10763-014-9615-0

Chapter 9
Comparative Study of Primary School Students' Values in Mathematics Learning in Ghana and Australia

Ernest Kofi Davis, Penelope Kalogeropoulos, and James Anthony Russo

9.1 Introduction

Values and valuing in mathematics education are usually categorised as general educational values, mathematical values, and mathematics educational values (Bishop, 1996). International studies suggest that mathematics educational values have the greatest impact on students' dispositions and experiences in mathematics education (Seah, 2018).

DeBellis and Goldin (2006) asserted that values are personal truths that motivate our short-term priorities and long-term decisions. The extent to which a value is embraced and prioritised is responsive to one's environment, and therefore value priorities continue to be examined and evaluated throughout one's life (Seah & Andersson, 2015). Values in mathematics are conative in nature; that is, a willingness or desire to maintain persistent effort to achieve maximum performance of an activity (DeBellis & Goldin, 2006). "Values provide the individual with the will and reason to maintain their course of action, despite challenges that might occur" (Hill et al., 2021, p. 6).

The literature suggests that ease of information sharing through technology has allowed the dissemination of mathematics pedagogical approaches around the world, resulting in a convergence of mathematics pedagogical practices (Seah et al., 2021).

E. K. Davis (✉)
University of Cape Coast, Cape Coast, Ghana
e-mail: ekdavis@ucc.edu.gh

P. Kalogeropoulos · J. A. Russo
Monash University, Melbourne, Australia
e-mail: penelope.kalogeropoulos@monash.edu

J. A. Russo
e-mail: james.russo@monash.edu

Y. Dede et al. (eds.), *Values and Valuing in Mathematics Education*,
https://doi.org/10.1007/978-981-99-9454-0_9

This shows that exploring practices across the context of mathematics education provides an opportunity for cross-cultural learning.

To set the context for the study, literature on values and culture, international comparison of mathematics education using TIMSS 2019 data, and the context of mathematics teaching and learning in Ghana and Australia have been provided to illuminate the research problem. The research method that was used to explore valuing in mathematics among primary school children from these two countries for the purpose of comparison, as well as the results and implications of the findings for practice, are presented.

9.1.1 Values and Culture

Values are generally socioculturally dependent, and "what we value reflects years of learning and influence from our historical experiences and social interactions" (Seah, 2018, p. 564). Values in mathematics education are equally positioned in literature as being socially and culturally driven (Bishop, 2008; Hill, et al., 2019; Seah et al., 2017). Hill et al. (2019), for example, observed from their study involving Pasifika minority students in New Zealand that valuing peer collaboration, practice, and family support were specific to this group of students and recommend the need for classroom culture and pedagogy to align with mathematics educational values of minority students in order to ensure equitable mathematics teaching. Even within a school context, valuing across grade levels, that is, primary, junior high school, and senior high school levels could differ based on the demands of schooling on students at each of these levels (Davis et al., 2019b; Zhang et al., 2016). Davis et al. (2019b), for example, observed from their study involving Ghanaian students that there was a gradual shift in the intensity of valuing of achievement from primary through junior high school to the senior high school level as a result of demands of high-stakes examinations on students.

There has been growing recognition of the impact of values in mathematics education on both cognitive and affective outcomes for students (Hill et al., 2021; Seah, 2018). This could explain the consistently high performance in international comparative tests of East Asian students since they share a common schooling experience with peers from other cultures. It has been suggested that the cultural values of East Asian students that have been internalised over the years as learners provide them with the determination to succeed and perform well in school mathematics (Seah & Andersson, 2015). At least two interrelated reasons have been put forward to explain this phenomenon. First, formal assessment and mastery are embedded into many Asian societies and have been linked to the cultural influence of Confucianism (Wang, 2017; Zhong & Xie, 2022). Second, competition between students for employment opportunities results in pressure from parents and students themselves to perform well in standardised high-stakes assessments that determine tertiary admission (Hill & Seah, 2023).

Context does not drive valuing in mathematics education among students only but also teachers. Corey and Ninomiya (2019) identified the influence of community values, such as the preparation of detailed lesson plans, material research, and emphasis on students' mathematical reasoning in instruction on valuing among Japanese teachers.

Comparative studies on values, particularly examining values across cultures and cultural groups, have been recommended in the literature to help carry out stronger comparative studies (Clarkson et al., 2019). Cross-cultural studies on values in mathematics education have been carried out among students (Dede, 2019; Seah et al., 2017; Zhang et al., 2016) and teachers (Dede, 2015). Studies have shown that student values differ across countries (Baba et al., 2013; Dede, 2019; Law et al., 2011; Lim, 2015; Zhang, 2016). Zhang et al (2016), for example, undertook a comparative study among primary school children from the Chinese Mainland, Hong Kong and Taiwan, and found that the students from each of the three groups valued "Achievement", "Relevance", "Practice", "Communication", "Information and Communication Technology", and "Feedback", which constituted the six main value structure dimensions, differently. Dede (2019) also carried out a study involving a comparison of valuing among Turkish, Turkish immigrant, and German students and found some differences in valuing among these categories of students. Dede (2019), for example, observed that "Consolidating" was valued by only German students, while "Practice" was valued by only Turkish students and attributed his observation to the sociocultural context of schooling in the two countries.

Results from cross-cultural studies on valuing among teachers show that, as with the students, the social and cultural context of schooling affects what teachers' value in mathematics education (Abdullah & Leung, 2019; Dede, 2015). Dede (2015), for example, found from his comparative study involving middle and high school teachers from Germany and Tùrkiye that the school levels of teachers between and within the two countries had an effect on valuing among the teachers. It appears evident that the social and cultural context of schooling including socio-mathematical norms of a country or community could have implications on what both teachers and students' value in mathematics education and ultimately their practices in mathematics. Abdullah and Leung (2019), argue that the values held by mathematics teachers in Brunei and those from Japan had an effect on the implementation of lesson studies. This suggests that the adoption of best practices across contexts and cultures cannot be done effectively without looking at the values that underpin these practices.

Developing countries in Sub-Saharan Africa such as Ghana continue to adopt best practices from developed countries but perform poorly in international assessment. There is therefore the need for research into comparative studies of valuing among students from developing countries, especially in Sub-Sahara Africa, and developed countries to ascertain the possible effect of value differences on the implementation of best practices adopted in mathematics education in developing countries. This chapter sets out to contribute to international literature in this area.

9.1.2 International Comparison of Math Education (TIMSS 2019 Data)

The latest iteration of the Trends in International Mathematics and Science Study (TIMSS) included nine statements on valuing of mathematics (Thomson et al., 2021). Year 8 students participating in TIMSS (although not Year 4 students) were asked the extent to which they agreed with each of these statements. These statements collectively comprised the Students Value Mathematics scale. Seven of these statements were the same as statements presented to students in the Values Alignment Study (Seah, 2020; see the Method section of the current chapter for the items). The two additional statements were:

- Learning mathematics will give me more job opportunities when I am an adult
- It is important to do well in mathematics.

Overall, it was found that 37% of students participating in TIMMS across all countries strongly valued mathematics, 47% somewhat valued mathematics, and 16% did not value mathematics. The ratio of Australian students in each of these categories was similar to the international average (38%, 48%, 14%). Australian male students (42%) were more likely to strongly value mathematics (34%) than female students (Thomson et al., 2021). Ghana, by contrast, did not participate in TIMMS in 2019. Interestingly, the country with the highest proportion of students that strongly valued mathematics was found in South Africa (68%), followed by Egypt (63%), Jordan (62%), and Morocco (60%) (Mullis et al., 2020); implying that students in at least some African countries place a comparatively high value on mathematics.

The current study builds on the data presented in TIMSS in two notable ways. First, it provides international comparative data for one country that did not participate in TIMSS (Ghana) and compares it to data collected in a country that did participate (Australia). Second, it presents data from primary school students (Years 3–6), whereas TIMSS focused on secondary students only (Year 8). A comparison study between the teaching strategies in Ghana and Australia could highlight what is aligned and what differs between the two countries. We also seek to find out if there are any particular manners in which students approach mathematics learning in their respective countries that have implications for their ongoing education and how this is the same or different in each country.

9.1.3 Context of Mathematics Teaching and Learning in Ghana

Ghana's pre-tertiary education system, like many countries in the world, is structured into two years of preschool education, six years of primary education, three years of junior high school education, and three years of senior high school/Technical and Vocational Education and Training (TVET) education. The study of mathematics is compulsory at all levels. The preschool and primary school levels are taught by generalist teachers who teach all subjects including mathematics, while the high school level is taught by subject teachers. These teachers are expected to train in mathematics either as their major or minor area at tertiary level.

Research in mathematics curriculum delivery in Ghana, in general, and mathematics teaching and learning, in particular, has attracted the attention of researchers in the past and continues to attract the attention of researchers (Ampiah, et al., n.d.; Davis et al., 2019a; Abenyega & Davis, 2015). A large body of research on the delivery of mathematics curriculum in Ghana continues to position the gap between the planned and the implemented curriculum in mathematics as one of the key factors that affect students' learning outcomes/attainment in the subject (Ampiah et al., n.d.; Davis et al., 2019a). Davis et al. (2019a), for example, observed from their study of curriculum implementation in mathematics that many of the topics that are identified as being difficult for students such as algebra, statistics and probability, were the topics students had very little opportunity to learn. Loss of instructional time through co-curricular activities, unavailability of required instructional materials, large class sizes, low mastery of the English language (the language of instruction), curriculum overload, and readiness of students were found to be among the major reasons for teachers' inability to cover all topics. This trend tends to create a cognitive deficit as students' progress from one grade to another since there is the automatic promotion of students from one grade level to another in Ghanaian government schools (Davis et al., 2022).

The quality and nature of teaching and learning in the Ghanaian mathematics classroom have also been often positioned in literature as being low and teacher-driven (Abenyega & Davis, 2015; Davis, 2018). Abenyega and Davis (2015), for example, found from the study involving the teaching of mathematics at the primary school level in Ghana that instructional language coupled with ineffective teaching approaches used by teachers excluded many students from mathematics lessons. Findings from other research on teaching and learning of mathematics in Ghanaian primary schools also suggest that the traditional school mathematics micro-culture, where the teacher is the centre of all major classroom decisions regarding what has to be learned, how it has to be learned, and what is assessed dominate the classroom discourse (Cobb & Bauersfeld, 1995; Davis, 2018). The context of curriculum delivery in mathematics has often resulted in low learning outcomes of students both at the national and international levels (Davis et al., 2022; Mullis et al., 2012), with gender disparities often reported (Boateng & Gaulee, 2019; Wrigley-Asante, et al., 2021).

Gender performance of students in mathematics has generally favoured males, with the situation worsening as one moves across grade levels (Boateng & Gaulee, 2019). This has resulted in the situation where females are generally underrepresented in mathematics-related disciplines and professions (Boateng & Gaulee, 2019; Wrigley-Asante, et al., 2021). However, for the past two and half decades efforts have been made at the national and local levels by the Ministry of Education and development partners such as the United Nations International Children's Emergency Fund [UNICEF] through policies and advocacies such as the organisation of vacation camps to encourage girls and women's participation in mathematics and mathematics related disciplines (Girls' Education Unit, 2002; UNICEF, 2018; National Gender Policy, 2015).

In 2018, the government of Ghana reformed the Ghanaian primary school mathematics curriculum from an objective-based system (in which the curriculum provided some specific learning outcomes/objectives expected to be achieved) to a standard-based one (in which there are national standards reflecting competencies that should be exhibited by learners at the end of each grade level) (Senk & Thompson, 2003), with the aim of transforming teaching and learning of the subject and improving the learning outcomes in the subject (MoE, 2018). This reform, despite its great intentions, has also come with some challenges such as the availability of textbooks, teaching and learning resources to support quality teaching and learning, and training of teachers to be able to fully implement the standard-based curriculum. While the information provided in this section does not seem to reflect values in mathematics, it provides context for the interpretation of results from this study.

9.1.4 Context of Mathematics Teaching and Learning in Australia

Previous to the inclusion of the four proficiencies, namely: understanding, reasoning, problem solving, and fluency, the mathematics curricula in Australia had been dominated by a focus on skills and techniques. Whilst the curriculum has incorporated proficiencies that are more relevant for the 21st-century skills, studies indicate that teaching in classrooms, especially at the secondary level, continues to privilege skills and techniques (Little & Anderson, 2015; Vale & Herbert, 2021). However, despite a national curriculum being introduced in 2010, education remains largely the responsibility of state governments in Australia (Stephens, 2014). Moreover, these various education systems tend to be characterised by decentralised structures, where principals are given substantial autonomy in terms of how they choose to organise the teaching and learning of curriculum content at the school level. This is further supported by national and state curricula being largely non-prescriptive. Consequently, Australian mathematics instruction is perhaps best distinguished by its diversity from school to school and system to system, with approaches varying from student-centred inquiry focussed approaches (Kalogeropoulos et al., 2021) to more teacher-orchestrated direct instruction (Ewing, 2011).

Given the socio-cultural nature of mathematics education, Australia, along with other countries, has focused on helping students to engage meaningfully with the future using mathematics knowledge, skills and dispositions (Seah et al., 2021). In a country like Australia, where parents, politicians and leaders want students to study mathematics to a high level and for students to leave school feeling that mathematics is important and valuable in their society, the proportion of students undertaking advanced mathematics courses has decreased and this is causing major concern (Attard & Holmes, 2020). A lack of student engagement with mathematics in school is one reason for this decline and can be explained by students thinking that mathematics is something you do (e.g. learn to regurgitate facts) rather than something you think about, imagine, argue or feel (Seah et al., 2016). It has been argued that the number one reason for the dislike, fear, and even hatred of mathematics is not the nature of mathematics itself, but the way the subject is portrayed and presented by the federal and state curricula and associated pedagogies (Seah et al., 2016).

Other than engagement in mathematics learning, it is also noticeable that other issues may still exist. Mathematics anxiety is a phenomenon that students and teachers exhibit, with females more likely to experience anxiety than males (Maloney et al., 2015). Also, in relation to gender, "Despite... initiatives [over the last 40 years or so], females' participation ... in particular in mathematics, from primary through tertiary education to the career level is still very low" (Leder, 2015, p. 149). A suggestion was made that since values and valuing are inner, stable qualities, then "if we can guide the development of students' and curriculum writers' valuing from whatever sorts of beliefs, attitudes and interests they possess, we should be able to better facilitate a more positive and productive view of mathematics learning, and also a more empowering and relevant approach to curriculum reform" (Seah et al., 2016, p. 14). Drawings from primary and secondary students in the city of Melbourne in Australia, were found to highly value fun, whole class interactions, out of class/ outdoor learning, challenge, group interaction, manipulatives, authenticity, quietness, hands-on, competition, visualisation and practice (Seah & Ho, 2009). In a small-scaled study, Seah (2010) identified 4 values associated by Australian students with different ability levels namely; examples, sharing, resources and multimodal representations during moments of effective mathematics learning with no apparent gender differences. This study may provide some insights as to whether the values indicated in previous studies are the same in current times and how it compares to data obtained in Ghana.

9.1.5 The Current Study

9.1.5.1 The Research Questions

This chapter reports on a study that used the VAS instrument to explore valuing among primary school children in Australia and Ghana. The research questions that guided this international comparative study are presented. The following questions were posed to guide the study:

- What are the differences/similarities, if any, in valuing in mathematics among primary school children from Australia and from Ghana?
- What are the differences/similarities, if any, in valuing in mathematics among primary school boys and girls from Australia and Ghana?

International literature on gender issues in mathematics has focussed very little on valuing among boys and girls. Again, although some studies have explored the effect of school levels, that is, primary, lower secondary, and upper secondary, on values and found differences in valuing in mathematics across grade levels (Davis et al., 2019b), no study that we can find has looked at valuing across the stage of schooling at the primary level, which constitutes the formative stage in children's education. In addition, international literature on values is informed largely by research in developed countries; including Ghana represents an attempt to explore the situation from a context in which mathematics learning is characterised by many challenges yet often remains under-researched. These research questions, therefore, provide an avenue to explore values in ways that bring new perspectives into what is already known in the literature. This will help improve mathematics education since the identification of what students value in mathematics education through research will inform mathematics curriculum development and delivery in Ghana. Mathematics curriculum development and delivery informed by what students value in mathematics education will provide the drive for the students to achieve in mathematics (Seah & Andersson, 2015).

9.2 Method

9.2.1 Participants

The authors administered this instrument themselves in the participating classrooms. Participants included 113 students from two Australian public primary schools (Year 3/4, n = 57; Year 5/6, n = 56), and 561 students from 11 Ghanaian public primary schools (Year 3/4, n = 279; Year 5/6, n = 282). The 3/4 indicates that these children were in a mixed year level classroom of students in years 3 and 4, a reasonably common occurrence in Australia and in Ghana. The research participants from Ghana were selected across the various context of schools namely, above-average, average and below average achieving schools in both rural and urban settings in one of the Regions in Southern Ghana. This region was chosen because the characteristics of students' performance in mathematics this region reflect the national picture, where the performance of students is generally low. Two Australian government primary schools, one in the Outer Eastern suburbs of Melbourne, Victoria (n = 88) and the other in the Western regional Victoria (n = 25) participated. The schools were neither advantaged nor disadvantaged relative to the Australian average, with an Index of Community Socio-Educational Advantage which placed them in the 54th and 42nd percentile respectively, relative to all Australian schools.

9.2.2 Instrument

The Value Alignment Study (VAS) instrument comprises two sections namely A and B (Seah, 2020). Section A of the VAS instrument requires students to consider a series of statements that reflected different aspects of their valuing of mathematics, and responding to each of these on a 4-point scale (disagree a lot, 1; disagree a little, 2; agree a little, 3; agree a lot, 4). The statements included:

- I think learning mathematics will help me in daily life
- Mathematics will assist me with my learning of other school subjects
- I need to do well in mathematics to get into the college or university of my choice
- I need to do well in mathematics to get the job I want
- I would like a job that involves using mathematics
- It is important to learn about mathematics to get ahead in the world
- My parents think that it is important that I do well in mathematics
- My family discusses mathematics at home
- My parents give me support when completing my maths homework/revision
- My parents expect me to do well in mathematics at school

Section A had 25 items in all. Section B was made up of six open-ended items that elicited information on valuing by students and their teachers in the actual classroom setting. Although the VAS instrument was created to explore value alignment, the Section A part of the instrument that was used in this study was chosen because the items addressed the research questions that guided the study.

In order to ensure that the instrument elicited valid responses, it was shared with experts in the area of values for their inputs. This was followed by pre-testing of the instrument in Australia. The sociocultural appropriateness of the items was evaluated before it was pre-tested in Ghana. Through the process, only item 15 was found to elicit invalid responses because the context did not reflect the experience of many of the Ghanaian primary school student participants. Item 15 served as an example for questionnaire respondents to nominate three attributes which each of them valued in mathematics learning, using the context of mobile phone selection instead of mathematics learning. It reads as follows:

> When choosing a new mobile phone, what do you think are important to you? Responses: Screen size. A big screen size is important to me because I work on my phone a lot; Weight. It is hard to walk around with a heavy phone in the pocket; Connectivity. A USB slot would be my dream. Transfers file efficiently.

Accordingly, it was modified to:

> When choosing a new shoe, what do you think are important to you? Responses: Size of the shoe. A which is bigger or smaller than the size of the foot will make its use uncomfortable; Colour. It is advisable to choose a colour that matches different dresses; Materials used. The quality of the shoe depends on the material used.

The validated instrument was then used to collect the data for the study.

9.2.3 Data Analysis

Data was cleaned and organised using Microsoft Excel, and then imported into SPSS v. 25 for analysis. Descriptive data was generated to compare and contrast student responses across countries. In addition, a series of independent sample t-tests were undertaken to establish whether statistically mean differences existed between student response profiles across countries. As sample sizes were notably different and the assumption of homogeneity of variance did not hold for nine of the ten comparisons, equal variances were not assumed. Moreover, to account for the fact that conducting multiple independent tests risks inflating type 1 error, a Bonferroni adjustment was applied to the statistical significance level.

9.3 Results

Table 9.1 presents the Likert scale data from the VAS for Australian students, whilst Table 9.2 presents the Likert scale data from the VAS for Ghanaian students. For each table, the items are ordered from the highest mean score to the lowest mean score. It is apparent from viewing both tables that the majority of students from each country agreed with each of the 10 statements. For Australian students, the most endorsed statement related to parents valuing of students' mathematics achievements, with almost all students agreeing that their parents think that it is important that they do well in mathematics, and three-fifths strongly agreeing with this statement. This item was similarly strongly endorsed by Ghanaian students. However, the most strongly endorsed statement by Ghanaian students, a statement that all Ghanaian student participants agreed with, and four-fifths strongly agreed with, was: "I think learning mathematics will help me in my daily life".

Approximately eight in ten Australian students and more than nine in ten Ghanaian students reported that they valued mathematics for instrumental reasons, specifically because it would: help them in their day-to-day life; support them to be successful in other school subjects; allow them to get into a desirable college or university course; or land them a job they wanted in the future. By contrast, there was evidence that Australian students at least were less inclined to value mathematics for reasons relating to interest or enjoyment, with only just over half of students agreeing that they would like a mathematically-oriented job in the future. This was not the case for Ghanaian students, for whom 93% agreed with the statement that they would like a job that involves using mathematics.

The least endorsed statements for Ghanaian students, and the second and third least endorsed statements for Australian students, related to whether parents provided the student with support when completing their mathematics homework, and whether the family discussed mathematics at home. Specifically, whereas most Australian and Ghanaian students strongly agreed that their parents had high expectations for their mathematics performance (53 and 60%), and that their parents thought that it was

Table 9.1 Student endorsement of statements from the VAS: Australia (Primary Students, n = 113)

Items	Agree (3 or 4) (%)	Agree a lot (4) (%)	Mean (SD) Overall	Mean (SD) Year 3/4	Mean (SD) Year 5/6	Mean (SD) Boys	Mean (SD) Girls
My parents think that it is important that I do well in mathematics	95	61	3.55 (0.63)	3.50 (0.66)	3.59 (0.60)	3.51 (0.65)	3.62 (0.57)
I think learning mathematics will help me in daily life	88	61	3.45 (0.80)	3.44 (0.76)	3.46 (0.86)	3.42 (0.78)	3.49 (0.85)
I need to do well in mathematics to get into the college or university of my choice	86	60	3.44 (0.80)	3.57 (0.74)	3.30 (0.84)	3.34 (0.84)	3.54 (0.73)
My parents expect me to do well in mathematics at school	87	53	3.34 (0.86)	3.29 (0.92)	3.39 (0.81)	3.16 (0.97)	3.58 (0.64)
I need to do well in mathematics to get the job I want	82	54	3.32 (0.86)	3.33 (0.84)	3.31 (0.88)	3.26 (0.88)	3.40 (0.84)
It is important to learn about mathematics to get ahead in the world	85	52	3.30 (0.89)	3.38 (0.84)	3.22 (0.95)	3.29 (0.93)	3.32 (0.87)
Mathematics will assist me with my learning of other school subjects	78	32	3.07 (0.79)	3.00 (0.81)	3.15 (0.76)	3.08 (0.77)	3.08 (0.80)
My parents give me support when completing my maths homework/ revision	64	26	2.68 (1.09)	2.80 (0.96)	2.56 (1.21)	2.61 (1.11)	2.76 (1.08)
My family discusses mathematics at home	56	28	2.67 (1.06)	2.75 (1.08)	2.58 (1.05)	2.67 (1.03)	2.69 (1.08)
I would like a job that involves using mathematics	56	22	2.59 (1.03)	2.68 (0.96)	2.50 (1.11)	2.66 (1.11)	2.50 (0.95)

important that they did well in mathematics (61 and 65%), relatively few strongly agreed that their families discussed mathematics at home (28 and 22%) or that their parents supported them to complete their mathematics homework (26 and 27%).

Overall, it appears that there is some evidence that, although the majority of students in both countries report positive valuing of mathematics, Ghanaian students were more likely to endorse certain statements on the VAS compared with their Australian counterparts. These aggregate differences are summarised visually in Fig. 9.1. Additional analysis applying a series of independent sample t-tests

Table 9.2 Student endorsement of statements from the VAS: Ghana (Primary Students, n = 561)

Items	Agree (3 or 4) (%)	Agree a lot (4) (%)	Mean (SD) Overall	Mean (SD) Year 3/4	Mean (SD) Year 5/6	Mean (SD) Boys	Mean (SD) Girls
I think learning mathematics will help me in daily life	100	81	3.81 (0.39)	3.84 (0.37)	3.79 (0.41)	3.74 (0.44)	3.89 (0.31)
My parents think that it is important that I do well in mathematics	99	65	3.62 (0.55)	3.61 (0.57)	3.64 (0.52)	3.53 (0.50)	3.68 (0.47)
I need to do well in mathematics to get into the college or university of my choice	97	64	3.61 (0.56)	3.58 (0.60)	3.63 (0.52)	3.61 (0.56)	3.61 (0.56)
My parents expect me to do well in mathematics at school	100	60	3.60 (0.49)	3.60 (0.49)	3.59 (0.49)	3.53 (0.50)	3.68 (0.47)
Mathematics will assist me with my learning of other school subjects	100	58	3.58 (0.49)	3.60 (0.49)	3.57 (0.50)	3.64 (0.48)	3.53 (0.50)
I need to do well in mathematics to get the job I want	96	60	3.56 (0.57)	3.57 (0.56)	3.55 (0.58)	3.51 (0.58)	3.61 (0.56)
I would like a job that involves using mathematics	93	53	3.42 (0.73)	3.48 (0.69)	3.35 (0.77)	3.30 (0.76)	3.54 (0.69)
It is important to learn about mathematics to get ahead in the world	90	50	3.37 (0.74)	3.23 0.82)	3.50 (0.63)	3.32 (0.74)	3.41 (0.75)
My parents give me support when completing my maths homework/revision	77	27	2.82 (1.07)	2.84 (1.06)	2.80 (1.07)	2.78 (1.04)	2.86 (1.10)
My family discusses mathematics at home	64	22	2.51 (1.19)	2.49 (1.22)	2.53 (1.17)	2.52 (1.14)	2.49 (1.25)

confirmed that Ghanaian students had a significantly more positive response profile than Australian students ($p < 0.05$) on four of the ten items:

- I think learning mathematics will help me in daily life
- Mathematics will assist me with my learning of other school subjects
- I would like a job that involves using mathematics
- My parents expect me to do well in mathematics at school

Moreover, there was evidence that there were some differences in the extent to which girls and boys valued mathematics, with a general trend of girls being more

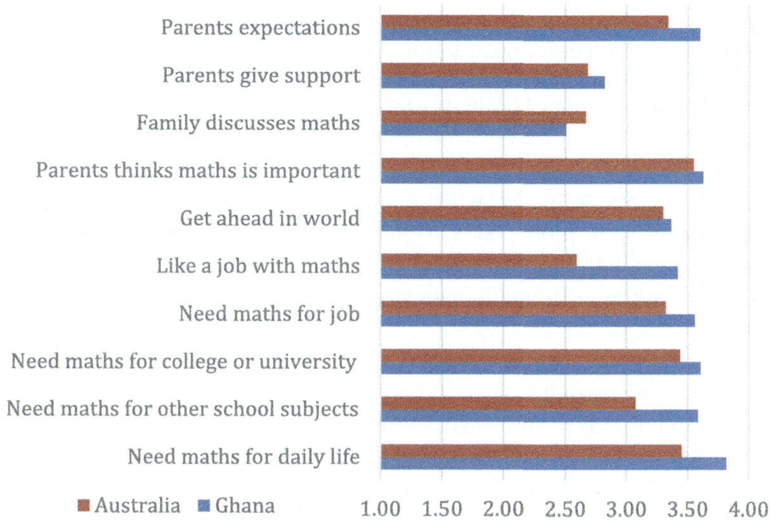

Fig. 9.1 Comparing mean scores on the VAS items: Australian versus Ghanaian students

likely to endorse VAS items than boys across both Australian and Ghanaian participant groups (see Fig. 9.2). However, a series of independent samples t-tests revealed that these differences only manifested as statistically significant in the substantially larger Ghanaian participant sample (given the associated increased statistical power). Specifically, Ghanaian girls were significantly more likely to endorse three statements than Ghanaian boys ($p < 0.05$), including:

- I think learning mathematics will help me in daily life

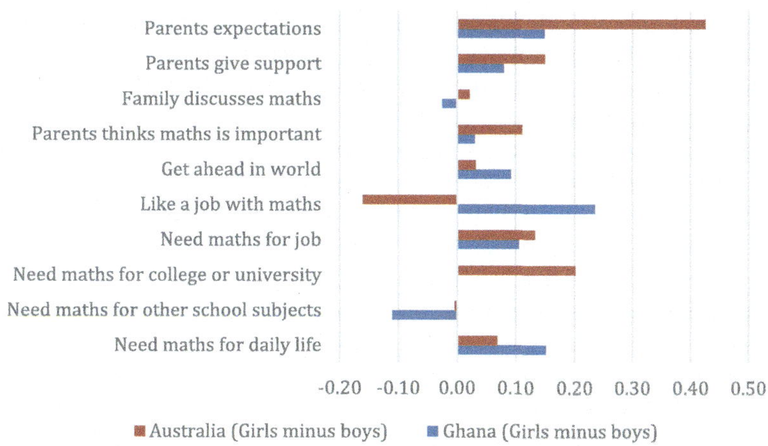

Fig. 9.2 Examining gender differences in mean scores on the VAS items: Australian versus Ghanaian students

- I would like a job that involves using mathematics
- My parents expect me to do well in mathematics at school

It is interesting that this last item regarding parental expectations also had the largest mean difference across gender amongst Australian girls and boys, and is in stark contradiction to the stereotype that parents are likely to have higher expectations of boys' performance in mathematics than girls. It is also particularly noteworthy that Ghanaian girls are more likely to agree that they would like a job involving mathematics than Ghanaian boys, given the additional stereotype that girls are less inclined to be interested in mathematical occupations and mathematics in general, compared with boys.

In contrast to gender, for the most part, stage of schooling did not appear to be a factor in influencing participants' values towards learning mathematics. Specifically, independent samples t-tests indicated no significant differences between Year 3/4 and Year 5/6 for Australian students for any of the items, whilst for Ghanaian students, the only difference was that Year 5/6 students were more likely to endorse the statement "It is important to learn about mathematics to get ahead in the world" compared with Year 3/4 students ($p < 0.05$).

9.4 Discussion

The majority of the VAS items were rated positively by primary school children, which suggests that despite their developmental stage, both Australian and Ghanaian students are aware of what is important in their mathematics learning as well as the expectation of their parents for their mathematics learning. Ghanaian primary school children's endorsement of parents valuing of students' mathematics achievement is consistent with earlier findings on values studies involving Ghanaian primary school children, which also revealed achievement as one of the main attributes primary school students' valued in their mathematics study (Davis et al., 2019b). The practice of ranking students based on their performance in examinations in various subjects including mathematics in Ghana and the prestige associated with high ranking and embarrassment associated with low ranking puts pressure on students to pass examinations, even at the primary school level. It could also be due to their awareness of the relevance of a pass in mathematics to their academic progression from the pre-senior high school level to the senior high school level in Ghana. It is, therefore, not surprising that the Ghanaian students valued achievement.

The relatively lower endorsement of parental and family support for the study of mathematics at home for Australian and Ghanaian students suggests an indirect involvement of a number of parents in their children's mathematics learning at home (Mejia-Rodriguez et al., 2020). While these parents may give their children the moral support they need to progress in mathematics, they appear not to be directly involved with mathematics homework neither do families of quite a number of children engage in mathematical discussion at home. This could be attributed to socio-economic and educational background of parents of the students who participated in this survey. In Ghana, many of the students who attend public/state-funded schools come from families with low socio-economic and educational backgrounds. Hence, the educational background of the parents and families might not permit them to engage directly in homework in mathematics or broader discussion in school mathematics. In the Australian context, a peculiar situation arises where many parents exhibit a certain degree of detachment from their child's mathematics education in specific terms. They also display limited enthusiasm for engaging in mathematical discussions within the household. Paradoxically, these same parents demonstrate a significant investment in their child's mathematical achievements. This phenomenon, previously documented in the literature, is not unique and has been observed elsewhere. For instance, Cao et al. (2007) discovered that even as Australian students perceived a decrease in parental encouragement, support, and assistance between Year 5 and Year 9, the expectations parents had regarding their children's academic performance remained relatively constant. This finding serves as a strong reminder that while teachers undeniably influence a student's mathematical values, parents equally possess a pivotal role in this regard (Harackiewicz et al., 2012). Nevertheless, the children's endorsement of valuing of their achievement by their parents suggests that parents' indirect involvement in their children's mathematics learning appears not to be enough to inform their value in mathematics learning. Mansour and Martin (2009) examined the role of parent involvement and support in students' academic disengagement and future intent. They found significant negative and positive links, respectively. Other studies have reported that greater parental involvement is associated with positive educational outcomes (Gonzalez, 2002; Pomerantz & Moorman, 2007) with more active involvement yielding the largest effects (Martin et al., 2012). Relatively less attention has been directed to the respective effects of mother/female caregiver involvement and father/male caregiver involvement. This is considered as an important area of future research considering the gender differences noticeable in this study.

It is not surprising to the researchers that the Australian and Ghanaian students valued the utilitarian value of mathematics and therefore strongly endorsed the statement "I think learning mathematics will help me in my daily life", and also valued mathematics for instrumental reasons because it will help them in their day-to-day life. This is because past and present mathematics curriculum has emphasized the utilitarian value of mathematics including its role in the daily lives of students as part of the rationale and overall aims and objectives of teaching and learning mathematics through the four proficiencies in the Australian curriculum and in Ghana (MoE, 2018, 2019, ACARA, 2019). For example, one of the aims of studying mathematics at the primary school level in the 2012 curriculum was to "help children to appreciate the value of mathematics and its usefulness to them..." (MoE, 2018, p. iv) and one of the aims of studying mathematics in the 2019 Ghanaian primary school curriculum, which is the revision of the 2012 curriculum, is to help learners to "become confident in using Mathematics to analyse and solve problems both in school and real-life situations" (MoE, 2019, vi). In an Australian curriculum context, numeracy is incorporated as a general capability and emphasises that "students become numerate as they develop the knowledge and skills to use mathematics confidently across other learning areas at school and in their lives more broadly" (ACARA, 2019). It is clear that each of these aims prepares the learners to value the role of mathematics in their daily lives. We, therefore, argue that the valuing of mathematics for utilitarian or instrumental reasons could be influenced by the aims of the primary school curriculum.

Again, Ghanaian students' value for jobs that involve using mathematics, may reflect their admiration and desire for jobs in fields such as engineering and medicine because of the remuneration and prestige associated with these jobs in Ghanaian society. This reflects the desire of most parents as well. In contrast, Australian students valued the statement "I would like a job that involves using mathematics" the least in the VAS. This difference possibly relates to the situated nature of values in a cultural context. Perhaps Australian parents do not proclaim such expectations or desires from their children, given that the egalitarian nature of the Australian society means that salaries for blue-collar jobs are comparable to those of professionals. Associated with this is the situation in Australia where one needs not go to a university to be a graduate in order for one to lead a reasonably comfortable lifestyle (Forsey, 2015).

The observation of Ghanaian students' positive valuing of mathematics might be explained by current academic literature, which suggests that countries that are often reported as performing poorly in mathematics achievement tests tend to score high in affective variables such as attitudes towards mathematics and mathematics self-concept (Mejia-Rodrigue et al., 2020). The significant differences in response rate for the items "I think learning mathematics will help me in daily life", "Mathematics will assist me with my learning of other school subjects" and "I would like a job that involve using mathematics" could be attributed to explicit exposure of the Ghanaian primary students to utilitarian values of mathematics in school. Why the positive valuing of mathematics by Ghanaian students does not translate into high achievement is not clear.

Our findings related to gender enrich current academic literature related to gender in mathematics education, given that it generally positions girls and women behind boys and men in a number of variables associated with mathematics (Mejia-Rodrigue et al., 2020). Literature suggests that boys are more likely to choose a STEM related career such as being an engineer or scientist, whereas girls seem to prefer jobs involving caring for people such as teaching or nursing (Chambers et al., 2018). However, it is perhaps not surprising that valuing in mathematics by girls is more positive than boys, given recent gender policies in STEM in Ghana have prioritised girls. As highlighted earlier in this chapter, there has been systematic effort to improve girls and women's participation in STEM programmes and careers for the past three decades. This has been achieved through mentoring and other programmes at all levels of education in Ghana that expose girls to opportunities in STEM and whip their interest compared with boys ("Report on the Science and Technology Seminar Series for Females", 2019). This might have reflected in more positive values of mathematics for girls compared with boys. By contrast to the situation in Ghana, and similar to the findings of our current study, previous research in the Australian context has found no significant gender differences in utility value (Watt, 2004, 2006).

9.5 Implications for Practice

The results from this study have implications for mathematics curriculum delivery. Specifically, it has implications for mathematics teaching and learning, and parental involvement in students' mathematics learning. Values in mathematics education, which constitute attributes of importance and worth that are internalised by students and teachers that provide them the will to maintain any course of action chosen in mathematics learning and teaching (Seah & Andersson, 2015), provide the drive for mathematics teaching and learning. Higher positive valuing of mathematics by girls as compared to boys provides a drive that could be leveraged on by teachers to boost their performance and participation in mathematics and mathematics related disciplines respectively. Mathematics teaching and learning could take on board what girls find important in mathematics education such that lesson delivery in mathematics will reflect what they value. This will help develop and maintain their interest in the subject and ultimately improve their achievement and participation in mathematics related disciplines.

Low performance of students in developing countries such as Ghana does not necessarily translate to negative valuing of mathematics. This suggests that even though literature positions classroom discourse in Ghana as being largely teacher-driven (Abenyega & Davis, 2015; Davis, 2018), the drive to succeed in mathematics is still very high among these primary school children. This positive valuing of mathematics among Ghanaian primary students offers some hope for the future in terms of students' performance in mathematics. Teachers would have to be made aware of the attributes that students' value in their mathematics learning so that they ensure that mathematics teaching and learning support what students' value. This will help students to develop self-drive to achieve in mathematics learning and become lifelong learners in mathematics.

The results highlight the need to pay attention to parental involvement in their children's/wards' mathematics education. The literature suggests a direct relationship between the level of parental involvement in their children/wards' education and the quality of their academic achievement (Lara & Saracostti, 2019). Lara and Saracostti (2019) observed from their study involving parents/guardians of primary school children in Chile that lower parental involvement translates to lower academic achievement. This calls for the need for teachers to engage parents in ways that will help them to value direct involvement in their children's/wards' mathematics learning and also equip them to do so. This will especially be important for parents of students who are at risk of dropping out of school and those with low educational backgrounds.

There is a need for studies to also explore valuing of parental involvement in children or wards' mathematics learning by parents/guardians, especially in Ghana where students' performance in mathematics has not been as good as expected. The results of the study reported in this chapter reflect the situation at the primary school level only, further research should be done at the junior high school and senior high school levels to ascertain the situation across grade levels. This will help ascertain, through international comparative study, how valuing among students in different context of schools evolve across grade levels and their implications for practice.

9.6 Limitations

Given the limited number of students involved in this study, especially from the Australian sample, it is difficult to generalise to wider contexts. The findings of this study were gathered from a small group of students and should therefore be replicated to include a larger, more diverse sample. This would also allow more sophisticated statistical techniques to be employed to address similar research questions. It would also be interesting and important to triangulate student's valuing in mathematics with classroom observations and student interviews to further validate our study's findings, and to deepen our understanding of these as well.

References

Abdullah, N. A., & Leung, F. K. S. (2019). Exploring teachers' values and valuing process in school-based lesson study: A Brunei Darussalam case study. In P. Clarkson, W. T. Seah & J. Pang (Eds.), *Values and valuing in mathematics education scanning and scoping the territory* (pp. 129–142). ICME-13 Monographs. Springer.

Australian Curriculum, Assessment and Reporting Authority (ACARA). (2019). *Australian curriculum: Mathematics*. Retrieved from https://www.australiancurriculum.edu.au/f-10-curriculum/mathematics/key-ideas/

Abenyega, J., & Davis, E. K. (2015). Exploring the intersection of English as an instructional language and inclusive pedagogy in primary mathematics classroom in Ghana. *International Journal of Whole Schooling, 11*(2), 45–64.

Ampiah, J. G., Davis, E. K., & Mankoe, J. O. (n.d.). *An investigation of provision of quality basic education in Ghana: A case study of selected schools in the Central Region*. Hiroshima, CICE, Hiroshima University. Available online at https://www.hiroshima_u.ac.jp/cice/kampala_Ghana.doc

Attard, C., & Holmes, K. (2020). It gives you that sense of hope: An exploration of technology use to mediate student engagement with mathematics. *Heliyon, 6*(1), e02945–e02945. https://doi.org/10.1016/j.heliyon.2019.e02945

Baba, T., Ueda, A., Osaka, N., Iwasaki, H., Kinone, C., Soeda, Y., & Shinno, Y. (2013). International comparative study "The Third Wave" regarding values in mathematics education I: General trends and comparative study among target groups [in Japanese]: Research in mathematics education. *Journal of Japan Academic Society of Mathematics Education, 19*(2), 127–140.

Bishop, A. J. (1996). How should mathematics teaching in modern societies relate to cultural values: Some preliminary questions. In: *Seventh Southeast Asian Conference on Mathematics Education* (Vol. 32).

Bishop, A. J. (2008). Teachers' mathematical values for developing mathematical thinking in classrooms: Theory, research and policy. *The Mathematics Educator, 11*(1/2), 79–88.

Boateng, F. K., & Gaulee, U. (2019). From studentship to academia: The academic female STEM trajectory in Ghana. *Journal of Underrepresented and Minority Progress, 3*(1), 67–86.

Cao, Z., Bishop, A., & Forgasz, H. (2007). Perceived parental influence on mathematics learning: A comparison among students in China and Australia. *Educational Studies in Mathematics, 64*(1), 85–106. https://doi.org/10.1007/s10649-006-9033-5

Chambers, N., Kashefpakdel, E. T., Rehill, J., & Percy, C. (2018). *Exploring the career aspirations of primary school children from around the world drawing the future.* Online: www.education andemploers.org.

Clarkson, P., Seah, W. T. & Pang, J. S. (Eds.) (2019). *Values and valuing in mathematics education scanning and scoping the territory.* ICME-13 Monographs. Springer.

Cobb, P., & Bauersfeld, H. (1995). *The emergence of mathematical meaning: Interaction in classroom cultures.* Lawrence Erlbaum Associates.

Corey, D. L., & Ninomiya, H. (2019). Values of the Japanese mathematics teacher community. In P. Clarkson, W. T. Seah & J. Pang (Eds), *Values and valuing in mathematics education* (pp. 89–102). ICME-13 Monographs. Springer.

Davis, E. K. (2018). Mathematics classroom discourse in typical Ghanaian public school: How does it look like? *The Oguaa Educator, 12*(1), 1–27.

Davis, E. K., Amponsah, M. O., Kwaah, C., Beccles, C., Howard, N. K., Nti, S. K., Takyi, A., & Abreh, M. K. (2019a). *An investigation into curriculum delivery in English language, mathematics and science at the basic school level in three selected districts in the Central Region of Ghana.* Cape Coast, Directorate of Research, Innovation and Consultancy, University of Cape Coast.

Davis, E. K., Carr, M. E., & Ampadu, E. (2019b). Valuing in mathematics learning amongst Ghanaian students: What does it look like across grade levels? In P. Clarkson, W. T. Seah & J. Pang (Eds.), *Values and valuing in mathematics education scanning and scoping the territory* (pp. 89–102). ICME-13 Monographs. Springer.

Davis, E. K., Ntow, F., & Beccles, C. (2022). Dichotomy between National Policy on progression of students in public junior high schools and realities of their performance: Cases of English language, mathematics and science. *SAGE Open,* 1–11. https://doi.org/10.1177/215824402211 23912

DeBellis, V. A., & Goldin, G. A. (2006). Affect and meta-affect in mathematical problem solving: A representational perspective. *Educational Studies in Mathematics, 63*(2), 131–147.

Dede, Y. (2015). Comparing primary and secondary mathematics teachers' preferences regarding values about mathematics teaching in Turkey and Germany. *International Journal of Science and Mathematics Education, 13*, 227–255.

Dede, Y. (2019). Why mathematics is valuable for Turkish, Turkish Immigrant and German students? A cross-cultural study. In P. Clarkson, W. T. Seah & J. Pang (Eds), *Values and valuing in mathematics education scanning and scoping the territory* (pp. 143–156). ICME-13 Monographs. Springer.

Ewing, B. (2011). Direct instruction in mathematics: Issues for schools with high indigenous enrolments: A literature review. *Australian Journal of Teacher Education, 36*(5), 65–92.

Forsey, M. (2015). Blue-collar affluence in a remote mining town: Challenging the modernist myth of education. *Ethnography and Education, 10*(3), 356–369. https://doi.org/10.1080/17457823. 2015.1051072

Girls' Education Unit (GEU). (2002). *Resource handbook for girls' education.* The Girls' Education Unit (Ghana).

Gonzalez, A. (2002). Parental involvement: Its contribution to high school students' motivation. *Urban Education, 75*, 132–135.

Harackiewicz, J. M., Rozek, C. S., Hulleman, C. S., & Hyde, J. S. (2012). Helping parents to motivate adolescents in mathematics and science: An experimental test of a utility-value intervention. *Psychological Science, 23*(8), 899–906. https://doi.org/10.1177/0956797611435530

Hill, J., Hunter, J., & Hunter, R. (2019). What do Pāsifika students in New Zealand value most for their mathematics learning? In P. Clarkson, W. T. Seah & J. Pang (Eds), *Values and valuing in mathematics education scanning and scoping the territory* (pp. 103–114). ICME-13 Monographs. Springer.

Hill, J. L., Kern, M. L., Seah, W. T., & van Driel, J. (2021). Feeling good and functioning well in mathematics education: Exploring students' conceptions of mathematical well-being and values. *ECNU Review of Education, 4*(2), 1–27. https://doi.org/10.1177/2096531120928084

Hill, J. L., & Seah, W. T. (2023). Student values and wellbeing in mathematics education: Perspectives of Chinese primary students. *ZDM–Mathematics Education, 55*(2), 385–398. https://doi.org/10.1007/s11858-022-01418-7

Howell, D. C. (2002). *Statistical methods for psychology* (5th ed.). Duxbury.

Kalogeropoulos, P., Roche, A., Russo, J., Vats, S., & Russo, T. (2021). Learning mathematics from home during COVID-19: Insights from two inquiry-focused primary schools. *Eurasia Journal of Mathematics, Science and Technology Education, 17*(5).

Lara, L., & Saracostti, M. (2019). Effect of parental involvement on children's academic achievement in Chile. *Frontiers in Psychology, 10*, 1464. https://doi.org/10.3389/fpsyg.2019.01464.

Law, H. Y., Wong, N. Y., & Lee, N. Y. L. (2011). The third wave studies of values in effective mathematics education: Developing students' mathematical autonomy in classroom learning. *The Mathematics Educator, 13*(1), 72–86.

Leder, G. C. (2015). Gender and mathematics education revisited. In S. J. Cho (Ed.), *The Proceedings of the 12th International Congress on Mathematical Education* (pp. 145–151). Springer.

Lim, C. S. (2015). Riding the third wave: Negotiating teacher and students' value preferences relating to effective mathematics lessons. In *Selected Regular Lectures from the 12th International Congress on Mathematical Education* (pp. 471–485). Springer Science + Business Media.

Little, J., & Anderson, J. (2015). What factors support or inhibit secondary mathematics pre-service teachers' implementation of problem-solving tasks during professional experience? *Asia-Pacific Journal of Teacher Education, 44*(5), 504–552.

Maloney, E. A., Ramirez, G., Gunderson, E. A., Levine, S. C., & Beilock, S. L. (2015). Intergenerational effects of parents' math anxiety on children's math achievement and anxiety. *Psychological Science, 26*(9), 1480–1488.

Mansour, M., & Martin, A. J. (2009). Home, parents, and achievement motivation: A study of key home and parental factors that predict student motivation and engagement. *Australian Educational and Developmental Psychologist, 26*, 111–126.

Martin, A. J., Anderson, J., Bobis, J., Way, J., & Vellar, R. (2012). Switching on and switching off in mathematics: An ecological study of future intent and disengagement among middle school students. *Journal of Educational Psychology, 104*(1), 1–18. Online: https://doi.org/10.1037/a0025988

Mejia-Rodriguez, A. M., Luyten, H., & Meelissen, M. R. M. (2020). Gender differences in mathematics self-concept across the world: An exploration of student and parent data of TIMSS 2015. *International Journal of Science and Mathematics Education, 19*, 1229–1250.

Ministry of Education (MoE). (2012). *National syllabus for mathematics (primary school 1–6).* Curriculum Research and Development Division (CRDD).

Ministry of Education (MoE). (2018). The Ghana Education Reform Agenda. Available online at https://www.moe.gov.gh/index.php/education-reform/

Ministry of Education (MoE). (2019). *Mathematics curriculum for primary schools.* National Curriculum and Assessment (NaCCA).

Mullis, I. V. S., Martin, M. O., Foy, P., & Arora, A. (2012). *TIMSS 2011 international results in mathematics.* TIMSS & PIRLS International Study Center.

Mullis, I. V. S., Martin, M. O., Foy, P., Kelly, D. L., & Fishbein, B. (2020). *TIMSS 2019 international results in mathematics and science.* Retrieved from Boston College, TIMSS & PIRLS International Study Center. Online: https://timssandpirls.bc.edu/timss2019/international-results/

National Gender Policy. (2015). *Ministry of gender and social protection.* Ghana.

Pomerantz, E., & Moorman, E. (2007). The how, whom, and why of parents' involvement in children's academic lives: More is not always better. *Review of Educational Research, 77*, 373–410. https://doi.org/10.3102/003465430305567

Report on the Science and Technology Seminar Series for Females in First Cycle Institutions (2019). *Phase II, held at College of Sciences*. Kwame Nkrumah University of Science and Technology (KNUST), from 20 to 21 June, 2019.Online: https://dfcentre.com/wp-content/uploads/2021/03/REPORT_Science-and-Technology-for-Girls-II-2019.pdf

Seah, W. T. (2010). The 'third wave': Values associated with effective learning of mathematics in Australian primary classrooms. *Research in Mathematics Education, 14*(2), 99–114.

Seah, W. T. (2018). Improving mathematics pedagogy through student/teacher valuing: Lessons from five continents. In G. Kaiser, H. Forgasz, M. Graven, A. Kuzniak, E. Simmt, & B. Xu (Eds.), *Invited lecturers from the 13th international congress on mathematical education* (pp. 561–580). Springer. https://doi.org/10.1007/978-3-319-72170-5_31

Seah, W. T. (2020). *Study 5 the values alignment study: Discussion paper*. Unpublished manuscript.

Seah, W. T., & Andersson, A. (2015). Valuing diversity in mathematics pedagogy through the volitional nature and alignment of values. In A. Bishop, H. Tan, & T. Barkatsas (Eds.), *Diversity in mathematics education: Towards inclusive practices* (pp. 167–183). Springer. Online at: https://doi.org/10.1007/978-3-319-05978-5_10

Seah, W. T., Andersson, A., Bishop, A., & Clarkson, P. (2016). What would the mathematics curriculum look like if values were the focus? *For the Learning of Mathematics, 36*(1), 14–20.

Seah, W. T., Baba, T., & Zhang, Q. (2017). The WIFI study students in Hong Kong and Japan. In J. W. Son, T. Watanabe, & J. Lo. (2017). (Eds.). *What matters? Research trends in international comparative studies in mathematics education, research in mathematics education* (pp. 333–354). Springer.

Seah, W. T., & Ho, S. Y. (2009). Values operating in effective mathematics lessons in Australia and Singapore: Reflections of pre-service teachers. In M. Tzekaki, M. Kaldrimidou & H. Sakonidis (Eds.), *Proceedings of the 33rd Conference of the International Group for the Psychology of Mathematics Education* (Vol. 5, pp. 57–64). PME.

Seah, W. T., Kim, H. J., & Kim, D. J. (2021). Reimagining Mathematics Education for the 21st Century in the 21st Century. *Journal of Educational Research in Mathematics, 31*(3), 393–404. https://doi.org/10.29275/jerm.2021.31.3.393.

Senk, S. L., & Thompson, D. R. (Eds.). (2003). Standards-based school mathematics curriculum. What are they? What do students learn? Lawrence Erlbaum.

Stephens, M. (2014). The Australian curriculum: Mathematics—How did it come about? What challenges does it present for teachers and for the teaching of mathematics? In Y. Li & G. Lappan (Eds.), *Mathematics curriculum in school education* (pp. 157–176). Springer.

Thomson, S., Wernert, N., Buckley, S., Rodrigues, S. O. & Grady, E., & Schmid, M. (2021). *TIMSS 2019 Australia. Volume II: School and classroom contexts for learning*. Department of Education, Skills and Employment. Online: https://research.acer.edu.au/cgi/viewcontent.cgi?article=1003&context=timss_2019

UNICEF. (2018). *Providing a safe and protective environment for the child: Additional training module & activity facilitation guide on adolescence, sexual and reproductive health, gender, and sexual and gender based violence (SGBV)*. UNICEF Ghana.

Vale, C., & Herbert, S. (2021). Secondary preservice teachers' anticipated objectives and practices for teaching mathematics. *Mathematics Teacher Education and Development, 23*(4), 5–28.

Wang, C. S. (2017). Restoring the authority of all-round development of education—Discussion on "transition from exam-oriented education to quality-oriented education." *Contemporary Teacher Education, 10*(1), 6–28.

Watt, H. M. G. (2004). Development of adolescents' self-perceptions, values, and task perceptions according to gender and domain in 7th- through 11th-grade Australian students. *Child Development, 75*(5), 1556–1574. https://doi.org/10.1111/j.1467-8624.2004.00757

Watt, H. M. G. (2006). The role of motivation in gendered educational and occupational trajectories related to maths. *Educational Research and Evaluation, 12*(4), 305–322. https://doi.org/10.1080/13803610600765562

Wrigley-Asante, C., Ackah, C. G., & Frimpong, L. K. (2021). Career aspirations and influencing factors among male and female students studying science technology engineering and mathematics (STEM) subjects in Ghana. *Ghana Journal of Geography, 14*(1), 85–103.

Zhang, Q., Barkatsas, T., Law, H., Leu, Y., Seah, W. T., & Wong, N. (2016). What primary students in Chinese Mainland, Hong Kong and Taiwan value in mathematics learning: A comparative analysis. *International Journal of Science and Mathematics Education, 14*, 907–924.

Zhong, C., & Xie, J. (2022). The dilemma and countermeasures for long-term implementation of the "double reduction" policy—From perspective of policy implementation networks. *China Academic Journal Electronic Publishing House, 07*, 64–72.

Chapter 10
Dominant Values and Value Shifts: The Valuing Process Seen from Curriculum Levels

Nor Azura Abdullah

10.1 Introduction

By focusing on the teachers' values, we only know how to effectively develop them as agents of change in their classrooms by understanding their orientations toward mathematics (Seah, 2018). Why do we need them as an agent of change? In Brunei's context, a small Muslim nation located in Southeast Asia, the nation has undergone an educational system reform starting in 2009. One of the reform aspects in the curriculum is geared towards a student-centered approach that focuses on students' thinking skills and higher mathematics processes in preparing them for life in the twenty-first century. One of the Ministry of Education's initiatives is to implement the change through the professional development of the teachers using Japanese Lesson Study.

Lesson study in Japan has been widely practiced and has become part of the school's professional development of teachers. As it was popularised as a successful practice for teacher learning, Brunei is also one of the countries that embarked on this novel approach. The interest in Lesson Study is a result of the new implementation of education reform in the tiny nation, known as SPN21 Sistem Pendidikan Negara Abad ke-21 translated as The National Educational System for the 21st Century in 2009. With the new educational system, updated curriculum, and syllabuses, the Ministry of Education initiated a Lesson Study to help convey and practice the new curriculum to teachers. The Lesson Study model is considered well-suited for assisting teachers in adopting the new student-centered approaches featured in the updated curriculum, as it emphasizes careful observation and examination of students' learning experiences.

Like Japan, Brunei has a core mathematics curriculum mandated by the Ministry of Education. However, unlike Japan, Brunei has yet to fully embrace the collaborative aspect of professional development and a classroom-based approach championed by

N. A. Abdullah (✉)
Universiti Brunei Darussalam, Gadong, Brunei Darussalam
e-mail: azura.abdullah@ubd.edu.bn

Y. Dede et al. (eds.), *Values and Valuing in Mathematics Education*,
https://doi.org/10.1007/978-981-99-9454-0_10

the teachers, which is the principle of Lesson Study. In Brunei, teacher professional development is typically practiced individually or through a self-upgrading system. Classroom observations are often synonymous with the assessment of teachers' practices. Therefore, introducing Lesson Study is a welcomed practice that may reposition teachers' traditional outlook of professional development as individual work, a top-down approach, and outside the classroom context to a collaborative effort, teacher-initiated and inside the classroom settings. The Ministry of Education included Lesson Study as one of its grand initiatives under its strategic planning for 2012–2017 as their commitment to promoting Lesson Study to enhance teachers' teaching quality and professional development (DPDR, 2012). To ensure the successful implementation of Lesson Study in Brunei, collaboration was established between key officers in the Department of Schools in the Ministry of Education, teacher educators, Lesson Study Head Mentors, and school leaders to oversee the practice (Khalid, 2016).

This study explored teachers' values associated with mathematics teaching that emerged from schools practicing Lesson Study. The research questions were: (i) What dominant values emerged in the two schools? and (ii) What factors influenced the schools' Lesson Study practices that contributed to developing their respective dominant values?

10.1.1 Lesson Study Adoption

Lesson Study process of collaborative planning, observing live lessons, and discussing the lessons taught can facilitate teachers to "create changes in teachers' knowledge and beliefs, professional community and teaching–learning resources" (Lewis et al., 2009, p. 286). By carefully studying lessons that they collaboratively create, practice, and evaluate, teachers may experience changes in terms not only of their content and pedagogical knowledge but also in their dispositions and the communal relationship among them. However, these changes might be influenced by the method of Lesson Study adoption in Brunei and how it contrasts with traditional professional development practices. Kennedy (2005) has described a range of Continuous Professional Development (CPD) models, which could provide valuable insight into the implementation process in Brunei, as illustrated in Table 10.1.

Furthermore, in a study looking at Lesson Study practices outside Japan, Fujii (2014) found that the practice needs to be more superficial and give the outcome as intended. The main component of Lesson Study practice that may be absent in its replication is the values tied to Lesson Study's features. He stated that the Japanese Lesson Study shows "that the consideration of educational values is always tied to, influenced by, and reflected in, the key features of Lesson Study" (Fujii, 2014, p. 78). This is illustrated in Fig. 10.1.

While Fujii (2014) did not explicitly specify the domain of the values, it might not be limited to the affective domain. This inference is based on his proposition that values influence teachers' instructional practices, which inherently involve cognitive

Table 10.1 The spectrum of CPD models (Kennedy, 2005, p. 248)

Model of CPD	Purpose of model	
The training model The award-bearing model The deficit model The cascade model	Transmission	Increasing capacity for professional autonomy
The standards-based model The coaching/mentoring model The community of practice model	Transitional	
The action research model The transformative model	Transformative	

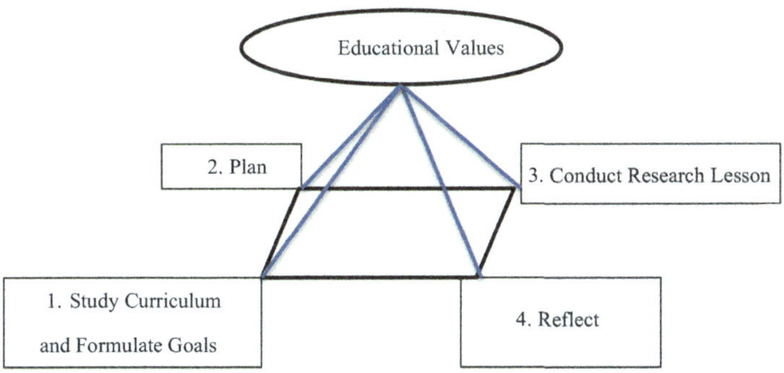

Fig. 10.1 Pyramid model of lesson study. *Note* Reprinted from implementing Japanese lesson study in foreign countries: misconceptions revealed (p. 78) by Fujii (2014)

processes. Seah (2008) acknowledged that viewing values solely as an affective construct could be limiting in the context of mathematics education. He proposed that valuing should be considered a conscious cognitive process involving making choices and decisions while recognizing that our internal emotions and assessments can influence what we value. In other words, valuing is a socio-cultural cognitive process that needs to account for both affective and cognitive aspects. Thus, there is a need for research values that are linked to teachers' decisions during Lesson Study.

10.1.2 Conceptual Framework

In this study, the values considered are whether they are espoused or enacted by the teachers during the teaching of mathematics. Espoused values are based on the teachers' talk when they are discussing with each other about the mathematics lessons or based on the interview responses with the researcher. The interview questions asked the teachers their perspectives on mathematics teaching and learning. The interview items also used hypothetical situations in mathematics lessons to make teachers' preferences in their practices apparent. Enacted values are values the researcher interpreted from the teachers' actions, direct and indirect verbal cues derived from their underlying beliefs and attitudes. According to Raths et al. (1987), for values to be formed, one's own beliefs and attitudes must go through several stages of valuing. This is done by looking at the beliefs and attitudes of the teachers since these affective and cognitive constructs can be considered indicators of values. At the earliest formation of valuing is the ability for a person to choose a belief or an attitude. Raths and colleagues explained that choosing can be further expanded to three processes: the ability to choose voluntarily, to choose from available options, and to choose carefully. The second stage of valuing is to prize the chosen belief or attitude. Prizing is a valuing process where an individual is seen to place importance on the chosen belief or attitude. The final process is to act on the prized belief or attitude. Looking at these stages of value formation and the Lesson Study process, the conceptual framework of this study was formed.

10.2 Methodology

This study follows two groups of mathematics teachers from two different schools, identified as School A and B, who were part of the Lesson Study initiatives by the Ministry of Education. The research design for this study was qualitative research involving multiple case studies. A naturalistic and descriptive narrative focused on teachers' learning experiences in situ. The researcher established non-participatory relationships with the participants in this study. Two schools with experience in Lesson Study participated for one whole academic year. The participants are a group of mathematics teachers in each school. Two Lesson Study cycles were done for each school at the beginning and toward the end of the academic year. All the phases of the Lesson Study were both video and audio recorded. Artifacts used during the sessions were collected through lesson plans, students' worksheets, teachers' observation forms, and classroom observation field notes. This study employed multiple observations, thus making it within-case and between-case observations (Gerring, 2006). This is illustrated in Fig. 10.2.

The school-based Lesson Study process can provide an explicit look into teachers' values and valuing process through teachers' decision-making of their instructional practices. A thematic analysis of the transcribed teachers' discussions and classroom observations was performed chronologically in the Lesson Study process, from planning meetings and research lesson implementations to post-lesson meetings. These processes also coincide with the intended, implemented, and attained curriculum stages. Specifically, teachers' manifestation of observable and inferred

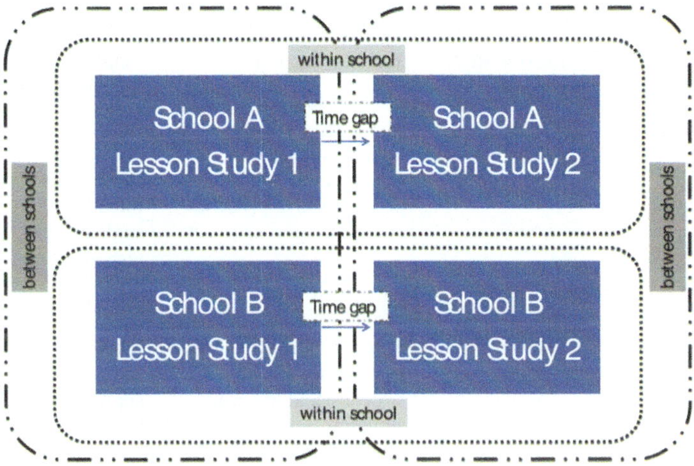

Fig. 10.2 Multiple-case study design. *Note* The representation of School A and School B lesson study cycles in a multiple-case study design

values was explored at these stages and termed as teachers' intended value indicators, implemented value indicators, and attained value indicators. These are shown in Table 10.2.

10.2.1 Schools Context

10.2.1.1 School A

School A is a government school located in the urban area in Brunei. It is a relatively large school with a student population of 600. This school was one of the pioneers in implementing Lesson Study on their own and continued to support the Ministry's initiatives with its involvement. Few of the mathematics teachers were the original members of the project, whereas the other mathematics teachers newly participated in lesson study. Their school leader, who was new to the school, was also unfamiliar with Lesson Study. The school leader authorized two Year 4 classes to participate in the study. Six mathematics teachers participated, consisting of two implementers and three observers. Regarding the highest qualifications, half of the group had a Diploma in Education, and the other half had a Bachelor in Education. Four of the teachers had teaching experience of more than 10 years, and the other two had more than 20 years. The topics for Lesson Study Cycle 1 were the Addition and Subtraction of Fractions and the Addition of Decimals for Lesson Study Cycle 2.

Table 10.2 Conceptual framework of the valuing process in lesson study

Valuing process	Lesson Study process	Teachers' actions	Value Indicators	Values
Choosing freely Choosing from alternatives Choosing after thoughtful consideration of the consequences of each alternative	Study curriculum and formulate goals	Teachers choose the mathematics lesson objectives and pedagogical approaches	Intended pedagogical approaches Intended students' responses	Espoused
Prizing and cherishing affirming	Plan lesson	Teachers produce a lesson plan with a chosen structure and learning activities	Intended pedagogical approaches Intended students' responses	Espoused
Acting upon choices	Conduct research lesson	Teachers act on the chosen learning activities and carry out the planned pedagogical approaches	Implemented pedagogical approaches Implemented students' learning activities	Enacted
Repeating	Reflect	Teachers discuss and highlight the intended and implemented learning activities and students' responses	Attained pedagogical approaches Attained students' responses	Espoused

10.2.1.2 School B

School A is a government school located in a residential urban area in Brunei. It is a relatively small school with a student population of 300. This school was one of the chosen schools for implementing a school-based Lesson Study identified by the Ministry of Education for its Professional Learning Community initiative in 2012. Few of the mathematics teachers were the original members of the project. During the study, a new school leader was in position and not familiar with the Lesson Study. The school leader only permitted one Year 4 class to participate. Nine mathematics teachers participated, consisting of one implementer and eight observers. Regarding highest qualifications, three observers had a Certificate in Education, two teachers had a Diploma in Education, three teachers with a Bachelor in Education, and one teacher with a Master's in Education. Three teachers were considered beginning teachers with less than 10 years of experience, which included the teacher implementer, two teachers with less than 20 years, and four teachers were considered senior teachers with more than 20 years of teaching experience. Similar to Lesson Study Cycle 1 from School A, the lesson focused on the Addition of Fractions, while Lesson Study Cycle 2 focused on the Area of Squares and Rectangles.

10.2.2 Coding Procedure

Value indicators were extracted from decisions made during teacher discussions before and after the lessons and during research lessons implementation. These transcribed talks were coded inductively and deductively in a repeated manner. Some of the teachers' conversations indicated their preference for students' actions. However, not all teachers' talks were clearly expressed. When this occurred, the researcher tried to infer the teachers' preferences based on teachers' interactions, among themselves or with students, using the coding scheme as a reference. Thus, the coding of the value indicators is done directly or by inferring the manifestation of values. Using deductive logic, the value indicators are based on the mathematical values coding by Bishop (2001), as presented in Table 10.3. Using inductive reasoning, a more grounded-theory procedure of constant comparative method (Creswell & Poth, 2018) was used. The study on values in mathematics teaching in Brunei was not done at the time of this study. Since teachers' values in Brunei are an unexplored territory, emergent themes were expected. This process involved the memoing process, open coding, and axial coding. The teachers' values in lesson planning and post-lesson discussion were referenced based on the thematic coding done. It started with open coding inductively to infer the emergent themes. Teachers' talk was transcribed, and open, axial coding was applied to the transcriptions.

Table 10.3 Mathematical values coding and teacher's decisions on instructional practices

Component	Values	Student actions
Ideological	Rationalism	i. Explain ii. Argue or iii. Show mathematical proof
	Objectism	i. Show ii. Use diagrams or iii. Concretizing mathematics ideas
Sentimental	Control	i. Understand the process of routine calculations or ii. Check their answers and justify them
	Progress	i. Explore ideas beyond given examples
Sociological	Openness	i. Present or ii. Defend their ideas with the whole class
	Mystery	i. Explore their imagination on the wonder of mathematical ideas

Note: Adapted from *What values do you teach when you teach mathematics?* by Bishop (2001)

10.3 Findings and Discussion

The study used the conceptual framework in Table 10.2 to analyze and synthesize the data. Based on the inductive and deductive data analysis, there are values identified from Schools A and B, where the values emerged to be prominent in terms of occurrences and constancy. These values are classified as dominant values. Both schools shared some dominant values, but they also had distinct differences. The same values are identified as shared dominant values, and the dissimilar values are classified as discrepant dominant values. The details of the dominant values shared and discrepant are shown in Table 10.4.

10.3.1 Dominant Values

The dominant values are the emergent values identified from the value indicators classified in the Lesson Study process in the Schools A and B cycles. The value indicators are the espoused and enacted values, the observable incidents based on teachers' interactions with each other or their students. The teachers' values are considered dominant based on the consistency of espoused and enacted values shown in the three stages of the Lesson Study. These recurring occurrences demonstrated the importance of such values based on the consistent and repeated presence in teachers' discussions and actions. To illustrate this, Table 10.5 offers a sample of School A teachers espoused and enacted values across different stages of the Lesson Study.

10.3.1.1 Shared and Discrepant Dominant Values

The shared dominant values are those values that are observed to be present in both schools. This indicates that teachers place these values in mathematics classrooms in their practices. In comparison, the discrepant dominant values explain the difference between the two schools of their dominant values. In a common curriculum, such as Brunei's mathematics curriculum, it is interesting to learn the common values practiced by the teachers and received by the students.

Table 10.4 Teachers' dominant espoused and enacted values

Shared dominant values	Discrepant dominant values	
Schools A and B	School A	School B
Rationalism Objectism Control Relational understanding Instrumental understanding Guided discovery	Seatwork, drill, and practice Students' own construction of knowledge	Teacher's exposition I-R-F interaction with students

Table 10.5 A sample of School A teachers' value indicators at different lesson study phases

Lesson study stage	Incident	Instructional strategy	Reason	Value indicator
Lesson planning discussion	Teaching addition and subtraction of fractions	Use of concrete materials to teach addition and subtraction of fractions	Bar models and pizza models to help students to 'see' the process of conversion of fractions	Valuing the mathematical value of Objectism
Lesson implementation	Lesson development on addition and subtraction of fractions	To use concrete manipulatives as an aid for lesson development and solve problems	To let students see and experience their own learning	Valuing the mathematical value of objectism
		To get students to present their work in front of the whole class	To let students communicate their thinking process verbally	Valuing communication from students Valuing the mathematical value of rationalism
Post-lesson Discussion	Focusing on Students' mistakes and misconceptions on addition and subtraction of fractions	Less time on manipulations of concrete or diagrammatic materials and more practise with calculations	It was too consuming to get students to see the concepts or discover the concept on their own	Valuing students' procedural skills Valuing end-product from students Valuing the mathematical value of control

Note This table provides a representative and not the complete list

In mathematics teaching, the lessons observed during the Lesson Study cycles have shown few shared dominant values between the two schools. These shared values are mathematical values of rationalism, objectism, and control; the mathematics educational values of relational and instrumental understanding and guided discovery. On the other hand, the values that stood out as being different were those related to mathematics educational values, such as instructions that value students providing fixed responses and the teacher delivering contents. While these were shared and discrepant values among the teachers, the data showed that there was a shift in focus toward other values during the process of valuing.

Based on teachers' discussions and lesson observations, the researcher inferred that, in terms of mathematical values, teachers from both schools suggested that they value both rationalism and objectism in their teaching. Rationalism value is illustrated when teachers planned to have students present their work to their peers as included in the lesson plans. In practice, this was observed in the classroom lessons where both teachers in Schools A and B encouraged their students to explain, discuss, and justify their mathematical workings. However, implementing such practices posed difficulties in both schools. While teachers encouraged students to present to explain and express their thinking processes in School A, they also incorporated Concrete-Pictorial-Abstract (C-P-A) based learning activities. Contrastingly, in School B, the main intended activity was for students to explain their thought processes. Yet, teachers there struggled to prompt students to articulate their mathematical understanding of concepts. As a compromise, they instead encourage students to demonstrate and explain their processes, whether through concrete methods, imagery, or symbolic representations. A commonality is that both schools value students communicating their thinking processes verbally. This emphasis is aligned with the advocacy of students' communication skills in educational reform planning as well as found practiced in the previous studies of Brunei's Lesson Study initiatives (Musa, 2015; Suhaili, 2018). The excerpts below showed the teacher in School A's struggle when getting students to articulate their reasoning when asked to reason for the addition of fractions involving the denominators.

63	T:	Okay, very good. I have a question. Okay, now, why is the denominator still the same?
64	S:	Because the denominator is the same-
65	Ss:	(mumbles)
66	T:	Why?
67	Ss:	(mumbles)
68	T:	Okay, how will we prove that the denominator still equals four?
69	Ss:	(mumbles)
70	S:	Because if you add 4 with 4, it will be different
71	T:	Will it be different? Equal to what?
72	Ss:	Eight
73	T:	Eight. Okay, now, does this come from one whole?
74	Ss:	Yes
75	T:	Okay, **can you use a diagram to explain** how you get three-quarters? Can you draw now? Draw. Can you draw now, okay? Explain how you get three quarters

Based on Bishop's (1988) and Aktas and Argün's (2018) definitions, teachers were observed to value control when they encouraged students to practice rote sequence and assigned a targeted outcome to develop their mathematical procedural skills. Teachers in School A were observed to value control by encouraging students to practice routine calculations, demonstrating a sense of urgency to develop students' understanding of procedural skills and target specific outcomes. Meanwhile, in School B, teachers adjusted their approach to several cycles of the initiate-response-feedback (I-R-F) pattern when students displayed their work with little or no explanation. This approach, classified as a sentimental value component of control by Bishop (1988), involved encouraging students to check and justify their answers with feedback provided by the teacher. Although both schools were observed to value control, School A focused on seatwork, drill, practice, and students' own construction of knowledge. In contrast, School B focused on the teacher's exposition and the I-R-F interaction with the students. The excerpt below shows one of the I-R-F interactions between the teacher and students.

61	T:	Okay, now, this one, we're using your fraction strips We will use your pizza fraction, who can do it in front? Okay, Student E
62	S.E:	(solving quietly in front)
63	T:	Show me which one is two out of eight
64	S.E:	(pointing to the fractions) (inaudible)
65	T:	Okay, show me three out of eight
66	S.E:	(pointing to the fractions) (inaudible)
67	T:	I need two out of eight. Okay, that is three. Two? Two out of eight?
68	S.E:	(pointing to the fractions) (inaudible)
69	T:	Anyone? Okay, right (inaudible)
70	T:	One, two, what's the fraction?
71	Ss:	Two out of eight. One, two, three. Three out of eight

Schools A and B intended their lesson activities to alternate between relational and instrumental understanding. During School A Lesson Study cycle 1, teachers aimed to promote students' relational understanding of adding fractions using a guided discovery approach, but this proved challenging for students. Teachers then switched to focusing on routine procedural steps for adding unlike fractions and reinforced this with drill and practice exercises. Teachers reflected on the difficulties of the relational understanding activities during post-lesson discussions, demonstrating a transition between relational and instrumental understanding values throughout the curriculum stages. Similarly, teachers in school B alternated between relational and instrumental understanding, but there was a disparity between the teacher implementer and teacher observers. Teacher implementer planned guided discovery activities to develop students' conceptual skills, while teacher observers preferred more direct activities to practice routine calculation. This resulted in the teacher implementer struggling to direct students to guided discovery activities and re-adjusted to exposition and initiate-response-feedback approaches. The study found it difficult to sustain intended pedagogical values in the implementation stage, which is possible with values clarification at each stage of the Lesson Study. Below are the excerpts from the teacher observers' discussion after the lesson implementation.

118	TO Nicole:	This student was still confused, listing the multiples {inaudible} to choose the right multiples to be used
119	TO Nikita:	
120	TO Nicole:	Yes. This shows that the students have not grasped the concept of fraction
123	TO Nikita:	That is why more explanation must be done

(continued)

(continued)

124	TO Nicole:	Aa, yes, so what shall be done (to improve)?
125	TO Nikita:	For equivalent fraction, (needs) more exposure and explanation (from the teacher)
415	TO Jenny:	Yes, no diagrams used. Mainly, we introduce to the students the LCM. First, they write that, unlike fractions, the denominator is different, so **how to change it**? That is why we need to introduce the LCM method
418	TO Nikita:	But it seems like we are stressing the use of the multiples in a fraction
419	TO Jenny:	How were the techniques being developed?
420	TO Nikita:	Need more exposure
421	TO Jenny:	Emphasize again **how to change unlike to like fractions**
422	TO Nikita:	And also the equivalent fraction
425	TO Jenny:	Emphasise more on how to find the answer
426	TO Nicole:	On the use of LCM on unlike fraction
427	TO Jenny:	Emphasize the use of LCM to change like to unlike fraction

10.3.2 Value Shift

In the case of School A's Lesson Study Cycle 1, value indicators shifted from planning to implementation to post-lesson discussion regarding the values of rationalism and objectism in teaching addition and subtraction of fractions. Teachers focused on the objectism value in the intended curriculum stage, prioritizing concrete activities such as paper folding and drawing fractional parts. The values of rationalism and objectism coexisted, but one was more emphasized than the other, as shown in Fig. 10.3.

Another case is where School A teachers intended for students to "see" the concept of adding or subtracting fractions through the C-P-A activities, with the value of students constructing their knowledge through discovery. However, in the implemented lessons, teachers reviewed the process by asking students to show their work and teachers aided by providing explanations rather than having students explain their mathematical thoughts as illustrated in Fig. 10.4.

Several factors might account for the instability and variability in teachers' values. Fujii (2014) previously noted that teachers' educational values could be influenced by the key features of Lesson Study. Another factor not directly related to Lesson

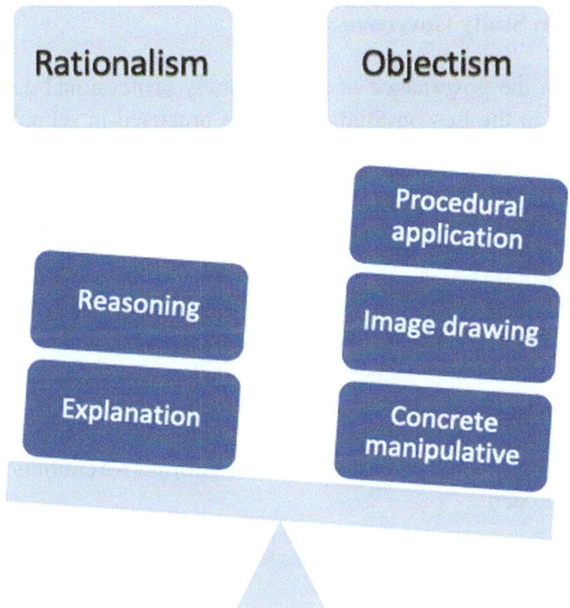

Fig. 10.3 Rationalism and objectism co-existing concurrently

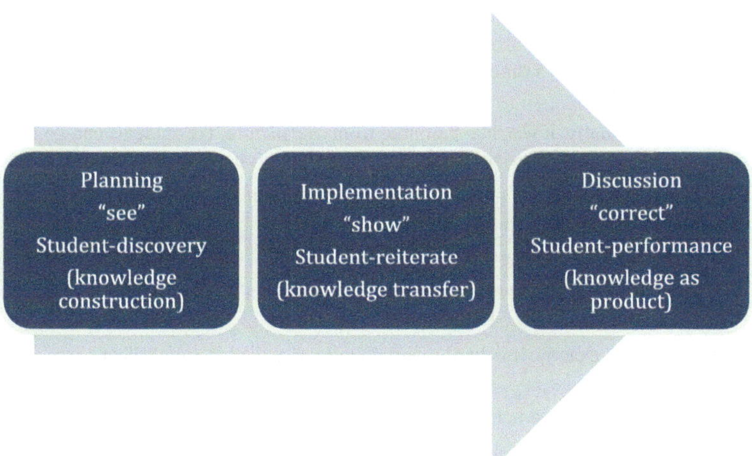

Fig. 10.4 School A's valuing process of C-P-A approach in teaching mathematics

Study is the classroom environment. Consequently, the prevailing values and shifts in the two schools might stem from distinct differences in their governance and conduct of lesson study activities and from variations in classroom environments.

10.3.2.1 Lesson Study Governance

The difference in the governance of Lesson Study professional development may have contributed to the Lesson Study approach practised in school. School A was given full autonomy by their school leader, where teachers were given the freedom to practice and conduct their Lesson Study with minimal interruption. Such a situation was not apparent for school B, where the school leader restricted the Lesson Study practices and placed importance on school matters over Lesson Study activities.

Based on their observed discussions, the teachers at School A were highly proactive. They would question each other and justify their opinions, particularly during lesson planning and post-lesson evaluations. The teachers acted as researchers, evaluating the instructional approaches they intended to implement and reflecting on student learning outcomes. As a result, referring to Kennedy's (2005) CPD model in Table 10.1, the Lesson Study approach adopted by School A fell somewhere between a transitional model and a promising transformative continuous professional development approach.

506	TI Melinda:	And for me, we cannot do this lesson in one day. I will do it for two days
507	TI Ida:	We were confident to do it in one day
508	TO Rosanna:	Luckily, yes. That's maybe (for) a good teacher, you know?
509	TI Melinda:	For me, I will do it in two days. Let us work on their concept, the concrete ones, pictorials…
510	TI Ida:	There was a time we did it straight away, but why did we discuss doing it this way?
511	TO Rosanna:	Actually, it was like that, like from one picture, the students can go to the next one, they can visualize it like an (inaudible) triangle -
512	TI Melinda:	But if they are smart, they can get it easily if we think about the average ones, the low ones. How?
513	TI Ida:	We are thinking about them
514	TO Rosanna:	(inaudible) That is why sometimes, they can't. It is complex for them to think about, like my 5A class
515	TI Melinda:	Yes, exactly. It is hard to tell them to visualize it, right? That's it
516	TI Ida:	Do you need help to visualize?
517	TO Rosanna:	To visualize is already hard

School B's Lesson Study approach combined the coaching/mentoring and community of practice models. The teachers primarily followed the coaching/mentoring model, where they exchanged opinions and compared their instructional practices. However, the Lesson Study practices extended beyond the one-to-one interaction of coaching/mentoring to include a group of teachers planning and teaching lessons. Despite this, Kennedy (2005) cautioned that passive learning in a community of practice model may occur, and this was inferred to take place in School B, as the teacher implementer was observed as a passive learner. Therefore, School B's Lesson Study approach can be classified as a transitional model of continuous professional development.

10.3.2.2 Lesson Study Activity

The paragraph describes the differences in the Lesson Study structure between School A and School B. School A had a more open, unguided focus on students" learning in their post-lesson discussions, with teachers focusing on evaluation and reflections. In contrast, School B had a more guided focus, with teachers using observation and post-lesson discussion forms to evaluate students" learning and make recommendations for improvement. Additionally, the nature and content of the conversations in their meetings differed, with School A being more informal and collaborative and School B being more formal and professional.

10.3.2.3 Classroom Environment

Students' communication skills—explaining their reasoning or mathematical thinking. Students have difficulty explaining their mathematical thinking beyond explaining the steps or procedures of their solutions. This may be due to their limitation of language usage. In Brunei, the medium of instruction is English, whereas the native mother tongue is Bahasa Melayu (Malay Language). Thus, it is a bilingual educational practice. Limited English language capability could hinder students' articulation of their thinking and reasoning. Based on the students' explanations, it seems that they highlighted their instrumental understanding. The steps and procedures of their mathematics solutions explain what they are doing as opposed to how they are doing it. With this limitation, teachers face difficulties in practicing their intended goals of getting students to explain the reasons or concepts of their work.

10.4 Conclusion

In conclusion, this study aimed to explore the dominant values that emerged among teachers from two schools during a lesson study professional development exercise. The findings indicated the presence of shared dominant values, as well as discrepant ones, and revealed a transition of these values from the intended to the attained curriculum levels. The shift in teachers' values appears to be influenced by two factors: the school governance, which dictates how lesson study was conducted, and the objectives of teacher professional development, which shaped the activities in the lesson study. These findings imply that values clarification within the Lesson

Study process teachers' professional development can be effectively carried out by manifesting and progressing their values. When interpreting the curriculum, teachers carry values that may align and conflict with the realities in the classroom setting. The lesson study practice allows observing value shifts or changes during the process. Thus enabling the progression of values clarification into practice. However, it is important to highlight that the limitation of this study is not to claim the dominant values of teachers for the whole nation. The preliminary exploration of the knowledge on the existence of dominant values shared and idiosyncratic values of the schools paves the way for further research to analyze the enabling and constraining factors of the teachers' existing values within the school and its dynamic relationship. Further research to understand these factors is a step for policymakers or educators to evaluate and advocate values in the mathematics curriculum.

Note This study is based on the author's unpublished PhD thesis submitted to Hong Kong University in 2021.

References

Aktas, F. N., & Argün, Z. (2018). Examination of mathematical values in the classroom practices: A case study of secondary mathematics teachers. *Education and Science, 43*(193), 121–141.

Bishop, A. (1988). *Mathematical enculturation: A cultural perspective on mathematics education.* Kluwer Academic Publishers.

Bishop, A. (2001). What values do you teach when you teach mathematics? *Teaching Children Mathematics, 7*(6), 346–349.

Creswell, J. W., & Poth, C. N. (2018). *Qualitative inquiry and research design: Choosing among five approaches* (4th Ed.). SAGE Publications.

Department of Planning, Development and Research [DPDR]. (2012). *The ministry of education strategic planning 2012–2017.* Ministry of Education.

Fujii, T. (2014). Implementing Japanese lesson Study in foreign countries: Misconception revealed. *Mathematics Teacher Education and Development, 16*(1), 65–83.

Gerring, J. (2006). *Case study research: Principles and practices.* Cambridge University Press.

Kennedy, A. (2005). Model of continuing professional development: A framework for analysis. *Journal of in-Service Education, 31*(2), 235–250.

Khalid, M. (2016). Reminiscing the lesson study journey in Brunei. In T. Ueyama (Ed.), *The lesson study seminar and practice 2014: Reflection on the lesson study practice in Brunei Darussalam* (pp. 100–104). Universiti Brunei Darussalam.

Lewis, C., Perry, R., & Hurd, J. (2009). Improving mathematics instruction through lesson study: A theoretical model and North American case. *Journal of Mathematics Teacher Education, 12*, 285–304.

Musa, J. (2015). *Lesson study report. Special programme: Implementation and monitoring unit.* Department of Schools, Ministry of Education.

Presmeg. (Eds.) *Critical issues in mathematics education* (pp. 239–253). Springer. https://doi.org/10.1007/978-0-387-09673-5_17

Raths, L. E., Harmin, M., & Simon, S. B. (1987). Selections from values and teaching. In P. F. Carbone (Ed.), *Value theory and education* (pp. 198–214). Krieger.

Seah, W. T. (2008). *Valuing values in mathematics education.*

Seah, W. T. (2018). Improving mathematics pedagogy through student/teacher valuing: Lessons from five continents. In *Invited lectures from the 13th international congress on mathematical education* (pp. 561–580). Springer International Publishing.

Suhaili, H. A. S. (2018). *Developing a school-based lesson study model for Brunei Primary mathematics teachers: A cultural historical activity theory perspective*. Unpublished PhD thesis. Universiti Brunei Darussalam.

Part III
Utilising the Values Perspective in Promoting and Sustaining Student Mathematical Wellbeing

Chapter 11
What Do Korean Students Value in Mathematics Learning? Insights into Mathematical Well-Being

Hee-jeong Kim, Won Kim, and Gima Lee

11.1 Introduction

International comparison studies, such as the Programme for International Student Assessment (PISA) (OECD, 2016, 2019a) and Trends in International Mathematics and Science Study (TIMSS) (Mullis et al., 2016), indicate that Korean students rank higher in mathematical achievements in comparison to students in other countries. However, their affective characteristics, namely interest, attitudes, and motivation towards learning mathematics, including valuing mathematics learning, are rated below the average of the countries of the Organization for Economic Corporation and Development (OECD) (Choe et al., 2014; OECD, 2019b).

Acknowledging this issue, both mathematics educators and policymakers in Korea have sought interventions to foster students' enthusiasm and appreciation for mathematics (Choe et al., 2014; Choe & Hwang, 2014; Ministry of Education, Korea, 2015; Lee, 2021). The Korean Ministry of Education has articulated a vision wherein students derive genuine pleasure from their mathematics learning experiences and nurture positive perspectives towards the subject (Ministry of Education, Korea, 2020). Tangible steps, such as establishment of mathematics-focused museums across the country and introduction of exemplary schools, aim to inculcate a deep-rooted value for mathematics in students, enriching their in-school and out-of-school learning experiences.

H. Kim (✉) · W. Kim
Korea University, Seoul, Korea
e-mail: heejeongkim@korea.ac.kr

W. Kim
e-mail: wonny00901@korea.ac.kr

G. Lee
Graduate School of Education, Korea University, Seoul, Korea
e-mail: lovejesus153@korea.ac.kr

Y. Dede et al. (eds.), *Values and Valuing in Mathematics Education*,
https://doi.org/10.1007/978-981-99-9454-0_11

Simultaneously, mathematics education researchers in Korea have been proactive. Researchers like Choi et al. (2020) and Kim and Ko (2018), have developed programs, undergirded by policy support, to support both students' mathematical achievement and their confidence in learning mathematics. Preliminary assessments of these interventions hint at their potential to uplift students' affective engagement with mathematics.

However, a significant gap persists. While there is an increased focus on and interest in understanding students' values and attitudes in the realm of mathematics education in Korea, there is a paucity of in-depth research into what exactly students value about mathematics learning and their associated mathematical well-being (MWB). While international studies, such as TIMSS 2015 (Mullis et al., 2016), spotlight the somewhat negative attitude of Korean students towards mathematics, their methodologies, often reliant on simplistic self-assessment scales, leave much to be desired. The methods employed in these studies are based on self-assessed questions on a scale of 1–4. For examples, "I enjoy learning mathematics," "mathematics is boring," "I like mathematics," "I usually do well in mathematics," "I learn things quickly in mathematics" are some examples of the typical survey questions (Yi & Lee, 2015). These surveys have enabled us to evaluate the overall patterns of students' self-conception on confidence of learning mathematics across countries. However, in-depth analyses of what students value and how they are creating those values in learning mathematics are limited. Research on MWB in connection with values, such as Clarkson et al. (2000) and Hill et al. (2022), needs to be expanded and what students value in learning mathematics needs to be extensively explored, particularly in the Korean context, where very high achievement outcomes with very low in confidence and value of learning mathematics.

The broader academic discourse acknowledges the multifaceted nature of *values*. Within mathematics education, unique features of values and valuing in mathematics teaching and learning include individuals' recognition of the importance and usefulness of their learning mathematics. The values are formed through learners' own learning experiences and these values play an important role in shaping attitudes and learning achievements (Askew et al., 2010; Bishop, 1988; Halstead & Taylor, 2000; Seah, 2005). They are intrinsically linked to students' mathematical well-being (MWB), encapsulated as a state of holistic contentment and functionality in the context of learning and applying mathematics (Seah et al., 2021).[1]

This chapter delves into the nuanced terrain of students' MWB. Guided by the reconceptualized MWB framework of Hill et al. (2022), we navigate the landscapes of Korean students' mathematical well-being, seeking to unravel the values they attach to their mathematical pursuits. In the following sections, we will re-articulate the MWB construct, ground it in empirical analysis, and furnish insights that illuminate

---

[1] Seah et al. (2021) refers students' MWB to "a sense of 'feeling good and functioning well' in relation to learning and using mathematics (p. 400)". In following this notion, our operational definition of student MWB is a sense as being achieved when students accomplish their goals and value-related aims while learning and doing mathematics.

the bridge between MWB and the values Korean students associate with learning mathematics.

11.2 Theoretical Background

11.2.1 Values of Mathematics Education

Research on values in mathematics education is often divided into three categories: general educational values, mathematical values, and mathematics educational values (Bishop, 1988; Pang et al., 2016; Seah, 2018). General educational values focus on the socialization function of education and the values of school's educational goals that contribute to cultural formation, including power distance within a social community, emotional preferences represented by gender, avoidance of uncertainty, future orientation, and inclusiveness of education methods or tools. Mathematical values refer to the value of mathematics as a discipline, including the ideology of the source's authority and power of mathematics, sentiment regarding the purpose of doing mathematics (stability of social order versus innovation), and the openness of mathematical knowledge from a sociological perspective (Bishop, 1988). Mathematics educational values are concerned with what is valued and important based on specific practices or norms that are formed during the process of teaching and learning mathematics (Atweh & Seah, 2008).

Early research in the field of values focused on general educational or mathematical values but did not specifically address mathematics educational values (e.g., Bishop, 1988; Eccles, 1983). That was until the first decade of the twenty-first century, when research on values expanded to include teaching, learning, and curricula, with an increasing interest in mathematical educational values (Atweh & Seah, 2008; Clarkson et al., 2000; Dede, 2006; Kalogeropoulos & Bishop, 2017; Seah & Bishop, 2000). For example, Seah (2005) focused on mathematics educational values in an international collaborative study, the Third Wave Project, which aimed to reveal the essence and attributes of values that affect mathematics learning. Consequently, the authors identified seven categories of mathematics educational values consisting of two opposing elements in each category (ability–effort, enjoyment–perseverance, process–result, application–calculation, fact and theory–ideas and practices, interpretation–exploration, and memory–creativity). With the expansion of the Third Wave Project research worldwide, awareness of values beyond these seven categories emerged. For example, in a study of Hong Kong and Japanese students, elements such as "exploration," "alternative approaches," "application," "mathematical identity," "creativity," "knowhow," and "discussion" were identified (Seah et al., 2017).

"What I Find Important (in my mathematics learning)" (WIFI), derived from the Third Wave Project, has been studied by collecting quantitative data on students' values from more than 20 countries (Seah, 2018). Korea began its research in earnest on values in mathematics learning through participation in this project. Specifically, Pang et al. (2016) conducted a survey of 409 sixth-grade elementary school students and 407 third-year middle school students (i.e., 9th grade) using a WIFI questionnaire to quantitatively compare and analyze values related to mathematics and mathematics learning, finding statistically significant differences. Further, other studies (Cho, 2018; Pang & Yim, 2019) have analyzed changes in the values of teachers and students by observing elementary school mathematics classes. These studies are important as they offer insights into promoting desirable values among students through mathematics education. In Cho's (2018) study, a teacher and students collaboratively developed and executed strategies for value negotiation, aligning their respective values in the process. As they engaged in this collaboration, the students began to understand that the scope of mathematics extends beyond mere calculations. They discovered that it also exists within artistic works and that they themselves have the ability to create such examples. This realization allowed them to appreciate the value of progress and they experienced joy in relation to doing mathematics.

Additionally, a new scale was developed by adapting the WIFI questionnaire to the Korean context to measure the mathematical educational values of pre-service and in-service teachers, resulting in the emergence of fun, problem-solving, expression, calculation, ability, and explanation as mathematical educational values (Yim et al., 2020). Finally, research trends have been analyzed by examining studies on mathematical values and mathematical educational values conducted globally, in terms of research period, project, country, analysis target (e.g., mathematical values, mathematical educational values), and research methods (Pang & Yim, 2019).

In sum, the exploration of values within mathematics education has undergone significant expansion, particularly with the notable contributions from Korea. This has deepened our understanding, shedding light on both universally shared values and those unique to specific contexts. Such insights accentuate the profound influence of values on shaping students' experiences, perceptions, and motivations in mathematics, reiterating the crucial role of values in making meaningful and engaging mathematics educational experiences. Nonetheless, there remains a gap in explicitly linking these values to students' content-specific well-being, even as students' well-being has become a central educational objective worldwide. While Clarkson et al. (2000) took initial steps to conceptualize MWB in relation to values, we delve deeper into this connection in the following section.

11.2.2 Mathematical Well-Being, Linked to Values

The concept of MWB, first introduced in the 2000s, refers to the subjective well-being that learners experience in the process of learning mathematics. It captures not only the essence of "feeling good and functioning well" as described by Huppert and So (2013, p. 839), but also encapsulates the positive state of functioning that emerges when students' classroom experiences resonate with their personal values (Hill & Seah, 2023). Clarkson et al. (2010), from the outset of their conceptualization of MWB, provided a foundational understanding of MWB by emphasizing three core domains: cognitive, affective, and emotional. The cognitive domain focuses on the knowledge and skills essential for performing mathematics in academic settings. Meanwhile, the affective domain delves deeper into the psychological aspects of learning, including learners' attitudes, values, and motivation, with a notable emphasis on learner motivation. Lastly, the emotional domain addresses the feelings, responses, and reactions students have towards mathematics learning. Considering Clarkson et al. (2010)'s framework, it becomes evident that for a holistic enhancement of MWB, it is imperative to nurture not just the cognitive domain, but also the affective and emotional domains.

While Clarkson et al. (2010)'s elaboration on the stages of achieving MWB was useful in the first instance, it did not offer sufficient insights into the mechanisms through which, for example, the development of pleasure of doing mathematics or growing appreciation of the value of mathematics could be achieved. As such, the framework was more descriptive than operational. Later work of Hill et al. (2021) speaks to this operational gap, by defining MWB in terms of value fulfilment theory. In other words, we say that achievement of students' MWB can be facilitated within the context of mathematics learning when students' values in the context of mathematics learning are fulfilled (Hill & Seah, 2023; Hill et al., 2021).

In more details, Hill et al. (2020) explored values related to MWB among 488 eighth-grade students in Melbourne, Australia. According to preliminary results, the values related to well-being were ranked in the order of relationship, participation, cognition, achievement, positive emotions, perseverance, music (listening to music while studying), and meaning. In their next study, Hill et al. (2021) developed the MWB model by analyzing and integrating the well-being model and its sub-factors on positive psychology and documents from the OECD. In this model, values related to well-being were categorized into seven areas: (1) a*ccomplishment* involving valuing achievement, goal attainment, confidence on math assignments, and test evaluations; (2) *cognition* involving valuing knowledge, skills, and necessary understanding for learning mathematics in school; (3) *engagement* involving valuing concentration, immersion, and deep interest or attention when learning math; (4) *meaning* concerning valuing the direction of math and feeling that math is valuable, useful, and purposeful; (5) *perseverance* referring to valuing driving force, grit, or exerting effort to achieve mathematics assignments or goals; (6) *positive emotions* referring to valuing positive feelings such as enjoyment, happiness, and pride when learning or doing mathematics and (7) *relationships*, that is, valuing supportive relationships or feelings of being valued, respected, cared for, feeling connected with other people, and valuing support from peers when it comes to mathematics. Hill et al. (2021) found statistically significant correlations between values and MWB in the first five areas of accomplishment, cognition, engagement, meaning, and perseverance. In a follow-up study, Hill and Seah (2023) termed these seven aspects *MWB ultimate values* while considering the importance of values for achieving MWB, arguing that there are *MWB instrumental values* that support ultimate values.[2] For example, valuing the teacher's explanation is viewed as supporting the ultimate value of *relationships*, while valuing challenges in mathematics is considered to support the ultimate value of *perseverance*.

In the field of mathematics education, MWB stands as a pivotal component of learners' subjective well-being. From the literature discussed above, our initial approach was to conceptualize MWB by intertwining values and the appreciation of learning mathematics. This encompasses cognitive aspects, a conative stance with an emphasis on a growth mindset, and positive emotional facets. However, this initial representation might appear fragmented. To provide a clear and more integrated perspective, we move towards a reconceptualization. Essentially, we aim to construct a coherent framework that aligns Hill's values within the dimensions proposed by Clarkson et al. (2010). This refined grid offers a systematic approach to understanding MWB and was instrumental in our analysis of data collected nationwide in Korea. We emphasize this step to ensure clarity and ease of comprehension in the next section.

11.3 Conceptual Framework

We synthesized MWB models from prior studies (Clarkson et al., 2010; Hill & Seah, 2023; Hill et al., 2021) to analyze the values that students attribute to learning mathematics and their association with MWB (see Table 11.1). The analytical framework, as proposed by Hill et al. (2021, 2022), integrates both the ultimate values crucial for achieving well-being and the instrumental values that support these primary values, elucidating the relationship between MWB and values. Meanwhile, Clarkson et al. (2010) posited a model suggesting that MWB extends beyond the cognitive aspects of learning mathematics, encompassing the comprehensive development of both affective and emotional domains. Drawing upon the strengths of each framework from these studies, the vertical axis of our analytical framework is segmented into three dimensions of MWB: cognition, conation, and emotion. The horizontal axis showcases the ultimate values essential for achieving MWB, along with examples of MWB instrumental values that reinforce these ultimate values, complete with their explanations and sources.

Building on this framework, we aim to define *ultimate values* as those perceived as having intrinsic worth, significantly impacting MWB, and considered universal regardless of culture and individual differences. In the same vein, *instrumental values* are defined as values used as means to achieve ultimate values; their contribution to well-being varies depending on the ultimate value they support, and these values can differ substantially among individuals.

[2] Although it is possible to consider both personal and sociocultural dimensions of values as in previous studies (e.g., Hill & Seah, 2023), this study narrows the discussion to learners' personal dimensions because the sociocultural dimensions of values (e.g., relationships) were not found in our tool in this study. This is described in the discussion section as a study limitation.

Table 11.1 Mathematical well-being with values framework (revised from Hill et al., 2022, p. 382)

MWB ultimate values	Descriptions	Source	Example of MWB instrumental values
Cognition			
Knowledge	Valuing the procedural process or remembering mathematical facts, concepts, and calculation e.g., memorization, repetition, and the speed and accuracy of calculations	Anderson and Krathwohl (2001), Bloom et al. (1956), and Clarkson et al. (2010)	Memory power, mental arithmetic ability, computational ability, formulas, memory, speed, accuracy
Comprehension	Valuing understanding mathematical facts and principles, understanding, translating mathematical expressions e.g., recognition, verification, classification, explanation		Comprehension, concept, vocabulary
Application	Valuing using acquired mathematical concepts and procedures in new situations e.g., planning, execution, problem-solving, utilization, application		Problem-solving, process, application, utilization
Analysis	Valuing identifying and distinguishing the situational context, mathematical facts, hypotheses, and conclusions e.g., differentiation, discrimination, organization		Exploratory ability, spatial ability, ability to analyze problem situations
Synthesis	Valuing connecting parts based on mathematical logic, making inferences, and solving problems e.g., connection, combination, composition, synthesis, inference, problem-solving		Ability to put parts together to form a new whole, ability to think, problem-solving capability, logical thinking, reasoning

(continued)

Table 11.1 (continued)

MWB ultimate values		Descriptions	Source	Example of MWB instrumental values
Evaluation		Valuing assessing and justifying mathematical ideas or problem-solving processes e.g., evaluation, criticism, argumentation, inspection, judgment		Proving, correcting wrong answers
Creativity		Valuing inventing something new and original e.g., exploration, design, development	OECD (2015)	Creativity, insight, imagination
Conation				
Engagement		Valuing concentration and deep focus while doing mathematics	Hill et al. (2021)	Concentration, curiosity, spontaneity
Perseverance		Valuing grit, passion, and hard work to learn mathematics	Duckworth (2016), and Hill et al. (2021)	Perseverance, effort, resilience
Mindset	Fixed	Valuing the unchangeable innate talents and intelligence in learning mathematics	Dweck (2006), OECD (2015)	Talent, intelligence
	Growth	Valuing beliefs in oneself on ability to do mathematics and capacity to do it proactively		Confidence, challenge, not giving up in the face of difficulties
Emotion				
Positive emotions		Valuing positive emotions such as fun, excitement, enjoyment, sense of achievement, and happiness experienced in learning mathematics	Hill et al. (2021)	Interest, enjoyment, fun, sense of accomplishment

Source OECD social, physical, psychological, and cognitive well-being (OECD, 2015)

11.3.1 Dimension 1: Cognition

The dimension of cognition was divided into seven levels of MWB ultimate values based on research on MWB (Clarkson et al., 2010), Bloom's taxonomy of educational objectives (Anderson & Krathwohl, 2001; Bloom et al., 1956), and the literature related to mathematics education and curriculum and assessment (National Council of Teachers of Mathematics (NCTM), 2000; OECD, 2019a). The level names are based on the framework proposed by Clarkson et al. (2010).

The first level of MWB ultimate values is Knowledge. At this level, the ability to recall mathematical facts or information is valued, as are calculations. While previous research (Anderson & Krathwohl, 2001) included "calculation" as part of level 3 "application," this study includes it in level 1. The cognitive demand levels required to perform procedures in mathematical tasks are distinguished as procedural without connections and procedural with connections, which involve understanding mathematical ideas and connecting them with procedures (Stein & Smith, 1998). However, in this study, "calculation" is defined as remembering and performing algorithms and is considered to have a lower-level cognitive demand.

At the second level, Comprehension, the ability to understand mathematical facts, principles, and laws is valued, along with the ability to interpret and translate mathematical representations (such as words, symbols, and images). In particular, the ability to interpret and translate representations such as terms, symbols, graphs, and diagrams in math problems is important not only for mathematical communication but also for extending mathematical concepts, and thus, included in the Comprehension level (Bloom et al., 1956; Janvier, 1987).

At the third level, Application, the ability to apply previously learned mathematical concepts and procedures in specific problem situations is valued. Developing application skills is important for mathematics learning and evaluation. In the future, mathematical knowledge and reasoning will be utilized to solve challenges faced in real life rather than simply acquiring knowledge. In addition, in Korea, mathematics plays the role of a gatekeeper in university entrance exams; therefore, students need to demonstrate their ability to apply and adapt generalized mathematical concepts and procedures to specific problem situations and to plan and execute problem-solving processes to achieve high scores.

At the fourth level, Analysis, the ability to identify and distinguish the constituent elements of a problem situation, including mathematical facts, hypotheses, and conclusions is valued. Bloom et al. (1956) defined this as the ability to decompose data into constituent elements, which helps learners to understand the structure of an organization. In mathematics education, Analysis is the ability to distinguish the elements, related mathematical knowledge, hypotheses, and conclusions that make up a problem situation. This ability involves not only understanding mathematical content but also perceiving, investigating, and analyzing the overall problem situation and its structure. Therefore, it is considered a higher cognitive level than Comprehension or Application.

At the fifth level, Synthesis, the ability to connect and integrate parts based on mathematical logic to create a logical structure for reasoning and problem-solving is valued. Bloom et al. (1956) emphasized creative behavior, describing Synthesis as going beyond connecting and integrating parts to produce original results. In our study, Synthesis and Creativity were operationally defined differently in the context of mathematics learning. Synthesis is the ability to logically connect mathematical facts such as definitions, axioms, theorems, and propositions to infer new facts or to solve problems by integrating mathematical facts and procedures. Emphasizing reasoning and problem-solving is important because students need to connect their existing knowledge and establish effective strategies to solve new problems. To infer new mathematical facts or to make decisions, they need to logically rearrange and reconstruct the given information to create a new logical system (NCTM, 2000).

At the sixth level, Evaluation, the ability to verify the validity of mathematical ideas or critically examine and evaluate the problem-solving process is valued. Bloom et al. (1956) described evaluation as the ability to judge the value of material for a given purpose. In mathematics education, evaluation involves establishing criteria for assessing the mathematical ideas and problem-solving processes provided by the mathematical community. The elements of Knowledge, Comprehension, Application, Analysis, and Synthesis serve this criterion. Thus, evaluation is considered a higher-order value that encompasses other values. To evaluate, one must first establish criteria and then judge the value of the material against these criteria.

At the seventh level, Creativity, the ability to devise innovative solutions by improving and combining existing ideas in challenging problem situations is valued. This is a higher level of Synthesis, which involves integrating provided information and designing unique ideas. It includes creativity to develop new ideas, insight to produce previously non-existing ideas, and the ability to express what has been imagined. Recently, in PISA, creative thinking has been valued and defined as "the competence to engage productively in the generation, evaluation and improvement of ideas, which can result in original and effective solutions, advances in knowledge and impactful expressions of imagination (OECD, 2019a).

11.3.2 Dimensions 2 and 3: Conation and Emotion

The conation dimension subdivides the MWB ultimate values into engagement, perseverance, and (growth) mindset. First, as an MWB ultimate value, engagement is related to the ability to concentrate and immerse oneself in mathematics learning. In well-being research, engagement is an essential element used to measure the structure of well-being; it is defined as being involved and interested in participating in any activity or the world itself. In a high level of immersion state, known as flow, thinking and emotion are lacking and the sense of time is forgotten. Such immersion has the characteristic of enjoying and pursuing the object rather than aiming for other purposes. In terms of mathematics learning, the value of engagement is achieved through pure curiosity about mathematics, voluntary participation, and concentrating in mathematics classes and during assignments (Hill & Seah, 2023; Kern et al., 2016; Seligman, 2011).

Second, perseverance is a value related to patience and passion for difficulties in mathematics learning. Kern et al. (2016) explain perseverance as an element of well-being, which means to persist, pursue goals despite difficulties, and accomplish them even if they are challenging or time-consuming. Recently, "the power to maintain perseverance and passion for long-term goals" has been conceptualized as grit (Duckworth et al., 2007), and attention has been drawn to the relationship between grit and subjective well-being in several research studies (e.g., Disabato et al., 2019; Lan & Moscardino, 2019; Lim & Yu, 2022).

Third, mindset refers to an individual's thoughts and attitudes, impacting their growth and development. It can be divided into growth or fixed mindsets. Students with a growth mindset see failure as an opportunity to learn and make efforts without giving up, while those with a fixed mindset tend to avoid challenges and choose easy tasks with a low risk of failure to avoid being evaluated as lacking ability (Dweck, 2006). Therefore, a student's mindset is closely related to their beliefs and confidence, which determine their behavior. Previous studies have shown that a growth mindset has a positive effect not only on cognitive factors such as academic achievement but also on affective factors such as motivation and enjoyment of the learning process, while a fixed mindset has a negative impact on those issues (Aronson et al., 2002; Dweck & Master, 2009; Dweck et al., 1995; Good et al., 2003).

Finally, the dimension of emotion is related to the value of positive emotions such as excitement, fun, and enjoyment in mathematics learning. Positive emotions affect life satisfaction in positive psychology research, and this is the first factor that has been shown to affect subjective well-being in recent well-being studies (Seligman, 2011). Kern et al. (2016) viewed happiness, cheer, and contentment as important factors of well-being among positive emotions in life.

Utilizing the grounded theory approach outlined by Miles et al. (2014), the conceptual framework in this study was refined through an iterative literature review process. The three researchers analyzed the data independently. Based on their findings, they engaged in continuous cross-checking and discussions. Given the nature of our approach, collaborative analysis took precedence over traditional quantitative measures of reliability. Further insights into our data analysis process will be elaborated in the subsequent section.

11.4 Methods

11.4.1 Participants and Contexts

This chapter is drawn from a larger project, *The International Mathematical Modeling Competency and Mathematical Values*. The goal of the larger project is to explore Korean students' mathematical modeling competencies and how they value the subject in their own learning of mathematics in comparison to students of Turkey. The written modeling tasks and value survey was distributed to 5th-, 6th-, 9th-, and 10th-grade students in both Korea and Turkey. Specifically, the value survey was developed from the WIFI project as well as drawn from the international collaborative study.

The participants in the survey were 193 elementary school students (5th and 6th grade) and 385 middle and high school students (9th and 10th grade) in Korea. To gain a general understanding of Korean students' values and modeling competencies, we recruited diverse schools and students[3] across the country, from metropolitan to rural areas. With the consent of the participants and that of their mathematics teachers and parents, 578 responses (from 193 elementary school, and 385 middle and high school students) to four open-ended questions on mathematical values in learning mathematics were obtained.

11.4.2 Survey Questions

This chapter made use of four contextualized open-ended questions that compose a part of the value survey mentioned above in order to explore Korean students' values in learning mathematics. In particular, the questions were designed to identify students' values that may not have been captured by the Likert-type questions used in other parts (Seah & Barkatsas, 2014; Pang & Seah, 2020). Specifically, these questions ask what it takes to become proficient in mathematics, as shown in Table 11.2. Though these questions seem to relate somewhat to aspects of cognition and accomplishment, most Korean students would be expected to answer what they find important and essential in learning mathematics in response to the questions because, like students from a few other East Asian countries (Zhang et al., 2016), their absolute goal of learning mathematics in the Korean context is to become good at mathematics through their learning. Therefore, the answers would imply their values in learning mathematics, which would be expected to illuminate their MWB in terms of the fact that values while learning mathematics are profoundly related to students' MWB (Hill et al., 2021).

11.4.3 Data Analysis

First, we extracted and coded words consisting of only nouns (or noun phrases) among parts of speech from the corpus using Excel spreadsheets because the responses of the majority of students within the corpus, in terms of frequency of appearance, were

[3] Only students (and their guardians or parents) who gave their consent on the form to participate in the project responded to the questionnaires and tasks. We do not believe that the few (N = 39) refusals for students to participate in the survey actually skewed the results in any way.

Table 11.2 Open-ended questions

Open-ended questions	Imagine that we are going to produce a magic pill. Anyone who takes this magic pill becomes very good at mathematics! What will you choose to be the top 3 ingredients of this magic pill? (Be imaginative, this main ingredient can be something we can touch and see, or something we can feel but cannot see)
Q1	Most important ingredient
Q2	Second most important ingredient
Q3	Third most important ingredient
Q4	I selected these ingredients because

nouns or noun phrases. Analyzing nouns typically provides the most significant information from within the corpus (Marcello et al., 2015; Yoon & Lee, 2019). Next, each noun's significance was analyzed linguistically and contextually by evaluating and discussing students' responses corresponding to the nouns through cross-checking by the three researchers. All the nouns were classified according to their confirmed significance into the MWB framework. Consequently, we divided the words into three categories comprising Cognition, Conation, and Emotion, which allowed us to obtain value-related words that reflected students' values of learning mathematics. Those not classified into these categories but linked to values that students thought of as significant in their (mathematical) learning were generally included in category of Others (for example, quick writing, stigma, and way to study). Lastly, to understand students' values in depth, the values were classified as more important or less important. To this end, we analyzed the frequency of these value-related words as the more frequently a specific word is mentioned, the more significant its relative values are.

11.5 Results

This chapter explored Korean students' MWB and the values of their mathematics learning from the responses of 578 students. These revealed their practical values based on the link between MWB and values. Consequently, three significant results were obtained to the following questions: (1) which words connect mathematical values and MWB; (2) what Korean students value most in the learning of mathematics and its relation to MWB in the cognition dimension; and (3) what Korean students value in their MWB in the conation and emotion dimensions.

11.5.1 Discovery of Words Regarding MWB and Values of Learning Mathematics

We discovered a variety of words referring to mathematical values while learning mathematics by analyzing students' responses according to our MWB framework. We discovered 97 value-related words showing and implicating values and successively classified them in accordance with our framework as presented in Table 11.1. We were further able to identify values that were considered more important than others to help us gain an in-depth understanding of students' relative values in the Korean educational context. The key results are outlined below.

First, 47 words related to Cognition were discovered. As depicted in Table 11.3, more words were included in the categories of Knowledge, Comprehension, and Application in comparison to other categories. Second, 31 words associated with Conation were discovered, and most of these belonged to the categories of perseverance (14) and mindset (14). Third, 7 words related to emotion were found and were associated with positive or satisfactory feelings. Last, there were 12 words that could not be categorized. However, because they were thought important by students, we categorized them into Others.

Consequently, 97 words were confirmed in the learning of mathematics. Analyzing the frequency of mention of these words enabled us to explore virtually which values students considered more important; this is discussed in the next section.

In sum, in the first examination of the student responses from the open-ended question, we uncover words reflecting mathematical values aligned with our MWB framework. 97 value-laden words were identified, offering insights into students' perceptions of values in the context of learning mathematics, especially within the Korean educational setting. A significant portion of these words, 47 in total, pertained to the Cognition dimension, with a dominant presence in the categories of Knowledge, Comprehension, and Application. Conation followed with 31 words, with Perseverance and Mindset being the most represented categories. Emotion-related words totaled seven, all evoking positive sentiments. An additional 12 words that did not fit neatly into our pre-defined categories were labeled as 'Others'. The frequency analysis of these words provides a deeper understanding of values students prioritize in mathematical learning, setting the stage for our subsequent discussions in the following sections.

11.5.2 What Korean Students Value and Their Perception of MWB

We identified Korean students' values and their perceptions of MWB by counting words written in the open-ended survey. As depicted in Table 11.4, words related to cognition were mentioned most frequently by the students, followed by words related

Table 11.3 Students' words related to values of learning mathematics

Dimension of MWB ultimate values		Instrumental values for MWB: words from data	Counts	Total
Cognition				
Knowledge		Memorization, mental arithmetic, calculation ability, formula application, mathematical knowledge, recollection, speed, mathematical rules, basic skill, practice, remembering, substitution ability, accuracy	13	47
Comprehension		Understanding, vocabulary, mathematical concepts, interpretation, observation ability, reading ability, literacy, mathematical principle	8	
Application		Application, flexibility, utilization, solution process, mathematical writing ability, explanation, commentary ability, organizing under conditions	8	
Analysis		Spatial perception ability, situation analysis ability, inquiry ability	3	
Synthesis		Thinking ability, problem-solving, logic, deduction, judgment, connecting	6	
Evaluation		Verification, correcting wrong answers, reasoning, calculation checking	5	
Creativity		Creativity, insight, ability to think diversely, imaginative ability	4	
Conation				
Engagement		Concentration, spontaneity, curiosity	3	31
Perseverance		Endurance, endeavor, sincerity, persistence, meticulousness, peace of mind, bravery, will, diligence, mental power, industriousness, calm nature, sedentary ability, passion	14	
Mind set	Fixed	Talent, smarts, ingenuity, intelligence, brain, IQ, sense	7	
	Growth	Confidence, self-reflection, grit, self-confidence, learning from mistakes, not giving up, challenging	7	

(continued)

Table 11.3 (continued)

Dimension of MWB ultimate values	Instrumental values for MWB: words from data	Counts	Total
Emotion	Enjoyment, fun, interest, happiness, aesthetics, serenity, sense of accomplishment	7	7
Others	Sensory experience, ingredient for doing well at school, action, ability to get the right answer, ability to penetrate, stamina, ability, method of study, quick writing, delicacy, contraction of solving, experience	12	12
Total			97

to conation and then, emotion. To be specific, the ratios of words related to cognition, conation, and emotion were 60.9, 35.7, and 1.6% respectively in the 5th grade context. Similarly, ratios of words related to cognition, conation, and emotion were 66.3, 30.5, and 2.3% respectively in the 9th grade context. Figure 11.1 shows the visual similarity between 5 and 9th grades. This suggests that students considered values related to cognition the most important among all the values of learning mathematics. This led us to mark the sub-categories (cognitive levels) in the cognition category that students considered more valuable (see Table 11.3 and Fig. 11.1).

As exhibited in Table 11.5 for the sub-category of our framework, words related to Knowledge (level 1) were most frequently mentioned, followed by Comprehension (level 2) and Application (level 3) in the 5th grade context. Similarly, in the 9th grade context, words related to Knowledge (level 1) were mentioned the most, followed by Comprehension (level 2) and Application (level 3).

Interestingly, value-related words from analysis (level 4) to Creativity (level 7) were rarely mentioned in the 5th grade context. However, words related to Synthesis (level 5) and Creativity (level 7) were mentioned relatively more frequently in the 9th grade context. Figure 11.2 shows this similarity and comparison between 5 and 9th grades visually, which is suggestive of two significant findings. First, both 5th and 9th graders focused relatively more on low-level cognition. Second, interest in high-level cognition was observed with grade progression, as Synthesis (level 5) and Creativity (level 7) were mentioned relatively more by 9th graders.

Sub-categories on which students focused more in the Conation category were observed. As seen from Table 11.6, words related to Perseverance and a Growth

Table 11.4 The number of value-related words mentioned by the students

Graders	Cognition	Conation	Emotion	Other	Total
5th	329 (60.9%)	194 (35.7%)	9 (1.6%)	10 (1.8%)	542 (100%)
9th	665 (66.3%)	307 (30.5%)	24 (2.3%)	10 (0.9%)	1006 (100%)
Total	994 (65.0%)	501 (32.0%)	33 (2.0%)	20 (1.0%)	1548 (100%)

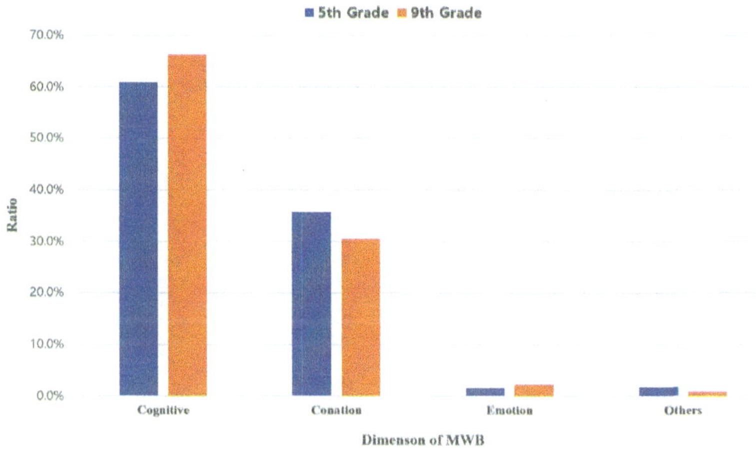

Fig. 11.1 Value-related words mentioned by the students in 5th and 9th grade

Table 11.5 The number of value-related words mentioned in the cognition category

Level	5th graders	9th graders	Total
L1 (Knowledge)	231 (70.3%[a])	383 (57.5%)	614 (61.8%)
L2 (Comprehension)	55 (16.7%)	93 (13.9%)	148 (14.9%)
L3 (Application)	27 (8.2%)	83 (12.4%)	110 (11.1%)
L4 (Analysis)	1 (0.3%)	5 (0.7%)	6 (0.6%)
L5 (Synthesis)	4 (1.2%)	52 (7.8%)	56 (5.6%)
L6 (Evaluation)	5 (1.5%)	6 (0.9%)	11 (1.1%)
L7 (Creativity)	6 (1.8%)	43 (6.4%)	49 (4.9%)
Total	329 (100%)	665 (100%)	994 (100%)

[a] % of responses in each category

Mindset were mentioned more frequently by 5th and 9th graders. Figure 11.3 visually presents this similarity between 5th and 9th grades in the Conation category. Therefore, Perseverance and Growth Mindset were considered more important by the students than other values. We indirectly confirmed that students consider effort and endeavor more important than innate talent.

Finally, only 7 value-related words linked to emotion were found that indicate being fully involved with positive or satisfactory feelings such as enjoyment and fun. Only 2% of the participants considered positive emotions important in learning mathematics.

In analyzing the values and perceptions of MWB among Korean students, our findings revealed a pronounced emphasis on cognitive aspects of learning mathematics. Words related to cognition overwhelmingly dominated the responses, particularly in the areas of Knowledge, Comprehension, and Application, which are labeled as

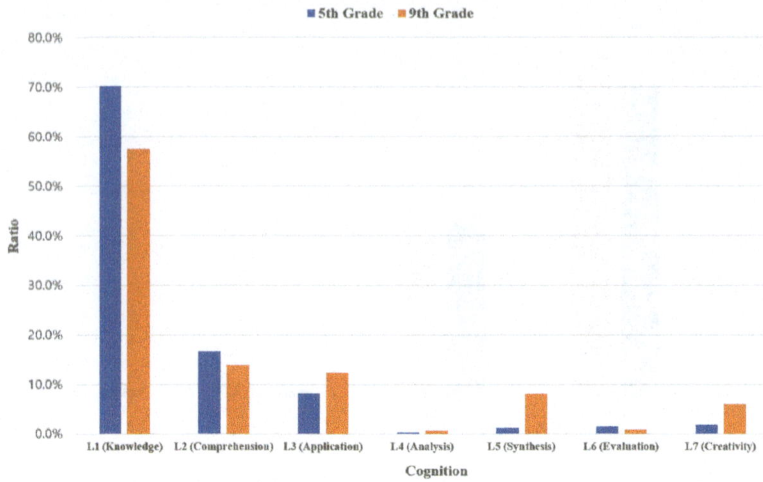

Fig. 11.2 Value-related words included in the cognition category of 5th and 9th grade students

Table 11.6 The number of value-related words in the conation category

Graders	Engagement	Perseverance	Mindset		Total
			Fixed	Growth	
5th	14 (7.2%)	81 (41.9%)	12 (6.1%)	87 (44.8%)	194 (100%)
9th	20 (6.5%)	127 (41.5%)	34 (11.0%)	126 (41.0%)	307 (100%)
Total	34 (6.8%)	208 (41.5%)	46 (9.2%)	213 (42.5%)	501 (100%)

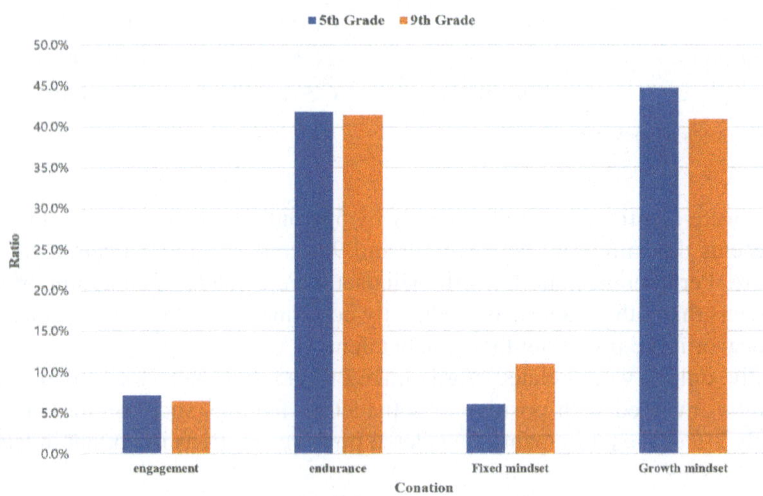

Fig. 11.3 Value-related words included in the Conation category

low level of cognition. While lower-level cognitive values were prioritized by both 5th and 9th graders, an increased interest in higher-level cognition such as synthesis and creativity were evident among the 9th graders. In the conation category, the prominence of perseverance and a growth mindset highlighted students' value in the importance of effort and perseverance over innate talent. Interestingly, only a small percentage of students emphasized the importance of positive emotions like enjoyment and fun in the learning process.

11.6 Discussion

The emphasis on lower cognitive skills, such as memorizing facts and rules and mental computation, present in Korean students' responses, highlights a significant trend in their approach to learning mathematics and what they value in learning mathematics. While these foundational skills are undeniably important, our results draw attention to the evolving cognition of the 9th graders, who begin to see the value in higher-order thinking skills like reasoning and creativity. This difference in perspective could be attributed to the nature of the Korean mathematics curriculum, where elementary stages primarily focus on computational strategies related to numbers and operations. In addition, this perception may be affected by the social context of mathematics education, such as parents' or teachers' value of mathematical learning or their MWB for children in different ages. This should be further investigated in a future study.

The observed emphasis on low cognitive load, despite its clear importance, raises questions about the potential implications for students' overall engagement and enjoyment in learning mathematics. Considering the relatively high academic achievements of Korean students alongside the TIMSS 2015 report showing that 58% of students expressed a dislike for learning mathematics, there is a pressing need for deep reflection on ways to support students in actively and enjoyably engaging with the high cognitive-demand tasks in the upcoming national curriculum. It is crucial to strike a balance where students not only achieve academically but also find joy and deep engagement in the learning process.

Encouraging students to participate with interest in such tasks can lead to profound satisfaction derived from immersion, consistent with our reconceptualized MWB framework. This perspective is grounded in the research and the *flow* theories put forth by the likes of Csikszentmihalyi (1990), Hill et al. (2022), Maslow (1962), and Seligman (2011), emphasizing the positive senses of individuals' experiences when deeply engaged cognitively and successfully navigating challenging tasks. Such experiences can lead to a deeper sense of fulfilment and a rekindled passion for the mathematics. As we reflected on our reconceptualized MWB framework, future curriculum developments and teaching methods should prioritize fostering this deep cognitive engagement while also making the learning experience more enjoyable for students.

Most interestingly, in our data, Korean students perceive that perseverance and a growth mindset are critical in learning mathematics. This result is in contrast with prior international comparison study results. PISA 2018 revealed that only 15.3%[4] of the students strongly refuted the notion that intelligent is a fixed trait, a result that ranks comparatively low among other OECD countries (OECD, 2019b). However, our data revealed that Korean students perceive that perseverance, efforts, perseverance, confidence, and challenge are key to improving mathematics achievement. This disparity prompts a re-evaluation of the underlying factors that drive such perceptions among Korean students. Is it possible that the pedagogical approaches in Korean education are fostering an environment that values persistent effort over predetermined abilities? Or perhaps the prominence of perseverance and growth mindset is indicative of a broader cultural shift, which students discerning that genuine progress in mathematics necessitates more than just innate talent, as an educational reform movement.

When considering the broader context of MWB, it becomes evident that cultivating well-being in mathematical learning is not merely about being happy without achieving high cognitive works. We believe that true MWB arises when students are deeply engaged in cognitively demanding tasks, navigating challenges, and encountering productive struggles (Blackwell et al., 2007; Henderson & Dweck, 1990). It is about equipping students with the resilience and mindset to persevere, fostering an environment where they believe in their capacity to grow and evolve. Therefore, by integrating and promoting the link between a growth mindset and mathematical well-being within our educational systems, we could be setting the stages and next steps for comprehensive mathematical advancement that goes beyond mere academic achievement.

By contrast, regarding the results on emotions, few students perceived positive emotions such as fun, excitement, interest, and happiness as important to achieve better results in mathematics. Considering that many students have a difficult experience in learning mathematics in the Korean educational context and must work hard (Pang et al., 2016), this result is important. It suggests that either few students think mathematics learning is accompanied by positive emotions or that few students consider mathematics learning to be fun and associated with happiness. Further investigations are needed into these positive emotions regarding MWB in learning mathematics in the Korean context.

As we coded the open-ended survey with a focus on well-being and values, we saw a possibility of a connection between the students' MWB and the values of doing and learning mathematics. This has also been confirmed by prior studies that probed the link between students' MWB and values of mathematics learning (Hill et al., 2021; Park & Peterson, 2008; Williams et al., 2015). This study contributes to the field both theoretically and empirically with its coding framework to investigate new constructs of students' MWB. We also provided new insights on how to foster

---

[4] About half of the Korean students disagreed or strongly disagreed with the following statement: "Your intelligence is something about you that you cannot easily change"; however, this was still low in comparison to other countries (OECD, 2019b).

students' MWB by offering higher-cognitive-demanding tasks and achievements through future-oriented mathematics, particularly in Korea as well as other Asian countries, where students' mathematical achievement is high, but MWB is relatively low.

Based on the findings of this study, there is a need to revamp the way mathematics is taught, especially in the elementary school curriculum, to foster MWB. The focus on formulae and effective solutions for mathematical problems in South Korea's educational curriculum may unintentionally limit opportunities for students to achieve a comprehensive understanding of mathematics and flexibilities and creativity for mathematical problem solving. The current trend of prioritizing memorization and quick calculations prevents students from developing both a deeper relationship within mathematics and inter-connections with other subject, such as STEM (Science, Technology, Engineering, and Mathematics). Along with improvements in the curriculum, teachers' instructional practices are also crucial. Teachers should be attentive and recognize moments when students are engaging in problem-solving, reasoning processes, mathematical modeling, and the high cognitive mathematics tasks they can discover the usefulness of mathematics in real-life situations. By noticing these moments and providing positive feedback, teachers might help students recognize the value of these experiences.

We note that this study has some limitations. With our survey, we cannot exclude the possibility that survey questions could have somewhat guided students' answers. Specifically, our contextualized questions that may have some association with aspects of cognition and accomplishment may have led students' answers to be somewhat related to cognitive values rather than emotional factors. Therefore, even though the result that cognitive factors are considered more important than other values in the Korean context seems to be consistent with those of previous studies in other East Asian countries (e.g., Hill et al., 2021; Zhang et al., 2016), this finding should be reconfirmed through follow-up studies.

11.7 Conclusion

Our study provides a unique lens into the Mathematical Well-being (MWB) of Korean students, revealing their values, perceptions, and challenges that shape students' mathematical learning. Central to our finding was the distinct emphasis students placed on cognitive aspects of learning and the value attributed to perseverance and a growth mindset. Such perceptions, particularly when juxtaposed against previous studies shed light on a potentially different trajectory of mathematics education in Korea. Korean students in this study exhibited a resilient orientation towards mathematical learning, emphasizing perseverance, effort, confidence, and the willingness to embrace challenges as essential to success in mathematical learning. Yet, our study also suggests there should be opportunities to make the learning experience more intuitive, enjoyable, and engaging for students with emotional support.

To move forward and put these findings into practice, teachers, as facilitators of learning, should be adept at not only making mathematics fun but also recognizing and nurturing moments of deep cognitive engagement, critical reasoning, and real-world applicability of mathematical concepts. Emphasizing high-cognitive-demand tasks that are not only challenging but also engaging and enjoyable can be pivotal. Positive facilitation and the promotion of a growth mindset during these moments can foster a deeper appreciation and enjoyment of mathematics among students. Such an approach aligns with the MWB framework and resonates with theories on positive engagement and immersion in learning.

Ultimately, this study's insights provide a foundation for a paradigm shift in mathematics education, particularly in regions where academic achievement is high yet MWB remains elusive. In addition, a significant discovery from this study was the identification of new constructs related to students' MWB, which adds an invaluable dimension to the existing body of research. By addressing these discovered constructs of MWB and restructuring educational approaches accordingly, we can aspire to foster a generation of students who find genuine satisfaction, enjoyment, and value in their mathematical learning.

References

Anderson, L. W., & Krathwohl, D. R. (2001). *A taxonomy for learning, teaching, and assessing: A revision of Bloom's taxonomy of educational objectives.* Longman Publishing.

Aronson, J., Fried, C. B., & Good, C. (2002). Reducing the effects of stereotype threat on African American college students by shaping theories of intelligence. *Journal of Experimenter Social Psychology, 38*(2), 113–125.

Askew, M., Hodgen, J., Hossain, S., & Bretscher, N. (2010). *Values and variables: Mathematics education in high-performing countries.* Nuffield Foundation.

Atweh, B., & Seah, W. T. (2008). *Theorizing values and their study in mathematics education.* In The Australian Association for Research in Education Conference, Fremantle, Australia. http://www.aare.edu.au/publications-database.php/271/theorizing-values-and-their-study-in-mathematics-education

Bishop, A. J. (1988). *Mathematical enculturation: A cultural perspective on mathematics education.* Kluwer Academic Publishers.

Blackwell, L. S., Trzesniewski, K. H., & Dweck, C. S. (2007). Implicit theories of intelligence predict achievement across an adolescent transition: A longitudinal study and an intervention. *Child Development, 78*(1), 246–263. https://doi.org/10.1111/j.1467-8624.2007.00995.x

Bloom, B. S., Engelhart, M. D., Hill, H. H., Furst, E. J., & Krathwohl, D. R. (1956). *The taxonomy of educational objectives, the classification of education goals, handbook I: Cognitive domain.* David McKay Company.

Cho, S. Y. (2018). Changing aspect of teacher and student's value in mathematics instruction. *Education of Primary School Mathematics, 21*(3), 273–287. https://doi.org/10.7468/jksmec.2018.21.3.273

Choe, S. H., & Hwang, H. J. (2014). The current situations of enhancing affective characteristics focused on the case of secondary school in Korea. *Communications of Mathematical Education, 28*(2), 235–253. https://doi.org/10.7468/jksmee.2014.28.2.235

Choe, S. H., Park, S., & Hwang, H. J. (2014). Analysis of the current situation of affective characteristics of Korean students based on the results of PISA and TIMSS. *Journal of the Korean School Mathematics Society, 17*(1), 23–43.

Choi, E. Y., Huh, N., & Ko, H. K. (2020). The effect of the communication-based math sharing school program on the affective domain. *Journal of Educational Culture, 26*(1), 397–423. https://doi.org/10.24159/joec.2020.26.1.397

Clarkson, P., Bishop, A. J., FitzSimons, G. E., & Seah, W. T. (2000). Challenges and constraints in researching values. In J. Bana & A. Chapman (Eds.), *Mathematics education beyond 2000* (pp. 188–195). Mathematics Education Research Group of Australasia.

Clarkson, P., Bishop, A. J., & Seah, W. T. (2010). Mathematics education and student values: The cultivation of mathematical wellbeing. In T. Lovat, R. Toomey, & N. Clement (Eds.), *International research handbook on values education and student wellbeing* (pp. 111–135). Springer.

Csikszentmihalyi, M. (1990). *Flow: The psychology of optimal experience.* Harper & Row.

Dede, Y. (2006). Values in Turkish middle school mathematics textbooks. *Quality and Quantity, 40*(3), 331–359. https://doi.org/10.1007/s11135-005-6133-8

Disabato, D. J., Goodman, F. R., & Kashdan, T. B. (2019). Is grit relevant to well-being and strengths? Evidence across the globe for separating perseverance of effort and consistency of interests. *Journal of Personality, 87*(2), 194–211. https://doi.org/10.1111/jopy.12382

Duckworth, A. L., Peterson, C., Matthews, M. D., & Kelly, D. R. (2007). Grit: Perseverance and passion for long-term goals. *Journal of Personality and Social Psychology, 92*(6), 1087–1101. https://doi.org/10.1037/0022-3514.92.6.1087

Dweck, C. S. (2006). *Mindset: The new psychology of success.* Random House.

Dweck, C. S., Chiu, C., & Hong, Y. (1995). Implicit theories and their role in judgments and reactions: A world from two perspectives. *Psychological Inquiry, 6*(4), 267–285. https://doi.org/10.1207/s15327965pli0604_1

Dweck, C. S., & Master, A. (2009). Self-theories and motivation: Students' beliefs about intelligence. In K. R. Wentzel & A. Wigfield (Eds.), *Handbook of motivation at school* (pp. 123–140). Routledge.

Duckworth, A. L. (2016). *Grit: The power of passion and perseverance.* Scribner.

Eccles, J. S. (1983). Expectancies, values, and academic behaviors. In J. T. Spence (Ed.), *Achievement and achievement motives: Psychological and sociological approaches* (pp. 75–146). W. H. Freeman.

Good, C., Aronson, J., & Inzlicht, M. (2003). Improving adolescents' standardized test performance: An intervention to reduce the effects of stereotype threat. *Journal of Applied Developmental Psychology, 24*(6), 645–662. https://doi.org/10.1016/j.appdev.2003.09.002

Halstead, J., & Taylor, M. (2000). Learning and teaching about values: A review of recent research. *Cambridge Journal of Education, 30*(2), 169–202. https://doi.org/10.1080/713657146

Henderson, V. L., & Dweck, C. S. (1990). Motivation and achievement. In S. S. Feldman & G. R. Elliott (Eds.), *At the threshold: The developing adolescent* (pp. 308–329). Harvard University Press.

Hill, J. L., Kern, M. L., Seah, W. T., & van Driel, J. (2020). A model of student mathematical wellbeing: Australian grade 8 students' conceptions. In M. Inprasitha, N. Changsri, & N. Boonsena (Eds.), *Interim Proceedings of the 44th Conference of the International Group for the Psychology of Mathematics Education* (pp. 273–281). PME.

Hill, J. L., Kern, M. L., Seah, W. T., & van Driel, J. (2021). Feeling good and functioning well in mathematics education: Exploring students' conceptions of mathematical well-being and values. *ECNU Review of Education, 4*(2), 349–375. https://doi.org/10.1177/2096531120928084

Hill, J. L., Kern, M. L., Seah, W. T., & van Driel, J. (2022). Developing a model of "mathematical wellbeing" through a thematic analysis of the literature. In M. Vollstedt, I. Kontorovich, K. Jones, & D. Sommerhoff (Eds.), *Proceedings of the 45th Conference of the International Group PME* (pp. 379–386). PME.

Hill, J. L., & Seah, W. T. (2023). Student values and wellbeing in mathematics education: Perspectives of Chinese primary students. *ZDM, 5*(22), 385–398. https://doi.org/10.1007/s11858-022-01418-7

Huppert, F., & So, T. (2013). Flourishing across Europe: Application of a new conceptual framework for defining flourishing across Europe. *Social Indicators Research, 110*(3), 837–861. https://doi.org/10.1007/s11205-011-9966-7

Janvier, C. (1987). Translation processes in mathematics education. In C. Janvier (Ed.), *Problems of representation in the teaching and learning of mathematics* (pp. 27–32). Lawrence Erlbaum Associates.

Kalogeropoulos, P., & Bishop, A. J. (2017). What is the role of value alignment in engaging mathematics learners? In A. Chronaki (Ed.), *Proceedings of the 9th International Mathematics Education and Society Conference* (Vol. 2, pp. 603–610). Mathematics Education and Society.

Kern, M. L., Benson, L., Steinberg, E. A., & Steinberg, L. (2016). The EPOCH measure of adolescent well-being. *Psychological Assessment, 28*(5), 586–597. https://doi.org/10.1037/pas0000201

Kim, H. K., & Ko, H. K. (2018). A case study on math clinical counseling based on connection with class. *Communications of Mathematical Education, 32*(2), 113–129. https://doi.org/10.7468/jksmee.2018.32.2.113

Lan, X., & Moscardino, U. (2019). Direct and interactive effects of perceived teacher-student relationship and grit on student wellbeing among stay-behind early adolescents in urban China. *Learning and Individual Differences, 69*, 129–137. https://doi.org/10.1016/j.lindif.2018.12.003

Lee, H. (2021). *A study on ways to improve students' mathematical affective achievement*. Report from the KOFAC (number: A22040003)

Lim, H., & Yu, B. (2022). The effects of grit on child subjective well-being. *Korean Journal of Sociology of Education, 32*(2), 187–212. https://doi.org/10.32465/ksocio.2022.32.2.007

Marcello, S. P., Hercules, A. D. P., Edilson, R., & Marcelo, L. (2015). An approach for text mining based on noun phrases. In N. S. Rui, C. J. Lakhmi, & J. H. Robert (Eds.), *Intelligent decision technologies* (pp. 525–536). IOS press.

Maslow, A. (1962). *Toward a psychology of being*. D Van Nostrand.

Miles, M. B., Huberman, A. M., & Saldana, J. (2014). *Qualitative data analysis: A methods sourcebook*. Sage.

Ministry of Education, Korea. (2015). *Mathematics curriculum*. Ministry of Education Notice 2015–74 [supplement 8].

Ministry of Education, Korea. (2020). *The plan for the mathematics education*. MOE.

Mullis, I. V. S., Martin, M. O., Foy, P., & Hooper, M. (2016). *TIMSS 2015 International results in mathematics*. Retrieved from http://timssandpirls.bc.edu/timss2015/international-results/

National Council of Teachers of Mathematics. (2000). *Principles and standards for school mathematics*. NCTM.

OECD. (2015). *What contributes to student well-being at school?* Retrieved from https://www.oecd.org/pisa/Well-being-Infographics.pdf

OECD. (2016). *PISA 2015 results (Volume I): Excellence and equity in education*. OECD Publishing.

OECD. (2019a). *PISA 2018 results (Volume I): What students know and can do*. Retrieved from https://www.oecd.org/pisa/publications/pisa-2018-results-volume-i-5f07c754-en.htm

OECD. (2019b). *PISA 2018 results (Volume III): What school life means for students' lives*. Retrieved from https://www.oecd-ilibrary.org/sites/bd69f805-en/index.html?itemId=/content/component/bd69f805-en

Pang, J., Cho, S. Y., & Seah, W. T. (2016). An investigation of what Korean students valued with regards to mathematics and mathematics learning: A study with sixth and ninth graders. *Journal of Elementary Mathematics Education in Korea, 55*(4), 467–484.

Pang, J., & Seah, W. T. (2020). Excellent mathematical performance despite "negative" affect of students in Korea: The values perspective. *ECNU Review of Education, 4*(2), 1–22. https://doi.org/10.1177/2096531120930726

Pang, J., & Yim, M. J. (2019). Perception and change of the values on mathematics learning by fifth graders through the teaching of a master teacher. *Journal of Elementary Mathematics Education in Korea, 23*(4), 405–435.

Park, N., & Peterson, C. (2008). Positive psychology and character strengths: Application to strengths-based school counselling. *Professional School Counselling, 12*(2), 85–92. https://doi.org/10.1177/2156759X0801200214

Seah, W. T. (2005). Negotiating about perceived value differences in mathematics teaching: The case of immigrant teachers in Australia. In H. Chick & J. L. Vincent (Eds.), *Proceedings of the 29th Conference of the International Group for the Psychology of Mathematics Education* (Vol. 4, pp. 145–152). PME.

Seah, W. T. (2018). Improving mathematics pedagogy through student/teacher valuing: Lessons from five continents. In G. Kaiser, H. Forgasz, M. Graven, A. Kuzniak, E. Simmt, & B. Xu (Eds.), *Invited lectures from the 13th international congress on mathematical education* (pp. 561–580). Springer.

Seah, W. T., Baba, T., & Zhang, Q. P. (2017). The WIFI study: Students' valuing of mathematics learning in Hong Kong and Japan. In J. W. Son, T. Watanabe, & J. J. Lo (Eds.), *What matters? Research trends in international comparative studies in mathematics education* (pp. 333–354). Springer.

Seah, W. T., & Barkatsas, T. (2014). *What Australian primary school students value in mathematics learning: A WIFI preliminary study*. Mathematics Education Research Group of Australasia.

Seah, W. T., & Bishop, A. J. (2000, April 24–28). *Values in mathematics textbooks: A view through two Australasian regions*. Paper presented at the 81st Annual Meeting of the American Educational Research Association, New Orleans, LA. (ERIC Document Reproduction Service No. ED 440 870).

Seah, W. T., Kim, H. J., & Kim, D. J. (2021). Reimagining mathematics education for the 21st century in the 21st century. *Journal of Educational Research in Mathematics, 31*(3), 393–404. https://doi.org/10.29275/jerm.2021.31.3.393

Seligman, M. (2011). *Flourish*. William Heinemann.

Stein, M. K., & Smith, M. S. (1998). Mathematical tasks as a framework for reflection: From research to practice. *Mathematics Teaching in the Middle School, 3*, 268–275. https://doi.org/10.5951/MTMS.3.4.0268

Williams, K. E., Ciarrochi, J., & Heaven, P. C. L. (2015). Relationships between valued action and well-being across the transition from high school to early adulthood. *The Journal of Positive Psychology, 10*(2), 127–140. https://doi.org/10.1080/17439760.2014.920404

Yi, H. S., & Lee, C. (2015). Analysis of longitudinal change of interest and self-confidence in mathematics among countries with high performance on TIMSS mathematics. *Journal of Educational Evaluation, 28*(2), 313–337.

Yim, M. J., Cho, S. Y., & Pang, J. (2020). Comparison of the mathematics educational values between pre-service and in-service elementary school teachers. *Communications of Mathematical Education, 34*(3), 277–297.

Yoon, T. Y., & Lee, S. Y. (2019). *Analyzing text through python*. Neulbom.

Zhang, Q., Barkatsas, T., Law, H. Y., Leu, Y. C., Seah, W. T., & Wong, N. Y. (2016). What primary students in the Chinese Mainland, Hong Kong, and Taiwan value in mathematics: A comparative analysis. *International Journal of Science and Mathematics Education, 14*, 907–924. https://doi.org/10.1007/s10763-014-9615-0

Chapter 12
The Role of Teacher Knowledge in Fostering Student Fulfillment of Values Crucial for Mathematical Wellbeing

Juan Zhong, Wee Tiong Seah, Yan Cao, and Ying Zhang

12.1 Background

Student disengagement with mathematics has been a persistent issue in mathematics education, and the COVID-19 pandemic had only accentuated this. While student participation in mathematics learning might be enthusiastic in the first few years of elementary schooling, our anecdotal experience in schools in a Chinese city has been that by the fifth grade, only about a quarter of students seem actively engaged and eager or willing to answer teacher-posed questions in class. This, in turn, has negative consequences in and beyond secondary mathematics education. There are serious economic and security implications arising from the consequent drop in the number of secondary school leavers electing to study mathematics and mathematics-intensive courses in tertiary institutions. These young citizens' disengagement can affect their numeracy skills and how the next generations view mathematics.

In most, if not all, instances disengagement with mathematics reflects a lack of wellbeing—that is, illbeing—in relation to learning the discipline. For example, disengagement during a learning activity can manifest itself in the form of task-withdrawing emotions such as distress, anger, frustration, anxiety and fear (Reeve, 2012), all of which are signs of illbeing. Fostering and maintaining mathematical wellbeing (MWB) is thus a priority in mathematics education (Clarkson et al., 2010; Hill et al., 2021a, 2021b).

J. Zhong (✉)
Chengdu Jinsha Elementary School, Chengdu, China
e-mail: 1597680554@qq.com

W. T. Seah
Faculty of Education, The University of Melbourne, Melbourne, Australia
e-mail: wt.seah@unimelb.edu.au

Y. Cao · Y. Zhang
Chengdu Qingyang Institute of Education Sciences, Chengdu, China

Y. Dede et al. (eds.), *Values and Valuing in Mathematics Education*,
https://doi.org/10.1007/978-981-99-9454-0_12

MWB was first conceptualized by Clarkson et al. (2010). Current, empirical studies in this area, however, led by Julia Hill (e.g. Hill et al., 2022) further developed this concept. In particular, Hill et al. (2022) conducted a scoping study to propose a MWB framework. This framework identifies seven ultimate values (UV) the fulfillment of which is necessary for positive MWB. For example, when a person values *accomplishment* but is unable to actualize it in their learning, this can affect their sense of wellbeing. Later, Hill trialed the model with students in Australia, China and New Zealand, which validated the seven UVs as being culturally robust. Nevertheless, some unique aspects of Chinese students' wellbeing and the values constituting it have been documented (Pan et al., 2022).

Considering the above, this current research study set to explore how classroom teachers (of mathematics) can successfully facilitate student fulfillment of UVs as a means of promoting (even) better MWB—and thus, more productive and meaningful mathematics learning—amongst students in China. That is, the intention here is not to explicitly investigate MWB change but to highlight the roles that cognitive and affective professional knowledge and expertise play in fostering the fulfillment of values.

Our devotion to fostering MWB, and to achieving this through facilitating the fulfillment of what students, teachers and others value in mathematics education, was largely sparked by the following experience in the first author's classroom.

12.1.1 An Experience with Ai Lin

It was 2021, when student questionnaires had just been collected in the first author's classroom, as part of a mapping of Chinese students' valuing associated with positive MWB. As had been planned for in the research design, the questionnaire returns were being analysed to identify representative respondents for interviews (the next stage in the research). During this exercise, the first author was surprised to find that a female student, Ai Lin, had indicated that she valued *verbal sharing* in class. This was because Ai Lin had always been a quiet student, one who would rarely volunteer to answer teacher questions publicly during mathematics lessons. The first author had in fact assumed that Ai Lin was an introvert and did not like to engage in public speaking. Ai Lin's questionnaire response also reflected a low MWB rating of 7 (the mean class score was 9.3) relative to her classmates', as well as a lack of confidence about her own mathematics learning. Feeling concerned, the first author had a conversation with Ai Lin, through which it became clear that her not volunteering to offer responses and views in class was rooted in a fear of answering wrongly, of teacher's potential criticism, and of peers laughing at her.

In order to support Ai Lin's learning, the first author became more watchful of Ai Lin's behaviour and feelings, as well as being more mindful of her needs. She reminded herself to praise Ai Lin for attainments, including those which were beyond mathematics performance (such as speed of completing given work, and quality of handwriting). The first author also engaged in individual conversations with Ai Lin, with the intention of encouraging her to speak up in class. She guided Ai Lin to reflect that nobody had ever been teased in class as a result of giving an incorrect response. She would also discuss with Ai Lin particular students' public responses in the mathematics class, leading her to realise that even though the students' responses were not entirely complete, they were never being laughed at by the teacher nor their classmates. The first author also made it more explicit to her class that 'the classroom is where mistakes especially need to take place', thereby encouraging all her students to feel free to share ideas verbally, not to be afraid of making mistakes, and acknowledging everyone's contributions and sharing.

Over time, Ai Lin felt encouraged and confident enough to raise her hand for the first time. The first author seized the opportunity to invite her to provide her response. Even though her first attempt did not deliver a complete answer to the mathematics question posed, the first author affirmed Ai Lin's change and progress, emphasizing to her that even though her response might not be perfect, that would not affect her improvement. Indeed, it is common for verbal responses to be incomplete. One year on, Ai Lin has now become a confident and regular contributor in mathematics lessons. She is no longer concerned about making mistakes, as she has grown to regard these as learning opportunities.

Through this intervention focus on Ai Lin over a year, the first author experienced how, by facilitating the student's valuing of such attributes of mathematics learning as *relationship, accomplishment* and *positive emotions*, the student's MWB can become more positive, leading to a more confident and engaged participation in her mathematics learning in class. Inspired by this, our focus, then, is on understanding more about the lesson activities which can help students fulfil values so that their respective MWB improves, while acknowledging the role of teacher professional knowledge and expertise in designing and delivering these activities. Thus, in this research, we went beyond assessing what students valued; we were interested in the mechanisms which were associated with improvements in students' MWB. This thus led to an extension of the initial mapping study in Chengdu (see Pan et al., 2022) that contributed to the Chinese MWB data, by one which is being reported in this chapter.

12.2 Literature Review

12.2.1 Values/Valuing in Mathematics Education

In preparation to discuss MWB, we first review values and valuing here, since we view wellbeing and values as intrinsically connected. This is best summarized by Tiberius' (2018) 'values fulfillment theory', which states that "well-being is served by the successful pursuit of a relatively stable set of values that are emotionally, motivationally, and cognitively suited to the person" (p. 13). In other words, a positive sense of MWB for a student comes about when they are able to fulfil their valuing of what they regard important in the context of mathematics learning. For example, if a student values *efficiency* in mathematics learning, and this student has opportunities in class to make use of any efficient solution approach, say, then this component of their MWB is being supported. On the other hand, if the mathematics teacher insists that all students are to solve mathematics problems using only the 'official' or textbook approach which may not be the most efficient one, for this student their MWB is under threat as they might feel restricted, frustrated or helpless.

Values refer to the attributes of personal being (e.g., *perseverance, kindness*), of interpersonal interactions (e.g., *friendship, collaboration*), of practices (e.g., *efficiency, creativity*), and of bodies of knowledge (e.g. *openness, truth*) which a person considers important. In the context of mathematics education, values were first conceptualized by Bishop (1988), who later proposed that these values as they are expressed in the Western classroom might be categorized in three ways: mathematical (pertaining to the discipline itself), mathematics educational (pertaining to the learning and teaching of mathematics), and general educational (pertaining to the wider set of values which school education aims to inculcate in students) (Bishop, 1996). He highlighted that values need not be learnt for their own sake (e.g., moral and ethical values), but also that culturally-based values are embedded in school disciplines and their pedagogies, and that paying attention to them has implications for students' successful learning.

Thus, in the context of mathematics education,

> valuing is defined as an individual's embracing of convictions in mathematics pedagogy which are of importance and worth personally. It shapes the individual's willpower to embody the convictions in the choice of actions, contributing to the individual's thrivability in ethical mathematics pedagogy. In the process, the conative variable also regulates the individual's activation of cognitive skills and affective dispositions in complementary ways (Seah, 2019, p. 107)

Over the last 15 years or so, research into values in mathematics education has largely been coordinated by a 21-system research consortium named 'The Third Wave Lab'; with the name emphasizing conation as the third perspective on (mathematics) learning and teaching, with the first two being cognition and affect. Amongst the various research studies supported by 'The Third Wave Lab', the 'What I Find Important (in my maths learning)' (WIFI) study mapped the values that students from around the world embraced in their respective mathematics classrooms (e.g., Aktaş et al., 2021; Österling et al., 2015; Zhang et al., 2016). Perhaps not surprisingly, a prominent finding has been that students from different cultures valued different attributes of mathematics learning, with no single value found to be embraced by students across each and every one of the 21 participating systems. This has given mathematics classrooms in individual systems a unique valuing signature, which reflects previous claims that "cultural practices, norms and traditions can be reproduced in the school classroom, via the values that students have internalized over time" (Seah, 2022, p. 18).

12.2.2 Mathematical Wellbeing

If a student's fulfillment of their own values brings about a sense of wellbeing, then how do we define MWB in such a way that it might be assessed? Our literature review reveals that in Julia Hill's writings (e.g. Hill et al., 2022), MWB refers to "feeling good and functioning well" (Huppert & So, 2013, p. 839) when one is learning or teaching mathematics.

Hill et al. (2020) conducted a combined deductive/inductive thematic analysis of 488 Grade 8 student data that were collected in Melbourne, Australia. This exercise confirmed that students' sense of wellbeing with mathematics learning was associated with the fulfillment of eight ultimate values (UVs). These UVs are, namely, *relationships, engagement, cognitive, accomplishment, positive emotions, perseverance, music*, and *meaning*. Examples of student-nominated learning characteristics which reflected these values are shown in Table 12.1.

Soon after, the UV *music* was dropped (see, for example, Hill et al., 2021a, 2021b), acknowledging that it effectively served the UVs *engagement* and *positive emotions* (see relevant codes in Table 12.1).

A small-scale questionnaire study was conducted in 2021 in Chengdu, China with 258 Grade 3 students (Pan et al., 2022) to find out if the same seven values dimensions might apply in a non-Western culture. That the values dimensions for positive MWB turned out to be similar to the one derived by Hill et al. (2022) provided an assurance to the validity and reliability of the MWB framework. Yet, as pointed out in Pan et al. (2022), there were differences amongst the learning characteristics constituting each of the values dimensions. For example, while there was no apparent mention of parents in Australian students' valuing of *relationships* in achieving MWB, this valuing of *relationships* amongst the Chinese students can be inferred from expressions such as 'when I get help from my parents in my study'. It was also found that the same teaching strategies can have different understandings in different cultural settings, thus potentially representing different values as well. For example, 'when I help my mathematics teacher out with chores' represented in China a fulfillment of the student valuing of *accomplishment*, but it would not be surprising to see students in the West interpreting such teacher requests with negative connotations.

It became clear from the literature review that little has been researched about how lesson activities might be designed to help students fulfil particular values, such that these fulfillments promote the students' sense of MWB. This led to the following Research Questions which we adopted for our study:

Research Question 1: For elementary school students in China who are experiencing negative MWB, which group(s) of ultimate values are especially not fulfilled?

Research Question 2: What is the nature of the teaching strategies which have been found to be effective in helping elementary school students in China fulfil values leading to positive MWB?

Table 12.1 Values and learning characteristics associated with mathematical wellbeing

Themes and codes	Student examples
Relationships	
Teacher support	A supportive or good teacher
Peer support	Having friends to help me
General support	When I get help for my learning
Engagement	
Interesting/hands on	Learning interesting stuff
Focused working	Being absorbed in my work
Independent/quietness	When it is quiet and I am by myself
Music (engagement)	Music helps me to concentrate well
Cognitive	
Understanding	When I understand the material
Accomplishment	
Good marks	When I do good in a test
Accuracy	When I get the answers right
General mastery	When successful at learning something
Completing tasks	When I complete my mathematics work
Confidence	When I'm really confident
Positive emotions	
Enjoyment/fun/happiness	If the mathematics class is enjoyable
Relax (no pressure)	When there is no pressure
Music (emotions)	Music to listen to, so enjoy it more
Perseverance	
Challenge	Having work I find challenging
Working hard/practice	When I work hard
Music (no reasoning)	Listening to music in class
Meaning	
Future skills	Skills that will help me in life
Real world relevance	I like it when problems relate to real life

12.3 Methodology

12.3.1 Data Sources

Data for this study were collected in Chengdu, the capital city of Sichuan province, located in south-western China. With an urban population of more than 9 million, Chengdu is one of the eight largest cities in China.

Student participants were selected from amongst the 87 students in two of the classes which took part in the earlier MWB study (Pan et al., 2022). The classes were located in a government elementary school serving the local suburban community. Students in Class A appeared to like mathematics learning less than their peers in Class B generally, according to the mathematics teacher teaching both classes. Indeed, when asked in the earlier questionnaire to rate the extent to which they liked their mathematics learning on a scale of 0 (total dislike) to 10 (very much like), the mean scores for Classes A and B were 8.99 and 9.83 respectively.

Two male and two female students were selected from each of the two classes, who were subsequently invited to take part in this study. In Class A, these were the male and female students with the lowest MWB scores in their questionnaire response in the previous MWB study. The same criterion was used to select the female student participants from Class B, but not for the identification of the male participants. This is because all the boys in Class B rated their MWB higher than 9 (out of 10), so it would be rather questionable to assume that the two boys with the lowest scores were experiencing negative MWB. Thus, the classroom mathematics teacher (who was the first author as well) identified two boys in Class B whom she had noticed to have been increasingly disengaged in class, and whose mathematics performance was slipping. In this chapter, the students would be identified by an alphabet representing their class (A or B) followed by a number (1, 2, 3 or 4).

12.3.2 Research Design

The first data collected were aimed at identifying what the unfulfilled values were, which was seen as possibly inhibiting the MWB of the students. This was achieved through a pre-intervention questionnaire survey. Based on this knowledge and understanding, the mathematics teacher then selected and executed teaching strategies in her daily mathematics lessons over a three-week period. The teacher was encouraged to introduce teaching strategies which were commonly seen in local classes, but which were known to actualize the targeted values. An example of such a teaching strategy might be to contextualise mathematics problems in real-life scenarios (more), thereby helping students to fulfil their valuing of *engagement* and/or *meaning*. The extent to which these teaching strategies were able to help students fulfil their values with regards to mathematics learning—and leading to positive MWB—were assessed in a post-intervention survey.

Our research design resembles what McKenney and Reeves (2019) called 'educational design research'. This is reflected in how the mathematics teacher's drawing upon her experience and knowledge to develop an intervention to improve her students' MWB is the context for this current study and is the source of theoretical understanding. The value of teacher professional experience and knowledge was earlier demonstrated in this chapter, in Ai Lin's case.

The intervention period of three weeks might be short, but the intention has been to determine at this stage of our ongoing research if values fulfillment can be meaningfully achieved within a short time so that students' MWB is improved. Follow-up assessment of the sustainability of any MWB change was to be conducted some six months later, and this is not the subject of the current chapter. After all, negative wellbeing can be stressful and distressing personally, leading to more issues if it contributes to a disengagement in students' mathematics learning, and thus a fast, accurate and effective intervention is important, from the practical perspective.

The intervention plan included a perusal of all student responses to the pre-intervention survey, to identify the factors—and thus which of the seven potential underlying unfulfilled UVs from Table 12.1—accounting for the less-than-ideal MWB. This information was mainly derived from Item 4 of the pre-intervention questionnaire (see Fig. 12.1). Based on this knowledge, the students would be grouped into intervention groups, with each group characterized by the same values that the students felt were unfulfilled and which were inhibiting their positive MWB. For example, there might have been a group of students whose valuing of *perseverance* was perceived individually to be unfulfilled for a variety of reasons, so they would come together as a group in the class. This approach to class organization allowed the classroom mathematics teacher to design teaching activities and approaches which allowed students to then be able to exercise and experience the valuing of particular attributes. Involving all the students for the intervention—instead of just the four student participants—in each class is not only an ethical and equitable research practice, but also allows for the student participants' responses to the intervention to be examined in a naturalistic context.

12.3.3 Data Collection

Data were collected in December 2022 over a three-week period. The mathematics topics being covered over this time in the Grade 5 classes were areas of composite figures, areas of irregular figures, and patterns. Although the school year had begun only three months prior, the classroom mathematics teacher had taught the same two classes in the previous year (2021–2022), which meant that the teacher-student and student–student relationships were already established.

As it turned out, the then prevailing COVID19 pandemic situation in Chengdu had meant that school lessons became home-based and online from the third week onwards. As such, only the first two weeks of the three-week intervention were conducted face-to-face in the school classroom. Similarly, the post-intervention MWB assessment was also conducted online.

Student Mathematical Wellbeing Survey – Pre-intervention

Name: ______________________ Class: _________ Date: ________

1. With regards to your mathematics learning, do you generally feel relaxed, happy and a sense of accomplishment?
(The larger the number, the greater level of you feeling relaxed, happy and sense of accomplishment)
0 1 2 3 4 5 6 7 8 9 10

2. Do you think that you are good at learning mathematics?
(The larger the number, the better you think you are good)
0 1 2 3 4 5 6 7 8 9 10

3. When you are learning mathematics, which of the following help to make you feel relaxed, happy and a sense of accomplishment?
(Select one or more)
(a) When others help me during mathematics learning
(b) Cooperating with friends during mathematics learning
(c) ____________________ during mathematics lessons
(d) When completing a mathematics learning task
(e) Being rewarded for good grades
(f) Solving daily life problems using mathematics knowledge
(g) Realising that mathematics learning is beneficial for life in the future

4. When you are learning mathematics, what are the conditions which would make you feel tired, powerless, depressed, sad, or lack of accomplishment?

Fig. 12.1 Pre-intervention questionnaire

12.3.4 Instruments

An instrument used for this study is a questionnaire survey. Specifically, a questionnaire in Chinese language was administered to the student participants before the intervention, and a similar one after. Figures 12.1 and 12.2 show the English translated version of these questionnaires.

Guided by Huppert and So's (2013) reference to MWB as "feeling good and functioning well" (p. 839), Items 1 and 2 of the questionnaires have been phrased to find out the extent to which students have been feeling good and functioning well respectively. This concept of MWB also underlies the initial mapping study (Pan et al., 2022). Together, these two items provide student responses that indicate the extent to which each of them was experiencing MWB (or not).

Item 3 of the pre-intervention questionnaire has been designed to help us identify learning moments which were associated with positive MWB. The choices available to the questionnaire respondents were drawn from Hill et al. (2020) to facilitate the specific identification of unfulfilled UVs, complemented by open-ended opportunities in one of the choices as well as Item 4. Cross-checking these with Item 4 responses

Student Mathematical Wellbeing Survey – Post-intervention

Name: ______________________ Class: _________ Date: ________

0. Has online lessons been better or worse for your mathematics learning? How do you feel about online lessons?

1. With regards to your mathematics learning, do you generally feel relaxed, happy and a sense of accomplishment?
(The larger the number, the greater level of you feeling relaxed, happy and sense of accomplishment)
0 1 2 3 4 5 6 7 8 9 10
Do you think that your score would be higher or lower right before online lessons began?

2. Do you think that you are good at learning mathematics?
(The larger the number, the better you think you are good)
0 1 2 3 4 5 6 7 8 9 10
Do you think that your score would be higher or lower right before online lessons began?

3. Do you think your scores for Items 1 and 2 are higher, lower or the same when you responded to them the last time? What are the reasons for the differences in scores?

4. Since the last time you responded to this questionnaire, what are the conditions which would make you feel relaxed, happy and a sense of accomplishment when learning mathematics?

5. Since the last time you responded to this questionnaire, what are the conditions which would make you feel tired, powerless, depressed, sad, or lack of accomplishment when learning mathematics?

Fig. 12.2 Post-intervention questionnaire

relating to factors of negative MWB, we were able to determine the conditions under which MWB flourished and when it was being threatened. Such information was harnessed to allocate the student participants into different intervention groups.

Item 4 of the pre-intervention survey, as well as Items 4 and 5 of the post-intervention survey, are intended to extract additional information to complement the information which is derived from the more restrictive Item 3 of the pre-intervention survey. In other words, these 4 similar items are intended to allow us to understand what contribute to positive MWB and MWB change, as well as the nature of the intervention activities.

A key item in the two questionnaires is Item 3 in the post-intervention survey, in which students were asked to comment on self-perceived differences in the ways they felt relaxed, happy and a sense of accomplishment, as well as how good they thought they were at mathematics learning. It is an open-ended item, and students' sharing here revealed key factors during the intervention period accounting for MWB change. These data allowed us to analyse the extent to which the teaching strategies (that were customised to students' values fulfilment needs) introduced in the mathematics lessons were useful, and how so.

The unplanned disruption in the third and last week of the intervention, in which classroom teaching became home-based and online, had introduced an additional factor for students' MWB change. As a consequence, a decision was made to introduce additional items in the post-intervention questionnaire. Item 0 asked student respondents in what ways the move to online lessons might have affected their respective mathematics learning. Additional questions were also added to the end of Items 1 and 2 to find out how (in their opinion) students' scoring might have been affected by the disruption to schooling.

The other 'instrument' used to collect relevant data was the informal interview, which took the form of one-on-one participatory conversations (Swain & King, 2022) between the classroom teacher and each of the eight student participants at any one time. When permissions were gathered from the students and their parents, they were informed that the interview conversations were meant to be similar to daily teacher-student exchanges, in order to minimize student-perceived disruptions to their (mathematics) learning experiences in class. They should also allow for authentic opinions to be shared by the student participants, while at the same time not taking up much additional time of the classroom teacher. The objective of the conversations was to seek for further information or explanations that became evident during the analyses of the questionnaire responses.

After the pre-test, the students were divided into groups according to the nature of the unfulfilled UVs. In the first week, the teacher used morning reading, recess, lunch break and other opportunities to engage with the eight student participants in one-on-one casual informal conversation, in order to better understand the intervention experience of each student. Based on the data from the pre-intervention questionnaire and from these casual informal conversations, the teacher developed an individual intervention plan. One week after the implementation of the intervention program, the teacher also used morning reading, recess, lunch break and other time to catch up with the eight students individually, to understand the effectiveness of the intervention mechanism.

The third data source was teacher observations. We collectively acknowledge that like all other teachers, the classroom teacher notices and observes how her students interact with one another and with her, and such information provide additional data with which the teacher interprets the eight student-participants' unfulfilled UVs, and which subsequently contribute to the shape and form of the interventions. To this end, the classroom teacher kept handwritten notes of her professional observations of the eight student participants during the three-week data-collection period, with the prompt being 'observations of what student participant A1 (or A2, ..., B4) does or say that is unique, unexpected, or out-of-the-ordinary'.

12.3.5 Data Analysis

The two sets of questionnaire data were entered into a spreadsheet to facilitate analysis. Since the sudden move to home-based, online lessons towards the end of the intervention period meant that the post-intervention questionnaire had to be administered online asynchronously, the formats of the questionnaire were different—hardcopy for the pre-intervention version, and Microsoft Word-based for the post-intervention exercise.

Given the small number of student participants, the classroom teacher was able to peruse and compare the information collected through the two questionnaires. She focused on identifying the group(s) of UVs which were not being fulfilled and which were inhibiting MWB development, especially Items 3 and 4 of the pre-intervention questionnaire, and Items 4 and 5 of the post-intervention one. Given that the students were too young to write their responses extensively, the classroom teacher made use of the informal interview opportunities to clarify what the students had written, complemented by her own observations of the relevant student participants' actions and behaviour in class if needed.

Since the student participants might not remember what their responses to Items 1 and 2 were in the pre-intervention questionnaire, their responses to Items 1 and 2 of the post-intervention questionnaire were not used to establish if MWB had improved, deteriorated, or stayed the same over the three weeks. The items were meant instead to draw students' attention to the two components of MWB, to 'warm up' their thinking and thoughts before they answered the rest of the questionnaires.

Item 3 of the post-intervention questionnaire provided the source of student information relating to reasons for any change in levels of MWB. Themes were identified through the process of text analysis.

12.4 Results

12.4.1 Factors Affecting MWB

Item 4 of the pre-intervention questionnaire asked respondents for the factors under which their MWB had been affected. Students A3, B1, B2 and B3 wrote about making mistakes when engaging in peer collaboration and discussions, or when their opinions in these situations were not accepted (see Appendix). On the other hand, the general response from students A1, A2, A4, B2, B3 and B4 could be categorized as not being good at mathematics learning and/or not being able to experience success at mathematics work.

This information above led the classroom mathematics teacher to group the eight student participants into three intervention groups, as shown in Table 12.2. It can be seen that two values dimensions underlie students' less-then-ideal MWB, namely, *relationships* and *accomplishment*.

Table 12.2 Composition of intervention groups

Factors	Examples of student responses	Intervention group members
Relationships (peer support)	Having friends to help me	Group 1: A3, B1
Accomplishment (good marks, accuracy, mastery, completing tasks)	When I do good in a test; When I get the answers right; When successful at learning something; When I complete my work; Assist the teacher to finish the task	Group 2: A1, A2, A4, B4
Relationship and accomplishment		Group 3: B2, B3

12.4.2 Intervention Approaches

As alluded to earlier, the intervention approaches were customized to the valuing needs of each of the intervention groups. Assisting students to fulfil what they value calls for an understanding of what relevant teaching strategies are available, and how they might be effective. The teacher was assured, however, that the teaching strategies did not need to be anything novel or hard-to-deploy, but could be those that were commonly seen or easily applied in mathematics classrooms (e.g. acknowledging students' learning progress). Thus, a feature of this three-week intervention program has been that it might not even be visible to peers of the eight student participants.

Students in Intervention Group 1 (i.e. students A3 and B1), for example, needed to feel that their valuing of *relationships* was fulfilled when they engaged in mathematics learning, so that this fulfillment could lead to a lift in their MWB. The teacher's plan was to talk to each of the two students individually, to identify what they perceived to be the issues in instances when they were working with peers during mathematics lessons. Both students A3 and B1 had perceived that whenever they worked in pairs or groups, their peers were not paying attention to their verbal contributions. Through the teacher-student conversations, and from the teacher's own observations, it was evident that both these two students tended to be impatient with saying something in collaborative work settings. They would often be the first to speak, and elaborating at length. Peers would often be unhappy with such domination of discussion time, and some would then interrupt A3 or B1's speeches.

The teacher's planned intervention for these two students was in the form of ongoing individual teacher-student interviews and feedback. The teacher continued to meet the two students individually, reminding each of them the importance of turn-taking and listening when working in groups. She reminded them to structure their verbal contributions in four parts, namely, "My opinion is … The reason(s) is … My approach to problem solve is … Does anyone have anything to add?" The teacher also made it a point to observe students A3 and B1 more closely during pair or group activities in class, so that she could offer each of them her feedback on their performances after class. While the teacher was able to execute her plan in the first two weeks when lessons were face-to-face, the move to home-based, online lessons in the third week had reduced opportunities for student–student interactions during mathematics lessons. As such, the intervention for students A3 and B1 was much reduced in the third and last week.

There were 4 students in Intervention Group 2, namely, students A1, A2, A4 and B4. The classroom teacher had planned the following teaching strategies to help these students fulfil their valuing of *accomplishment*:

1. Selection of mathematics questions in homework tasks and in teacher's classroom questions in which the cognitive demands matched the students' perceived readiness at the moment
2. Opportunities for success
3. Teacher's timely critique
4. Praise and encouragement for students' changes
5. Teacher to better understand where students find difficulty learning content, so as to provide learning assistance timely
6. Teacher's follow up with individual students.

As such, during the intervention period (and thereafter), the teacher was more mindful of these students' learning progress, so that the questions she posed to them in class were at or slightly higher than the respective perceived levels of understanding. In so doing, these questions were within the Zone of Proximal Development of each student, which also allowed them to experience success and a sense of accomplishment. Note that the teacher was not posing easy questions to them, or questions which were easy relative to the students' levels of mathematical understanding. For example, when the teacher was probing her students for an approach to finding the area of the given (composite) shape, her questions ranged from the relatively easy "how might we find the area of this shape?" to "explain how we can find the area of this shape". The latter question was posed to the higher achieving students who needed relatively greater challenge to feel accomplished. In particular, when any of the six students (i.e. students A1, A2, A4, B2, B3, B4) was answering questions in front of their peers in class, the teacher was also mindful of being more liberal with affirmation and encouragements. During individual seatwork in class, the teacher would also pay more attention to these four students' work, so as to provide assistance in a more timely manner when needed. These intervention strategies were also able to be carried out in the week when lessons were home-based and online.

Students B2 and B3 were designated to Intervention Group 3, as both students appeared to require learning experiences which embraced both *relationships* and *accomplishment*. As these two values define Groups 1 and 2, the classroom teacher's plan was to adopt pedagogical practices which combine those used in the two other groups respectively. Through the conversations which the classroom teacher made with students B2 and B3, it was established that neither student believed that there was any difficulty with peer communication, except for their repeated accounts of how their opinions were often ignored by their peers and how they felt frustrated as a result. Follow-up conversations with their groupmates revealed that students B2 and B3 would often be amongst the last in the group to contribute, have no new content to add to the group discussion, and talked at length. Consequently, peers have found it hard to recognise the thrust of their message.

Students B2 and B3 were later able to reflect on this feedback, with both expressing that they were often keen to provide comprehensive and valid summaries. Integrating peers' opinions and comments into what they had in mind would explain why their arguments seemed lengthy and arduous. For these two students, the intervention then included advice by the classroom teacher for them to adopt two new practices. First, students B2 and B3 were encouraged to contribute early on during group discussions, and ending with a sentence such as, "now I shall listen to what others have to say, and I will add only if I have developed new ideas". Secondly, the two students were also discouraged from repeating what peers have said, unless they could improve on these.

In addition, during the third week of home-based, online lessons, the classroom teacher also gave these students more opportunities to answer teacher-posed questions. There was also more attention on providing positive feedback and encouragement to them when their mathematics work was being assessed online (Table 12.3).

Table 12.3 Intervention approaches

Intervention Group	Intervention Face-to-face	Online
1 (A3, B1)	Individual teacher-student coaching and feedback	
2 (A1, A2, A4, B4)	Selection of mathematics questions in homework tasks and in teacher's classroom questions in which the cognitive demands matched the students' perceived readiness at the moment Opportunities for success Teacher's timely critique Praise and encouragement for students' changes Teacher to better understand where students find difficulty learning content, so as to provide learning assistance timely Teacher's follow up with individual students	No change
3 (B2, B3)	Teacher matches difficulty level of learning tasks (e.g., homework, classroom questions) to perceived student readiness, providing students with opportunities for success Being more cognizant of where the students might find difficulties when learning, so that timely and relevant assistance can be provided One-on-one interaction	Provide these students with more opportunities to answer teacher-posed questions, mindful of their mathematical readiness Attention to providing more positive feedback and encouragement during online assessments

12.4.3 Post-intervention MWB

Item 3 of the post-intervention questionnaire (see Fig. 12.2) provides information on whether the teacher's adjustments to her lesson delivery (i.e. the intervention) had resulted in students experiencing more positive MWB, where MWB refers to "feeling good and functioning well" (Huppert & So, 2013, p. 839). The responses from the eight students are summarised in the Appendix. From this table, we can conclude that both components of MWB (i.e., feeling good and functioning well) had improved for students A1, A3, A4 and B2. For students A2 and B3, the 'feeling good' component was perceived to be similar before and after intervention, whereas 'functioning well' was boosted. Note that for student B3 here, her response was impacted by the switch to home-based, online learning, since she wrote that she was "happier in class since there is interaction. Online lessons can affect progress'.

On the other hand, for student B1, the 'feeling good' component apparently suffered but the quality of 'functioning well' seemed to have been held. Thus, the three-week intervention period appeared to have only affected students' MWB negatively for students B1 and B4, and affecting only the 'feeling good' component. Yet, reading these two students' feedback, the cause of these perceived drop in MWB was related to the last week of lessons being delivered online remotely: "We can't answer questions when learning online, so I have not been able to discuss with peers" (student B1), and "no sense of accomplishment when learning online" (student B4).

In other words, it is reasonable to say that the three-week intervention program, during which the classroom mathematics teacher facilitated students' fulfillment of MWB values, had generally improved the quality of student MWB, with half the student participants reporting improvements across both components of MWB.

For the group of student participants in this Chengdu research who valued *accomplishment*, it appears that an increase of MWB was facilitated through engaging with learning tasks that matched students' cognitive readiness. This was observed regardless of the students' initial levels of MWB. For these students, it was very important to be able to complete learning tasks and to correctly answer questions.

The next important factor regulating students' MWB was the quality of their interactions with peers. This aspect of their (mathematics) learning experience was however severely impacted by the nature of home-based, online lessons, leading to a drop of MWB that was observed in the last week of the intervention.

Thus, between the valuing of *accomplishment* and of *relationships* (both of which were key to MWB development and maintenance), the Chinese students in Chengdu prioritized the former in the form of personal accomplishment. This was not just suggested by the data, but also supported by teacher daily observation. Indeed, some of the students held the view that accomplishment brings about relationships; an excellent accomplishment in mathematics learning provides one with more say and more weight within peer groups. For these students, accomplishment was not derived from mathematics scores alone. In fact, a great proportion of this accomplishment came about from being successful in completing mathematics learning tasks, which included being correct in answering/speaking in front of the class, independently solving problems, and completing mathematics activities. The classroom teacher's targeted intervention plan which was based on an understanding of each student's readiness to learn the concepts related to areas of composite shapes, areas of irregular shapes, and patterns had meant that they got to experience positive emotions during their learning experiences. As such, they also felt that they were being—and becoming—good at mathematics learning.

These in no way suggest, however, that the students' relationships with one another were not developed. In fact, the prevailing relationships were strong and advanced. The students were generally confident when interacting with peers in these groupings, and they were also respectful of diverse perspectives and views expressed by their peers.

For students whose MWB was negatively impacted by difficulties in fulfilling the valuing of *relationships*, the two main factors were related to self-indulgence and to issues with mathematical reasoning. Student indulgence in themselves has meant that they seldom heard what their group members were saying, leading the latter to not feel respected, thus the negative reactions. The other factor related to students not being able to explain or reason mathematically in ways which were concise and clear, resulting in mathematical expressions which were unnecessarily long-winded, and leading to difficulties in understanding amongst their peers.

12.5 Discussion

It became clear through this current study that for students whose valuing of *accomplishment* needs a boost, this can come about through weaving into the mathematics lessons more opportunities for the students to experience success. This does not mean, however, a compromise in the difficulty level and/ or demands of mathematical tasks for students. Instead, based on the teacher's understanding of each student's cognitive readiness, teaching strategies were chosen such that they positioned them in their respective ZPD, and engaged them in productive struggle (NCTM, 2014). In this current study, these teaching strategies were expressed through classroom questions, groupwork, and individual mathematics work. Through these teaching strategies, students experienced success and thus, fulfilled their valuing of *accomplishment,* a UV that is highly valued by students (Hill et al., 2021a).

For those students who perceive that their valuing of *relationships* is not fulfilled, it is useful to work with the students concerned to identify the source(s) of negative emotions, so that teachers can guide these students to reflect, analyse and adjust. This is a value that is important to help students realise, since culturally diverse (including the ethnic Chinese) students had been found to regard *relationships* to be the most significant amongst the seven MWB UVs (Hill et al., 2021b).

The (Westerner) reader may be surprised that many of the pedagogical strategies/ approaches adopted in the intervention phase in this study were not already part of the teacher's repertoire. After all, one might expect that a teacher providing students with opportunities for success, or acknowledging and praising students for positive changes, are examples of what is commonly believed to be 'best practice' rather than part of intervention activities (as we saw above). Yet, in the Chinese school education system, the traditional emphases in mathematics lessons are on students' mastery of content, associated with deep understanding and being able to solve problems efficiently (Zhou et al., 2023). While students are expected to perform well in assessments, this expectation does not translate to students feeling accomplished. In fact, it is a cultural virtue to remain humble, to continue to scale new heights and to better one's own performance. "On the other hand, with relatively broad and deep content, and large class sizes, the teaching pace in Chinese mathematics classroom is always fast. Teachers have limited time available to deal with students' unexpected productive thinking" (Zhou et al., 2023, p. 1191). Indeed, the classroom teacher in this study represents a new generation of Chinese teachers who subscribe to a more holistic view of what optimal education quality should look like, taking into consideration not just cognitive demands but also affective and conative ones. This generation of teachers share views emerging in the West that the fostering of positive MWB is proactive towards the prevention of negative affect, thereby promoting positive cognitive growth.

As suggested by Table 12.1, a variety of teacher classroom strategies will be required to support students' fulfilment of all seven UVs. That they are all commonly employed by teachers in mathematics classrooms, rather than being novel approaches which require teacher professional learning, might suggest that teachers with 'good practices' would have been supporting students' fulfilment of the seven UVs already. This may indeed be the case with exemplary teachers of mathematics, exemplary in the sense that their students' MWB has been fostered optimally such that they are performing well in the discipline. However, it is likely more probable that teachers do not always teach in ways which support the fulfilment of all seven UVs, so that their students' MWB is affected in one way or another. For this group of teachers, which we believe most members of the profession belong to, identifying the UV(s) which is not yet supported for the class and/ or for individual students would inform how additional classroom strategies should be incorporated into their day-to-day mathematics teaching to support students' MWB.

Quite clearly, such intervention demands a lot from the classroom teacher. The teacher's immediate response to a student's request for assistance is crucial to the success of the intervention exercise. In the interest of sustainability, the number of targeted students needs to be kept small. After all, the effects of the intervention were evident in the three weeks adopted in this current study (Sect. 12.4.3), thereby opening up opportunities for other students with low MWB to benefit from such intervention approaches as well.

The classroom teacher's ability to help her students value *accomplishment* and *relationships* in this current study was also a function of her cognitive and affective knowledge and expertise. She drew on her professional experience to provide students with opportunities to gain and to feel accomplished and related, both cognitively and affectively. As we saw above in Sect. 12.4.2, cognitive strategies included advising students how to turn-take, how to structure their verbal contributions in four parts, and selecting questions based on teacher understanding of students' individual

ZPD. Those of the affective nature included providing students with opportunities to experience success, as well as ensuring that teacher praise and encouragement were given when students deserved the boost.

On the practical dimension, this current study provides teachers with pedagogical ideas to facilitate students' fulfillment of *accomplishment* and *relationships*, and perhaps also providing clues to how the other five UVs might also be facilitated in other classes. At a more philosophical level, the findings here demonstrate how cognitive and affective knowledge/ expertise can and do promote conative variables such as values (Seah, 2022), just as conative variables also promote cognitive and affective ones.

Although these findings above hold great promise for further understanding the construct that is mathematical wellbeing, that the study was conducted with only one teacher in one school has meant that generalisations cannot be assumed. Another limitation of this study relates to the self-reporting nature of student feedback regarding their MWB; it was through the students' questionnaire responses that unfulfilled UVs were identified, for example.

12.6 Conclusion

In this current research study, less-than-ideal levels of MWB amongst elementary school students in Chengdu were found to associate with students who were not able to fulfil the values of *accomplishment* and *relationships* (Research Question 1). An intervention which was informed by student data was executed over a three-week period, although the then prevailing COVID-19 pandemic had unexpectedly led to the third week of lessons being home-based and conducted online. Nevertheless, the targeted adjustments to the classroom teacher's practices had led to an improvement of the student participants' sense of mathematical wellbeing. These adjustments would vary according to the contexts in which the valuing of particular UVs of MWB was found wanting, drawing on teachers' cognitive and affective knowledge and expertise with regards to effective mathematics pedagogy.

Thus, with regards to Research Question 2, both cognitive and affective knowledge and expertise have proven useful in introducing the targeted adjustments needed to facilitate student valuing that are in turn crucial for enhancing MWB (see Sect. 12.5). The current study has shown how these two forms of knowledge were applied to help students fulfil a conative variable, that is, values (of *accomplishment* and *relationships*).

Importantly, this study has demonstrated that positive MWB is associated with students being able to fulfil *all* seven valuing of *accomplishment*, *cognitive*, *engagement, positive emotions, perseverance, meaning,* and *relationships*. Thus, in this study, the two students in Intervention Group 1 (i.e. students A3 and B1) who felt that their valuing of *relationships* was not actualized reported that their MWB was not optimal. On the other hand, the more important message for teachers is that they can facilitate students' fulfillment of the seven MWB values through the deployment of teaching strategies which are commonly seen in mathematics classrooms. There is no need for special, novel teaching strategy. There is no need for planned mathematics lessons to be re-organised. There is also no need for 'official' intervention programs.

We encourage more of similar intervention studies to be conducted to examine if—and how—this plays out when students' MWB can become more positive with a fulfillment in the valuing of any of the other five value dimensions, that is, *engagement, cognitive, positive emotions, perseverance,* and *meaning.* We are also eager to find out the extent to which the intervention approach adopted here remains effective and sustainable in the longer term.

Appendix: Responses to Item 3 of Post-intervention Questionnaire

Student	Gender	More relaxed, happier and greater sense of accomplishment?	Reasons?	Better at learning mathematics?	Reasons?
A1	M	Yes	When I raise my hand to speak, I become happier	Yes	Am now able to respond to teacher's questions
A2	M	Same	Answering questions	Yes	Active interactions
A3	F	Yes	Am able to reason/think to answer questions lately, thus a sense of accomplishment	Yes	Am able to reason/think to answer questions lately, thus a sense of accomplishment
A4	F	Yes	There are now more classroom activities, mathematics learning has become more interesting, I volunteer to answer questions more, I have firmer grasp of my knowledge, thus I am happier now when learning	Yes	There are now more classroom activities, mathematics learning has become more interesting, I volunteer to answer questions more, I have firmer grasp of my knowledge, thus I am happier now when learning

(continued)

(continued)

Student	Gender	More relaxed, happier and greater sense of accomplishment?	Reasons?	Better at learning mathematics?	Reasons?
B1	M	No	We can't answer questions when learning online, so I have not been able to discuss with peers	Same	We can't answer questions when learning online, so I have not been able to discuss with peers
B2	M	Yes	Learning is easier when online. Questions are also easier to answer	Yes	Due to recent revisions, and I have also learnt Olympiad mathematics
B3	F	Same	Am happier in class since there is interaction. Online lessons can affect progress	Yes	Am happier in class since there is interaction. Online lessons can affect progress
B4	F	No	No sense of accomplishment when learning online	No	Learning has become more relaxing, and I have also mastered more

References

Aktaş, F. N., Akyıldız, P., & Dede, Y. (2021). Students' mathematics educational values at religious vocational middle schools: A cross-sectional study. *Education, 3*(13), 1–13. https://doi.org/10.1080/03004279.2021.1981421

Bishop, A. J. (1988). *Mathematical enculturation: A cultural perspective on mathematics education*. Kluwer Academic Publishers.

Bishop, A. J. (1996). How should mathematics teaching in modern societies relate to cultural values—Some preliminary questions. In N. D. Tri, P. T. Long, C. Comiti, D. R. Green, E. Southwell, & J. Izard (Eds.), *Seventh Southeast Asian conference on mathematics education* (pp. 19–24). Vietnamese Mathematical Society.

Clarkson, P., Bishop, A., & Seah, W. T. (2010). Mathematics education and student values: The cultivation of mathematical well-being. In T. Lovat & R. Toomey (Eds.), *International handbook on values education and student well-being* (pp. 111–136). Springer.

Hill, J. L., Kern, M. L., Seah, W. T., & van Driel, J. (2020). A model of student mathematical well-being: Australian Grade 8 students' conceptions. In M. Inprasitha, N. Changsri, & N. Boonsena (Eds.), *Interim Proceedings of the 44th Conference of the International Group for the Psychology of Mathematics Education* (pp. 273–281). PME.

Hill, J. L., Kern, M. L., Seah, W. T., & van Driel, J. (2021a). Feeling good and functioning well in mathematics education: Exploring students' conceptions of mathematical well-being and values. *ECNU Review of Education, 4*(2), Article 2. https://doi.org/10.1177/2096531120928084

Hill, J., L Kern, M., van Driel, J., & Seah, W. T. (2021b). The importance of positive classroom relationships for diverse students' well-being in mathematics education. In D. Burghes & J. Hunter (Eds.), *Mathematics Education for Sustainable Economic Growth and Job Creation* (1st ed., pp. 76–89). Routledge. https://doi.org/10.4324/9781003048558-7

Hill, J. L., Kern, M. L., Seah, W. T., & van Driel, J. (2022). Developing a model of mathematical wellbeing through a thematic analysis of the literature. In C. Fernández, S. Llinares, A. Gutiérrez, & N. Planas (Eds.), *Proceedings of the 45th Conference of the International Group for the Psychology of Mathematics Education* (Vol. 2, pp. 379–386). PME.

Huppert, F. A., & So, T. T. C. (2013). Flourishing across Europe: Application of a new conceptual framework for defining well-being. *Social Indicators Research, 110*(3), 837–861. https://doi.org/10.1007/s11205-011-9966-7

McKenney, S., & Reeves, T. C. (2019). *Conducting educational design research* (2nd ed.). Routledge.

NCTM. (2014). *Principles to actions: Ensuring mathematical success for all.* National Council of Teachers of Mathematics.

Österling, L., Helena G., & Andersson, A. (2015). Balancing students' valuing and mathematical values. In S. Mukhopadhyay & B. Greer (Eds.), *Proceedings of the Eighth International Mathematics Education and Society Conference* (Vol. 3, pp. 860–872). MES8.

Pan, Y., Zhong, J., Hill, J. L., & Seah, W. T. (2022). Values which facilitate mathematical wellbeing of Chinese primary school students: A preliminary study. In *Twelfth Congress of the European Society for Research in Mathematics Education (CERME12), TWG08(13)*. https://hal.archives-ouvertes.fr/hal-03745622

Reeve, J. (2012). A self-determination theory perspective on student engagement. In S. L. Christenson, A. L. Reschly, & C. Wylie (Eds.), *Handbook of research on student engagement* (pp. 149–172). Springer. https://doi.org/10.1007/978-1-4614-2018-7_7

Seah, W. T. (2019). Values in mathematics education: Its conative nature, and how it can be developed. *Research in Mathematical Education, 22*(2), 2. https://doi.org/10.7468/JKSMED.2019.22.2.99

Seah, W. T. (2022). Values-based mathematics teaching: Promoting mathematical performance, competencies and wellbeing. *Proceedings of the Singapore National Academy of Science, 16*(01). https://doi.org/10.1142/S2591722622400026

Swain, J., & King, B. (2022). Using informal conversations in qualitative research. *International Journal of Qualitative Methods, 21*, 16094069221085056. https://doi.org/10.1177/16094069221085056

Tiberius, V. (2018). *Well-being as value fulfilment: How we can help each other to live well.* Oxford University Press.

Zhang, Q., Barkatsas, T., Law, H. Y., Leu, Y.-C., Seah, W. T., & Wong, N.-Y. (2016). What primary students in the Chinese Mainland, Hong Kong and Taiwan value in mathematics learning: A comparative analysis. *International Journal of Science and Mathematics Education, 14*(5), Article 5. https://doi.org/10.1007/s10763-014-9615-0

Zhou, J., Bao, J., & He, R. (2023). Characteristics of good mathematics teaching in China: Findings from classroom observations. *International Journal of Science and Mathematics Education, 21*(4), 1177–1196. https://doi.org/10.1007/s10763-022-10291-5

Chapter 13
To What Extent Are Students Fulfilling Their Values and Thriving in Mathematics Education?—The Case for Victoria, Australia

Julia L. Hill, Margaret L. Kern, Wee Tiong Seah, and Jan van Driel

13.1 Introduction

Anna, a Year 8 student, is a few months into the new school year. Usually, an articulate and dedicated mathematics student, this year Anna has found herself feeling frustrated, her attention is waning, and she is beginning to question the value mathematics serves for her future aspirations. Anna believes she learns mathematics most effectively when she is working with her peers, sharing ideas, and learning from others; when she feels her teacher know, respect, and care for her; and when she feels like mathematics is meaningful to her. However, this year, Anna has to sit in straight rows, facing the board with minimal peer interaction; her teacher appears to care more about student achievement then developing a relationship with the students, and uses a direct instructional approach that feels disconnected to the world around her. Thus, through her teacher's actions and pedagogical practices, Anna's mathematics educational values (Bishop, 1996) are no longer being fulfilled, contributing to her negative feelings and disengagement in mathematics. That is, Anna's mathematical wellbeing (MWB) has diminished, and she is beginning to experience mathematical 'illbeing'.

The cultivation of student wellbeing in schools has become a global priority (e.g. United Nations, n.d.). Unfortunately, the adolescent period shows a near universal

J. L. Hill (✉) · M. L. Kern · W. T. Seah · J. van Driel
Faculty of Education, The University of Melbourne, Melbourne, Australia
e-mail: hill.j@unimelb.edu.au

M. L. Kern
e-mail: peggy.kern@unimelb.edu.au

W. T. Seah
e-mail: wt.seah@unimelb.edu.au

J. van Driel
e-mail: j.vandriel@unimelb.edu.au

Y. Dede et al. (eds.), *Values and Valuing in Mathematics Education*,
https://doi.org/10.1007/978-981-99-9454-0_13

decrease in wellbeing, which is steeper than during any other life period, with the sharpest decline occurring during the primary to secondary school transition (Hill et al., 2022a, 2022b). Student wellbeing is increasingly on government agendas, including New Zealand, Australia, and the United States, with millions in funding recently being allocated to support wellbeing initiatives in schools (e.g. Clark, 2023). However, most investments and explorations focus on a global measure of wellbeing, and lack specificity and relevance to individual subject domains including mathematics. There are problems with this global approach; for example, wellbeing can differ across contexts (e.g. environments), and across school subjects. Also, a global approach prevents teachers from articulating the language and principles of wellbeing to make wellbeing relevant to their pedagogies and classrooms.

Many students do not enjoy mathematics (Attard, 2012; Larkin & Jorgensen, 2016). Indeed, it is one of the most unpopular and anxiety-triggering subjects which students face at school (Grootenboer & Marshman, 2015). We suggest that this illbeing arises in part from students' subject-specific values not being fulfilled. Indeed, many core values that support wellbeing more broadly, such as close relationships, a sense of enjoyment, or meaning (Seligman, 2011), are often not being fulfilled in mathematics education, especially in secondary years. That is because teachers often use teacher-centric direct instruction (Hunter et al., 2016); overlook social ways of working (Grootenboer & Marshman, 2015); and classes lack meaningful learning opportunities, leaving students wondering "when will I ever use this?".

If we hope to improve how students feel and function in their mathematics class, we must first understand what students value in the subject. Previous studies posit that primary and secondary students across several countries espouse seven core or 'ultimate values' (UVs) (i.e. accomplishment, cognitions, engagement, meaning, perseverance, positive emotions, and relationship) that support their positive learning experiences and wellbeing in mathematics (Hill et al., 2020). Then to improve students' feelings and functioning in mathematics, we should explore how successful students are at fulfilling these UVs, since high value fulfilment contributes to high MWB. In this chapter, we measure how students prioritize the seven UVs; their experiences of these seven UVs; and the difference between these (i.e. values and experiences), which is students value fulfilment—or their MWB. We first provide background on a values-based understanding of wellbeing, and then present a model of MWB that we developed through prior work (Hill et al., 2020, 2022a, 2022b). Next, we describe our quantitative study which includes testing the psychometrics of the measures, and then, once validated, we report on students' levels of MWB. We end with a discussion of the implications and applications for research and practice, including practical examples that educators can incorporate into their mathematics classroom to cultivate students' MWB.

13.1.1 A Value Fulfilment Model of Student Wellbeing

Values provide a blueprint for wellbeing (Tiberius, 2018). Values, however, can differ across persons, cultures, and contexts (Alexandrova, 2017). Values are the things we consider deeply important, act as a compass guiding our decisions and actions in life, and are standards upon which we judge how well our lives (and learning) are going (Halstead, 1996; Tiberius, 2018). For example, a student valuing peer support would likely seek out collaborative learning experiences in the classroom, and judge that their learning was going well if given repeated opportunities to work alongside peers. We define student wellbeing in terms of value fulfillment theory (Tiberius, 2018), which suggests that an individual's experience of wellbeing depends on the extent to which their values are fulfilled. Specifically, Tiberius asserts: "our lives go well to the extent that we peruse, and fulfill or realize, our appropriate values…your life goes badly to the degree that you live a life that has little value fulfillment" (p. 34). Value fulfillment theory is helpful because it gives us signposts about how to improve one's wellbeing, by understanding what students value and then signaling what teachers and students might do (e.g. the pedagogies, classroom climate, class positioning) to help fulfil these values.

Tiberius distinguishes between UVs, or core values, and instrumental values (IVs). UVs are at the highest level—the things valued for their own sake, which are generally consistent across individuals and contexts. For example, people from most cultures tend to value close relationships, accomplishments, and a sense of meaning. However, certain cultures (e.g. collectivist, individualist) may prioritize these UVs in a different hierarchy (e.g. people from individualist societies, such as Australians, Swedes, often prioritize the positive emotions UV, like happiness, more so than collectivist societies, such East Asians (Joshonloo et al., 2021). Underneath these UVs are IVs, which are the things valued to achieve the UVs. These IVs can vary widely between individuals (Tiberius, 2018). For example, the fulfillment of close relationships (an UV) might be achieved through different IVs, such as team sports, playing video games with friends, working collaboratively in class. In this chapter we focus on both IVs (e.g. peer support, achieving high marks on tests) and the corresponding seven UVs in mathematics, which have previously been identified as important for students' MWB (Hill et al., 2022a, 2022b).

13.2 The Development of Our MWB Model and Explorations of MWB

We define MWB as the fulfillment of core or UVs within the mathematics learning process that is accompanied by positive feelings (e.g. enjoyment) and functioning (e.g. engagement, accomplishment) in the subject (Hill et al., 2022a, 2022b). Our definition combines value fulfillment theory (Tiberius, 2018) with hedonic (i.e. subjectively felt aspects of happiness) and eudemonic (i.e. optimal functioning) wellbeing perspectives (Huppert & So, 2013). The MWB model, upon which the current study is based, was developed by Hill et al., (2022a, 2022b). The authors undertook a scoping review, thematically analyzing 40 mathematics education values publications aiming to uncover how students' IVs in mathematics aligned with UVs proposed in the wellbeing literature. Hill et al., (2022a, 2022b) discovered 90 unique inductive IVs espoused by mathematics students which could be deductively categorized into seven UVs: accomplishments, cognitions, engagement, meaning, perseverance, positive emotions, and relationships (see Table 13.1 for descriptions of these). Source models for these UVs included Seligman's (2011) PERMA model (Positive emotions, Engagement, Relationships, Meaning, and Accomplishment); Kern's et al. EPOCH model (Engagement, Perseverance, Optimism, Connectedness, Happiness); the Organization for Economic Co-operation and Development (OECD) framework of student wellbeing in schools (OECD, 2015); and an earlier MWB model.

Several studies across multiple countries have explicitly explored students' MWB using the model developed by Hill and Kern et al., (2022a, 2022b). For example, Year 3 Chinese students rated and nominated IVs most important for their MWB, including good grades (fulfilling UV accomplishment); fun (UV positive emotions); interest and mathematics themed videos (UV engagement); and teacher praise (UV relationships) (Hill & Seah, 2023). Australian Year 8 students described factors that made them feel good and function well in mathematics, revealing 21 unique IVs that could be categorized into the seven UVs, the most frequent being positive classroom relationships, engagements, then cognitions (Hill et al., 2020). New Zealand Year 3 to 8 students experienced the UVs relationships, accomplishments, and meaning the most, and the UV engagement the least, whilst students' experiences across all seven UVs declined as the grade level increased (Hill et al., 2022a, 2022b).

13.3 Gender Differences and Values in Mathematics Education

Research studies show both similarities and differences in relation to the values espoused by males and females in mathematics education (e.g. Wigfield et al., 1997). For example, Wong (1995) asserted that females valued collaboration and males valued problem solving and competition in mathematics education. Likewise, Barkatsas et al. found males valued problem solving processes with mathematical understanding, meaningfulness, effort and practice (e.g. working on lots of examples) whereas girls valued autonomy, having their voices heard, and mathematical discourse. Whilst females and males espoused similar values concerning the importance of mathematical achievement (Wigfield et al., 1997) females attributed their success more to teacher and peer support than males (Riegle-Crumb et al., 2006). Notably, despite several studies showing females espouse greater relational values in mathematics, such as family support, peer collaboration and respectful classroom

Table 13.1 Our model of mathematical wellbeing

Ultimate values in maths education	Description	Source	Example instrumental values in maths education
Accomplishment	Valuing achievement, reaching goals, confidence or mastery completing mathematical tasks and tests	PERMA	Accuracy, high marks, goals, confidence
Cognitions	Valuing knowledge, skills, and/or understanding required to do mathematics at school	MWB, OECD	Efficiency, recall, prior knowledge, understanding
Engagement	Valuing concentration, absorption, deep intertest, or focus when learning/doing mathematics	PERMA, EPOCH	Attention, interesting work, novel learning, autonomy
Meaning	Valuing direction in mathematics; feeling mathematics is valuable, useful, worthwhile or has a purpose	PERMA	Maths agency, real world links, utility, task value
Perseverance	Valuing drive, grit, or working hard towards completing a mathematical task or goal	EPOCH	Challenging maths, perseverance, practice and hard work
Positive Emotions	Valuing positive emotions when learning/doing mathematics e.g., enjoyment, happiness, or pride	PERMA, EPOCH, MWB	Minimal anxiety, fun, safe climate, pride
Relationships	Valuing supportive relationships; feeling valued, respected and cared for; connected with others; or supporting peers in mathematics	PERMA, EPOCH, OECD	Belonging, group work, family support, teacher explanations, teacher warmth and care, peer support

Note PERMA (Seligman (2011): *P* positive emotion, *E* engagement, *R* relationships, *M* meaning, *A* accomplishment; EPOCH: *E* engagement, *P* perseverance, *O* optimism; *C* Connectedness; *H* Happiness; *OECD* Organization for Economic Co-operation and Development (OECD, 2015); *MWB* mathematical wellbeing

relationships (e.g. Hill & Hunter, 2023; Wong, 1995), they often simultaneously experience lower perceived social support, use group work less, and feel less involved than males in the mathematics classroom (Eccles, 2011; Samuelsson & Samuelsson, 2016). Based on these findings we conjecture that girls' valuing of relationships (an UV) is often not being entirely fulfilled in mathematics. In this study we investigate these gender difference in value fulfillment in mathematics education.

13.4 Best Practice Guidelines for Assessing Student Wellbeing

Jarden et al. (2021) proposed best practice guidelines concerning wellbeing assessment with students, which we focus on here. First, assessment should follow appropriate psychological processes like careful planning (e.g. identify goals), appropriate data collection (e.g. using qualitative and/or quantitative measures), processing and coding the data (e.g. appropriate statical analyses), and communicating the data (e.g. written journals, conferences). Applied here, we considered our goals (i.e. developing a survey to measure value fulfillment and MWB), method of data collection (i.e. using Likert survey responses), and methods of communication (e.g. this book chapter).

Secondly, assessment should be psychometrically sound (i.e. with reliability and validity tested and affirmed). Validity means the assessment can be relied upon to provide consistent and reliable data (Jarden et al., 2021). Validity ensures assessment effectively measures what it is supposed to measure. Reliability indicates that items within sub-scales are consistent with one another. Considering these principles, this chapter reports on the psychometric properties of our MWB survey, before describing the results of the survey.

Thirdly, the comprehensiveness of surveys must be balanced with survey length. That is because students may lose focus or not complete lengthy surveys impacting on the reliability and validity of the data (Jarden et al., 2021). Here we decided on a short survey to capture the domains of MWB as briefly as possible, reducing burden on participants and schools while still validly capturing their MWB.

13.5 The Current Study

The current study first tests the psychometrics of a MWB survey. We then report on the importance of students' values (i.e. how important the seven UVs are to students); students' experiences of these values in the context of learning mathematics (i.e. how successful students are at actioning these UVs); and students' personal value fulfillment (i.e. the difference between the importance of and experience of these UVs). Specifically, we focus on the following research questions:

1. How important are the seven UVs and what are students' experiences of these seven UVs in mathematics education?
2. Does gender relate to the importance of and experiences of these seven UVs?
3. To what extent are students on average fulfilling their valuing of these seven UVs?
4. Does gender relate to the fulfillment of students' valuing of these seven UVs?

13.5.1 Participants

Participants included 488 Year 8 students (233 female; 255 male) aged between 13 and 14 years. This age group was purposely selected because student wellbeing tends to decline as students transition from the primary to secondary school years (Hill et al., 2022a, 2022b); thus, understanding factors supporting students MWB for this age group is pertinent. Also, students in Year 8 have experienced at least one full year of secondary mathematics instruction. Student participants attended one of nine secondary schools (two Catholic; three private/independent; four Government/public) located throughout urban Melbourne and the surrounding regional cities of Victoria, Australia. Schools ranged from medium to high socio-economic status. Students self-identified themselves as Australian (71%), Asian (14%), European (6%), Indian/Pakistani (6%), Indigenous Australian (2%), North/South American (2%), or Middle Eastern (1%). Most students (80%) were Australian born, whilst 20% were born overseas. Data was collected during class time, between April and December in 2019.

13.5.2 Measures

The measure consisted of a Qualtrics-based survey featuring demographic, free-response, and Likert style questions. Here we report on students' responses to 44 Likert questions split across two sections (the wording of questions in Sections A and B is summarized in Table 13.2).

Section A asked students about their values, specifically the relative importance of the 22 IVs and corresponding seven UVs (see Table 13.2). Given that all seven UVs were likely valued to some extent by most students, we were interested in the hierarchy of these UVs. For example, *It is important to me that I have help and support from my maths teacher/s when I need it* measured the degree of importance of the IV teacher support for students.

Section B measured students' experiences of these same 22 IVs and corresponding seven UVs. For example, *I have help and support from my maths teacher/s when I need it* measured students' experiences of teacher support (an IV) in the mathematics classroom.

Using a digital device, students responded to the 44 randomized survey questions using a sliding bar scale ranging from 0 ('not like me at all') to 10 ('completely like me'). Labels appeared only on the endpoints, as recommend by the OECD (2013a) guidelines. The 11-point scale, labels, and sliding scale were adapted from the PERMA Profiler (Butler & Kern, 2016). Because measures of wellbeing and values tend to skew towards the positive end (Butler & Kern, 2016) we decided to use an 11-point scale, as opposed to a standard 5-point scale, to reduce data skewedness and increase variability (Dawes, 2008).

Table 13.2 Seven ultimate values and corresponding survey question wording

Ultimate values	Item #	Instrumental values	Section A wording: importance	Section B wording: experiences
Accomplishment	A1	Progress	It is important to me that I am making progress towards accomplishing my goals in maths	I feel like I am making progress towards accomplishing my goals in maths
	A2	Goals	In maths it is important to me that I am achieving the important goals I have set for myself	In maths I feel like I am achieving the important goals I have set for myself
	A3	Test achievement	It is important to me that I can handle my maths tests	I feel like I can handle my maths tests
	A4	Assessment achievement	It is important to me that I can handle my maths assignments	I feel like I can handle my maths assignments
Cognitions	C5	Understanding	In maths it is important to me that I understand what I have been learning	In maths I understand what I have been learning
	C6	Maths skills	It is important that I have the maths skills that I need to complete my work in maths	I have the maths skills that I need to complete my work in maths
	C7	Problem solving	It is important that I am able to solve the maths problems that we need to do	I am able to solve the maths problems that we need to do
Engagement	E8	Flow	When I am doing maths it is important to me that I get completely absorbed in what I'm doing	When I am doing maths I get completely absorbed in what I'm doing
	E9	Dedication	It is important to me that I feel dedicated to my maths learning	I feel dedicated to my maths learning
	E10	Interest	When I am learning new things in maths it is important to me that I feel deeply interested	When I am learning new things in maths I feel deeply interested
Meaning	M11	Purposeful	It is important to me that I feel like my maths learning has a purpose and is meaningful to me	I feel like my maths learning has a purpose and is meaningful to me
	M12	Worthy	It is important to me that I feel like my maths learning is valuable and worthwhile	I feel like my maths learning is valuable and worthwhile

(continued)

Table 13.2 (continued)

Ultimate values	Item #	Instrumental values	Section A wording: importance	Section B wording: experiences
	M13	Utility	It is important that my maths learning gives me a sense of direction in my life	My maths learning gives me a sense of direction in my life
Perseverance	P14	Completing work	In maths it is important to me that I finish whatever I begin	In maths I finish whatever I begin
	P15	Persistence	It is important that I keep at my maths work until I am done with it	I keep at my maths work until I am done with it
	P16	Effort	It is important to me that I work hard at my maths learning	I work hard at my maths learning
Positive Emotions	PE17	Fun	When I am doing maths it is important to me that I have a lot of fun	When I am doing maths I have a lot of fun
	PE18	Happiness	When I am doing maths it is important to me that I feel happy	When I am doing maths I feel happy
	PE19	Optimism	When I am doing maths it is important to me that I feel positive	When I am doing maths I feel positive
Relationships	R20	Teacher support	It is important to me that I have help and support from my maths teacher/s when I need it	I have help and support from my maths teacher/s when I need it
	R21	Peer support	It is important to me that I have friends that support me with my maths learning when I need it	I have friends that support me with my maths learning when I need it
	R22	General support	When I have a problem with my maths learning it is important to me that I have someone to help me	When I have a problem with my maths learning I have someone to help me

Because our seven UVs were deductively derived from other wellbeing models, we adapted the wording of existing validated wellbeing surveys (e.g., Butler & Kern, 2016). For example, the EPOCH survey measures adolescent wellbeing across five domains, four of which (i.e. engagement, perseverance, connectedness, and happiness) are included in our MWB model, shown in Table 13.1. Subsequently, the EPOCH survey item *I am a hard worker* measuring perseverance, was adapted here to *I work hard at my maths learning*; likewise, *I feel happy* (measuring positive emotions) was adapted to *When I am doing maths, it is important to me that I feel happy*.

Each of the seven UVs was associated with three survey items, except accomplishment, which had four items. To get a composite score for each UV we calculated the means of the three or four survey items.

13.5.3 Psychometric Testing of the Survey

We used R software (version 4.2.2) with the lavaan (Rosseel, 2012) and psych (Revelle, 2015) packages to test for validity and reliability respectively. First, for survey validity, we used confirmatory factor analysis (CFA) to evaluate the model, shown in Fig. 13.1, first for the values, then for experiences, along with calculating the latent correlation between values and experiences. Model fit was tested using the root mean square error of approximation (RMSEA), the standardized root mean residual (SRMR), the Tucker Lewis Index (TLI), and the Comparative Fit Index (CFI). Acceptable model fits are represented by RMSEA and SRMR values being below 0.08, TLI and CFI values being above 0.9 (Hooper et al., 2008).

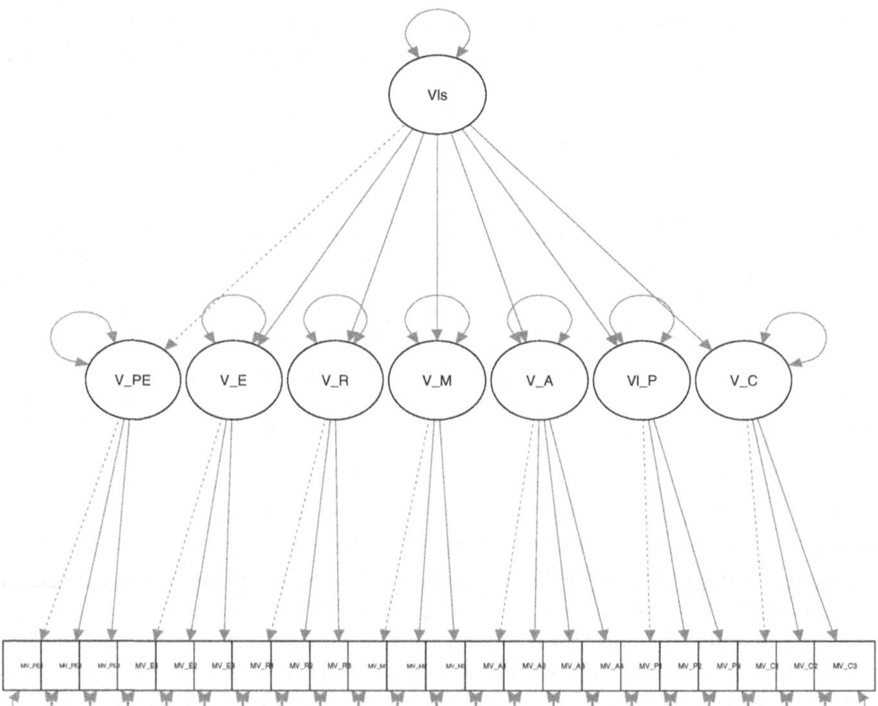

Fig. 13.1 Estimated factorial model, tested with importance then with experiences

13.6 Results

13.6.1 Psychometric Properties of the Survey

Confirmatory factor analysis supported our seven-factor model, for the two models (i.e. value importance, and the experience models), with one exception. R21 (*I have friends that support me with my maths learning when I need it)* demonstrated compromised fit to the model ($\lambda_{R2} = 0.51$). Removing this item improved model fit. Table 13.3 summarizes the latent factor loadings, and indictors of model fit across the values importance and experience models, with Fig. 13.1 illustrating the estimated models. Test of invariance showed invariance across gender was supported (Δ CFI < 0.01) indicating the measures (i.e. importance and experiences) are directly comparable across gender. This meant our analysis could continue with direct comparisons of means and loadings.

Internal consistencies across importance and experiences are summarized in Table 13.4. Across both importance and experiences all seven UVs items showed a high level of reliability. Across both, relationships was the least reliable UV. However, this was expected because we purposely increased item validity by broadening how we measured relationships (i.e. incorporating teacher, friends, and general support) at the expense of high internal reliability. All importance items were found to correlate with the experience items. Gender correlated only with importance ratings for cognitions, and relationships because females valued these aspects higher than males.

In sum, our results indicated that our measures validly and reliably measure students' valuing and their experiences of these seven UVs in mathematics.

13.6.2 Value Importance and Experience Survey Results

Because our measures were both valid and reliable, we turned to our research questions. We consider the importance (Section A) and the student experiences (Section B) of these values, including mean scores, standard deviations, data spread at the group level (RQ1), and separately by gender (RQ2), as summarized in Table 13.5, and illustrated in Figs. 13.2 and 13.3.

Across the whole sample (RQ1) for the values importance section (Section A, top of Table 13.5) students rated the UV cognitions as most important, followed by UVs relationships, accomplishments, perseverance, meaning, positive emotions, and lastly engagement. However, all seven UVs skewed towards being important ($M > 7.10$), which is expected given other wellbeing surveys show a similar skewing towards the positive end (Butler & Kern, 2016; Zeng & Kern, 2019). For the experiences section of the survey (Section B, bottom of Table 13.5), students rated that they experienced the cognitions UV the most, followed by relationships, perseverance, accomplishments, meaning, engagement, and lastly positive emotions. Also, the less experienced UVs (i.e. meaning, engagement and positive emotions) had

Table 13.3 Latent factor loadings and fit indices for the importance and experiences confirmatory factor analyses (see Table 13.2 for survey questions and see Fig. 13.1 for estimated model)

Factor/item		Importance	Experiences
Positive emotion			
PE17	λ_{PE1}	0.78	0.89
PE18	λ_{PE2}	0.88	0.94
PE19	λ_{PE3}	0.89	0.88
Engagement			
E8	λ_{E1}	0.83	0.87
E9	λ_{E2}	0.87	0.86
E10	λ_{E3}	0.85	0.85
Relationships			
R20	λ_{R1}	0.85	0.72
R21	λ_{R2}	0.72	–
R22	λ_{R3}	0.87	0.84
Meaning			
M11	λ_{M1}	0.87	0.89
M12	λ_{M2}	0.86	0.87
M13	λ_{M3}	0.73	0.80
Accomplishment			
A1	λ_{A1}	0.85	0.84
A2	λ_{A2}	0.81	0.79
A3	λ_{A3}	0.78	0.81
A4	λ_{A4}	0.79	0.86
Perseverance			
P14	λ_{P1}	0.84	0.85
P15	λ_{P2}	0.83	0.84
P16	λ_{P3}	0.86	0.75
Cognitions			
C5	λ_{C1}	0.88	0.87
C6	λ_{C2}	0.86	0.91
C7	λ_{C3}	0.84	0.83
Latent factor			
Positive emotions	λ_{PE}	0.77	0.92
Engagement	λ_{E}	0.92	0.99
Relationships	λ_{R}	0.81	0.69
Meaning	λ_{M}	0.85	0.91
Accomplishment	λ_{A}	0.99	0.90
Perseverance	λ_{P}	0.93	0.92

(continued)

Table 13.3 (continued)

Factor/item		Importance	Experiences
Cognitions	λ_C	0.91	0.83
Model fit			
N		472	483
RMSEA		0.098	0.087
RMSEA 90% confidence interval		0.092, 0.103	0.081, 0.093
SRMR		0.053	0.052
χ^2 (df = 202)		112.27	849.19
CFI		0.902	0.928
TLI		0.887	0.917

Note Confirmatory factor analysis estimated using the Lavaan package (version 0.6.13) in R (version 4.2.2). *RMSEA* Root mean square error of approximation, *SRMR* Standardized root mean residual, *CFI* Comparative fit index, *TLI* Tucker Lewis index

Table 13.4 Internal reliability scores across the value importance (A) and experiences (B) sections

Survey	Seven ultimate values						
	A	C	E	M	P	PE	R
Section A: importance							
Cronbach's α	0.88	0.90	0.89	0.86	0.88	0.89	0.85
Guttman's λ_6	0.88	0.85	0.84	0.82	0.84	0.84	0.80
Minimum split half (λ_4)	0.93	0.81	0.79	0.82	0.81	0.81	0.76
Maximum split half (ß)	0.79	0.79	0.79	0.75	0.80	0.76	0.71
Section B: experiences							
Cronbach's α	0.89	0.90	0.89	0.89	0.85	0.93	0.76
Guttman's λ_6	0.88	0.86	0.85	0.85	0.80	0.90	0.76
Minimum split half (λ_4)	0.92	0.82	0.80	0.81	0.78	0.84	0.76
Maximum split half (ß)	0.85	0.80	0.79	0.80	0.78	0.82	0.76

Note Minimum and maximum split halves are based on 10,000 random draws across the data, tested using the psych package (Revelle, 2015) in R. *A* accomplishments, *C* cognitions, *E* engagement, *M* meaning, *P* perseverance, *PE* positive emotions, *R* relationships

larger inter-quartile ranges than the other experienced UVs (i.e. cognitions, relationships, perseverance, accomplishments), as shown in Fig. 13.2, suggesting these lower less experienced UVs also tended to be rated more negatively across the Year 8 cohort. Notably, Fig. 13.2 shows students on average (the mean ratings are represented by the x symbols) rated the importance of the seven UVs (shaded boxes) higher than their experiences (solid boxes) of these seven UVs, and especially for positive emotions which showed the highest overall mean difference between value importance and experience.

Table 13.5 Overall importance and experience descriptive survey data for each of the seven UVs, for the whole sample and separately by gender

Section A: importance descriptive data

UV	Whole sample				Males				Females			
	M	*SD*	Min	Max	*M*	*SD*	Min	Max	*M*	*SD*	Min	Max
A	7.76^{3}	1.95	0	10	7.67^{3}	2.13	0	10	7.87^{3}	1.73	0	10
C	8.32^{1}	1.84	0	10	8.14^{1*}	1.99	0	10	8.50^{1*}	1.65	0.33	10
E	7.10^{7}	2.15	0	10	7.21^{7}	2.26	0	10	6.98^{6}	2.03	0	10
M	7.44^{5}	2.07	0	10	7.40^{6}	2.22	0	10	7.48^{4}	1.91	0	10
P	7.46^{4}	2.09	0	10	7.43^{4}	2.24	0	10	7.48^{4}	1.92	2	10
PE	7.39^{6}	2.26	0	10	7.36^{5}	2.26	0	10	7.41^{5}	2.28	0	10
R	8.13^{2}	1.95	0	10	7.80^{2**}	2.22	0	10	8.49^{2**}	1.54	1	10

Section B: experiences descriptive data

	Whole sample				Males				Females			
UV	*M*	*SD*	Min	Max	*M*	*SD*	Min	Max	*M*	*SD*	Min	Max
A	6.97^{4}	2.13	0	10	7.13^{3}	2.25	0	10	6.80^{4}	1.99	0	10
C	7.50^{1}	2.07	0	10	7.62^{1}	2.21	0	10	7.36^{2}	1.90	0	10
E	6.20^{6}	2.39	0	10	6.41^{6*}	2.50	0	10	5.98^{6*}	2.25	0	10
M	6.62^{5}	2.38	0	10	6.73^{5}	2.56	0	10	6.51^{5}	2.16	0	10
P	7.05^{3}	2.13	0	10	7.08^{4}	2.30	0	10	7.03^{3}	1.93	1	10
PE	5.82^{7}	2.56	0	10	6.03^{7}	2.63	0	10	5.59^{7}	2.45	0	10
R	7.38^{2}	2.00	0	10	7.33^{2}	2.14	0	10	7.44^{1}	1.84	1.33	10

Note UV seven ultimate values, *A* accomplishments, *C* cognitions, *E* engagement, *M* meaning, *P* perseverance, *PE* positive emotions, *R* relationships. *M* mean scores, *SD* standard deviations. 1–7 = ranking of each UV from highest to lowest scores across groups
* $p < 0.05$, ** $p < 0.01$

Concerning gender differences in value importance and experiences (RQ2), as shown in Fig. 13.3, females rated all the UVs, especially cognitions and relationships, as more important than males, except for the engagement UV, which males rated significantly higher. The importance ratings for three UVs (i.e. cognitions, perseverance, and relationships) also showed a lower spread for females compared to males, indicating that females were less likely to rate these UVs as low as males, with UVs cognitions and relationships showing statistically significant differences. Notably, despite females rating most UVs as more important, they also reported less experiences of these UVs compared to the males (except for experiencing relationships, which was rated higher by females than males). For instance, across all UVs combined, males scored 0.23 points higher (out of 10) for their experiences of these UVs in mathematics than females, especially for their experience of positive emotions, which was 0.44 point higher for males than females. Only for the experiencing of relationships did females score slightly higher than males (mean score difference − 0.11). In summary, females generally attribute higher importance to

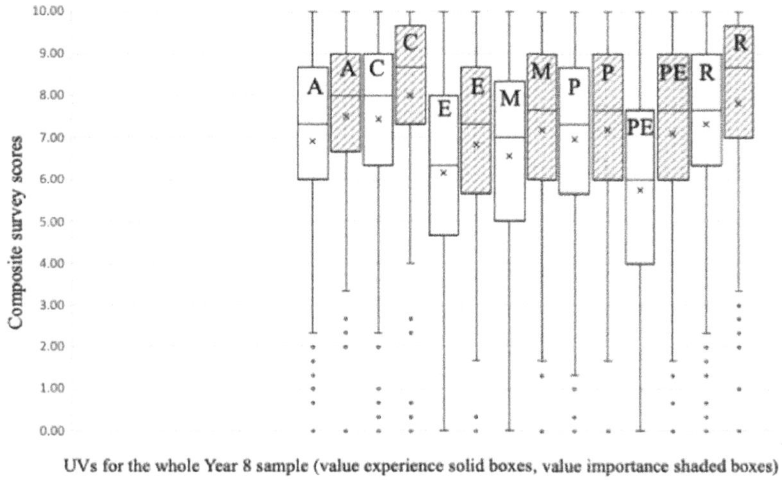

Note. A = accomplishments, C = cognitions, E = engagement, M = meaning, P = perseverance, PE = positive emotions, R = relationships, boxplot center line represents median scores, x = mean score values

Fig. 13.2 Boxplots of value importance and experiences for the whole Year 8 sample

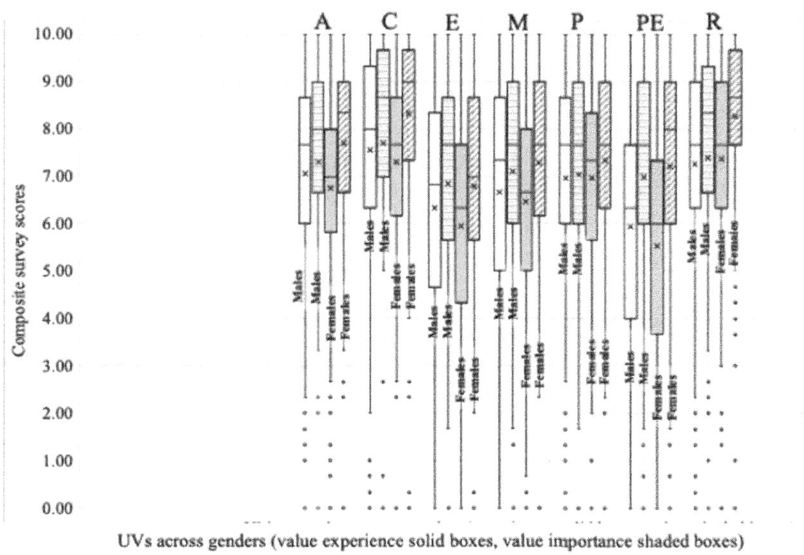

Note. A = accomplishments, C = cognitions, E = engagement, M = meaning, P = perseverance, PE = positive emotions, R = relationships, boxplot center line represents median scores, x = mean score values

Fig. 13.3 Boxplots of value importance and experiences across genders

the seven UVs but also reported being less successful experiencing these UVs in the mathematics classroom.

13.6.3 Students' Personal Value Fulfilment

Because wellbeing is context specific and value dependent (Alexandrova, 2017; Tiberius, 2018), different students may attribute higher importance to certain values. For instance, research studies show females often place higher importance on relational aspects of learning mathematics than males (Hill & Hunter, 2023). We purport that MWB is most accurately captured by measuring students' success at fulfilling their own personal values. Thus, we also explored the students' levels of personal value fulfilment by subtracting the students mean values importance scores, from the scores measuring their experiences of these values, for the whole sample (RQ3) and across genders (RQ4). A higher positive mean difference indicates a greater difference between the level of importance attributed to the UVs and students' success experiencing that UV. In other words, they are less successful fulfilling their own values. As summarized in Table 13.6, we found that across the whole sample, the mean difference was by far the greatest for positive emotions, meaning students were least successful fulfilling this UV. Conversely, students were most successful fulfilling their valuing of perseverance.

Table 13.6 Students' personal value fulfilment, or the average mean differences between the value importance and experience scores

UV	Whole sample				Males				Females			
	MD	*SD*	Min	Max	*MD*	*SD*	Min	Max	*MD*	*SD*	Min	Max
A	0.81	1.78	–6.67	9.33	0.56**	1.80	–6.67	6.33	1.07**	1.73	–5.00	9.33
C	0.81	1.99	–9.67	7.67	0.51**	2.12	–9.67	7.67	1.13**	1.78	–5.67	7.33
E	0.92	1.70	–5.00	8.33	0.85	1.69	–5.00	8.33	1.00	1.71	–5.00	6.67
M	0.82	1.73	–5.00	8.33	0.67*	1.74	–5.00	8.33	0.99*	1.70	–5.00	8.33
P	0.41	1.41	–3.00	7.00	0.37	1.38	–3.00	6.33	0.45	1.45	–3.00	7.00
PE	1.60	2.35	–8.00	8.67	1.38*	2.08	–5.33	8.67	1.84*	2.58	–8.00	8.67
R	0.74	1.88	–6.00	8.67	0.46**	1.79	–6.00	8.33	1.05**	1.92	–3.67	8.67

Note UV seven ultimate values, *MD* mean difference, *SD* standard deviation, *A* accomplishments, *C* cognitions, *E* engagement, *M* meaning, *P* perseverance, *PE* positive emotions, *R* relationships
* $p < 0.05$, ** $p < 0.01$

Across genders, males and females similarly were the most successful in fulfilling their valuing of perseverance and least successful fulfilling positive emotions. Yet a striking difference was that females were considerably less successful fulfilling their values for all the seven UVs, suggesting that female students report lower MWB than do males, overall and across the seven UVs.

13.7 Discussion

In this study we introduced a survey designed to measure students' valuing, experiences, and fulfilment of seven UVs previously shown to support students' positive wellbeing in mathematics education (e.g. Hill et al., 2020). Our survey was found to be both valid and reliable. Findings indicated Australian Year 8 students rated being most successful experiencing their valuing of the UVs cognitions and relationships, and least successful for UVs engagement and positive emotions. However, because not all students prioritize their values in the same hierarchy, we argue that students' MWB is most accurately captured by their value fulfillment—being the difference between the students' own personal values and their experiences of these same values. We found individual students were the most successful fulfilling their personal valuing of UV perseverance and the least successful fulfilling the UV positive emotions. Whilst female students valued most of the UVs as more important than males, the females also reported being less successful at experiencing and fulfilling these UVs.

Overall, cognitions and relationships were rated the highest, in terms of both importance and experience suggesting these aspects reflect strengths of these Year 8 students. Students often see mathematics as cognitively challenging, difficult, progressive, and abstract, and thus often feel anxious about being left behind (Hill et al., 2021a, 2021b). Thus, it was encouraging that students, on average, felt competent with their mathematical skills, knowledge, and understandings; and they felt socially supported in the classroom. Feeling competent is a core psychological need supporting overall wellbeing (Ryan & Deci, 2017); it is also linked to mathematical performance and procedural knowledge (Schukajlow et al., 2023). Close relationships (another core psychological need) (Ryan & Deci, 2017) between students and teachers matter, and belonging supports many positive mathematics learning outcomes like achievement, engagement, enjoyment, (e.g. Averill, 2012; Schukajlow et al., 2023).

Conversely, the students rated their experiences the lowest for the UVs engagement and positive emotions, signaling areas that potentially undermine students' overall MWB. The pervasive negative emotions and disengagement in mathematics are global challenges. For example, the international PISA studies reported that 30% of students around the world felt helpless and stressed when doing mathematics (OECD, 2013b). A New Zealand study found positive emotions dropped from being the highest to the lowest rated UV from Years 3 to Year 8, whilst engagement remained the second lowest rated UV (out of the seven UVs) across year levels (Hill et al.,

2022a, 2022b). The majority of Israeli primary students also noted that they generally did not enjoy learning mathematics (Markovits & Forgasz, 2017). Whilst numerous studies demonstrate that student engagement in mathematics education often declines as students progress through school, with the sharpest decline in engagement over the primary to secondary school transition (Grootenboer & Marshman, 2015; Martin et al., 2015). The nature of these two UVs (i.e. positive emotions and engagement) also suggests that like UV perseverance, there might well be an interactional effect between some UVs and MWB. That is, low levels of experiencing positive emotions and engagement—as well as low levels of fulfilling positive emotions and perseverance—might affect MWB negatively, which in turn drives these levels further downwards, potentially creating a spiraling impact on MWB, and thus the quality of mathematics learning.

An important finding are the gender differences across all measures. The fact that female students rated most UVs as more important yet achieved lower experiences of these values, and consequently, lower value fulfilment than males might, indicates that females are more likely to experience mathematical illbeing. It is well publicized that females hold more negative attitudes and emotions; are more mathematically anxious; and are underrepresented in mathematic subjects and courses at school and beyond (e.g. Frenzel et al., 2007; Guo et al., 2015; Samuelsson & Samuelsson, 2016; Wang & Degol, 2013). Studies (e.g. Hill & Hunter, 2023) show female students often espouse more social IVs in mathematics education compared to males. Might it be the case that while these gender-unique IVs serve the same UVs, there have been fewer opportunities for them to be fulfilled, leading to lower MWB? Also, our findings suggest that female students may have higher expectations (i.e. having higher values) for the conditions that support their mathematics learning than males. Thus, females' values are perhaps harder to fulfil (e.g. higher values may require more effort or resources to be fulfilled), whilst females may also face more challenges fulfilling their values (e.g. females values' may often clash with the pedagogical values and thus be harder to fulfil) compared to males. Perhaps female students need greater support (social and emotional), more encouragement, and enjoyable learning experiences than males to fulfil these higher expectations and values. Added to this earlier research female students often perceive receiving less support, social interactions, and involvement than boys in the mathematics classroom (Eccles, 2011; Samuelsson & Samuelson, 2016). To enhance female students' MWB teachers should better acknowledge the unique values and higher expectations (or valuing) of female students, particularly the social aspects, and then incorporate pedagogies to support these values in the mathematics classroom.

Students' MWB is most accurately represented by students' success fulfilling their own values (Tiberius, 2018). For example, two mathematics students may espouse similar UVs, yet one student (e.g. from a collectivist culture) may prioritize social aspects over meaningful learning, another student (e.g. from a more individualist culture) may prioritize accomplishment over social learning opportunities (Hofstede, 2011). These students prioritize these UVs differently; thus, it makes sense to capture both UVs and the associated IVs through which these UVs are being fulfilled. Our findings indicated that Australian Year 8 students (i.e. the whole sample, males, and females) were most successful fulfilling their valuing of perseverance (see Table 13.6). This suggests that students might have found it easiest to actualize aspects of perseverance, which they also found important and valued, including practicing, sustained effort, and seeking out challenging mathematics. Having the agency to persevere thus elevates a student's sense of MWB. That this UV seems to be the easiest to fulfil should be understandable, since it is initiated by a student themselves as long as they value it. That is, fulfilling the valuing of perseverance might be less dependent on external factors and more dependent on internal factors. In fact, amongst the seven UVs, perseverance is perhaps the only one which students can accomplish and fulfil on their own accord. For example, fulfilling relationships relies on others' support, and thus is outside one's locus of control. The challenge for teachers, then, would be to nurture in students a sense of acknowledging that perseverance is an important attribute to possess and develop within oneself. In this regard, the development of growth mindsets or mathematical mindsets could be a useful starting point.

Students were also least successful fulfilling their valuing of positive emotions, suggesting that this presents as a key barrier to students experience of more positive MWB. While this may imply that Year 8 mathematics lessons in Australia are unsuccessful in promoting positive emotions (e.g., joy, fun), it is also probable that student expectations and valuing of positive emotions are too high. One thing which is quite clear, however, from the sample IVs in Table 13.1 is that this failure to fulfil the students' valuing of positive emotions is very likely reflected in the widespread experience of (mathematics) anxiety amongst students in Australia. In the absence of systems-level data, we have to rely on PISA 2012 findings (Thomson et al., 2013), which indicated that one-quarter to one-third of 15-year-old students in Australia were experiencing mathematics anxiety. There is reason to believe that this situation has become more severe in the last few years, as our 2019 data suggest. There is also the compounded effect of long periods of home-based online mathematics learning between 2020 and 2021 because of the COVID-19 pandemic. All of these factors signal the need for schools and the wider community to work harder in addressing mathematics anxiety and in promoting students' experience of positive emotions during their respective learning experiences. Reducing mathematics anxiety is important because mathematics anxiety can persist in adulthood, also students are more likely to underperform, act out, and disengage when experiencing mathematics anxiety (Dowker et al., 2016).

13.8 Implications and Limitations

Our findings point to several practical implications whilst we also note several limitations of this study. For example, Likert type surveys provide consistency and reduce the complexity of highly subjective constructs (like wellbeing), yet they can also miss important aspects of one's experience. For example, the UV relationships may encompass countless sources of support (e.g. family, tutors, teacher aides) not included in our survey. Also, surveys cannot adequately capture the richness of cultural values. Further studies might include a cultural component to the survey for more culturally diverse populations. Our sample size was relatively small and included only Year 8 students who were also culturally homogenous. Future investigations might test the survey with a more culturally diverse sample and include both primary and secondary students.

Our study has shown that whilst the seven UVs are still considered important for most students, the extent to which these UVs are valued is different for different groups of students (e.g. females). This has implications for other wellbeing surveys that rarely consider students' values, because a more accurate representation of one's wellbeing is by measuring values fulfilment (Alexandrova, 2017; Tiberius, 2018).

Our MWB model and survey might be helpful for schools to quickly assess where students are going well and what is helpful, and also where they could benefit from extra attention in mathematics education. Such assessments send a message to students that MWB matters and that the school cares about students' feelings and perspectives (Jarden et al., 2021). Teachers might use our survey with students to help signal what aspects of the teacher's practice and pedagogy that requires more attention (i.e. UVs scoring the lowest in value fulfilment).

The seven UVs are valued to some extent by most students and thus it is important teachers emphasize all the seven UVs in the mathematics classroom. For example, teachers might begin the school year with a checklist of the seven UVs, selecting to focus on specific UVs in each school term. Given the UV relationship was considered most important by most students, which aligns with earlier MWB studies (e.g. Hill et al., 2020; Hill et al., 2022a, 2022b), teachers could consider to make this UV central to every lesson (see Table 13.7 for some examples). Interestingly, Attard's (2012) Framework for Engagement emphasizes how pedagogical relationships (similar to what student participants here valued and experienced the most) lead to enhanced engagement in mathematics lessons (which the same student participants valued and experienced the least). Teachers may choose to emphasize an additional UV that was identified as needing attention through the survey and then over each school term rotate across other the UVs. Table 13.7 provides some ideas for teachers which have been shown to improve student wellbeing and learning experiences mathematics or more broadly across all subjects. Different activities suit different students and teachers. Thus, teachers might experiment to see what activities work best in their classroom. However, further research is required to identify pedagogical activities which specifically support students' MWB across these seven UVs while teaching mathematics at the same time.

Finally, teachers should incorporate all seven UVs into the 'taught' (i.e. explicit teaching) and 'caught' (i.e. implicit norms and practices) mathematics curriculum in ways which allow these to be fulfilled as well. For example, signposting lessons that focus on positive emotions (taught curriculum) whilst also modelling how to remain optimistic during challenges in mathematics (caught curriculum).

Table 13.7 Example activities to help students fulfil the seven UVs in mathematics

Ultimate value	Strategy	Example activity
Accomplishment	Learning from failures	Ask students to think about a situation that went really wrong for them in mathematics. What happened? Think through the details of the situation without judgement. What did they do? What were their actions? What were the outcomes and what did they learn from the situation? What would you do differently in the future? By processing our failures, students can learn and improve
Cognitions	Instructional support	Using effective and constructive feedback by answer the following three questions: (1) *Where am I going?* Reminding students of the focus of the task/assignment etc.; (2) *How am I going?* Giving specific and clear information about students' performance; (3) *Where to next?* Provide individually specific advice to help students improve on particular areas in mathematics (Hattie & Timperley, 2007)
Engagement	Listening to music in class	Ask students to create a class play list of music that inspires them. Play this music during class to enhance concentration, enjoyment, and engagement with mathematics (Hill et al., 2021a, 2021b)
Meaning	Contextualizing mathematics to the bigger ideas	Connect the learning material to students interests and future goals (Burns et al., 2022); connecting mathematical knowledge by making connections to and between bigger real-world ideas (Hunter et al., 2016)
Perseverance	Exploring the strength of perseverance in mathematics	Students consider '*what is perseverance*?" Provide examples of individuals who exemplify the strength of perseverance. Students explore their own experiences of mathematical perseverance, e.g., *what's a challenge in mathematics you have overcome using perseverance?* Students create a timeline of this strength identifying instances in which perseverance in mathematics was helpful (Oppenheimer et al., 2014)

(continued)

Table 13.7 (continued)

Ultimate value	Strategy	Example activity
Positive emotions	Addressing mathematical anxiety	Younger students read about and empathize with characters in texts who are experiencing similar mathematical anxiety challenges (e.g., the Math Curse by Jon Scieszka). Engage in discussions and activities to help students draw out the key experiences of the character and link these to the emotions they may be feeling. Older students might use deep breathing exercises or expressive writing where they write down all negative mathematical emotions without judgement (Buckley, 2020)
Relationships	Knowing the students	At the beginning of all mathematics lessons teachers ask students about their week/weekend/ important events, and learning more about students' goals and interests in school and outside school (Burns et al., 2022)

13.9 Conclusion

Tiberius (2018) asserts that we can achieve high wellbeing when we become aware of and then fulfil our values. Applied to mathematics education students MWB might be enhanced when students' values relating to mathematics learning are fulfilled. To achieve this teachers' might draw on pedagogies and practices that align with and thus fulfil their students' values. The challenge in many countries is to improve the pervasive inequalities, anxieties, disengagement, poor student participation and so forth in mathematics (e.g. Grootenboer & Marshman, 2015). We contend that focusing on student values and MWB might improve some of these negative experiences and trends. This chapter provides a contribution by demonstrating value fulfilment—and thus MWB—can be validly and reliably assessed using a brief survey. It also highlights the gender differences in value fulfillment which may partly explain why females often experience mathematics more negatively than males. Our study provides an important first step in moving towards improving student MWB by understanding what aspects of teaching and learning in mathematics education are in greatest need of attention.

References

Alexandrova, A. (2017). A philosophy for the science of well-being. *Oxford University Press*. https://doi.org/10.1093/oso/9780199300518.001.0001

Attard, C. (2012). Applying a framework for engagement with mathematics in the primary classroom. *Australian Primary Mathematics Classroom, 17*(4), 22–27.

Averill, R. (2012). Caring teaching practices in multiethnic mathematics classrooms: Attending to health and well-being. *Mathematics Education Research Journal, 24*(2), 105–128. https://doi.org/10.1007/s13394-011-0028-x

Bishop, A. J. (1996). How should mathematics teaching in modern societies relate to cultural values: Some preliminary questions. In *7th Southeast Asian Conference on Mathematics Education* (vol. 32, June 3–7).

Buckley, S. (2020). *Issue in the teaching of mathematics: Mathematics anxiety.* Victorian Department of Education. https://www.education.vic.gov.au/Documents/school/teachers/teachingresources/discipline/maths/MTT_Mathematics_Anxiety.pdf

Burns, E. C., van Bergen, P., Leonard, A., & Amin, Y. (2022). Positive, complicated, distant, and negative: How different teacher-student relationship profiles relate to students' science motivation. *Journal of Adolescence, 94*(8), 1150–1162. https://doi.org/10.1002/jad.12093

Butler, J., & Kern, M. L. (2016). The PERMA-Profiler: A brief multidimensional measure of flourishing. *International Journal of Wellbeing, 6*(3), 1–48. https://doi.org/10.5502/ijw.v6i3.526

Clark, S. (2023). *Funding for schools and student wellbeing.* Parliament of Australia. https://www.aph.gov.au/About_Parliament/Parliamentary_departments/Parliamentary_Library/pubs/rp/BudgetReviewOctober202223/FundingSchoolsStudentWellbeing

Dawes, J. (2008). Do data characteristics change according to the number of scale points used? An experiment using 5-point, 7-point and 10-point scales. *International Journal of Market Research, 50*(1), 61–104. https://doi.org/10.1177/147078530805000106

Dowker, A., Sarkar, A., & Looi, C. Y. (2016). Mathematics anxiety: What have we learned in 60 years? *Frontiers in Psychology.* https://doi.org/10.3389/fpsyg.2016.00508

Eccles, J. (2011). Gendered educational and occupational choices: Applying Eccles et al. model of achievement related choices. *International Journal of Behavioral Development, 35*(3), 195–201. https://doi.org/10.1177/0165025411398185

Frenzel, A. C., Pekrun, R., & Goetz, T. (2007). Girls and mathematics—A "hopeless" issue? A control-value approach to gender differences in emotions towards mathematics. *European Journal of Psychology of Education, 22*(4), 497–514. https://doi.org/10.1007/bf03173468

Gaspard, H., Dicke, A.-L., Flunger, B., Brisson, B. M., Häfner, I., Nagengast, B., & Trautwein, U. (2015). Fostering adolescents' value beliefs for mathematics with a relevance intervention in the classroom. *Developmental Psychology, 51*(9), 1226–1240. https://doi.org/10.1037/dev0000028

Grootenboer, P., & Marshman, M. (2015). Mathematics, affect and learning: Middle school students' beliefs and attitudes about mathematics education. *Springer.* https://doi.org/10.1007/978-981-287-679-9_4

Guo, J., Marsh, H. W., Parker, P. D., Morin, A. J. S., & Yeung, A. S. (2015). Expectancy-value in mathematics, gender and socioeconomic background as predictors of achievement and aspirations: A multi-cohort study. *Learning and Individual Differences, 37*, 161–168. https://doi.org/10.1016/j.lindif.2015.01.008

Halstead, J. M. (1996). Values and values education in schools. In J. Halstead & M. Taylor (Eds.), *Values in education and education in values* (pp. 3–14). The Falmer Press. https://doi.org/10.4324/9780203973554

Hattie, J., & Timperley, H. (2007). The power of feedback. *Review of Educational Research, 77*(1), 81–112. https://doi.org/10.3102/003465430298487

Hill, J. L., Bowmar, A., & Hunter, J. (2022). A preliminary study exploring the mathematical wellbeing of grade 3 to 8 students in New Zealand. In M. Vollstedt, I. Kontorovich, K. Jones, & D. Sommerhoff (Eds.), *Proceedings of the 45th Conference of the International Group Psychology in Mathematics Education* (pp. 387–394). PME. https://rua.ua.es/dspace/handle/10045/126571

Hill, J. L., & Hunter, J. (2023). Examining the mathematics education values of diverse groups of students. *International Journal of Mathematical Education in Science and Technology, 54*(8), 1641–1633. https://doi.org/10.1080/0020739x.2023.2184280

Hill, J. L., & Seah, W. T. (2023). Student values and wellbeing in mathematics education: Perspectives of Chinese primary students. *ZDM- Mathematics Education, 55,* 385–398 https://doi.org/10.1007/s11858-022-01418-7

Hill, J. L., Kern, M. L., Seah, W. T., & van Driel, J. (2020). A model of student mathematical wellbeing: Australian grade eight students' conceptions. In M. Inprasitha, N. N. Changsri, & N. Boonsena (Eds.), *Proceedings of the 44th Conference of the International Group Psychology in Mathematics Education* (pp. 273–281). PME. https://www.igpme.org/publications/current-proceedings/

Hill, J. L., Kern, M. L., Seah, W. T., & van Driel, J. (2021a). Feeling good and functioning well in mathematics education: Exploring students' conceptions of mathematical wellbeing and values. *ECNU Review of Education, 4*(2), 349–375. https://doi.org/10.1177/2096531120928084

Hill, J. L., Kern, M. L., van Driel, J., & Seah, W. T. (2021b). The importance of positive classroom relationships for diverse students' well-being in mathematics education. In D. Burghs & J. Hunter (Eds.), *Mathematics Education for Sustainable Economic Growth and Job Creation* (pp. 76–89). Routledge. https://doi.org/10.4324/9781003048558-7

Hill, J. L., Kern, M. L., Seah, W. T., & van Driel, J. (2022). Developing a model of "mathematical wellbeing" through a thematic analysis of the literature. In M. Vollstedt, I. Kontorovich, K. Jones, & D. Sommerhoff (Eds.), *Proceedings of the 45th Conference of the International Group Psychology in Mathematics Education* (Vol. 2, pp. 32–39). PME. https://www.igpme.org/publications/current-proceedings/

Hofstede, G. (2011). Dimensionalizing cultures: The Hofstede model in context. *Online Readings in Psychology and Culture, 2*(1). https://doi.org/10.9707/2307-0919.1014

Hooper, D., Coughlan, J., & Mullen, M. R. (2008). Structural equation modelling: Guidelines for determining model fit. *Electronic Journal of Business Research Methods, 6*, 53–60. https://doi.org/10.1093/hesc/9780192896599.003.0012

Hunter, R., Hunter, J., Jorgensen, R., & Choy, B. (2016). Innovative and powerful pedagogical practices in mathematics education. In K. Maker, S. Dole, J. Visnovska, M. Goos, A. Bennison, & K. Fry (Eds.), *Research in mathematics education in Australasia 2012–2015* (pp. 213–234). Springer. https://doi.org/10.1007/978-981-10-1419-2_11

Huppert, F. A., & So, T. T. C. (2013). Flourishing across Europe: Application of a new conceptual framework for defining flourishing across Europe. *Social Indicators Research, 110*(3), 837–861. https://doi.org/10.1007/s11205-011-9966-7

Jarden, A., Jarden, R., Chin, T. C., & Kern, M. L. (2021). Assessing wellbeing in school communities. In M. L. Kern & M. L. Wehmeyer (Eds.), *The Palgrave handbook of positive education* (pp. 297–324). Springer. https://doi.org/10.1007/978-3-030-64537-3_12

Joshanloo, M., van de Vliert, E., & Jose, P. E. (2021). Four fundamental distinctions in conceptions of wellbeing across cultures. In M. L. Kern & M. L. Wehmeyer (Eds.), *The Palgrave handbook of positive education* (pp. 675–703). Springer. https://doi.org/10.1007/978-3-030-64537-3_26

Larkin, K., & Jorgensen, R. (2016). 'I hate maths: Why do we need to do maths?' Using iPad video diaries to investigate attitudes and emotions towards mathematics in Year 3 and Year 6 students. *International Journal of Science and Mathematics Education, 14*(5), 925–944. https://doi.org/10.1007/s10763-015-9621-x

Markovits, Z., & Forgasz, H. (2017). "Mathematics is like a lion": Elementary students' beliefs about mathematics. *Educational Studies in Mathematics, 96*(1), 49–64. https://doi.org/10.1007/s10649-017-9759-2

Martin, A. J., Way, J., Bobis, J., & Anderson, J. (2015). Exploring the ups and downs of mathematics engagement in the middle years of school. *The Journal of Early Adolescence, 35*(2), 199–244. https://doi.org/10.1177/0272431614529365

OECD. (2013a). *OECD guidelines on measuring subjective well-being.* OECD Publishing.

OECD. (2013b). *PISA 2012 results: Ready to learn—Students' engagement, drive and self-beliefs* (Volume III). OECD Publishing.

OECD (2015). *What contributes to student well-being at school?* OECD Publishing.

Oppenheimer, M. F., Fialkov, C., Ecker, B., & Portnoy, S. (2014). Teaching to strengths: Character education for urban middle school students. *Journal of Character Education, 10*(2), 91–105. https://doi.org/10.4135/9781529714654.n7

Revelle, W. (2015). *Package 'psych'*. http://personality-project.org/r/psych.manual.pdf

Riegle-Crumb, C., Farkas, G., & Muller, C. (2006). The role of gender and friendship in advanced course taking. *Sociology of Education, 79*, 206–228. https://doi.org/10.1177/003804070607900302

Rose, T., Joe, S., Williams, A., Harris, R., Betz, G., & Stewart-Brown, S. (2017). Measuring mental wellbeing among adolescents: A systematic review of instruments. *Journal of Child and Family Studies, 26*(9), 2349–2362. https://doi.org/10.1007/s10826-017-0754-0

Rosseel, Y. (2012). Lavaan: An R package for structural equation modelling. *Journal of Statistical Software, 48*, 1–36.

Ryan, R. M., & Deci, E. L. (2017). Self-determination theory: Basic psychological needs in motivation, development, and wellness. *Guilford Publications*. https://doi.org/10.1521/978.14625/28806

Samuelsson, M., & Samuelsson, J. (2016). Gender differences in boys' and girls' perception of teaching and learning mathematics. *Open Review of Educational Research, 3*(1), 18–34. https://doi.org/10.1080/23265507.2015.1127770

Schukajlow, S., Rakoczy, K., & Pekrun, R. (2023). Emotions and motivation in mathematics education: Where we are today and where we need to go. *ZDM—Mathematics Education, 55*(2), 249–267. https://doi.org/10.1007/s11858-022-01463-2

Seligman, M. E. P. (2011). *Flourish*. William Heinemann.

Thomson, S., de Bortoli, L., & Buckley, S. (2013). *PISA 2012: How does Australia measure up? The PISA 2012 assessment of students' mathematical, scientific and reading literacy*. Australian Council for Educational Research.

Tiberius, V. (2018). Well-being as value fulfillment: How we can help each other to live well. *Oxford University Press*. https://doi.org/10.1093/oso/9780198809494.001.0001

United Nations. (n.d.). *Sustainable development goals*. U.N. www.un.org/sustainabledevelopment/

Wang, M. T., & Degol, J. (2013). Motivational pathways to STEM career choices: Using expectancy–value perspective to understand individual and gender differences in STEM fields. *Developmental Review, 33*(4), 304–340. https://doi.org/10.1016/j.dr.2013.08.001

Wigfield, A., Eccles, J. S., Yoon, K. S., Harold, R. D., Arbreton, A. J., Freedman-Doan, C., & Blumenfeld, P. C. (1997). Change in children's competence beliefs and subjective task values across the elementary school years: A 3-year study. *Journal of Educational Psychology, 89*(3), 451–469. https://doi.org/10.1037/0022-0663.89.3.451

Wong, N. Y. (1995). *The relationship between Hong Kong students' perception of their mathematics classroom environment and their approaches to learning: A longitudinal study* [Doctoral dissertation, The University of Hong Kong]. Proquest Dissertations.

Zeng, G., & Kern, M. L. (2019). The Chinese EPOCH measure of adolescent wellbeing: Further testing of the psychometrics of the measure. *Frontiers in Psychology*. https://doi.org/10.3389/fpsyg.2019.01457

Part IV
Applying the Values Perspective to Teaching Problem Solving

Chapter 14
Categories and Their Relationships Among Socially Open-Ended Problems

Takuya Baba, Isao Shimada, Yuichiro Hattori, and Hiroto Fukuda

14.1 Introduction

Since the outbreak of the COVID-19 pandemic, new cases are being reported daily. Although this trend has been analysed scientifically, the decision to declare a state of emergency cannot solely rely on a mathematical solution. This exemplifies a 'trans-science problem', which can be queried scientifically but cannot be answered only by science (Weinberg, 1972). In addition to the number of cases and other scientific facts, several false rumours and extensive demagogy surround the pandemic. The dichotomy of true/false and the existence of various ideas reflecting different interests collectively make the situation even more complicated. In such a situation, it is necessary to make not only mathematical and scientific decisions but also social ones through the reconciliation of interests. That is, societal shifts do not seek only one answer but rather require optimal or preferable solutions based on values (Keeney, 1992; Science Council of Japan, 2007).

In mathematics education, considering the relationship between problem-solving and social and value aspects, Brown (1984) highlighted the emergence of values in mathematical problem-solving and the importance of considering such values. However, the expression 'problem-solving' tends to emphasis 'solving' a problem

T. Baba (✉)
Hiroshima University, Higashihiroshima, Japan
e-mail: takuba@hiroshima-u.ac.jp

I. Shimada
Formerly of Nippon Sport Science University, Tokyo, Japan

Y. Hattori
Okayama University, Okayama, Japan
e-mail: hattori.yuichiro@okayama-u.ac.jp

H. Fukuda
Okayama University of Science, Okayama, Japan
e-mail: hfukuda@ous.ac.jp

Y. Dede et al. (eds.), *Values and Valuing in Mathematics Education*,
https://doi.org/10.1007/978-981-99-9454-0_14

while treating values as noise and unnecessary when solving mathematical problems (Iida et al., 1994). Returning to the initial idea of problem-solving, Polya (1945) proposed four principles: understanding the problem, devising a plan, carrying out the plan, and looking back on mathematical problem-solving. Here, solving the problem corresponds to carrying out a plan; if we include generating a strategy, it also includes devising a plan. Understanding the problem and looking back at problem-solving in Polya's proposal (1945) extend beyond solving a problem in a narrow sense and allow us to connect the social dimension with mathematical problem-solving. Indeed, Lesh and Zawojewski (2007) distinguished a new type of problem-solving from traditional problem-solving in proposing model-eliciting activities. Here, problem-solving does not limit itself to 'solving' the problem but includes pre- and post-problem-solving activities. The foregoing study revitalises Polya's work (1945) and emphasises the development of a mathematical model and interpreting the problem situation after a solution has been found. In this context, Shimada and Baba (2016) proposed a particular type of problem-solving involving model development and problem interpretation. Students develop mathematical models based on their values when solving socially open-ended problems (Baba, 2007, 2009).

Additionally, Shimada (2017) developed some 'socially open-ended problems' and delivered experimental lessons from the perspective of the mathematical modelling process. In other words, the lessons flow as students develop mathematical models based on their values, discuss them, and select their values and models after discussion.

To further develop these ideas (Shimada, 2017; Shimada & Baba, 2016), a research team of seven members was formed in 2016, and this team engaged with socially open-ended problems from the perspective of critical mathematics education under the thematic study of the Japan Society of Mathematics Education (Baba, 2016, 2017, 2019, 2020).This team found that such problem-solving can be realised from the pre-primary school to the tertiary level, and some emerging values are related to equity, which is crucial in modern society.

These efforts have clarified the characteristics and possibilities of socially open-ended problems in mathematics education. However, three issues still require further investigation:

(1) systematisation of problem categories;
(2) arrangement of problems, values, and mathematical models according to educational stages from pre-primary to tertiary education;
(3) identification of competence that can be nurtured through (1) and (2).

Kridel (2010) described the concretisation of the organisation of an entire curriculum in terms of scope and sequence. Oliva (2009) also noted eight components of a curriculum, grouped into two categories: scope and sequence. The former denotes the breadth of the curriculum: 'The contents of any course curriculum or grade level constitute the scope for that course…' (Oliva, 2009, p. 417), and the latter denotes the arrangement of topics—'the order in which the organising elements or centres are arranged…' (Oliva, 2009, p. 428). That is, issue (1) is related to scope and issue (2) to sequence.

This study engages with issue (1), which corresponds to systematisation. In particular, this means identifying possible characteristics and relationships among those categories. Through this, we can determine the characteristics of a mathematical activity within each category and ensure the completeness of the curriculum.

To date, Shimada (2017) has identified four problem categories in practice: sharing, selecting, rule-making, and planning. Shimada and Baba (2022) later revised the categories to sharing, selecting, rule-making, and planning/designing and compiled 32 socially open-ended problems for primary schools. However, these categories were developed inductively after almost 10 problems were produced, and the other problems were subsequently classified. In this process, we experientially explore the possible existence of different categories and problems in each category. It is thus important to consider these categories systematically. The approach adopted in this study involves first reconsidering existing categories from the perspective of the six universal activities. The above categories, such as sharing and selecting, are mathematical activities embedded in various social contexts, and thus, they might be related to the six universal activities (Bishop, 1988).

Here, a brief explanation of some key words must be provided. 'Problem-solving' has been a centre of discussion in mathematics education-based practice and research for decades, and the term has various meanings. In this study, the problems considered involve containing social context and engaging values in solving them. This type of problem is called a 'socially open-ended problem (Baba, 2007, 2009)'. Here, solving a socially open-ended problem follows the steps of mathematical modelling (Shimada, 2017). It corresponds with a modelling cycle (Organisation for Economic Co-operation and Development [OECD], 2018), and the mathematical solution and values are closely connected. This mathematical solution is called a 'mathematical model'. 'Mathematical modelling' is a process of developing such a mathematical model based on social values in correspondence with the given situation of the problem. Thus, we expect that mathematical modelling tasks from previous research may contain a social context that requires developing mathematical models based on values.

The research questions (RQs) addressed in this study are as follows:

RQ 1: What are the characteristics of the categories of socially open-ended problems?

RQ 2: What are the relationships among them?

RQ 3: Do socially open-ended problems exist among tasks in mathematical modelling?

RQ 4: Are the existing categories, such as sharing, selecting, rule-making, and planning/designing, sufficient to cover all the tasks found in major previous studies on mathematical modelling?

As stated above, values have traditionally been avoided as noise in mathematical problem-solving. The current research addresses this research gap and systematises an approach to social open-ended problems for day-to-day teaching. More concretely, it aims to consider the characteristics and relationships among existing problem categories while considering a method of creating a new socially open-ended problem based on the research on mathematical modelling, potentially containing socially open-ended problems. This endeavour will enrich students' learning in a 21st-century context by coherently utilising mathematical problem-solving and values.

14.2 Research Engagement with Socially Open-Ended Problems

Before engaging with the RQs, we describe the development of theories and practices regarding socially open-ended problems below.

14.2.1 Pre-stage

Before developing the idea of a socially open-ended problem, Iida et al. (1994) first identified values within problem-solving in Japan. They indicated that some problem-solving contained values in its solution. One such problem is the melon-sharing problem. This problem states that three teams are competing, and they must share 10 melons as a prize according to the points each team receives. Sharing can be accomplished in several ways, such as equal sharing, sharing in proportion to the points, and winner takes all, according to the students' values. Here, values seem to have appeared by chance. For example, they happen to divide the melons as a prize equally while exploring different ways of dividing, not because they aim at equality. On the other hand, in a socially open-ended problem, students are prompted to provide mathematical models based on social values. Thus, the word 'value' in the current study is not the same as in the study conducted by Iida et al. (1994). Importantly, this study noted that values and mathematical solutions are connected.

14.2.2 Stage I: Proposal of the Initial Idea

Baba (2007, 2009), based on the approach adopted by Iida et al. (1994), proposed a socially open-ended problem as a new type of problem-solving approach necessary in a society with diversified values. Its objective, problem, and method are compared with those of the renowned open-ended approach (Shimada, 1977) in Table 14.1. To differentiate these, 'mathematically' is added as a prefix to 'open-ended problem'.

As a socially open-ended problem, the 'melon-sharing problem' (Iida et al., 1994) can be used to induce students' values such as equality, proportionality, and winner's domination, as well as mathematical models based on those values. Ways of sharing depend on the values of each student. Baba (2009) noted that sharing activities have

Table 14.1 Comparison of two types of open-ended problems (Baba, 2007, p. 12)

	Mathematically open-ended problem	Socially open-ended problem
Objective	To nurture mathematical thinking	To nurture mathematical thinking and judgement based on mathematical thinking and associated social values
Problem	To allow mathematically diverse solutions	To allow mathematically diverse solutions and associated social values
Method	Discussion of mathematically diverse solutions and their generalisation and symbolisation	Discussion of mathematically diverse solutions and the associated social values

been pretty common among human beings throughout history. As such, this can be regarded as a cradle for mathematical ideas.

Students' values naturally emerge in solving socially open-ended problems at the beginning when they are reminded of daily experiences. Their attitudes afterward may depend on how their teacher treats students' mathematical models based on social values. If the teacher welcomes such ideas, the students are willing to come up with various mathematical models based on social values. Otherwise, they will stop talking about values and models. These emerged values create students' feelings of closeness between problem-solving and their daily life. In this process, they encounter and appreciate others' values, as well as other diverse values, which are fundamental in today's multi-value era and society. In the socially open-ended problem, a positive treatment of these values aims at cultivating proactive attitudes of problem-solving in the present and future and relating problem-solving to life in society.

14.2.3 Stage II: Development of Experimental Lessons

A Japanese mathematics lesson is characterised as a structured problem-solving approach (Stigler et al., 1999, pp. 36–41). This means that it basically follows stages such as 'Reviewing yesterday's lesson', 'Presenting the problem for the day', 'Working on the problem individually', 'Working on the problem in groups', and 'Summarising the main point'. These stages generally apply to most mathematics lessons in Japan regardless of the contents to be taught.

Shimada (2017) developed lessons using socially open-ended problems for primary schools, following these stages of Japanese mathematics lessons. Here, after the problem is presented, individual students work on its solution. Owing to the nature of a socially open-ended problem, they develop their own mathematical models based on their values. The combinations of social values and mathematical models are diverse, some typical cases among them are picked up by the teacher for discussion by the whole class. After the discussion, students are individually asked to choose their values and models. Some of them modify their values and models. Others maintain their own but refine them. This lesson process is similar to the modelling process (OECD, 2018).

These lessons provide a concrete reflection of the initial idea of a socially open-ended problem and widen its applicability. The development of these problems has led to the inductive and experiential identification of four categories and the concretisation of values and mathematical models (Shimada, 2017; Shimada & Baba, 2022). For example, in the problem 'Hitting the target', three balls are thrown at a target (Fig. 14.1). While two of the thrown balls hit 3-point and 5-point areas, the remaining ball hits the border between the 1-point and 3-point areas. In this case, participants are asked: 'How would you score this?' As a solution, students provide various points, such as 1, 2 [(1 + 3)/2], 3, or 4 [1 + 3], for various reasons. Typically, the reasons and thus values are categorised into kindness to the first grader and impartiality to all participants; the corresponding models are shown in Table 14.2. This lesson does not aim to obtain only one set of definite models and values but rather to interpret the problem situation, develop mathematical models with embedded values, and discuss their interpretations.

At a school cultural festival, your class offers a game of hitting a target with three balls. If the total score is more than 13 points, you can choose three favorite gifts. If you score 10 to 12 points, you get two prizes, and if you score 3 to 9 points, you get only one prize. A first grader threw a ball three times and hit the target in the 5-point area, the 3-point area, and on the border between the 3-point and 1-point areas. How do you give the score to the student?

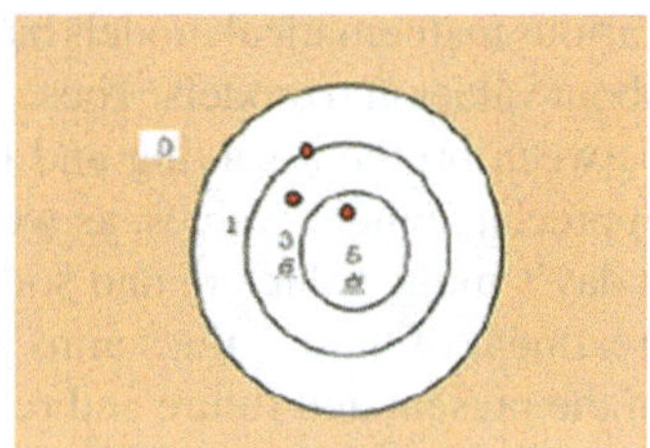

Fig. 14.1 Hitting the target (Shimada, 2017)

Table 14.2 Students' mathematical models and associated social values at the beginning stage ($N = 38$; Baba & Shimada, 2019, p. 178)

Mathematical models	Associated social values	Percentage of explicit social values
a. 5 + 3 + 3	Kindness to the first grader (specific person)	92.9 (13/14)
b. 5 + 3 + (3 + 1)		100.0 (1/1)
c. 5 + 3 + 3 + 1 + 1		100.0 (1/1)
d. 5 + 3 + 2		100.0 (2/2)
e. 5 + 3 + 2	Fairness and equality to the whole participant (all students)	0.0 (0/9)
f. 5 + 3 + 1		0.0 (0/10)
g. 5 + 3 + 3		0.0 (0/1)

Note In the column for the percentage of explicit social values, the fractions in parenthesis showed the number of students who wrote a particular mathematical model and expressed the social values explicitly against the total number of students who wrote a particular mathematical model

14.2.4 Stage III: Theoretical Foundation

The socially open-ended problem is regarded as a new type of problem for mathematics education, particularly in 21st-century society, and it can be characterised in terms of diversified values and trans-science problems (Weinberg, 1972). Society provides and enjoys alternatives owing to technological advancements, but sometimes, it experiences ethical dilemmas, such as those related to energy, food, and environmental issues. These represent a group of problems also called Ethical, Legal, and Social Issues and/or Socio-science Issues. In the above socially open-ended problem (Hitting the target), values are expressed by the participants, such as kindness to a specific group of people, equity, and impartiality.

In correspondence with such social change, there exist other approaches to problem-solving. When a problem is solved, the focus is not solely on arriving at an answer but also on developing a model presenting an interpretation of the problem context. Socio-critical modelling (Barbosa, 2006; Dede et al., 2021) and ethnomodelling (Rosa & Orey, 2010) have been proposed to interpret and examine the problem context, which is associated with the complexity of this highly technological society. They form a bridge between mathematicality and sociality.

Based on this trend and to extend the possibility of socially open-ended problems, a research team has been formed since 2016 to explore common properties and provide a theoretical foundation for such problems (Table 14.3). The team has conducted joint research combining the socially open-ended problem and the theory of Critical Mathematics Education (Skovsmose, 1994) to address the above social change under the title 'Research on Emergent Themes' during the Spring Session of the Japan Society of Mathematics Education from 2016 to 2020 (Baba, 2016, 2017, 2019, 2020).

Through this engagement, the team has identified the comprehensiveness of solutions, social justice embedded in the solutions, and exemplarity of sample socially open-ended problems for lessons. Here, the comprehensiveness of the solution means that mathematics and other factors cannot be isolated from the solution

Table 14.3 Research team for socially open-ended problems and Critical Mathematics Education

Name	Affiliation (as of 2016)	Role
Takuya Baba	Hiroshima University	Team Leader
Shimada Isao	Nippon Sport Science University	Primary Education
Kubo Yoshihiro	Hokkaido University of Education	Tertiary Education
Nakawa Nagisa	Kanto Gakuin University	Pre-primary Education
Kosaka Masato	Fukui University	Pre-primary Education, Relation with Science Education
Hattori Yuichiro	Kochi University	Secondary Education
Fukuda Hiroto	Hiroshima University	Secondary Education, Statistics Education

(Baba, 2020). For example, students at an early stage cannot clearly identify mathematical models but rather intermix mathematical models, language, and emotions (Nakawa & Kosaka, 2019). Another type of comprehensiveness appears in the later stages of education. For example, university students consider the cost sharing of a thesis binding. They consider several factors, such as page numbers, relationships among friends, and individual wishes. They intensively discuss what fairness is to all members (Kubo, 2019). The former can be called introductory comprehensiveness in the sense of having an introductory position within one's own education. The latter can be called extensional comprehensiveness in the sense of extending one's position after receiving a school education. As pre-primary and primary students think about problem-solving within their close environment, they expand their scope of thinking to the larger society and think about social issues, based on their previous experiences as they grow and move to the next school stage.

Social justice is rarely discussed in mathematics lessons in Japan; however, socially open-ended problems would enable students to address the complications between ensuring equity and being kind to the disadvantaged. The exemplarity (Skovsmose, 1994) of one specific and/or different socially open-ended problem is discussed regarding the appearance of certain common values across different grades and problems. Furthermore, some team members have extended part of this idea to propose socio-critical open-ended problems (Hattori et al., 2021). However, the systematisation of problem categories has not yet been clarified. Thus, this study clarified these problem categories.

14.3 Characteristics and Relationships Among Problem Categories

This section aims to answer RQs 1 and 2 by identifying characteristics and relationships among existing categories of socially open-ended problems.

14.3.1 Methodology for Consideration of Characteristics and Relationships Among Socially Open-Ended Problems Categories

The existing categories are inductively and experientially developed. To understand these categories more deeply and identify their characteristics and relationships, six universal activities (Bishop, 1988) are chosen to answer RQ 1. There are two reasons for this. First, these universal activities exist in any culture, and the analysis of problem categories may hold implications beyond cultural borders. For the same reason, the semantic meanings in the dictionary and other leading studies are also added, and their implications are drawn. Second, the problem categories are expressed in verbs and represent mathematical activities. Thus, consideration of these categories in relation with universal activities is expected to yield further insight into the characteristics as mathematical activities.

Regarding RQ 2, the relationships among the problem categories are considered in terms of activity and objects to act upon. There are two reasons for this. The first is related to the problem categories as mathematical activity. It is important to know the objects to be acted upon in a mathematical activity. The second reason is that we experientially learned that some relationship exists between some categories. For example, sharing and designing may be related to each other when we consider and design a method of sharing.

However, the researchers are aware of the limitations of this approach as well. Although the six universal mathematical activities (Bishop, 1988) are among the most important research topics in this area, we may be able to identify other characteristics if we employ another reference. An important point is that simply counting and measuring may not produce social values, but social value may emerge if we conduct activities of counting and measuring for sharing. This point should be investigated further in the future.

14.3.2 Existing Categories of Socially Open-Ended Problems

Problem-solving in Japan has been discussed and practiced for many years, both explicitly and implicitly. Currently, it targets three points: acquisition of mathematical knowledge and skills; growing mathematical thinking, judging, and expressing abilities; and nurturing mathematical dispositions such as interests, willingness, and attitudes. To achieve these three points, the Course of Study clearly states the importance of mathematical activities utilising mathematical views and thinking. 'To grasp the phenomenon by paying attention to numbers, quantities, geometrical shapes, and their relations, think logically based on reason, and deliberate it integratively and extensively' (Ministry of Education, Culture, Sports, Science and Technology [MEXT], 2017a, p. 23; 2017b, p. 21).

Broadly speaking, there are two types of problem-solving processes, one focusing on day-to-day and context-rich problems and the other on mathematical problems. In both, the problem-solving process is characterised by the emergence of the question, problem formulation, understanding the problem, planning a problem-solving strategy, executing the plan, reviewing this, and the emergence of a new question, (MEXT, 2017a, p.72; 2017b, p. 59). This process is thought to be influenced by four stages of problem-solving (Polya, 1945). However, again, values are not taken up even in these types of problem-solving. Problem-solving is emphasised only from the perspective of mathematical values and thus generalisation and logicality are appreciated, but not social values. That is why it is essential to solidly develop a new approach to problem-solving utilising social values. That is the uniqueness of this approach.

Through the above activities, Shimada and Baba (2022) stated a revised four hypothetical categories—*Sharing, Selecting, Rule-making*, and *Planning/Designing*—and compiled 32 socially open-ended problems. Each category was explained, and some example given as follows (Shimada & Baba, 2022, pp. 27–28):

> Sharing: To consider diverse values such as impartiality and thoughtfulness, and mathematical models, by 'sharing' a finite number of objects among people. Critically consider the intention to share in a given situation, such as policy.
> An example is 'Number of couches required' (Hattori et al., 2021) as shown below (Fig. 14.2). Here, the number of couches is decided depending upon values such as fairness, equality, and economy.
> Selecting: To consider diverse values and mathematical models through 'selecting' objects (e.g., car, mobile phone, foods), people (e.g., sports representatives, voters), and jobs (e.g., teachers, doctors). Critically consider the intention to select in each situation, such as policy.
> An example is 'Purchasing a mobile phone' (Hattori et al., 2021), as shown below (Fig. 14.3). Depending upon the values such as saving money, unlimited usage, and balanced management, the mobile phone plan is chosen.
> Rule-making: To consider diverse values, such as impartiality and thoughtfulness, and mathematical models through 'rule-making (system making)'. Critically

There are 30 children. How many couches, each seating up to four people, should be available?

Fig. 14.2 Number of couches required

Our teacher is thinking of purchasing a new mobile phone. He would like to purchase it from one of the following three companies.

Plan＼Companies	A	B	C
Amount of data that can be used monthly and fees	25GB for JP¥9000	12GB for JP¥5000	6GB for JP¥3000
Calling charges	Free	JP¥50 per minute	JP¥150 per minute*

* The upper limit of call charges of Company C is JP¥9000, with no additional call charge beyond this.

The monthly charge of each company is calculated as: Monthly charges = Monthly data usage charge + Call charges. In this case, which company's plan would be the most economical for our teacher?

Fig. 14.3 Purchasing a mobile phone

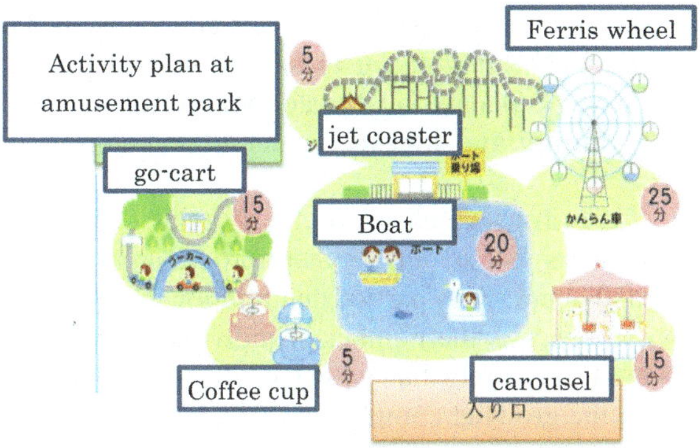

Waiting time	
Vehicle	Waiting time
Coffee cup	5
Go-cart	10
Jet coaster	10
Boat	0
Ferris wheel	20
carousel	15

※This problem is developed by revising the problem in grade 3 mathematics textbook volume 1 (Sawada et al., 2012, p.37).

Fig. 14.4 Activity plan at an amusement park

consider the intention of the rule (system) in a given situation, such as policy. An example is hitting the target, as shown above (Shimada, 2017).

Planning/Designing: To consider diverse values and mathematical models by 'planning/designing' an event in the future by arranging elements of time and space. Critically consider the intention to plan/design in a given situation, such as policy.

An example is 'an activity plan at an amusement park' (Shimada, 2017). While the time taken for each vehicle ride is shown in the picture (Fig. 14.4), the waiting time is given below the table and all transfer times from one vehicle to another are assumed to be 5 min. Students make an activity plan to enjoy riding vehicles for the time between 10:00 and 11:30.

14.3.3 Consideration of Problem Categories from the Perspective of Universal Activities

Socially open-ended problems represent situations that have been addressed throughout the history of humanity, for example, sharing a limited number of objects, such as food and resources, among a group of people. Here, such situations are turned into socially open-ended problems in the context of today's society. To reconsider existing categories of socially open-ended problems, this study employed the perspective of six universal activities (Bishop, 1988), a dictionary, and representative literature to understand the nature of human activities and the characteristics of the categories.

First, we considered the sharing category. According to the Oxford Learner's Dictionary (https://www.oxfordlearnersdictionaries.com), *sharing* means 'to have, use or experience something at the same time as somebody else' and 'to have part of something while another person or other people also have part'. This means that objects, events, and people are held simultaneously among several people. Throughout history, human beings have needed to collaborate for survival, and thus sharing is an important strategy. **Counting and measuring** serve as premises for sharing.

According to Bishop (1988), '**Counting**, which is so closely related to trade, wealth, employment, property, and status in a society, is therefore strongly related to the social values in that group, and accuracy is part of that relationship' (p. 27). One significant difference among societies is the number of individuals, which is related to their environment. 'We can also see that with the growth in sizes of numbers in the community and with the growing complexity of societies, more and more complex number systems have evolved' (Bishop, 1988, p. 26). Societies have developed number systems, which are a foundation for sharing activities. Further, Bishop (1988) stated that '**Measuring** relates to ideas like "more than" and "less than" since the need for measuring only occurs if phenomena are to be compared' (p. 35) and 'measuring is deeply embedded in economic and commercial life. As well as involving numerical features, therefore it undoubtedly has a strong social aspect' (p. 37). In this sense, measuring is related to social activity, and sharing uses such activity to develop relationships with other members of society. Bishop also mentioned that valuing accuracy and rarity depends on the society in which the individual lives. Today's societies typically contain many individuals and the technology which forms a core of these societies requires extremely high accuracy, but it still depends on the context. Depending upon the society and context, we may **design and explain** why and how we share.

Second, according to the Oxford Learner's Dictionary, *selecting* means 'to choose somebody/something from a group of people or things, usually according to a system'. For example, human beings face many options when selecting an object to eat, a person to play with, a school to attend, a place to live, and so on. Simply speaking, the existence of options represents the core of this activity and the richness of our lives.

To choose, one may consider and compare options. If the comparison is quantitative, then it is deeply related to the **measuring** activity noted above; 'Comparing more than two or three objects develops another idea, that of ordering' (Bishop, 1988, p. 35). Ordering objects and selecting one among them is practiced in daily life. However, in daily life, one may decide not to quantify, as expressed in the following quote 'Gay and Cole bring another aspect to our attention, namely what can be counted in Kpelle culture, and they sensitise us to their taboos associated with number' (Bishop, 1988, p. 27). Therefore, **counting and measuring** activities are important steps towards quantification and are thus closely related to sharing and selecting activities.

In today's societies, social choice theory and collective choice theory address collective choices (Sakai, 2013). Social choice theory contains logical aspects associated with a mathematical model. Here, choices in society can be made based not only on individual interests and impressions but also on the collective will. We further consider a method of sharing and selecting among other people. This is related to

designing and explaining, but the primary activities are sharing and selecting. In other words, designing and explaining belong to the secondary level, which is also called the meta-level, and is explained in the next section.

Third, according to the Oxford Learner's Dictionary, *rule* means 'a statement of what may, must or must not be done in a particular situation or when playing a game.' Thus, rule-making means developing such rules. For example, Shimada and Baba (2022) considered socially open-ended problems, such as games, sports, and taxes. Many of these are related to games and play.

Bishop (1988) stated that '**Playing** may seem initially to be a rather strange activity to include in a collection of activities relevant to the development of mathematical ideas, until one realises just how many games have mathematical connections' (p. 42). Interestingly, playing is not required for survival. Rather, Huizinga (2018) referred to humans as 'Homo-Ludens', referring to the performance of activities besides those needed for survival. 'Certainly imitation, or the modelling of reality, is a feature of many games, and has much importance for use here' (p. 45). In this sense, rules of play may model societal rule, or vice versa. Bishop further stated, 'Games are often valued by mathematicians because of their rule-governed behaviour which it is said, is like mathematics itself' (p. 45). Mathematics thus shares some characteristics with games.

In addition, humans are social animals who form groups and live together in societies that require rules to maintain order (Engeström, 2014). Some kind of rules can be considered for **any other activities** in our societies. For example, a tax can be interpreted as a rule that determines how to share resources and make life meaningful in society.

Finally, according to the Oxford Learner's Dictionary, *planning* means 'to make detailed arrangements for something you want to do in the future' and *designing* means 'to decide how something will look, work, etc., by drawing plans, making computer models, etc.' In some ways, these concepts are related and can even be used interchangeably.

Bishop (1988) described **locating and designing** jointly thus: 'Where the "locating" activities refer to positioning oneself and other objects within the spatial environment, the activities of designing concern the "manufactured" objects, artifacts and technology which all cultures create for their home life, for trade, for adornment, for warfare, for games and for religious purposes' (p. 38). Designing is related not only to actual production but also to abstract thinking, since 'designing concerns abstracting a shape from the natural environment' (p. 39). Thus, the category of planning/designing addresses the allocation of resources in terms of time and space to fulfil a certain purpose in life. **Any other activities** can be combined with this planning/designing activity.

The table is to be looked at horizontally. Each category is examined one by one in relation with the six universal activities. Some universal activities are found to be essential for a particular category. For example, a sharing activity regards counting and measuring as fundamental and thus as directly related activities. In addition, a sharing activity becomes a little more intricate so that we may consider various conditions to design and explain how to share with others. In this way, designing and

Table 14.4 Problem categories about six universal activities

Problem category	Six universal activities					
	Counting	Measuring	Locating	Designing	Playing	Explaining
Sharing	○	○		△		△
Selecting	△	○		△		△
Rule-making	△	△	△	△	○	△
Planning/designing	△	△	○	○	△	△

※ Developed by the authors (○ and △ indicate direct and indirect correspondences, respectively)

explaining activities are regarded as secondary compared to the above fundamental activities. The triangle (△) marks such an activity as secondary or indirect.

From these, it is found that these four categories were associated with the six universal activities. In fact, all of these are interrelated to some degree. For example, in the melon problem, we designed rules for sharing melons. Here, sharing, rule-making, and designing are involved. However, this problem is centred on melon-sharing. Thus, the primary activity is sharing, while other activities may be secondary or indirect. Regarding RQ 1, the characteristics of four categories are identified in relation with universal activities and the result is summarised in Table 14.4.

14.3.4 Relationships Among Problem Categories from the Perspective of Objects

As shown in the previous section, some activities belong to the primary level while others belong to the meta-level. Since meta-level activities act on the primary level as an object, the primary level can be called an object level. For example, we may consider a socially open-ended problem of designing how to share certain objects. Let us consider two examples of the sharing category. The first is the melon problem (Iida et al., 1994). We can share melons according to the points that each group obtained, although there are a few different ways of sharing. The second is the tax system, which requires a far more complicated consideration of the income, number of family members, number of handicaps, and the amount of damage incurred during the typhoon. Here, sharing and designing can be considered at different levels. Sharing belongs to the object level, and designing how to share belongs to the meta-level of sharing activity (Table 14.5).

Table 14.5 Levels of activities

Meta-level activity	Activity	Designing
	Objects	Sharing
Object-level activity	Activity	Sharing
	Objects	Number of melons

In this section, four problem categories are reconsidered in terms of the activity and objects to be acted upon. Since the beginning, human beings have acted on their environment and invented tools, with today's most sophisticated tools, such as computers and ICT, used to extend human capacity. In other words, we can think of a 'tool for tools' by thinking about the usefulness and/or effectiveness of the tool. Essentially mathematical activities work at multilaminar levels. However, for the learners the objects of activity remain in the environment at large, including people, objects, and the order of such acts, called events.

(1) The sharing activity in the *Sharing* problem category identifies the number and quantity of people and objects through counting and measuring and proceeds to divide such numbers and quantities equally, proportionally, or in even more complicated ways. In such cases, a rule for sharing may be considered. In this sense, rule-making also belongs to the meta-level.
(2) The selecting activity in the *Selecting* category can be regarded as an option instead of sharing with others. Not only a particular object but also the environment can be an object of selection. Regardless, selection by society requires a logical explanation. A leading researcher in management studies (Keeney, 1992) has compared alternative-focused thinking and value-focused thinking. The former involves selecting an option among existing alternatives and is sometimes criticised for not considering favourable alternatives. The latter aims to create alternatives based on preferred value (González et al., 2018). Keeney (1992) emphasised the latter, value-focused thinking in modern management and the development of alternatives based on such value.
(3) Rule-making activities in the *Rule-making* category can be formulated at the meta-level when sharing and selecting activities are repeated many times. Another important aspect of rule-making is related to playing (Bishop, 1988). This is related to how games and social activities are conducted smoothly while maintaining interest and are thus part of the affective domain. Engeström (2014) noted that rule-making is important for human beings and society to function smoothly.
(4) Designing activity in the *Planning/designing* category is limited to geometrical aspects in Bishop (1988); however, in this study, it also involves designing objects, events, time, and space. Such activities may have spacious, pleasurable, and economic properties. Planning/designing also has the characteristics of meta-activity for sharing, selecting, and rule-making. How to share, select, and make rules are the targets of planning and designing.

Table 14.6 shows the relationships among the four categories as an answer to RQ2.

Table 14.6 Relationships among problem categories

Meta-level activity	Activity	Planning/designing	Planning/designing	Planning/ designing
	Objects	Sharing, selecting, rule-making	Selecting, rule-making, designing	Selecting, rule-making
Object-level activity	Activity	Sharing, selecting, rule-making	Selecting, rule-making, designing	Selecting, rule-making
	Objects	Number and quantity (person, object)	Environment (layouts, objects, shapes)	Events

14.4 Categorisation of Tasks in Mathematical Modelling Research

Based on the above discussion, RQs 3 and 4 were explored in relation to the creation of further socially open-ended problems. Possible tasks were chosen from modelling research to confirm whether they could be converted to socially open-ended problems and grouped into the four categories.

14.4.1 Methodology for Task Selection and Categorisation

As described previously, solving a socially open-ended problem involves mathematical modelling. Therefore, two groups of papers on mathematical modelling were chosen. The criteria for selection were international representation and systematic collection. The first is an anthology of research papers by the international mathematics education research community, which focuses on mathematical modelling, namely the International Community of Teachers of Mathematical Modelling and Applications (ICTMA). The ICTMA has published two books: *Mathematical Modelling Education in East and West* (Leung et al., 2021) and *Mathematical Modelling Education and Sense-making* (Stillman et al., 2020). We anticipated that these books would contain tasks in the practice section and selected five papers from the fifth chapter in the first volume, 'Teaching Practice', and nine papers from the second chapter in the second volume, 'Research into, or Evaluation of, Teaching Practice in Mathematical Modelling Education'. In total, 14 papers were extracted for analysis. The second source of data was the special issue of the International Mathematics Education Research Journal, *ZDM*, which features teaching practice on mathematical modelling. The theme of the 2018 special issue was empirical research on teaching and learning mathematical modelling; the issue contained 26 papers. Thus, the target of the current analysis was 40 papers in total.

Each paper was examined as to whether it contained a mathematical modelling task. Here, 'modelling task' refers to a problem setting which requires a modelling activity. These tasks were analysed from two perspectives. One was the type of mathematical modelling that each task contained. Here 'type' refers to either one of three types of modelling: descriptive modelling, which captures existing reality in a given extra-mathematical domain (Niss, 2018); normative modelling, which creates reality in this domain (Niss, 2018); and socio-critical modelling, which critiques the role of mathematics in society and mathematical modelling itself in society (Gibbs, 2019). Descriptive modelling includes, for example, activities to solve the problem of mathematically finding the height of a tree by drawing a right triangle and using a similarity ratio. Normative modelling includes, for example, activities that consider 'per unit quantity' by dividing the number of people by the area in order to grasp a new reality that is more crowded. Socio-critical modelling includes, for example, activities in which students discuss the mathematical errors involved in the social content of a newspaper article (Barbosa, 2006). Some of socio-critical modelling may be normative modelling. In other words, these are sometimes intertwined. However, the authors identify materials that seem to more emphasize the characteristic to critique the role of mathematics in society and modelling itself as socio-critical modelling materials.

To summarise the method of analysis, 40 papers were first examined to determine whether they contain mathematical modelling tasks. Second, the modelling tasks were grouped according to the types of mathematical modelling and to the problem categories of socially open-ended problems to which they belong. The analysis result is arranged accordingly (as shown in Table 14.7). As complete objectivity cannot be guaranteed by classification, the interobserver agreement rate of 80% or more, which was obtained by the third and fourth authors carrying out the classification work and collating their results, is deemed to guarantee the reliability of the analysis.

The coding and review processes are described here in more detail. First, the third and fourth authors examined how to judge the inclusion of a mathematical modelling task. As a result, they agreed to the condition that a figure or a text in the task must explicitly mention mathematical modelling. Next, the following conditions were developed for the classification of problem categories for socially open-ended problems. Descriptive modelling is modelling that only captures real-world situations; normative modelling not only captures real-world situations but also creates a new world; and socio-critical modelling is modelling centred on a critical examination of the role of mathematics in society. To ensure their adequate mastery before the review, we first conducted a preliminary review with a couple of randomly selected tasks. Subsequently, independent reviews were conducted.

This approach has the following limitations. Modelling tasks are extracted only from mathematical modelling research, and this is found to be a prominent way of developing new socially open-ended problems. However, this method may narrow our scope of the possibility of socially open-ended problems from that of developing problems from scratch or from other problem-solving tasks. For example, categories of sharing and rule-making were not found, and this may be due to following this approach. This is a limitation of research as well as future issues.

Table 14.7 Sample: result of task analysis

Article	Summary of teaching material	Type of modelling	Category
Greefrath et al. (2018)	A realistic situation presents itself: piling up straw bales in a field. The height of the stacks was questioned	Descriptive modelling	–
Ärlebäck (2020)	The task is a basketball free throw. Divide the students into three teams and ask them to each make 10 basketball free throw from three different distances and determine 'which team is the best' based on bar charts, the total frequency, mean, mode, box plots, and standard deviation	Normative modelling	Selecting
Czocher and Hardison (2021)	The task is to create scenarios in which a veterinarian aims a tranquillising dart at a monkey. Other conditions are left open-ended, and students must determine and model variables such as the length of the straight path from the vet's gun to the monkey, initial velocity of the dart, angle of the gun, and so on	Socio-critical modelling	Planning/ designing

14.4.2 *Interpretation of Mathematical Modelling Tasks in Terms of Four Categories*

The 40 articles were analysed using the above-mentioned method. Some studies did not contain any concrete tasks, whereas others contained several tasks. Thus, the number of tasks was 61, greater than the number of papers. The interobserver agreement rate between the third and fourth authors was 96.7% (agreement on 59 tasks out of a total of 61 tasks), which guarantees the reliability of this analysis. The result of the classifications of these tasks is presented in Table 14.8.

These tasks were categorised into descriptive modelling (30), normative modelling (25), and socio-critical modelling (6). Thus, there were more cases of descriptive and normative modelling than socio-critical modelling. This chapter focuses on normative modelling and socio-critical modelling because descriptive modelling

Table 14.8 Summary of analysis results

Modelling type	Descriptive modelling	Normative modelling				Socio-critical modelling			
Problem category		1	2	3	4	1	2	3	4
No. of tasks	30	0	8	0	17	0	1	0	5

⁎ The problem category is indicated thus: 1, sharing; 2, selecting; 3, rule-making; and 4, planning/ designing

aims to capture existing reality, normative modelling to create reality that is not currently available, and socio-critical modelling to criticise mathematical models and modelling in society. That is, the latter two types seek not only one solution but also an optimal solution, and thus values tend to appear in the solution; thus, they are related to socially open-ended problems. In particular, socio-critical modelling agrees with the idea of a socially open-ended problem in terms of its aims.

Shimada (2017) identified a unique set of four properties of socially open-ended problems: emphasis on the social context, authenticity of the problem, conditionality of the problem, and treatment of the problem. Each task is examined if it satisfies each of the four properties. Among these properties, the fourth property indicates the treatment of the problem. That is, tasks contained in the above two modelling types, normative modelling and socio-critical modelling, can be converted into socially open-ended problems, depending upon the treatment of the problem in the classroom. For example, the task 'The Basketball Penalty Competition' in Ärlebäck (2020) is not devoted to interpreting existing reality but aims to create reality (the best team) and is thus judged as normative modelling. In achieving this aim, each student can freely interpret 'best team', select a suitable indicator (herein a representative value), and decide how to use these indicators based on their social values. In this sense, the task satisfies the third property and authenticates the problem situation because of its similarity to real life (the second property). This reflects the social context (the first property). Thus, all four properties are present in this task. Therefore, this chapter focuses only on tasks which contain either of two modelling types, normative modelling or socio-critical modelling. Next, this chapter analyses them in terms of problem categories. The analysis revealed that all 31 tasks belonged to the selecting or planning/designing category; no tasks belonged to the categories of sharing and rule-making. Furthermore, planning/designing was more common than the other categories. However, tasks that belong to the planning/designing category are also related to the other three categories because of object-meta-level properties mentioned in the previous chapter. To clarify this point, the tasks found in normative modelling and socio-critical modelling were scrutinised. Three tasks found in normative modelling are considered here.

【Outline of Task 1 judged to be normative modelling】 (Burkhardt, 2018).

The task was to prepare and justify advice for the town council on the best way to reduce accidents within a specified budget, given the cost of various improvements, based on a custom-tailored database of 120 reports on road accidents in a small fictional town.

【Outline of Task 2 judged to be normative modelling】 (Stender, 2018).

The task was to investigate the best positions of bus stops for the entire public transport system in the city of Hamburg.

【Outline of Task 3 judged to be normative modelling】 (Burkhardt, 2018).

The task was to devise, schedule, run, and practically evaluate classroom games.

These three tasks were considered to belong to the planning/designing category: designing measures to prevent traffic accidents, establishing the locations of new bus stops, and designing classroom games. Because these represent normative modelling, a preferable reality will be designed or created. However, these tasks also involved other categories, such as sharing, selecting, and rule-making. The task of exploring measures to prevent traffic accidents contains a sharing aspect at the object level because it concerns sharing a budget under budgetary constraints. The task of selecting the location of a bus stop from among infinite places involves the selecting category at the object level. Since the options are infinite, the selection of location is not focused on alternatives but is rather value-focused thinking. The task of planning/designing a game requires making rules and thus involves the rule-making category at the object level. From these results, it is clear that these tasks belong to the planning/designing category at the meta-level and to either of the three other categories at the object level. Alternatively, the planning/designing category uses sharing, selecting, and rule-making as objects and planning/designing as an activity. The aim can be known from the task. For example, the first is 'to prepare advise on the best way' and the aim of this task is to design the best way but not to share the budget.

Similarly, the tasks found in socio-critical modelling were analysed in the same way. Three such tasks are considered below.

【Outline of Task 1 judged to be socio-critical modelling】 (Czocher & Hardison, 2021).

The task was to create scenarios in which a veterinarian aims a tranquillising dart to shoot a monkey. Other conditions were left open-ended, so that students determined and modelled variables such as the length of the straight path from the vet's gun to the monkey, initial velocity of the dart, and gun angle.

【Outline of Task 2 judged to be socio-critical modelling】 (Yvain-Prébiski, 2021).

Botanists have discovered and sketched an exotic tree every year since 2013. They believe that the tree will reach its full size by 2023. They want to build a greenhouse to protect the trees. The task was to help predict the size of the trees in 2023.

【Outline of Task 3 judged to be socio-critical modelling】 (Orey et al., 2020).

The task was to design and construct roller carts. In this task, it was expected that students would consider the mass, circumference, and diameter of wheels, the number of wheels, geometric shapes of the fixed and movable parts of the structure, and so on.

Each task involved design methods and procedures, in this case with respect to dart shooting, greenhouses and roller carts, and it was expected that mathematical modelling would be critically examined after several modelling cycles. These tasks belong to the category of planning/designing and socio-critical modelling, which criticise the mathematical models and mathematical modelling extant in society. As stated above, the category of planning/designing has the property of being at the meta-level, and either of the other categories is at the object level. For example, when designing roller carts and considering their production cost, the problem can also belong to the sharing category at the object level by considering the cost of each part within the scope of a limited budget. Another example is that a veterinarian designs a method and procedure for shooting an anaesthetic dart into a monkey. It is a design activity at the meta-level and a selecting activity at the object level. This is because the veterinarian must consider how to select the necessary variables among distance between them and the monkey, initial speed of the dart, shooting angle, and so on. Here, students are expected to select the appropriate methods and procedures. The aim is to design methods; therefore, it is not alternative-focused but rather value-focused thinking in mathematical modelling terms. The final example is the designing of a greenhouse, which requires the prediction of tree growth. In Yvain-Prébiski (2021), students proposed a greenhouse design contingent on a variable speed that depended on sunshine and watering. Rules pertaining to sun exposure and watering can be designed to solve this problem. Thus, problem belongs to the rule-making category at the object level.

From the above, the answers to RQ3 and RQ4 can be stated as follows. Tasks from modelling research can be treated as socially open-ended problems, and the modelling types are normative modelling and socio-critical modelling. Tasks within normative and socio-critical modelling have a strong tendency towards falling under the category of planning/designing, but at the same time they include aspects of other categories, such as sharing, selecting, and rule-making at the object level. This means that these tasks identified as convertible to socially open-ended problems may be grouped under one or a few of the four problem categories. Simultaneously, this also means that mathematical modelling tasks that involve normative modelling and socio-critical modelling can be utilised as a new socially open-ended problem.

14.5 Conclusion

In conclusion, the following findings emerged from the current study. First, inductively and experientially developed categories (Shimada, 2017; Shimada & Baba, 2022) are characterised in relation with six universal activities. Second, some categories belong to the meta-level, while others belong to the object level. In particular, the planning/designing category belong to both the object level and meta-level. Third, mathematical modelling tasks in normative and socio-critical modelling have the potential to be converted into socially open-ended problems. Finally, these tasks can be grouped into either one or a few of four categories of socially open-ended problems. In particular, tasks in normative and socio-critical modelling are biased toward the category of planning/designing, and partially fall within the categories of sharing, selecting, and rule-making.

These findings show the potential for desired approach that can combine mathematical problem-solving and values. While mathematics education used to emphasis students' formation of mathematical knowledge and skills, the proposed approach enriches students' ability of interpreting the problem, which may emerge in their daily life, and making a judgement through mathematics. In other words, they realise connections between social values and mathematical models within the societal context, and that is a new possibility of mathematics education. In particular, as stated in the beginning, it is very important for students to know others' values and appreciate the existence of diverse values in a society with value pluralism. Here, this new possibility can be interpreted in terms of four points.

- Students weak in mathematics may feel encouraged by giving their own social values and mathematical models. In that sense, they find it easy to participate in the mathematics lesson.
- If socially open-ended problems are treated coherently, students' ability to interpret society in terms of mathematics, as stated in mathematical literacy (OECD, 2018), is enhanced. The problem categories can provide more comprehensive ways of dealing with socially open-ended problems.
- Treatment of students' emerged social values and mathematical models requires teachers' competence to treat them appropriately.
- Teachers are able to develop teaching materials with reference to the framework and sample problems.

There are some issues for future research to develop a comprehensive framework of socially open-ended problem from pre-primary education to tertiary education:

- Further consideration of the possibilities of any other problem categories, as stated above.
- Creation and practice of more socially open-ended problems and accumulation of students' responses for validation of this approach of socially open-ended problems.
- Longitudinal research to accumulate data from the same group. The values and mathematical models which appear in the primary school may form a foundation for values and mathematical models for further education, and for citizenship after graduation. As they grow and advance to next educational stage, students will encounter problems with a larger scope, such as social equity (Baba, 2023).
- Comparative research across countries regarding social values and mathematical problems utilising the same framework. This will explain the characteristics of social values within the same cultural group.

Acknowledgements We would like to thank Editage (www.editage.com) for English language editing. Regarding the research funding, this work was supported by KAKENHI (19K14219, 21K18511, 21K02470, 22K13717).

References

Ärlebäck, J. B. (2020). A case study of tensions and challenges arising as a Swedish upper secondary teacher designs and implements a model development sequence on statistics. In G. A. Stillman, G. Kaiser, & C. E. Lampen (Eds.), *Mathematical modelling education and sense-making. International perspectives on the teaching and learning of mathematical modelling* (pp. 139–150). Springer.

Baba, T. (2007). Primary mathematics education in a society and times with multiple values. *Journal of Japan Society of Mathematics Education, 89*(10), 20–27. (in Japanese).

Baba, T. (2009). Analysis of socially open-ended problems in mathematics education from the perspective of values. *Journal of JASME Research in Mathematics Education, 15*(2), 51–57. (in Japanese).

Baba. T. (2016). Fostering critical thinking skills through mathematics education. *Proceeding of 4th Spring Session* (pp. 95–96). Japan Society of Mathematics Education (in Japanese)

Baba, T. (2017). Fostering critical thinking skills through mathematics education: Consideration of critical mathematics education (Skovsmose, 1994) from Japanese context. In *Proceeding of 5th Spring Session* (pp.199–200). Japan Society of Mathematics Education (in Japanese).

Baba, T. (2019). Fostering critical thinking skills through mathematics education. In *Proceeding of 7th Spring Session* (pp. 1–2). Japan Society of Mathematics Education (in Japanese).

Baba, T. (2020). Fostering critical thinking skills through mathematics education: Comprehensive discussion using exemplarity. In *Proceeding of 8th Spring Session* (pp. 129–130). Japan Society of Mathematics Education (in Japanese).

Baba, T. (2023). Curriculum of mathematics education to nurture critical mathematical literacy. *Proceedings of the Annual Meeting of JSSE, 47*, 205–206. (in Japanese).

Baba, T., & Shimada, I. (2019). Socially open-ended problems for enriching student learning with mathematical models and social values. In P. Clarkson, W. Seah, & J. Pang (Eds.), *Values and Valuing in Mathematics Education. ICME-13 Monographs* (pp. 171–183). Springer.

Barbosa, J. C. (2006). Mathematical modelling in classroom: A socio-critical and discursive perspective. *ZDM Mathematics Education, 38*(3), 293–301.

Bishop, A. J. (1988). *Mathematical enculturation: A cultural perspective on mathematics education.* Kluwer Academic Publishers.

Brown, S. I. (1984). The logic of problem generation: From morality and solving to de-posing and rebellion. *For the Learning of Mathematics, 4*(1), 9–20.

Burkhardt, H. (2018). Ways to teach modelling-A 50 year study. *ZDM Mathematics Education, 50*(1–2), 61–75.

Czocher, J. A., & Hardison, H. L. (2021). Attending to quantities through the modelling space. In F. K. S. Leung, G. A. Stillman, G. Kaiser, & K. L. Wong (Eds.), *Mathematical modelling education in east and west: International perspectives on the teaching and learning of mathematical modelling* (pp. 263–272). Springer.

Dede, Y., Akçakın, V., & Kaya, G. (2021). Mathematical, mathematics educational, and educational values in mathematical modeling tasks. *ECNU Review of Education, 4*(2), 241–260.

Engeström, Y. (2014). *Learning by expanding: An activity-theoretical approach to developmental research* (2nd ed.) (Yamazumi, K., et al. 1999, Trans.). Cambridge University Press (in Japanese).

Gibbs, A. M. (2019). *Socio-critical mathematical modeling and the role of mathematics in society.* Doctoral dissertation. Florida Institute of Technology.

González, O., Baba, T., & Shimada, I. (2018). Value-focused thinking in the mathematics classroom: Engaging students in decision-making through socially open-ended problem solving. In B. Rott, G. Törner, J. Peters-Dasdemir, A. Möller, & Safrudiannur (Eds.), *Views and beliefs in mathematics education: The role of beliefs in the classroom* (pp. 55–68). Springer International Publishing.

Greefrath, G., Hertleif, C. & Siller, HS. (2018). Mathematical modelling with digital tools: A quantitative study on mathematising with dynamic geometry software. *ZDM Mathematics Education, 50*(1–2), 233–244.

Hattori, Y., Fukuda, H., & Baba, T. (2021). Development of socio-critically open-ended problems for critical mathematical literacy: A Japanese case. *Journal of Educational Research in Mathematics, 31*(3), 357–378.

Huizinga, J. (2014, originally published in Germany in 1933, translated by Satomi, G. in 2018). *Homo ludens: A study of the play-element of culture.* Koudansha-Academic (in Japanese).

Iida, S., Yamashita, A., Sumi, M., & Komori, A. (1994). Research on recognition of values in open end problem in mathematics education (1): Summary and preliminary report. *Research Journal on Kyushu Mathematics Education, 1*, 32–43. (in Japanese).

Keeney, R. L. (1992). *Value-focused thinking: A path to creative decision making*. Harvard University Press.

Kridel, C. (Ed.). (Original version 2010, Nishioka et al., translated and edited 2021). *Encyclopedia of curriculum studies*. SAGE.

Kubo, Y. (2019). University students' attitudes on relevance in critical thinking. In *Proceeding of 7th Spring Session* (pp. 3–10). Japan Society of Mathematics Education (in Japanese).

Lesh, R., & Zawojewski, J. (2007). Problem solving and modelling. In F. K. Lester (Ed.), *Handbook of research on mathematics teaching and learning* (2nd ed., pp. 763–804). Information Age Publishing.

Leung, F. K. S., Stillman, G. A., Kaiser, G., & Wong, K. L. (Eds.). (2021). *Mathematical modelling education in east and west: International perspectives on the teaching and learning of mathematical modelling*. Springer Nature.

Ministry of Education, Culture, Sports, Science and Technology (Hereafter MEXT). (2017a). *Course of study for primary school*. MEXT (in Japanese).

MEXT. (2017b). *Course of study for junior secondary school*. MEXT (in Japanese).

Nakawa, N., & Kosaka, M. (2019). Children's social values for critical thinking in mathematical activities in pre-school education. In *Proceeding of 7th Spring Session* (pp. 27–34). Japan Society of Mathematics Education (in Japanese).

Niss, M. (2018). Advances in research and development concerning mathematical modelling in mathematics education. In *Proceedings of EARCOME8, 1* (pp. 26–36). EARCOME.

Organisation for Economic Co-operation and Development [OECD]. (2018). *PISA 2022 mathematics framework (Draft)*. OECD.

Oliva, P. F. (2009). *Developing the curriculum* (7th ed.). Allyn & Bacon.

Orey, D. C., Rosa, M., & Soares, R. B. (2020). Mathematical modelling as a learning environment to transform a street activity into a sport practice. In G. A. Stillman, G. Kaiser, & Ch. E. Lampen (Eds.), *Mathematical modelling education and sense-making: International perspectives on the teaching and learning of mathematical modelling* (pp. 199–219). Springer.

Polya, G. (1945: Reprinted 1990). *How to solve it* (Penguin 2nd ed.).

Rosa, M., & Orey, D. (2010). Ethnomodelling as a pedagogical tool for the ethno-mathematics program. *Revista Latinoamericana De Etnomatemática, 3*(2), 14–23.

Sakai, T. (2013). *Invitation of social choice theory: Science on voting and majority decision*. Nihon-Hyoron-sha (in Japanese).

Sawada, T., et al. (2012). *Grade 3 mathematics textbook volume 1*. Kyouiku Shuppan (in Japanese).

Science Council of Japan. (2007). *[Proposal] Integration of knowledge—Towards science for society*. Science Council of Japan (in Japanese).

Shimada, S. (Ed.). (original version 1977, new version 1995). *Open-ended approach in mathematics*. Toyokan Shuppan (in Japanese).

Shimada, I. (2017). *Arithmetic and mathematics education and diverse values: Approaches through socially open-ended problems*. Toyokan Shuppan (in Japanese).

Shimada, I., & Baba, T. (2016). Transformation of students' values in the process of solving socially open-ended problems (2): Focusing on long-term transformation. In: *Proceedings of the 40th Conference of the International Group for the Psychology of Mathematics Education* (Vol. 4, pp.187–194). PME.

Shimada, I., & Baba, T. (Eds.). (2022). *Open-ended approach in mathematics lesson to enrich diversified values and mathematical viewing and thinking*. Meijitosho Shuppan (in Japanese).

Skovsmose, O. (1994). *Towards a philosophy of critical mathematics education*. Springer Science+Business Media Dordecht.

Stender, P. (2018). The use of heuristic strategies in modelling activities. *ZDM Mathematics Education, 50*(1–2), 315–326.

Stigler, J. W., & Hiebert, J. (1999). *The teaching gap: Best ideas from the world's teachers for improving in the classroom*. The Free Press.

Stillman, G. A., Kaiser, G., & Lampen, C. E. (Eds.) (2020). *Mathematical modelling education and sense-making: International perspectives on the teaching and learning of mathematical modelling*. Springer Nature.

Weinberg, A. M. (1972). Science and trans-science. *Minerva, 10*(2), 209–222.

Yvain-Prébiski, S. (2021). Didactical adaptation of professional practice of modelling: A case study. In F. K. S. Leung, G. A. Stillman, G. Kaiser, & K. L. Wong (Eds.), *Mathematical modelling education in east and west: International perspectives on the teaching and learning of mathematical modelling* (pp. 305–315). Springer.

Chapter 15
A Study of Japanese and Australian Students' Mathematical Values in Problem-Solving

Miho Yamazaki

15.1 Introduction

Mathematics was invented by people, and their values have influenced its development (e.g., Kline, 1953). To do mathematics creatively, it is important to recognise people's values regarding mathematics. In mathematics education, students need to share these values to engage in proactive and creative mathematical activities. Because values are conative in nature (Seah, 2019), they help students concentrate on learning mathematics. Their connection to mathematics development helps students develop their mathematical knowledge.

Bishop (1988) argued that mathematics is not a value-free subject and that values are taught together in mathematics teaching, even if teachers are unaware of it. Bishop (1988) proposed three pairs of mathematical values that emerge from mathematics: rationalism-objectism, control-progress, and openness-mystery. These values are related to the social developments by which mathematics has developed and are fundamental aspects of Western mathematics. The particularly relevant pair of values that are driving forces for creative mathematical activity are control and progress. Control is a static desire to use mathematics to explain the natural and human-made environments. Progress is a dynamic feeling that the unknown can become known in mathematics. These values are related to feelings and attitudes that the mathematical culture has driven and reinforced.

For students to do mathematics creatively, teachers need to teach the mathematical values of control and progress in a balanced manner. Moreover, students need to appreciate these values in a balanced way (Bishop, 1988). Bishop (1988) proposed a curriculum based on projects which are time-consuming personal research based on historical situations and exemplify the values of control and progress. This curriculum is intended to focus on these values by dealing with projects related to society in the

M. Yamazaki (✉)
Teikyo University, Itabashi, Japan
e-mail: m_yamazaki@main.teikyo-u.ac.jp

Y. Dede et al. (eds.), *Values and Valuing in Mathematics Education*,
https://doi.org/10.1007/978-981-99-9454-0_15

past, at present and in the future. In other words, control and progress values are conveyed through studying how mathematics is used in society. However, conveying the control and progress values in a balanced manner can come from projects and mathematics content because mathematical content is associated with several mathematical values. Dede et al. (2021) found that several mathematical values could be expressed by focusing on mathematical modelling tasks, and it may be possible to balance control and progress values through a single mathematical content. However, there have been no proposals in mathematics education on how to convey control and progress values in a balanced manner through a single mathematics content.

In Japan, mathematics is viewed as a creative activity. There is an emphasis on the values that people should be aware of when doing mathematics creatively, and the mathematical values of simplification, clarification, and unification have been proposed (Nakajima, 1971). Simplification makes concrete events easier to express, work with, and think about by mathematizing them. Clarification shows correct thought and correct relationships by logical thinking and expression. Unification captures different things in a unified way so that a single form is logically valid. In varied expressions, these values have been written into the goals of mathematics curricula in the past and recent years in Japan about the type of students they wish to nurture through mathematics education. Nakajima (1971) assumed that mathematical creation could be achieved by recognising these values and feeling that it is unpleasant not to do so. Even seemingly complex events in a problem-solving process can be simplified by expressing them using mathematical equations. It is possible to clarify an event's causes by considering it logically. Then, it is possible to develop further thinking by unifying the situation in which the idea has been applied to make it applicable to other situations.

Thus, it is necessary to appreciate the three mathematical values for the problem-solving process to become a creative activity. Japanese researches have revealed that these mathematical values appear in problem-solving contexts (e.g., Tsubomatsu, 2011) and that students consider these values in comparing various solutions for problem-solving (e.g., Hagiwara, 2007). But no research has considered problem-solving to be a product of mathematical culture. Problem-solving is an activity in mathematical culture, and developing problem-solving in one situation to apply to other situations involves learning that engages the sense of the control values and, subsequently, the progress values. In other words, problem-solving can be positioned as mathematical content that may convey control-progress values in a balanced manner. Therefore, it may be possible to link control-progress values and the mathematical values of simplification, clarification, and unification to capture the problem-solving process, but no study has been done.

This study aims to clarify what is needed for practice to convey both control and progress values in problem-solving in a balanced manner, which does not end with control values but can be felt even in progress values. To solve this problem, the following two issues are considered. First, the mathematical values proposed by Bishop (1988) and Nakajima (1971) are derived by focusing on the properties of mathematics, which are ideological concepts, and it is unclear what values students hold concerning the properties of mathematics. Therefore, students' values concerning the properties of mathematics in problem-solving need to be clarified. A survey is conducted to capture students' values in problem-solving situations. Second, the mathematical values proposed by Bishop (1988) and Nakajima (1971) are derived from the properties of Western mathematics and are inherent in mathematics content, such as problem-solving methods. This raises the question of whether mathematical values can be conveyed to students simply by studying mathematics content or whether they are influenced by the culture in which students' study and live. Therefore, suggestions are obtained on what is needed to implement practices that allow students to be creative in problem-solving and to feel control-progress values, considering the values identified in the first task. A cross-cultural comparison of students' values was conducted, focusing on students who study the same curriculum but live in different cultures. Specifically, students in Japan and Australia studying the Japanese mathematics curriculum are surveyed. The students living in Australia in this study were studying the Japanese curriculum in addition to the Australian curriculum.

The study shows how using problem-solving can draw students into values related to creativity, control, and progress in mathematics. The study identifies students' values related to the properties of mathematics and offers suggestions for how students can be creative in problem-solving. The novelty of this study is that it compares values by focusing on students from two cultures learning mathematics using the same materials. Australia is an ethnically diverse country with diverse values (e.g., Han and Seah, 2019) whereas Japan is not. While similar values may be fostered by learning mathematics using the same materials, students' cultures may influence mathematics learning. In this sense, both factors are considered in this study.

15.2 Theoretical Framework

15.2.1 Mathematical Values

There are three types of values related to mathematics education: mathematical values, mathematics educational values, and general educational values (Bishop, 2001, 2008). Mathematical values are related to and have developed as knowledge of Western mathematics. Mathematics educational values are related to the norms in classrooms in which mathematics education is conducted. General educational values are not limited to mathematics education but arise from the demands of socialisation by society. All values are influenced by society, and various properties and objects are valued in different countries and cultures (e.g., Clarkson et al., 2019). Among these values, the mathematical values are those that students aim to appreciate through mathematics education to understand mathematics more deeply. Therefore, this study focuses on mathematical values to determine what students consider worthwhile in mathematics. The mathematical values proposed by Bishop

(1988) for Western mathematics are well known. They represent valuable properties of Western mathematics and are implicitly or explicitly believed in and sustained by people in a society. These values are inherent in mathematics and should be conveyed to students, primarily through mathematics education.

In Japan, where Western mathematics is used in mathematics education, values that should be consciously pursued have been proposed from a different perspective to that of Bishop (1988). Nakajima (1971) proposed three value types that should be pursued for students to engage in creative activities: clarification, simplification, and unification. From the standpoint that creative activities should be the aim of mathematics education, students need to think mathematically, engage in tasks, and produce new things with the feeling that they are compelled to pursue these three values. The word 'creative' used here refers to creation in the sense that students feel as if they have invented a new idea on their own and, as a result, acquire new content based on their need to work on the task. These values represent the important properties of mathematics as perceived by those who do mathematics creatively. These values, also inherent in mathematics, correspond to mathematical values. Students are expected to appreciate these values through mathematics education.

15.2.2 Mathematical Values and Problem-Solving in the Japanese Curriculum

Problem-solving can have a single answer or solution or be open-ended. In Japan, it is important to consider and compare various solutions because there are many ways to arrive at a solution, even for a single answer (e.g., Hino, 2015). To compare and summarise children's ideas, it is important to reflect on problem-solving ideas from various perspectives, not only for a deeper understanding of ideas but also to increase children's motivation to pursue ideas (Koto et al., 1992). Thus, learning mathematics to compare ideas and consider their values is important. Mathematical values function as a point of view for organising ideas in problem-solving. Depending on the perspective from which they are organised, there will be differences in how they are characterised and, consequently, in the new content created.

The fact that a problem can be solved and new content created by developing the ideas used in the solution represents the mathematical values of control and progress. Nakajima's (1971) mathematical values of clarification, simplification, and unification drive creative activities. Therefore, simplification, clarification, and unification values may be associated with the values of control and progress. Solving a problem is related to control, made possible by simplification. By clarifying and organizing the viewpoints of various solution methods and unifying them, a previously unknown problem situation will become known.

However, besides the abovementioned mathematical values, it is natural for students to become aware of other mathematical values. Creative activities can be conducted through the complex involvement of various mathematical values. In this study, practices that can convey a balance between control and progress values in problem-solving are the focus. Thus, the study is limited to the mathematical values of clarification, simplification, and unification, as well as ideas related to these values.

15.2.3 Learners' Mathematical Values in Problem-Solving

Understanding students' values aids good mathematics education, as values are a conation component and influence powerful motivating forces (Seah, 2019). Positive conative states foster students' well-being and, thus, performance in mathematics learning experiences. Focusing on students' values can improve their well-being in mathematics learning (Clarkson et al., 2010). Recent research supports teaching mathematics to align with students' values (Hill et al., 2022; Kalogeropoulos and Clarkson, 2019) to discuss students' diverse values in the classroom and increase positive feelings and engagement with mathematics.

In Japan, student values are important factors in learning. One of the goals of mathematics education in Japan today includes the development of "human nature and the ability to pursue learning" (Ministry of Education, Culture, Sports, Science and Technology, 2017, p. 18), which aims for students to learn mathematics with values, which influence mathematics learning, including cognition and thinking, and motivate students to learn. Students should appreciate the intrinsic value of mathematics because it enriches the human spirit. For example, a student who enjoys the mathematical values of unification will not learn anything by seeing different things as different but will be able to see other things as having the same structure and organize them.

Even though the goal is for students to appreciate the valuable properties of mathematics, they develop their own values not only in mathematics class but also in their culture and at home. Therefore, it cannot be said that students' values regarding the properties of mathematics are exactly the same as their conceptual values; it is natural for differences to exist among students. Therefore, learners' mathematical values are used in this study to emphasise that they are students' values related to mathematics' properties. Since mathematics education can be regarded as an enculturation process (Bishop, 1988) and students are recreators of the values of mathematical culture, their mathematical values are like, yet somewhat different, the mathematical values embedded in mathematics. Therefore, mathematical values and learners' mathematical values may differ, yet are connected. This allows us to observe the pair of mathematical values—control and progress—and their details, including subtle differences among learners.

15.2.4 Choice and Learner Mathematical Values

Students' values affect their conscious choices on how to act and can be determined by their choices. Krathwohl et al. (1973) proposed a taxonomy of internalisation in the affect domain, describing valuing at its third level. At this level, valuing through internalising specific values is consistent and stable. Learners' mathematical values are students' internalisation of the valuable properties of mathematics, appearing as consistent and stable behaviour by students.

When consciously 'selecting things of value', the criterion of choice is what the student perceives as valuable. Then, items that satisfy the criteria are selected. At this point, the criterion, the thing perceived as valuable, is what satisfies the student's desire through experience prior to the conscious act of selection. In the conscious act of selecting something of value, the students perceive both the criterion and the selected item as valuable. These can be regarded as representations of students' mathematical values. Therefore, learners' mathematical values can be captured using the criterion of choice and what is chosen in the conscious act of choosing 'what is important'.

15.3 Methods

15.3.1 Method for Capturing Learner Mathematical Values in Problem-Solving

This qualitative study is based on answers to a questionnaire survey in which students were asked to choose items. The learners' mathematical values are captured by analysing their responses. To explore learners' mathematical values, a questionnaire was designed in which students were asked to select the most mathematically important idea from solutions containing different ideas. In asking students to select the 'most mathematically important idea' by comparing several solution methods, they were reminded of what they perceived as valuable in mathematics, which became the criterion for selection, and the idea satisfying this criterion was selected. Thus, in describing the reasons for the choice, the value perceived by the student was mentioned, and the expression of the learner's mathematical values could be captured.

The questionnaire was developed for mathematical problem-solving. Three solution methods of varying qualities were set as choices. Learners' mathematical values were expected to emerge in the answers to 'mathematically good solution' questions in an open-ended manner. However, because this study is concerned with progress values achieved through creative activities in the problem-solving process, the aim was to capture the mathematical values of students concerning the idea of solutions of different qualities that lead to creative activities. Thus, the three solution methods differed in the quality of problem-solving and related to control and progress values, presenting them as student options. For example, if a student chose a solution method because 'it is a method that can be used in other cases', their mathematical values could be captured as 'applicability', which is less a matter of what is important in mathematics education and more what is inherent in mathematics. This expresses the value associated with covering not only the current problem but also other cases. The values are connected to Bishop's (1988) mathematical control values from the student's perspective, making it possible to closely examine them based on the characteristics of the selection criterion and selected object.

Since this is a theoretical assumption, it is possible that students would have answered without much consideration, and their mathematical values would not be expressed in the reasons for their choices. To minimise this risk, the purpose of the study was explained to the students before the survey, and they were asked to think carefully about their responses. They were asked whether they understood the idea of the choice, whether they thought the idea was good, and why. Students who made choices without due care or understanding of the solution idea and whose answers and reasons were inconsistent were excluded from the analysis. Six students selected the idea of the solution as a good idea, even though they did not understand it, and were inconsistent in their evaluation of the goodness of the solution and the statement of reasons for their choice of solution. They were excluded from the data analysis.

15.3.2 The Cultural Differences Influenced Learner Mathematical Values

This study focused on students studying the same mathematics curriculum living in different countries: Japan and Australia. Students in both locations learned Japanese mathematics content from Japanese teachers obtained from Japanese textbooks produced by the same textbook company. The Japanese government approved the textbooks. The main differences between students in the two countries were the place where they studied Japanese mathematics, the type of school in which they studied Japanese mathematics, the amount of time they spent studying Japanese mathematics (on weekdays or only on Saturdays), the qualifications of their Japanese mathematics teachers, and the culture in which they lived. The cultural differences in this study are due to Australia being an ethnically diverse country with diverse values and Japan not. Although there are Japanese schools in Australia, the students who participated in this study attended local schools on weekdays in Australia and were assumed to be more influenced by Australian culture than students who attended Japanese schools in Australia.

The students in this study who live in Japan attended a public elementary school in Tokyo located in a lush green area away from the central business district and studied Japanese mathematics textbooks almost every weekday. Their schooling was compulsory. Mathematics teachers were university or graduate school graduates with teaching licences.

The students in this study who lived in Australia were local students, including Japanese, Australians, and people of other nationalities. Some had studied mathematics in places other than Japan and Australia. They studied the content of Japanese mathematics textbooks only on Saturdays at a supplementary school in Victoria, the second-largest state in Australia, located in a lush green area. On weekdays, they studied mathematics content conforming to the Australian curriculum (Australian Curriculum Assessment and Reporting Authority, 2017) or the Victorian curriculum (Victorian Curriculum and Assessment Authority, n.d.) at their elementary schools in Victoria. Teachers of Japanese mathematics in supplementary schools were native Japanese speakers who were not required to have a teaching licence. Teachers used handouts based on Japanese mathematics textbooks to teach mathematics at the supplementary schools.

15.3.3 Research Ethics

The survey was administered with the prior consent of the teachers at the school and was communicated and explained to the students' parents as appropriate. Students participating in the survey were informed prior to the survey and told that their answers were not related to their marks at school and that they were expected to answer the questions only to the extent that they were able. Students were asked to respond to the survey if they were able to cooperate. Students who did not respond to the survey were considered to have refused to participate and were excluded from the overall data count. Student names were coded using numbers to remove personal identifiers from the data. Japanese students were prefixed with SJ and numbered consecutively using three digits, such as SJ001. Australian students were prefixed with SA and numbered sequentially, such as SA001. All data were stored electronically, and the students' responses' originals were kept locked.

15.4 Overview of the Survey

15.4.1 A Problem and Solutions in a Questionnaire

The mathematical problem-solving task was to count the number of marbles when the number of marbles increases regularly. Starting with the initial stage of the number of marbles, the first to third stages are illustrated, with the student asked to find the number of marbles in the fourth step (Fig. 15.1). This task could be answered by counting the numbers one by one, but it can also be answered simply by taking a mathematical view; there is diversity in the solution methods, and clarification can sort out the differences in solution ideas; meanwhile, unification can deepen understanding of the task phenomenon. Therefore, this task is appropriate

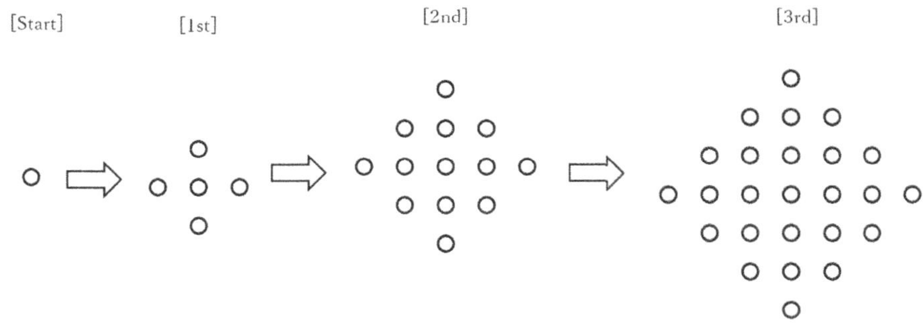

Fig. 15.1 The mathematical problem-solving task

for capturing mathematical values. The three solutions represent different ways of thinking (Fig. 15.2). Students compared the mathematical differences between these alternatives, consciously recalled what was valuable to them, and selected the 'most mathematically important' idea. The choices were presented using the word 'idea' to emphasise to the students that they should pay attention to ideas in each solution.

Idea 1 involves drawing and counting marbles. This method focuses on alignment symmetry to simplify counting the number of marbles in a circle, creating one 5 × 5 square and four clusters of triangles containing four marbles. Idea 2 is a way to find the pattern of how many more marbles there would be by focusing on the gradual change in the shape of the arrangement of the marbles. This method uses the number of marbles that increase in multiples of 4, that is, 4, 8, 12, and so on, and infers that the subsequent increase is 16. Idea 3 focuses on the gradual change in the shape of the marble arrangement and the mechanism for increasing the number of marbles. In this method, one marble in the initial stage is positioned at the centre, and an increasing number of marbles is seen as four triangular clusters with the same number of marbles around it. The number of marbles in each triangular cluster is 1 in the first step, $1 + 2$ in the second step, and $1 + 2 + 3$ in the third step; the next step is estimated as $1 + 2 + 3 + 4$. As the number of marbles increases as the steps progress, if the number of steps is x and the number of marbles at that step is y, the number of marbles can be captured using $y = (1 + 2 + \cdots + x) \times 4 + 1$.

Every idea has a simple and clarified explanation written with a diagram and an equation that leads to a correct answer. For example, idea 3 responds to the value of simplicity, as it aims to capture the number of marbles by linking it to the number of steps. It also responds to the value of clarification, as the number of steps is used in the formula for the number of marbles. In terms of unification, any solution idea can be applied to steps other than the fourth step in the problem or the case of different arrangements of figures. However, idea 3 explains how the number of steps is related to the number of marbles based on the four-directional symmetry of the square, and this mechanism can be used to consider the relationship between the number of steps and the number of marbles in the case of other shapes arranged in pentagons or hexagons, for example, by using the same idea (Fig. 15.3). Idea 3 is very useful for solving the problem related the relationship between the number of steps and

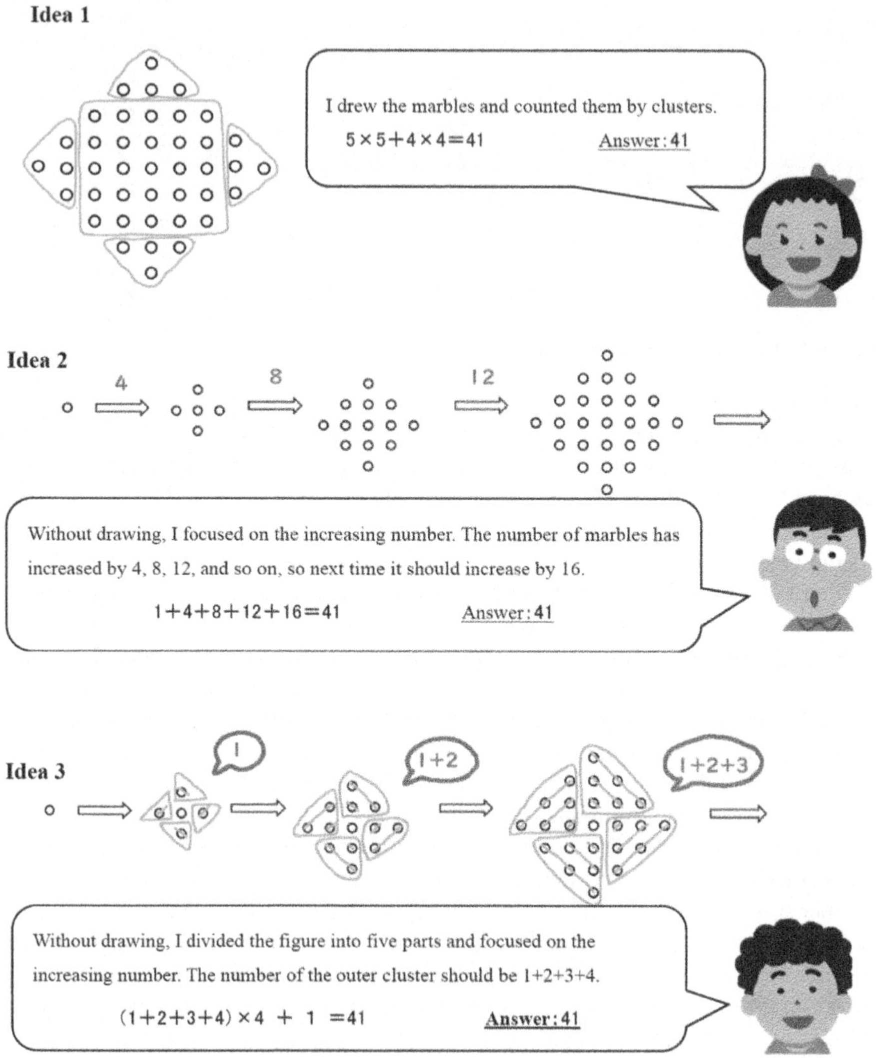

Fig. 15.2 Three different solutions

the number of marbles. Idea 3 can integrate and capture different phenomena, such as lining up in a square and lining up in other shapes, using the same mechanism, marking progress in how phenomena are perceived.

Bishop's (1988) mathematical values can be conveyed through activities like those used in this study. For example, the values of rationalism can be conveyed by rationally thinking about how the number of marbles increases according to rules. In contrast, the values of objectism can be conveyed by thinking about the characteristics of how the number of marbles increases using diagrams. The values of control

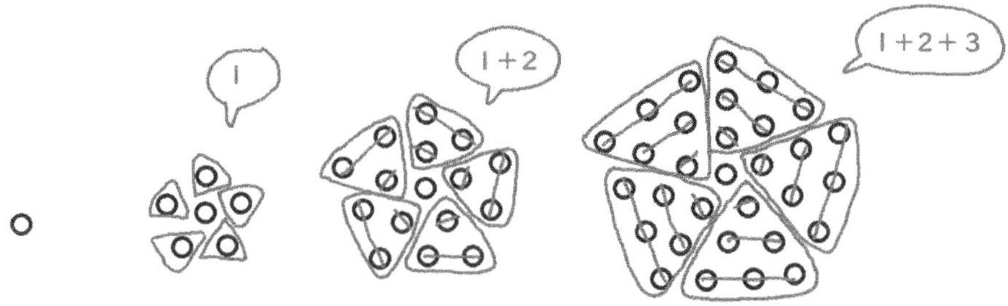

Fig. 15.3 Examples of pentagonal arrangements with the same structure

appear in explaining the number of marbles with an expression, and the values of progress appear in expanding the view adapted to the four sides of a square to the unknown problems of pentagons and hexagons. The value of openness can be expressed by explaining ideas used to solve a problem. The values of mystery can be expressed by there being many other ways of thinking about solving a problem. This study focuses on the control and progress values among mathematical values to capture how students value mathematics in problem-solving.

15.4.2 *Questionnaire*

After showing the solutions to the three ideas, the participants were asked to indicate (1) whether they understood idea 1; (2) whether they thought idea 1 was mathematically good, and (3) why they thought it was a mathematically good idea (or not). The same three questions (4, 5, and 6) were asked for idea 2, and the same three questions (7, 8, and 9) were asked for idea 3. For item 10, the participants were asked to choose the most mathematically important idea from ideas 1 to 3. For item 11, they were asked to describe the reasons for their choice. The Appendix x.1 shows the English version of the questionnaire.

Items 2, 5, and 8 asked whether each idea was mathematically good, maybe the same as the statement of reason in item 11, which compares the three ideas and then makes a choice, or they may be different statements. This is because it is necessary to use one common criterion to compare the three ideas. In contrast, an additional criterion may be used each time when judging whether each idea is mathematically good. The selection criteria for judging whether each idea is mathematically good may differ from those used for comparing the three ideas. Because learners' mathematical values are expressed when students consciously choose, their responses are analysed by item 11, for which students choose between the three ideas. However, as items 2, 5, and 8, which ask about each idea, and item 11 may be linked, the answers to items 2, 5, and 8 are referred to only when they are related to the content of item 11.

15.4.3 The Mathematics Curriculum Learned by Participants in Japan and Australia

The study participants were third- to sixth-grade students living in the Tokyo area of Japan and third- to ninth-grade students living in Victoria, Australia. The study participants in both countries used Japanese mathematics textbooks. In these textbooks, problem-solving is not a separate unit but is presented as a text problem in mathematics content units. In Japan, the method of dividing numbers into groups of the same number, as in idea 1, is used in the 2nd-grade multiplication unit in elementary schools (Fujii and Majima, 2020a). Students are taught to think about the meaning of the numbers in the multiplication equation by connecting them to clusters of the same number represented in the diagram. Problem-solving involving variables, such as those discussed in this study, appears in the 4th-grade textbook in a mathematics unit titled 'Investigating How Things Change', in which students consider how quantities change (Fujii and Majima, 2020b). In this unit, students examine how variable Y changes when variable X changes by organising it in a table and using a diagram. In problem situations in which a specific value of Y is required, the following methods are introduced: focusing only on how variable Y changes and finding a way to increase the value of Y (idea 2), finding an equation relating variables X and Y (idea 3), and counting by drawing a specific diagram (idea 1).

However, in Australia, there is no fixed textbook for mathematics; teachers select and use materials and printouts that conform to Australian or Victorian curricula. Therefore, referring to the Australian curriculum for mathematics, patterns, and algebra are positioned as sub-strands within the Content Strand of Numbers and Algebra. Patterns and algebra specify that writing, following, and creating number patterns should be taught in mathematics education and that students should make tables of changing events and look for patterns. For example, in a student book adapted from the Australian curriculum (Harris, 2018; Turner, 2018a, 2018b), some activities look for patterns in the sequence of numbers created by four arithmetic operations until the 4th grade. In the 5th grade, the book contains problems in finding the relationship between the number of steps and the number of blocks, and methods (e.g., idea 2) to look for patterns of increase by looking only at the number of blocks are discussed. This book also addresses the problem of determining the number of blocks required for specific steps. In the 6th grade, the search for rules regarding the number of sides in a regular polygon and the number of lines of symmetry (which may lead to idea 3) is discussed.

15.4.4 Data Collection

The survey was conducted during August–October 2022. Before the survey start, the students were informed of the purpose of the study and the survey outline and asked to fill in their honest opinions. Valid data were collected from 312 elementary school students in Japan: 71 in the 3rd grade, 62 in the 4th grade, 96 in the 5th grade, and 83 in the 6th grade. Valid data were collected from 125 students in Australia: 31 in the 3rd grade, 24 in the 4th grade, 30 in the 5th grade, 15 in the 6th grade, and 25 junior high school students in the 7th grade and above. The survey was administered to more than one grade to observe their differences and whether the content studied in each grade significantly impacted the responses.

15.4.5 Data Analysis

The collected response forms were assigned numbers so that individuals could not be identified, and the responses were organised into a list so that the contents of each number ('the most mathematically important idea' and 'reason it was the most mathematically important idea') were linked. This list was used to conduct a quantitative analysis of the data characteristics.

The number of responses for each selected idea was counted for each type. To capture differences between Japanese and Australian in selecting idea, a Chi-square test was performed as needed. An open coding procedure was used for the selection. First, keywords were extracted from the reasons for selection mentioned by all students. Next, if the keywords were similar, they were organised into labels. If several labels were similar, they were grouped. Then, one group was set as the main code, and one label as the subcode, and the reasons for the selection were classified. For example, the student in SA395 chose idea 3 by answering, 'If the question is "Answer the Xth number", I can answer it immediately and accurately'. In this case, we extracted three keywords: the Xth number, immediately, and accurately. Then, by combining the keywords with similar keywords in other students' response data, we quickly and accurately labelled each, as in the other case. These were later used as subcodes. Finally, we grouped the label unification, effect, and clarification. These three codes later became the main codes.

The validity of the keywords extracted from the student's reasons for selection is that the keywords in the reasons for selection are at least what the student considered 'mathematically important'. Because the students considered the contents of the keywords to be 'mathematically important', they selected an idea that satisfied the keywords. In other words, both the selected ideas and the keywords written in the reasons for the selection were matters that the students considered 'mathematically important'. Therefore, the learner's mathematical values were captured by extracting, classifying, and organising such keywords. Keywords described in negation form were extracted.

To ensure the reliability of the analysis, the same procedure was performed on 21 November 2022, 25 January 2023, and 27 January 2023. The order in which the students were analysed changed each time. Consequently, different parts of the analysis were examined to refine the analysis.

15.5 Findings

15.5.1 Differences Between Japan and Australia in the Type of Selected Solutions

The following is a breakdown of the types of solutions selected by the students as their most important ideas (Tables 15.1 and 15.2): The columns of each table indicate the students' grade level and the rows of the type of idea selected. For Australian students, grades 7, 8, and 9 also participated in the survey; however, because of the small total number of students, the figures are for all three grades combined. In total, 124 Australian and 307 Japanese students provided valid data.

Six Japanese students were required to settle on one solution and select two. One 6th-grade student chose ideas 1 and 2. One 4th-grade student and one 5th-grade student chose ideas 1 and 3, respectively. Three 4th-grade students chose ideas 2 and 3. One Australian 3rd-grade student who selected idea 1 evaluated it without understanding, as did one Japanese 3rd-grade student who selected idea 1, three 3rd-grade students who selected idea 3, and one Japanese 4th-grade student who selected idea 3. These students were excluded from the analysis.

The Australian students who participated in this study were likelier to choose idea 2 as their most important idea. These students accounted for approximately 69.4% of the total. These characteristics were similar throughout the school year. Among the Japanese students who participated in the survey, about 51.8% chose idea 2 as the most important idea, but ideas 1 and 3 were also present in some proportions. These characteristics were similar throughout the school year.

Table 15.1 Breakdown of the types of ideas selected by the students in Australia (n = 124)

	3rd	4th	5th	6th	7th	Total	Percentage (%)
Idea 1	5	2	4	1	3	15	12.1
Idea 2	18	20	20	13	15	86	69.4
Idea 3	5	1	5	1	6	18	14.5
n.d.	2	1	1	0	1	5	4.0
Total	30	24	30	15	25	124	100

Table 15.2 Breakdown of the types of ideas selected by the students in Japan (n = 307)

	3rd	4th	5th	6th	Total	Percentage (%)
Idea 1	17	10	35	18	80	26.0
Idea 2	32	38	44	45	159	51.8
Idea 3	8	15	17	21	61	19.9
n.d.	10	2	1	0	13	4.2
Total	67	65	97	84	313	101.9

15.5.2 Categories of Reasons for Selection

The main codes were classified into six categories by organising students' reasons for their choices: simplification, clarification, unification, technique, feeling, and effect. The simplification category is derived from an easy way of expressing, working, and thinking. The clarification category comes from the ability to explain something by attributing it to a few essential things to understand it correctly. The unification category is based on the assumptions of other problematic situations. The technique category focuses on the techniques used in each solution method, such as counting, grouping, and focusing on changes. The feeling category is an evaluation based on conformity with the students' sensory perceptions. The effect category evaluates the educational context in which abilities are gained through solutions. Sample examples of each main code are listed (Table 15.3). In every case in which ideas 1, 2, or 3 were selected, at least one response classified as simplification, clarification, or unification was included. Some students' responses were sorted into multiple codes.

The most common reasons Australian students gave for choosing idea 2 were simplification and clarification, followed by effect. Specifically, idea 2 was often chosen for such reasons as being the simplest, easy to do, or easy to understand, followed by quick, which is often chosen for the speed of the solution. The most common reason Japanese students chose idea 2 was technique, followed by simplification and effect. Specifically, for example, students often evaluated idea 2 because of its focus on the difference between variables and increasing patterns, followed by the response that it was easy to make an equation and quick.

The most common reasons for selecting idea 3 were effect, simplification, and unification for Australian students and technique, simplification, and unification for Japanese students. Idea 3 was expressed as an equation relating the number of steps to the number of marbles. Australian students focused on the fact that the answer could be obtained by applying the number of steps and evaluating it based on its simplicity and speed. Meanwhile, idea 3 focused on the fact that the number of marbles increased in all four directions because of the symmetry of the square, and the Japanese students focused on the fact that they found a relationship between the number of steps and the

Table 15.3 List of examples for each main code

Main code	Sample
Simplification	It is the simplest and the easiest to figure out without having to do too much in my mind. (SA370)
Clarification	Because it is easy to understand $1 + 2 + 3$. (SJ13)
Unification	It is easier to think of a larger number of steps because the seventh would be $(1 + 2 + 3 + 4 + 5 + 6 + 7) \times 4 = 28 \times 4 = 112$. (SJ360)
Technique	It is a way of finding a pattern. (SJ46)
Feeling	Because I did it at the start, unlike the other ones. (SA152)
Effect	Ideas like Solution 2 would help me to be more aware of all sorts of things. (SJ185)

number of marbles by focusing on the groups and evaluated it based on its technique and simplification. In addition, unification, the value for which idea 3 was evaluated in both Japan and Australia, applies to other step-number situations. For example, the Japanese student SJ200 stated that the reason was 'because I can quickly find the answer even if it is the 100th number', and the Australian student SA395 stated, 'because if I am asked to answer the Xth number, I can quickly and correctly answer the question'. Thus, other problems were assumed to be the case for other steps.

15.6 Discussion

15.6.1 A Consideration Related to Selected Idea 2

The two countries have similarities and differences regarding the types of ideas selected as most important. The most significant percentage of students in both countries chose idea 2. The difference is that an extremely large percentage of Australian students chose idea 2, whereas a certain percentage of Japanese students chose ideas 1 and 3. These results were similar throughout the school year in both Australia and Japan, suggesting that the results are not a consequence of the mathematics content of a particular grade level but instead of what students are learning in the mathematics curriculum in each country. In fact, a Chi-square test was conducted about whether students who answered item 11 selected idea 2 or not. The results revealed a significant disparity between the Australian and Japanese students ($\chi 2(1) = 11.61192548$, $p = 0.000655303$) (Table 15.4). The reason for these results may be the cultural influence of the differences in what is covered in mathematics education in each country.

Turning to the mathematics content, there are differences between Japan and Australia. In Japan, ideas 1, 2, and 3 are found in textbooks and are familiar to students. All of these ideas have been taught in math classes on weekdays and are studied carefully in class; therefore, it is assumed that they are recognised and selected as important ideas by students. Meanwhile, mathematics in the Australian curriculum includes methods that focus on the way the variable Y increases, such as idea 2 in the pattern and algebra sub-strands, and these methods were studied in weekday classes in Australia. Therefore, idea 2 is recognized as important and was selected more often. In both countries, the content of weekday mathematics classes might have influenced the types of ideas that the students selected as most important. However,

Table 15.4 Breakdown of the students who selected idea 2 or not in Australia and Japan

	Australia	Japan	Total
Selecting idea 2	86 (70.59322)	159 (174.4068)	245
Not selecting idea 2	33 (48.40678)	135 (119.5932)	168
Total	119	294	413

it is possible that something other than mathematical content influenced these results, such as teaching methods, the position of mathematics as a subject, and mathematics educational values. The relationship between the idea type and these other factors remains a topic for future research.

There were also differences between Japanese and Australian students in their reasons for choosing idea 2. The most important criterion for selecting an idea was the students' values, which functioned in the problem-solving process. Australian students' choice of idea 2 was strongly influenced by the values of simplification and clarification, while Japanese students' choices were strongly influenced by the values of technique. The reason for the differences in the values expressed as described above may be the cultural influence of differences in the content of mathematics education in each country.

In the Australian curriculum, students are expected to construct a table of values to record two possible number patterns (Australian Curriculum Assessment and Reporting Authority, 2017). In weekday classes in the Australian curriculum, Australian students focus on the variable Y and look for patterns to see what kind of changes occur, or work on tasks to find the value of Y for a specific value of X. Therefore, for example, in terms of finding a pattern, simplification is better than complexity to find a specific value of Y, and the value of clarification is considered more important. By contrast, the Japanese curriculum stipulates that expressing a figure in a formula, considering the relationship expressed in the formula, and investigating the characteristics of change or correspondence between two things should be taught in mathematics education (Ministry of Education, Culture, Sports, Science and Technology, 2017). For this reason, it is considered that Japanese students strongly value and engage in the technique of focusing on the relationship between variables X and Y and expressing the relationship in equations.

15.6.2 A Consideration Related to Selected Idea 3

While 61 students in Japan (19.9% of the students in Japan) and 18 students in Australia (14.5% of the students in Australia) chose idea 3 for various reasons, 9 students in Japan and 4 students in Australia chose idea 3 specifically because of its applicability to other situations. The reason given by these students was one characteristic: they were not only solving the fourth step but also recalled the case of a more significant number of steps and integrated them into the same situation. These students did not think that solving the 4th step alone was sufficient; instead, they recalled the case of an increased number of steps and integrated them into the same situation. In other words, these students appreciate the value of finding the value of Y in any case by considering that the same relationship can be used for various variables X. This is regarded as an expression of the mathematical value of the control.

Meanwhile, because idea 3 expresses the mechanism of the relationship between the number of steps and the number of marbles, we can consider applying this idea not only to other problems involving the number of steps but also to problems involving other shapes by keeping these mechanisms in mind. In other words, students can perceive unknown problems as known problems if they have the same mechanism of relations. In this sense, this idea could be viewed as an expression of progress values in mathematics. However, there was no reference to other shapes in the reasons why the students chose idea 3. In other words, no student in this study judged idea 3 to be important from the progress viewpoint. This indicates that what is necessary for students to be able to perceive progress values is to understand that even the premise of a problem can be considered a changeable factor and that the problems can be viewed in an integrated manner. In this example, it is necessary not only to consider the number of steps in the problem as a variable but also to consider the given figure, which is the premise of the problem, as a variable and to view the problems of other figures in an integrated manner so that they could be included as part of the same set.

Students bring their own values to mathematics. If these values correspond to the progress values proposed by Bishop (1988), they can relate unknown problems to known ones and systematically construct knowledge. If the values created by the students correspond to the unification proposed by Nakajima (1971), it is possible to learn mathematics creatively based on the knowledge already built by the students. Mathematics is not mysterious knowledge discovered by great mathematicians that must be received but is a product created by people in a culture. Students are expected to appreciate the values of Bishop (1988) and Nakajima (1971) and be able to embody them in mathematics.

15.7 Suggestions for Practice

Idea 2 was selected most often by the students; idea 3 could lead to progress values. Based on the above discussion of student selection of these ideas, some suggestions for practice are offered.

Based on the discussion of students choosing idea 2, the instruction content may influence the choice of ideas they consider most important. The content taught on weekdays in mathematics education is perceived by students as worthwhile learning because it is carefully taught over time. What is taught and focused on in terms of weekday mathematics education content helps convey to students what mathematical ideas are valuable and what is valued in mathematics. Taking time to teach mathematics content significantly impacts the development of students' values. To make students recognise the importance of ideas, such as idea 3, and to make them value thinking creatively and making progress, it is effective to spend time in mathematics classes and organise lessons that focus on the idea and emphasise the value of creativity. For students, ideas mentioned over time in class are more effective in fostering a sense of value about the content than those seen or introduced only once in a book. The results of this study reaffirm the importance of teacher practice.

Based on the discussion of the students who chose idea 3, solving problems is an objective in problem-solving, but it is difficult to achieve progress through creative thinking if students only solve one required problem, which is the goal of the problem-solving process. Students are expected to understand not only phenomena in terms of variables but also the mechanism of the relationship between variables X and Y, and then to integrate phenomena with the same mechanism, thereby making progress in understanding the phenomena of the problems. In this study, both the Japanese and Australian students tended to regard as valuable the fact that they could derive answers to the questions they were asked and the ability to derive answers to questions at other steps of the process. In both Japan and Australia, students need to appreciate the value of developing an integrated view through problem-solving processes to engage in activities associated with progress values.

The following suggestions are offered on the teacher instruction required for students to obtain a sense of progress values. First, teachers should discuss various solution methods for content in weekday mathematics classes. In particular, idea 3, which focuses on the relationship between variables X and Y, and clearly explains the mechanism of these relationships, should be discussed in class. This is because the mechanism clarifies the structure of the arrangement, provides a framework for viewing the arranged marbles, and can be used for any arrangement with the same structure. Second, in problem-solving, students should not stop finding the answer to a question but rather spend time considering the solution's value after obtaining it. Teachers should discuss various ideas with students and share the values of ideas that integrate the required problem with other problems of figures that have not yet been faced and convince students that these ideas can be used for other figure arrangements, such as idea 3. These practices are expected to help students acquire an essential view of how the marbles are arranged and how the number of marbles increases, and of feeling that problems requiring the same structural arrangement are no longer unknown problems. This practice is related to Bishop's (1988) progress values.

15.8 Limitations and Further Research

This study focused on problem-solving situations and captured students' mathematical values through their responses to the survey questions. I set up a problem-solving situation that can be viewed from a variable perspective, with three options of different ideas. By doing so, I was able to capture what ideas students valued and for what reasons, and I was able to make practical suggestions. However, there are limitations to this study, and further research is needed.

First, the values expressed by the students are limited to problem-solving situations dealing with variables such as those in this study. This does not necessarily mean students express values in problem-solving situations or choices. Students' values are stable, which means they would reflect the same values if given the same problem and choices. The students' values captured in this study were relatively superficial, depending on the questions and choices. To deepen understanding of students' values, it is necessary to collect and analyse multiple datasets, for example, by giving students a variety of questions and open-ended answer choices.

Next, students' values were identified as far as the survey could capture them. This was based on the assumption that the students understood the survey's intent, thought carefully about the survey questions and options, and answered the questions by confronting themselves. Although the survey and designed questions were explained to students to help this assumption hold true, the results of this study are limited. To capture students' values more accurately and in more detail, researchers should be in the classroom, understand the classroom culture, get to know the students well, and combine other factors that may need to be investigated in the future.

Acknowledgements This study was supported by JSPS KAKENHI Grant Number JP19K14222. This work was supported by JSPS KAKENHI Grant Number JP19K14222. Any opinions expressed herein are those of the author and do not necessarily represent the views of JSPS KAKENHI.

Appendix 1 The English Version of the Questionnaire

(1) Circle your condition and let me know your view about idea1.

1. Do you understand the idea? Yes ・ No

2. Do you think this is a mathematically good idea? Yes ・ No

(Please give reasons…)

(2) Circle your condition and let me know your view about idea2.

1. Do you understand the idea? Yes ・ No

2. Do you think this is a mathematically good idea? Yes ・ No

(Please give reasons…)

(3) Circle your condition and let me know your view about idea3.

1. Do you understand the idea? Yes ・ No

2. Do you think this is a mathematically good idea? Yes ・ No

(Please give reasons…)

(4) Which idea do you think is the most mathematically important?

Idea 1 ・ Idea 2 ・ Idea 3

(Please give reasons in detail…)

Thank you for your cooperation.

References

Australian Curriculum, Assessment and Reporting Authority. (2017). *Foundation to year 10 curriculum: Mathematics (8.4)*. https://www.australiancurriculum.edu.au/f-10-curriculum/mathematics/

Bishop, A. J. (1988). *Mathematical enculturation: A cultural perspective on mathematics education*. Kluwer Academic Publishers.

Bishop, A. J. (2001). Educating student teachers about values in mathematics education. In F.-L. Lin & T. J. Cooney (Eds.), *Making sense of mathematics teacher education* (pp. 233–246). Kluwer Academic Publishers.

Bishop, A. J. (2008). Teachers' mathematical values for developing mathematical thinking in classrooms: Theory, research and policy. *The Mathematics Educator, 11*(1/2), 79–88.

Clarkson, P., Bishop, A. J., & Seah, W. T. (2010). Mathematics education and student values: The cultivation of mathematical wellbeing. In T. Lovat, R. Toomey, & N. Clement (Eds.), *International research handbook on values education and student wellbeing* (pp. 111–135). Springer.

Clarkson, P., Seah, W. T., & Pang, J. (2019). Scanning and scoping of values and valuing in mathematics education. In P. Clarkson, W. T. Seah, & J. Pang (Eds.), *Values and valuing in mathematics education* (pp. 1–10). Springer.

Dede, Y., Akçakın, V., & Kaya, G. (2021). Mathematical, mathematics educational, and educational values in mathematical modeling tasks. *ECNU Review of Education, 2*(2), 241–260.

Fujii, T., & Majima, H. (Eds.) (2020a). *Atarashii sansuu 2*. Tokyo Shoseki.

Fujii, T., & Majima, H. (Eds.) (2020b). *Atarashii sansuu 4*. Tokyo Shoseki.

Hagiwara, R. (2007). Tayou na hasso ga kitaidekiru kadai to kodomo no hanno no jissai [Children's reactions to tasks where various ideas are expected]. In *Proceedings of the 40th Research Conference of Japan Society of Mathematical Education*, pp. 235–240.

Han, C. D., & Seah, W. T. (2019). What Australian students say they value most in their mathematics learning. In G. Hine, S. Blackley, & A. Cooke (Eds.), *Mathematics education research: Impacting practice (Proceedings of the 42nd Annual Conference of the Mathematics Education Research Group of Australasia)* (pp. 332–339). MERGA.

Harris, G. (2018). *Australian targeting maths: Year 6 student book*. Pascal Press.

Hill, J. L., Ker\n, M. L., Seah, W. T., & Van Driel, J. (2022). Developing a model of mathematical wellbeing through a thematic analysis of the literature. In C. Fernández et al. (Eds.), *Proceedings of the 45th Conference of the International Group for the Psychology of Mathematics Education* (Vol. 2, pp. 379–386). PME.

Hino, K. (2015). Comparing multiple solutions in the structured problem solving: Deconstructing Japanese lessons from learner's perspective. *Educational Studies in Mathematics, 90*, 121–141.

Kalogeropoulos, P., & Clarkson, P. (2019). The role of values alignment in levels of engagement of mathematics learning. In P. Clarkson, W. T. Seah, & J. Pang (Eds.), *Values and valuing in mathematics education* (pp. 115–127). Springer.

Kline, M. (1953). *Mathematics in Western culture*. Oxford University Press.

Koto, S., & Niigata Sansu Kyoiku Kenkyukai. (Eds.) (1992). *Sansuka tayou na kangae no ikashikata matomekata [Methods of utilizing and summarizing various manners of thinking in elementary mathematics class]*. Toyokan.

Krathwohl, D. R., Bloom, B. S., & Masia, B. B. (1973). *Taxonomy of educational objectives, The classification of educational goals. Handbook II: Affective domain*. David McKay Co.

Ministry of Education, Culture, Sports, Science, and Technology. (2017). *Shougakkou gakushu shidou youryou. [Course of study for elementary schools]*. Online: https://www.mext.go.jp/content/20230120-mxt_kyoiku02-100002604_01.pdf

Nakajima, K. (1971). On 'mathematical thinking' in mathematics education in Japan. *Journal of Japan Society of Mathematical Education: Supplementary Issue, Reports of Mathematical Education, 53*, 50–56.

Seah, W. T. (2019). Values in mathematics education: Its conative nature, and how it can be developed. *Research in Mathematical Education, 22*(2), 99–121.

Tsubomatsu, A. (2011). Shogakkou sansuka ni okeru souzouotekina gakushusido: Sentaisho to tentaisho no dounyu no jyugyo wo toshite [Creative learning instruction in elementary school arithmetic: Through an introductory lesson on line and point symmetry]. *Journal of Japan Society of Mathematical Education, 93*(10), 2–9.

Turner, G. (2018b). *Australian targeting maths: Year 5 student book*. Pascal Press.

Turner, G. (2018a). *Australian targeting maths: Year 4 student book*. Pascal Press.

Victorian Curriculum and Assessment Authority. (n.d.). *Victorian curriculum: Foundation-10*. Online: https://victoriancurriculum.vcaa.vic.edu.au/

Part V
Values and Socio-cultural Contexts

Chapter 16
Analysing Young Japanese Children's Social Values, Mathematical Values and Mathematical Expressions

Nagisa Nakawa

16.1 Introduction[1]

Globally, preschool mathematics education have flourished. Its importance is increasing (Björklund et al., 2020) as evidenced by a series of books on numbers, statistics and young children's cognition in preschool mathematics education, including those related to, for instance, the International Congress of Mathematical Education (ICME) and the Congress of the European Society for Research in Mathematics Education (CERME). They include studies focusing on cognitive and affective aspects and other sociocultural perspectives (Nakawa et al., 2023; Radford, 2022; Tirosh et al., 2020). However, there are a limited number of studies treating young children's values, making it worth examining, as proposed by Nakawa (2019).

In this chapter, the author would like to take the social and mathematical values suggested by Shimada and Baba (2022), paying attention to the preschool setting that focuses on the primary mathematics education stage. They identified three types of values—mathematical, social and personal. They stated that "conciseness, clarity and integration are considered as mathematical values" and "social values refer to values shared by the many people in a society and can be said to be expressed in relation to others" (Shimada & Baba, 2022, p. 21). In their book, a variety of examples are proposed in the Japanese context, based on their four problem classifications—distribution, rulemaking, selection, and planning and designing. For fostering mathematical and social values at the pre-primary level, it is crucial to see how young children

[1] This paper is based on three presentations made at the Japanese Society for Mathematics Education—Verification of Early Childhood Mathematics Education from Critical Mathematics Education (Nakawa, 2017), An Examination of the Social Values Expressed by Young Children in Mathematical Activities in Pre-school Education (Nakawa and Kosaka, 2019) and Examining Exemplarities in Pre-school Education in Relation to Critical Thinking (Nakawa, 2020).

N. Nakawa (✉)
Kanto Gakuin University, Yokohama, Japan
e-mail: nagisa@kanto-gakuin.ac.jp

Y. Dede et al. (eds.), *Values and Valuing in Mathematics Education*,
https://doi.org/10.1007/978-981-99-9454-0_16

at the preschool stage behave and act in a similar situation to the ones at the primary level. Among the problems proposed by Shimada and Baba (2022), considering the developmental stage of preschool children, particularly 5–6-year-olds, rulemaking and distributive activities could be modified and implemented. Among distributive activities at the primary level, children are asked how they decide to divide a concrete object, such as a melon among "people of different ages". Among rulemaking activities in primary school, they are asked to decide what kind of rules are valid in a particular situation for "children of different ages". These two contexts can be easily understood by young children.

Therefore, this chapter reports on social values of young children based on the observations made during activities related to rulemaking and distributive behaviour. The research questions were as follows:

(1) What social values do young children hold in rulemaking and distribution activities?
(2) What types of mathematical values are observed in connection with social values in these activities?

16.2 Literature Review

16.2.1 Shimada and Baba's (2022) Mathematical, Social and Personal Values and Democratic Competences

This study focuses on the social values suggested by Shimada and Baba (2022), specifically in the primary mathematics education stage, and the three identified social value types—mathematical, social, and personal. They stated that "conciseness, clarity and integration are considered mathematical values" (Shimada & Baba, 2022, p. 21). For instance, a distribution activity offers children the opportunity to think about dividing objects equally, which is related to the fundamental idea of division and fractions. "Social values refer to values shared by the many people in a society and can be said to be expressed in relation to others" (Shimada & Baba, 2022, p. 21). Examples include compassion for others and values of equality and fairness. Personal values are those which apply to individuals alone. They are expressed in situations where the individual thinks independently and does not interact with others. For example, "when purchasing a new car, the choice depends on what is valued by the individual" (Shimada & Baba, 2022, p. 21).

According to Shimada and Baba (2022), the first mathematical values are often expressed in mathematics classes. For preschool children, mathematical values may be related to fundamental ideas of calculations, or the attribution perspective which focuses on a certain standard, such as the shape or size of objects. Social and personal values are also related to mathematical models. They state that mathematical values and models are strongly interrelated. This means that social values could have mathematical model representations. For example, values, such as equality and fairness, are expressed in a situation and children construct mathematical models accordingly. For instance, in this problem, "5 children sit down on a bench, surrounding a table. There are 36 children in the class. How many benches do we need to prepare?", some children would express their ideas using mathematical expressions, such as $35 \div 5 = 7 \ldots r1$, and would, therefore, answer that they need 8 benches. Other children could think of $36 \div 4 = 9$ and therefore their answer would be that they need 9 benches.

Thus, to solve the problem, children utilized mathematical models to explain their ideas. In addition, mathematical models can make social values more explicit through students' articulations. For example, a state of equality and fairness is ambiguous without explanation and often mathematical models can be used to communicate and clarify such equality and fairness. Even if social values are the same, mathematical models may be expressed in a variety of ways. Therefore, presenting a model clarifies how children perceive these values. In other words, the use of models can clarify an individual's social values. For instance, in the above problem, the second mathematical model probably shows that children believe in the importance of equality, deciding to reduce the number of children who share one desk. Therefore, the mathematical model emphasises these children's social values. Subtle transformations in social values can be inferred using mathematical models. The subtle transformations of children's values are analysed in this study by focusing on mathematical model transformations and their reasons. Shimada and Baba (2022) suggested that it is easier to see whether the transformation is explained by a mathematical model than to capture social value transformation.

When mathematical modelling appears in the activities of young children, it can be examined based on the above statements, especially in Japanese preschools where young children are yet to learn arithmetic and mathematics. Nakawa (2019) found that while examining the activities and then the statements of young children as they worked out a solution, their primitive mathematical thinking became clear; the mathematical value of rationalism. Therefore, following Nakawa (2019), this study focuses on young children's mathematical thinking, even when they cannot find mathematical models, and examines how their fundamental thinking may lead to mathematical modelling or values.

Skovsmose (1994) discussed the key concepts of critical mathematics education. One of them focused on the democratic competencies proposed by Giroux (1989). Thus, schools should educate pupils to become critical citizens who take risks, challenge themselves and whose actions make a difference in society. To foster the foundation of such democratic competencies among children, this study emphasizes the importance of group discussion and communication among them.

16.2.2 Preschool Education in Japan

Research on preschool mathematics education is gradually increasing in Japan and the cognitive and affective characteristics of young children are becoming progressively clearer (Matsuo & Nakawa, 2019; Nakawa, 2019). Nakawa (2019) dealt with role-playing in preschool education and discussed the relationship between social and mathematical values in young children's mathematical activities. The study argued that young children had values of fairness and were influenced by social factors, such as life situations, teachers and friends. Furthermore, it found that young children were able to think and express themselves logically and mathematically in various activities, which confirmed the potential and importance of dialogue in mathematical development, even for those who may not have received formal mathematical education. However, research from the perspective of critical mathematics education has not been conducted in Japan, which is the perspective the author addresses in this chapter.

16.3 Methodology

16.3.1 Study Focus

Multiple case studies were conducted in a private preschool in Japan, to perform a qualitative analysis of the interactions captured by children's narratives (Creswell & Poth, 2018). The author was not able to select Japanese preschools purposively as Japan does not rank preschools according to academic performance, nor do preschools offer subject-based learning. Hence, a new approach had to be developed to examine whether children's statements and actions were related to different values and democratic capacities. Informal discussions were held with the supervising teachers who understood the characteristics of each child to analyse and understand the phenomena that occurred. Research permission was obtained from the preschool and children's guardians.

Our study focuses on distribution and rulemaking, using reasonable situations for young children to work on, which were similar to previous research conducted in primary schools (Shimada & Baba, 2022). It examines how their social and mathematical values are expressed using multiple case studies. As this was a new trial involving young children, two data collection rounds, Cases 1 and 2, were undertaken and data were collected in December 2018 and August 2019, respectively. In the two cases, two different sets of children, aged 3–6 years, participated. The number of children varied in each case (shown in Table 16.1). The focal study subject was how 5–6-year-old children negotiate and reach a consensus within a group. The presence of children of different ages in each group may affect their emerging values (e.g., their body is smaller, so we will give them a smaller portion of the cake), especially in distributing activities. The focal perspective of the two cases was authenticity (Vos, 2018).

Table 16.1 Number and age of participants in Cases 1 and 2

	Rulemaking	Division	
	Target game	Continuous quantity	Discrete quantity
Case 1	11 children (5 boys and 6 girls), aged 5–6 years	Group A	Group C
		6 children (4 boys and 2 girls), consisting of four older children and two younger children	6 children (3 boys and 3 girls), consisting of four older children and two younger children
		Group B	Group D
		7 children (3 boys and 4 girls), consisting of four older children and three younger children	6 children (4 girls and 2 boys), consisting of four older children and two younger children
Case 2	9 children (5 boys and 4 girls), consisting of three 3/4-year-olds, three 4/5-year-olds and three 5–6-year-olds	Group E	16 children (one boy and 15 girls), all were 5/6-year-old
		5 children (2 boys and 3 girls), consisting of three 5–6-year-old girls, one 5–6-year-old boy and one younger child	
		Group F	
		6 children (4 boys and 2 girls), consisting of two 5–6-year-old girls and four younger children from a different class	

16.3.2 Participants

Table 16.1 presents the number, gender, and age of the participants in both cases. In each case study, the number of children varied according to the purpose and content of the case study. For example, in Case 1, the target game was free to have several children; therefore, 5–6-year-old children were free to join the activity. However, the division activities followed Shimada's (2017a, 2017b) case study where a melon was divided among family members of different ages, that is, grandfather, father, mother, children, and a baby. Thus, the division activities at preschool were planned similarly. Hence, children of different ages were gathered for the study. The author intended to see the oldest children's actions as the younger ones were not able to express their feelings verbally.

16.3.3 Activities in Case 1

16.3.3.1 Target Game

A target game board was set up on the floor (Fig. 16.1).

Fig. 16.1 Target setting

Each child threw a sticky ball three times and points were awarded based on where the ball landed: centre circle = 5 points, middle circle = 3 points, outer circle = 1 point and outside the circle = 0 points.

After the children practised the game repeatedly, they were given the following task to discuss.

A 3-year-old child threw the ball three times. The first throw hit the five points area, the second throw the three points area, and the third three/one point boundary line. What is the child's total score?

16.3.3.2 Paper Cake Distribution (Continuous Quantity)

The children were divided into two groups and each group was given the following task to perform.

We have a cake (A4 paper, coloured, round and flat), which is drawn on a piece of paper. We want to share it with everyone, so we want you to use scissors to divide it. Let us consider how to divide it.

16.3.3.3 Distribution of Oranges (Discrete Quantity)

The same group (Group C) that performed the second task participated in the third task where nine oranges were used. The participants were asked the following questions.

We want to share these oranges, but how should we do it? Shall we discuss and share them?

16.3.4 Activities in Case 2

16.3.4.1 Target Game

The target game was the same as Case 1 (Fig. 16.1). The gameplay, rules and scoring scheme, too, were the same. In Case 1, except for a few children, everyone enjoyed the game, however they had difficulty discussing the hypothetical situation. Therefore, in Case 2, the children were included in the discussion regarding the setting of the game rules.

16.3.4.2 Distribution of Round and Elongated Cakes (Continuous Quantity)

The different age groups were divided into four groups of 6 children.

One group was given a round cheesecake and three groups were given a long, thin roll cake, to work on the following task. Share the cake so that everyone is happy and don't let anyone feel unhappy. There are younger children from the Peach Class, so please discuss among yourselves and make sure that everyone agrees. You can decide how you will divide the cake.

16.3.4.3 Distribution of Bananas (Discrete Quantity)

This was an accidental investigation that occurred when all the bananas were to be eaten together after an activity using bananas.

There were 16 children and only 15 bananas. As the children were engaged in dividing and distributing activities and had a good foundation in group discussion, the teacher suggested that they discuss and distribute the bananas in a way that was agreeable to everyone.

16.4 Results

The following sections outline the outstanding and important characteristics related to social and mathematical values identified during the activities. In the following exchange processes, T refers to the teacher; CA, CB, CC, etc. refer to individual children; and Cc refers to several children answering together.

The author underlined parts of the excerpts considered to be focal points that would be discussed later in the chapter, and the numbers (i)–(xiv) inserted next to them are used for cross-reference purposes.

16.4.1 Case 1: Results

16.4.1.1 Target Game

The teacher and children went over the rules of the game together and the teacher demonstrated the game once. Target games were played per child for three times and the total number of points obtained were written on a chart, using numbers and tapes. Once everyone finished the game, they were asked about the task.

> T: I was wondering what I would do if the Peach class (three-year-old class) were here … What should we do about the score then? If we have small children from the Peach class ... is there anything you think we should change?
> CA: Oh, good idea! (Goes to the circle and changes the scores on the stickers).
> T: Oh, I see. You mean it gets easier?
> CA: Yes...
> CB: Good...
> T: We will switch the 3- and 1-point marks. What about the 0-point mark?
> CA: As it is...
> T: As it is ... Can I ask you why? Why did you switch things up?
> CA: You know, the Peach class (the younger children's class), maybe, you know, they're going to get one point; so, I feel sorry for them; so, I'm going to give them 3 points here. (i)
> T: I see. It might not get this far. It might only get to the 1-point mark. So, I'll give you 3 points then. How about that? Is that all right, everybody?
> Cc: Yes...
> T: And when other children do it? Would it be just the same?
> Cc: Yes...
> T: Well, if you don't mind me asking, how many throws did I just make?
> Cc: 3.
> T: So, after throwing it three times, for the third time, when it landed between these marks, how would we score it?
> CA: I'll give you 0 points for hitting this mark...
> T: What would you do if someone from the Peach class goes over the line? The first time I got 5 points, the second time I got 3 points, the third time I got 4 points and the fourth time I got 5 points. For the last time, when it touched the line here, what should I do? What is the best way to do it?
> CA: Oh, good! You know, I think maybe, the Peach class, even from here, you can go over there like this, so I think we should make it three here.
> T: What if, when you throw it, it goes here (the boundary of the area for 5 points and 3 points)? (Pointing to the curve)
> CA: If it lands here (the boundary of the area for 5 points and 3 points), I'll give you 5 points! Oh, 4 points! (Pointing to the curve)
> Cc: Good, good. (ii)
> T: Okay, so after the 5, there is a 4 here. Why is there a 4 here?

CA: Because, you know, when you get 0 points here, you know, I feel sorry for you, so I tell you, you know, you get 4 points here. (iii)
T: Okay, so, let's make this area 4 points...

The conversation related to the children's values, especially compassion towards others. However, children's values were not observed to be more than expected. Children found it difficult to make logical conversations in the provided situation.

16.4.1.2 Paper Cake Distribution (Continuous Quantity)

In Group A (Table 16.1), the following conversation took place regarding dividing the paper cake.

T: What should we do? How should we divide the cake?
CA: With a knife, like this, like this (gestures to cut the cake per its diameter and divide it). (iv)
T: Oh, yes, I'll bring a toy knife. Oh, I see. So that's how. You'll cut it with a knife. I see...
CB: Can we cut from the beginning?
CA: Teacher, he says that first, he is going to cut it.
T: Well, you first have to draw lines on the parts to cut and CA (a younger child), your elder brother is cutting that for you, too.
CB: Teacher, the strawberries are going to crack; is that okay?
T: Oh, I see. There are strawberries right where you will cut the cake.
CA: *Pakkan* (the sound of cracking strawberry)! It's cracked.
CB: It's alright, it's alright. Now, quarter, *pakkan*, *pakkan* (making the sound of the cake cracking).
CB: 1, 2, 3, 4, 5, 6
CB: 1, 2, 3, 4, 5, 6 (Everyone is looking at the cake), being cut into 7 pieces.
CC: Teacher, we're down to 7.
T: Wow, it was divided into 7!
CA: Then why don't you just cut another one?
CB: Why? If you do it like this, you're going to have 9 pieces...
CB: Oh, let's practice once (pretending to cut a piece of toy cheese; cutting it into 6 equal pieces, pressing against the cake base, using his hand as if he had a knife).
CC: You have 1, 2, 3, 4, 5, 6. 6!
CC: How would you do it?
CB: Put it down like this...
CC: Ah!
CB: 1, 2, 3, 4, 5, 6. 6!
T: Just as well.
CB: Then cut it with scissors. All right? There are now 6 pieces. Because there are 2 pieces, they are on the line; so, first cut them in half. (v)
CC: Now it's cracked perfectly! Next time, it's going to be difficult.
(CA and CC are in the oldest class).

The act of division proceeded in Group A, with a boy, CB, from the eldest class taking the lead. They tried to cut the cake according to what they had initially drawn on the paper cake with a pencil. They tried to divide the cake into six equal parts by referring to their cake toys. Several children used their hands in different situations to imitate cutting the cake and it can be assumed that cake division, as seen in real life, influenced their speech and behaviour. The pieces differed slightly in size because the children had used their hands to simulate the cake cutting.

In Group B, a girl, CD, from the eldest class, summarised the group's opinions. She made a cutting gesture with her hand; however, she did not pay attention to the size of the pieces after she had cut them in half. One half was divided into three halves and the other half was divided into four halves of different sizes. At the cutting stage using scissors, the following conversation took place.

CD: What if we cut 3 over here and 4 over here—1, 2, 3, 4, 5, 6, 7?
CE: Then you won't get here.
CD: What is here? 1, 2, 3, 4, 5, 6, 7.
CD: If you cut here like this, it will look like this? (Looking at the teacher).
T: If it's good enough for everyone, why not?
CD: Is that right?
Cc: OK...
CD: (Cutting the cake in half with scissors)
CD: What if we do this one for 3 and this one for 4? And then, why don't we just turn it around with this? (vi)

In this conversation, CD confirmed with the teacher that they all wanted to try cutting the cake at least once. The teacher encouraged the group, especially 5–6-year-olds, to reach a consensus. The other children from the older class, who understood the concept, agreed to cut the cake one piece at a time with scissors and repeated the distribution process. They were not concerned about the size of the pieces, instead, they focused on those who used or did not use scissors. At the end of the session, the children cut the cakes and each ate a piece.

16.4.1.3 Distribution of Oranges (Discrete Quantity)

In Group C (Table 16.1), the following conversation took place regarding the change of the unit of division.

T: There are nine oranges here; a lot of oranges. We've played enough and we may be a bit full, but we'll all have oranges. I would like to divide them but how should we do it?
CA: Separate them one by one, all together...
T: So, can we talk about it and divide them up?
CE: 2 and 5 and…
CB: Wait a minute, just a minute, just a little bit; there are 9.
CA: Oh, what if we split them into 4 and 3?

CB: How about 6? So, if we take out these 4, these 4…?
CF: Why, and then 1, 2, 3, 4 and 5?
CB: Oh, I see, these 3, that makes 6.
CB: I've got 6.
CA: It would be 6.
CB: And then, because of this again...
CA: 1, 2, 3, 4, 5, 6.
CB: Yes, done. (CB tries to give the other 3 oranges to the teacher.)
T: You don't want them?
CB: We don't want them and this one too.
CA: We now have 6 oranges.
CB: Can I eat one?
T: Is there one orange for each of us? Then, you can share them, too, since you're here. (Giving the other 3 oranges to the children)
Cc: Eh, eh, eh! (They try to give all three oranges as leftovers for the teacher.)
T: How do we separate them?
CB: There are oranges there! (Pointing to the other oranges put on a desk)
T: You know, the oranges placed over there cannot be given out.
CF: Yes, we can all peel them and then divide the fruit amongst ourselves. (vii)
Cc: I see.
CB: Then each of us has one.
CA: You eat from yours first.
CB: Yes, I know. I know. (They form a circle and each one eats one portion.)

After eating the first one, the children started to divide the three oranges. CB, CE and CF, holding the second orange, peels it. CB gives the smaller of the halves to CF. CE gives a portion to the smaller child, and CD and CA give the roughly halved one to CC. Finally, when the teacher asks how they had divided them, CB replies, "I divided them in half".

CB: We split them up.
T: How did you divide the oranges?
CB: Like this (pointing and counting other children in turn).
CB: CE shared his portion with CD. CA gave his portion to CC. We all split our portions.

After this conversation, the author asked about the difference between this activity and the one where the cake was divided unequally; however, no clear answer was provided.

In the distribution behaviour of Group D, all members received one piece each, similar to Group C. However, the remaining three oranges were divided equally (equal division), with the girl who seemed to lead the action, distributing a piece to each child. The idea of "half" was not the same as in Group C. The girl acting as the leader returned the remaining portions to the teacher. The reasoning is explained below.

R: Do you still want everyone to have the same number?

Cc: Yes.
R: Why?
CG: Because, you know, if you have one or two more, the other friends will think that it's nice and they want more. (viii)
CH: Yes, I think so too. (Other children nodding)

16.4.2 Case 2: Result

In Case 2, the activities did not differ much from Case 1. The main difference was with the continuous quantity activity: the cakes for distribution were real and of different shapes (a cylindrical cheesecake with a low height and a long, cylindrical roll cake) and in the discrete quantity activity, the number of bananas was insufficient.

Specifically, T1 and T2 denote teachers and R denotes the author.

16.4.2.1 Target Game

Children, under the teacher's direction, discussed among themselves and formed two teams with three children each. Each group comprised children of different ages. The teacher encouraged the children to discuss the rules and positions for throwing the ball that could convincingly satisfy everyone, including the younger children. However, the children, in this Case compared to the children in Case 1, did not actively engage in conversation. The teacher asked the children to gather to make it easier to talk to them and asked them to think about the starting position. They were guided to stand on the line closest to the target circle to check the distance. All the children were placed in the same position. The teachers let the children sit on the floor and continued talking.

T1: What should we do with the children from the Peach Class (the younger children's class)? Where can they throw it from?
CA: At the front?
T1: Well? Let's write the rules then...
CA: Around here (Pointing to an area closer than the line on which they stood).
T1: Okay, so we're going to go around here? (Asking the others if they are fine with the suggestion)
CA: Is that all right with everybody?
CB (The Peach Class): (nodding).
T1: What will you do?
CB: (nodding).
CA: Here (Pointing to the nearest line).
T1: Do you want to throw it closer? How's that? Do you like that?
CB: Yes.
T1: Okay, let's do that. Can I ask you to move the line a bit? (She tries to move the line forward.)

CA (Older child): Do you want to try it? (Reapplying the tape in the front for the younger ones)
CB (The Peach Class): (nodding).
CA (Older child): He said that he'll try it once... (ix)

The older children were allowed to throw from anywhere because some wanted to throw from different places. Some children suggested that the exact score on the tape should follow that on the inside of the tape. After these rules were decided by the children, they started playing the target game.

16.4.2.2 Distribution of Round and Elongated Cakes (Continuous Quantity)

Six children in each group were given a roll of cake and a knife each and the environment was structured such that the cutting and distribution of the rolls could be performed freely by the group. In the first group, agreements were made as follows:

CC: (Pointing to a small child) Can you cut it?
CD: Cut your own and if there are any leftovers, the next child will cut it. OK?
CC: Yes...
CD: Is that good enough?
CS: (Nodding). (x)
CD: Can I cut it? (Holding a knife).
CD: I think it's about this much (Pointing to the cutting point) here...
CD: Is that good enough?

After the conversation, CD, who took the lead, asked the other children in the group one by one, "Is that okay?", and the others agreed. CD cut the cake into six equal pieces. The cakes were cut one after the other, with the elder children taking over for the younger children. The cutting procedure consisted of one person cutting the cake and then handing the knife over to the next child. From the outset, there was no discussion about the portion sizes and it was decided that each person would cut a thin slice, with any leftover pieces being "refilled". It was observed that as the children cut the cakes, the children from the older class checked and divided them equally, regardless of their size. Following is an example of the attention paid to the size:

CC: (CE tries to cut it.) That's too big!
CS: Big! (In a voice of discontent) Big! (xi)
CC: Make it tiny!

For the last slice that was left over, the children agreed to cut it into six equal pieces. The other group cut the pieces first in sequence, starting from the end, dividing them into six equal pieces and matching the size of the other pieces to the first six pieces. The children reached a consensus as shown below.

CE (in the older class): Let's divide it into six equal parts. Watch out! (To the other children in the older class). Let's think first ... six equal parts? (Pretending to cut 6 rolls of cake with her hands).
CF: There are 1, 2, 3, 4, 5, 6. Six children are here.
CE: Yes (Pretending to cut the cake rolls with his fingers). I can divide it into 6 pieces with my fingers.
CG: (Pretending to cut the cake rolls with her fingers) How many are there? 1, 2, 3, 4, 5, 6? (xii)
CE: But it's not safe for children from the Peach Class to cut, so children from the Lily class (the older class) will cut for you. Do it this way, Peach class. It's easier if I cut it together with them. Can you cut by yourself alone? (Addressing the children from the Peach class).
CG: (Nodding)
R: When you decide how to cut it, you can cut it. Who cuts it doesn't matter.
CE: But you all want to try cutting it, don't you? Do you want to try it?
CC: (Nods or raises hands.) (xiii)
CE: Who wants to cut it, raise your hand.

The children cut the cake into six parts and ate each piece together.

16.4.2.3 Distribution of Bananas (Discrete Quantity)

The children asked each other whether the "6-year-olds would eat" or "who won't eat". They suggested that "one of them has to leave" and "someone must leave". They asked each other, "How many people usually eat bananas?" Although no conclusions were reached, several children insisted that they wanted to eat a banana. One boy said that he did not have to eat it.

CJ: You don't want to eat a banana?
CK: I don't want to eat a banana.
CJ: Okay, CK is out.
CK: (Stands up).
CS: Thank you, thank you!
CH: (To CK) You should stay here...
CJ: But CK will get out of here.
CT: You can leave if you want to. (xiv)
CJ: So, who doesn't like bananas?
CL: No one... (Silent)
CS: No one... (Silent)
CM: What? Why? Are you full?
CK: Yes...
CS: (Chuckles)
CJ: Then, get out of here.
CS: Out, out. Bye.
CK: (Walking away from the room)

T: Are you sure, CK?
CN: He's full.
T: You're out, CK.
CJ: Why don't you say thank you later?

As CK went out, each of the other participant received a banana and everyone, excluding CK, ate their bananas.

16.5 Discussion

16.5.1 Relationship Between the Expression of Social Values and Authenticity in Young Children

In Case 1's target game, there was no active discussion among the children. One boy (CA) actively expressed his opinion and said that he would change the score because he felt sorry for the younger children ((i)–(iii)) and the other children agreed with him. The social value of considering others was observed. In the cake-cutting activities of Case 1, Groups A and B initially tried to share the cake equally. Group A tried to distribute the cake fairly ((iv) and (v)), while Group B tried to be fair in the act of cutting with scissors, rather than considering the size of the cake ((vi)). During the distribution of oranges, in both groups, young children thought that it was important to divide the number of oranges equally, regardless of their age, gender or body size ((vii) and (viii)). Thus, equality was regarded as important.

Shimada (2017a) found that in the context of rulemaking and distribution, social values, such as equality, fairness, caring and diversity, were demonstrated by primary school children. In Case 1, only compassion was found in the rulemaking activity ((i)–(iii)) and only equality was found in the distribution activity ((v)–(viii)). In addition, in both case studies these values were formed mainly by young children who had the loudest voices in the groups. Children seemed to find rulemaking activities more difficult to understand and commit to as they had to imagine the situation, compared to distribution activities, which were easier for them to grasp based on the teacher's descriptions.

The authenticity of the context significantly influenced the results. In Case 1's target game, the children did not appear to be engaged which seemed to be because they were asked about a virtual situation. In the same game in Case 2, different groups showed different fairness values in various situations. Comparing the distribution activities in Cases 1 and 2, the paper cake cutting in Case 1 did not involve real eating; hence, it was inferred that fairness was shown in different situations—at the time of cutting. The girls in Group B in Case 1 showed that the young children were more likely to divide the cake equally when an actual cake was used for division.

Case 2 was intentionally designed with more emphasis on authenticity than Case 1, and it seems that because of this, both cases showed different results. In Case 1's target game the children had to imagine a situation with younger children, whereas in Case 2, several younger children were present, which made it easier for the 5–6-year-olds to grasp the scene (refer to the excerpts in 16.4..2.1). In making the rule in Case 2, caring for the smaller children became a priority; hence, the older children decided to let them throw from a closer position ((viii)); however, score differences were not discussed. For 5–6 year olds, freedom to throw from anywhere was ensured, rather than throwing from the same place. This was not observed in Case 1. Moreover, they discussed the game more actively than in Case 1.

The results of the distribution of roll cakes and round cakes were similar. However, the method of cutting was slightly different. In the roll cake distribution, the children decided to cut the cake into smaller pieces one at a time from the end ((x)) and divide the final excess into six equal pieces. Another example of building consensus was the 5–6-year-old children coming up with a plan and reaching a consensus with the younger children. Consensus was reached about cutting, with each child cutting once. As in Case 1, equality emerged as an important factor for the children ((xiii)). Regarding the size, in case of the paper cake, although there was some mention of dividing it equally, the children did not pay much attention to the size, since it was made of paper. In Case 2, however, size was strictly observed by children ((xi)). Although the conversations during the round cakes and elongated roll cakes were not addressed here, the children had previously divided a cake into equal parts, using their hands as knives in gestures ((xii)). However, dividing a round cake into six equal parts was difficult and, in practice, the cake was divided into eight equal parts. In the distribution of the bananas, while equal division was emphasised, since the number of bananas was one less than the number of total students, the children who wanted to eat asked the boy who did not want to eat to "Leave the room"((xiv)).

16.5.2 Comparison of Young Children's Expression of Social Values

The results of Cases 1 and 2, considering the representation of social values in primary schools in Shimada (2017b), are compared and summarised in Table 16.2.

This study investigated the social and mathematical values of young children. The results of Cases 1 and 2 differed. In the rulemaking activity, older children showed compassion for younger children in both cases. However, the desire "to throw from anywhere" was clear in Case 2, which was not observed in Case 1. This situation led to the expression of respect for the younger children; however, even though the older children showed compassion towards the younger ones, they stated that "We also want to throw balls as near as the younger ones do to get higher scores." This may have reflected a childish attitude, which is reasonable at the developmental stage of 5–6 years. There are several situations in preschool education settings in Japan where older children must help and care for younger children (Nakawa, 2019). In the target school, children of different ages intermingled. In some cases, teachers

Table 16.2 Comparison of social values expressed in Cases 1 and 2

	Rulemaking	Division		Mathematical model
	Target game	Continuous quantity	Discrete quantity	
Primary school[a]	Compassion, equality, diversity	Equality, fairness, compassion, and diversity	No examinations	Existence
Case 1	Compassion	Equality	Equality	None
Case 2	Compassion for others and one's own desire and feelings	Equality	Showing their wishes	None

[a] The social values in this row are derived fromShimada (2017b)

actively involved older children with younger children to provide support. Indeed, compassionate behaviours are valued in Japanese schools (Shimada & Baba, 2022). However, when considering the developmental stages of children aged 5–6 years, these two feelings are simultaneously manifested in wanting to do things and the feeling of compassion.

In the distribution activities, the focus was on fairness and equality, even after considering children's different ages and body sizes. The most important value was the sharing of a cake equally. In the case of the bananas, unlike the other activities, the discussion among the older children was more about "eating their own food" than about being considerate of others. The feeling of sharing equally was given priority, despite no suggestions about sharing the banana, indicating that both fairness and one's own feelings and desires were valued. Therefore, the two opposing values appear differently in different situations. This shows that children were context-dependent.

Moreover, in these results, the social values of young children manifested differently if authenticity was emphasised. In Case 1, equality and fairness were emphasised while the children appeared to value their own feelings. In Case 2, the way the real cakes were divided differed from the cake division in Case 1, with a focus on the amount of cake and an obsession with dividing the cakes equally. Young children thought about how to divide the cake in their own way and divided it in a way that was agreeable to all. This shows that even for young children, whether the scene is real or virtual, influences their decisions and behaviour.

16.5.3 Representation of Mathematical Concepts

As stated by Nakawa (2019), no mathematical models discussed in primary school were found in social value situations in preschool (Table 16.2). This may be partially because preschool education in Japan does not provide formal mathematics teaching and learning. Meanwhile, social values that may have been formed by young children inside and outside the school were observed. For example, the understanding of arithmetic quantities and their related skills and the importance of verbal expressions and gestures were observed. During the distribution activities, gestures showed a mathematical concept and understanding of the division concept; several children simulating division by hand. This implies that although young children cannot express themselves verbally, gestures convey their intention and understanding of mathematical concepts.

During the distribution of the roll cake in Case 2, the children were observed dividing the cake into six equal parts by hand and focusing on its size. The division method was used to ensure equality for the benefit of young children by first cutting out six pieces from one cake and then dividing the remainder into six equal parts. Although it seemed difficult to divide them strictly into six equal parts, dividing them equally was a matter of awareness and action. Especially, in dividing the remainder into six equal parts, they changed the unit of quantities to be divided, which related to the fundamental idea of division and units. These findings are of special interest as they provide the basis for learning fractions and division that are taught in the second year of primary school. In the distribution of oranges, children counted the number of people in the group, repeatedly counted the number of oranges and used the concept of "half" during the speech, indicating that, as stated in Nakawa (2019), rationalism was expressed through mathematical values. Although the relationship between social values and mathematical modelling seen in Shimada (2017b) was not observed, the differences between the characteristics of these young children and children in primary school and beyond could be a major feature of preschool mathematics education.

16.5.4 Integrated Summary and Consideration of the Result

The analysis of Cases 1 and 2 shows that rulemaking can be difficult for young children in imaginary situations; however, in realistic situations, such as Case 2, children were able to verbalise and communicate ideas through words and gestures. Regarding distribution, an authentic situation, including real and concrete objects, allows the observation of multiple children's opinions and active involvement. In addition, equality was emphasised among children. However, despite equality and sympathy for others, a feeling different from sympathy was expressed in the children's desire for their opinions to be respected, according to their developmental stages. Kubo (2017) stated that consensus-building, based on the creation of options and consideration of their validity, is crucial for fostering democratic competence. Similarly, this study provided evidence that young children could make their own decisions through communication.

The implementation of various cases from a democratic perspective in mathematics education in Japan has been extremely limited. However, in response to the demands of modern times, the recent courses of study are oriented towards a new mathematics education system that incorporates application-oriented aspects and the traditional structure-oriented approach. Furthermore, cross-curricular learning is being developed; hence, it is crucial to take mathematics education one step further. From early childhood education, in the context of early mathematical thinking, listening and cooperation, which are the foundations for developing democratic competence, must be practised.

In addition to language, the children's behaviour in this study, in particular their gestures, has been used to clarify their mathematical abilities and skills. In other words, the children's gestures reveal aspects of their informal learning before they transition to formal education as they can be used to simulate and divide equally, even if the behaviour is not precise, and learn how to divide objects.

Regarding the children's behaviours, especially within these small groups, the children engaged in dialogues based on the thoughts they had developed, including the support of teachers and adults. In this dialogue, the children used mathematical skills and knowledge that they acquired through daily life, even if the mathematical modelling seen in primary school and beyond was not observed, as described by Nakawa (2019). The author speculates that this may lead to what Skovsmose (1994) calls the basis for democratic competence, that is, the direct use of arithmetical means during discussions. Furthermore, these children did communicate through words and gestures. This study, eliciting these two types of communication, showed that the children's verbal expressions and gestures might be considered to identify their existing mathematical reasoning.

This study also emphasised the importance of capturing the children's negotiation process. Activities regarding rulemaking and division can provide young children with the opportunity to observe social values according to context and group membership and help develop skills that lead to democratic competence to talk, negotiate and reach a consensus.

Regarding the composition and development of mathematical models, although the children made full use of mathematical ideas (equal parts, halves, etc.) in the tasks, they had not reached the stage of composing mathematical models due to the absence of formal lessons. However, this can be of practical value for the development of social values in preschool education. This study showed that young children have their own way of convincing and discussing through mathematical ideas and gestures, which is different from adults and primary school students. Not many studies regarding gestures and conversation processes in early childhood have focused on mathematics learning, except for Nakawa et al. (2023).

This study's findings have implications for teaching. It provides recommendations regarding providing preschool mathematics education. Japanese children in this study were at a developmental stage when they were starting to engage in reflection and reflective thinking. Although some children know that equality, assistance, and cooperation are important owing to daily life experiences and guidance from adults, at the developmental stage (i.e., 5–6 years), only their desires are expressed. Hence, providing opportunities for children to share their thoughts about other children's opinions while listening to other children's ideas is an ongoing necessity in preschool education, even when children may not be able to consciously do so. Young children expressed their thoughts by relating gestures to language. This was also observed and discussed by Nakawa et al. (2023). Therefore, it is considered effective for teachers to actively use gestures, and pay attention to children's gestures in conjunction with language, particularly in situations with concrete objects.

16.6 Conclusions

This chapter considered two research questions:

(1) What social values do young children hold in terms of rulemaking and distributive behaviour?
(2) What types of mathematical values are observed in connection with social values?

In rulemaking activities, compassion was observed as a social value, while, in distribution activities, equality was observed. However, the quality of equality was different depending on the authenticity of the situation. Within distribution activities, the change of focal units was observed, which related to mathematical values.

This study has several limitations, owing to the uniqueness of Japanese preschool education. As there is no formal mathematical learning in Japanese preschools, young children's mathematical knowledge and skills must be promoted informally. There would likely be large differences in the abilities of young children depending on their family environment. Differences with countries that have a formal preschool education may need to be considered and examined. In addition, since there are other examples in Shimada and Baba's (2022) framework, it is necessary to consider the values that can be seen (or not seen) in the other classifications. The author would like to examine the characteristics and differences between preschoolers and first graders when considering the early childhood/primary school connection. In future work, it would be worthwhile to examine other cases and design other tasks (e.g., Shimada & Baba, 2022) in a preschool mathematics education setting.

In the future, the author plans to implement this approach in more preschool educational institutions. The results of the present study are influenced by the characteristics possessed by the groups of young children. At the preschool education stage, in mathematics, the content of the discussion is not necessarily opposed to the expression of social values that are not related to mathematical content from a comprehensive human resource development perspective. Academically, it is necessary to discuss whether mathematical modelling is possible at the preschool education stage and if so, in what activities and contexts could it be expressed. In terms of comparative mathematics education, the author plans to examine the observable differences when the same tasks are performed in different countries.

References

Björklund, C., van den Heuvel-Panhuizen, M., & Kullberg, A. (2020). Research on early childhood mathematics teaching and learning. *ZDM, 52*, 607–619.

Creswell, W. J., & Poth, N. C. (2018). *Qualitative inquiry and research design: Choosing among five approaches* (4th ed.). Sage Publications.

Giroux, H. A. (1989). *Schooling for democracy: Critical pedagogy in the modern age*. Routledge.

Kubo, R. (2017). Democratic ability and critical thinking in mathematics education. In *Proceedings of the 5th Spring Conference of the Japanese Society for Mathematics Education* (pp. 201–208). The Japanese Society for Mathematics Education (in Japanese).

Matsuo, N., & Nakawa, N. (2019). Preschool children's understanding of length and area measurement in Japan. In U. T. Jankvist, M. van den Heuvel-Panhuizen, & M. Veldhuis (Eds.), *Proceedings of the Eleventh Congress of the European Society for Research in Mathematics Education* (pp. 2311–2319). Freudenthal Group & Freudenthal Institute, Utrecht University, Netherlands and ERME.

Nakawa, N. (2017). An examination of early childhood arithmetic education from critical mathematics education. In *Proceedings of the 5th Spring Conference of the Japanese Society for Mathematics Education* (pp. 225–232). The Japanese Society for Mathematics Education (in Japanese).

Nakawa, N., & Kosaka, M. (2019). Examining young children's social values in mathematical activities of preschool education. In *Proceedings of the 7th Spring Conference of the Japanese Society for Mathematics Education* (pp. 225–232). The Japanese Society for Mathematics Education (in Japanese).

Nakawa, N. (2020). Examining exemplarities in pre-school education in relation to critical thinking. In *Proceedings of the 8th Spring Conference of the Japanese Society for Mathematics Education* (pp. 147–154). The Japanese Society for Mathematics Education (in Japanese).

Nakawa, N., Uegatani, Y., Otani, H., & Fukuda, H. (2023). Young Japanese children's subjectification and objectification through the lens of joint labor in a mathematical activity at a preschool: A case study. *Early Childhood Education Journal*.

Nakawa, N. (2019). Mathematical values through social values: A number activity in a Japanese kindergarten. In P. Clarkson, W. T. Seah, & J. S. Pang (Eds.), *Values and valuing in mathematics education* (pp. 157–169). Springer.

Radford, L. (2022). *The theory of objectification: A Vygotskian perspective on knowing and becoming in mathematics teaching and learning*. Brill Academic Publications.

Shimada, I. (2017a). *Arithmetic and mathematics education and diverse values: Approaches by social open-ended issues*. Toyokan (in Japanese).

Shimada, I. (2017b). The possibility of developing critical thinking skills through social open-ended problems. In *Proceedings of the 5th Spring Conference of the Japanese Society for Mathematics Education* (pp. 217–224). The Japanese Society for Mathematics Education (in Japanese).

Shimada, I., & Baba, T. (2022). *Open-ended approach in mathematics lesson to enrich diversified values and mathematical viewing and thinking*. Meiji Books.

Skovsmose, O. (1994). *Towards a philosophy of critical mathematics education*. Kluwer Academic Publishers.

Tirosh, D., Tsamir, P., Levenson, E. S., & Barkai, R. (2020). Setting the table with toddlers: A playful context for engaging in one-to-one correspondence. *ZDM, 52*, 717–728.

Vos, P. (2018). "How real people really need mathematics in the real world"—Authenticity in mathematics education. *Education Sciences, 8*(4), 195.

Chapter 17
Teaching Math and Preserving Culture: The Intersection of Values in Indonesian Pedagogy

Ifada Novikasari, Ahmad Muttaqin, and Noor Elebiary

17.1 Introduction

Indonesia is rich with cultural diversity; thus, it faces the challenge of preserving its unique identity and heritage. McLean (2002) suggests knowledge of where students come from or what is called local culture can be a source, not only for local communities, but also for other cultures. Thus, there is a potential for local cultures to become a knowledge base for making mathematical connections. Introducing a culture can be through universal topics such as money, time, and measurement of objects (Iliev & D'Angelo, 2014; Malloy & Malloy, 1998). Real life situations based in a particular culture and their relationships can sometimes be expressed using mathematics (Haines & Crouch, 2007). Interestingly mathematics is also a product of culture (Bishop, 1988). Bedewy et al. (2022) suggested that the design of mathematics learning using appropriate aspects of the students' culture can be a strategy for building knowledge in various environments.

Seah and Wong (2012) believe, apart from cognitive factors, there are also other dimensions that influence student learning success, namely the affective dimension. A few studies have been conducted on culture in mathematics learning in relation to values (Dede et al., 2021; Hunter, 2021). One of them suggest that values are a crucial component in mathematics education and indeed values are always taught in mathematics learning (Dede, 2013). Lim and Ernest (1997) suggest that some of the values

I. Novikasari (✉) · A. Muttaqin
UIN Prof. K.H. Saifuddin Zuhri Purwokerto, Purwokerto, Indonesia
e-mail: ifa_da@uinsaizu.ac.id

A. Muttaqin
e-mail: ahmadmuttaqin@uinsaizu.ac.id

N. Elebiary
University of Howler, Kurdistan, Iraq
e-mail: nae3636@aol.com

Y. Dede et al. (eds.), *Values and Valuing in Mathematics Education*,
https://doi.org/10.1007/978-981-99-9454-0_17

that can be developed in the mathematics classroom include accuracy, systematic and rational thinking, as well as social and cultural values (e.g., cooperation, fairness, and appreciation).

In mathematics education there is a transfer of values (Seah & Bishop, 2003) that occurs at all levels of education, from elementary to tertiary institutions. Daher's (2020) study in a teacher education program shows that using appropriate mathematics activities with pre-service teachers can encourage them to build values including creativity, criticality, and metacognition. In practice, the values possessed by teachers influence the choice of pedagogy they use in their teaching (Hill et al, 2021). This study examines what values are contained in the design of local culture-based mathematics pedagogy for pre-service teachers. In this activity, prospective teachers explore the diversity of local cultures in Indonesia and then relate them to mathematical concepts in the form of teaching materials.

17.2 Values and Culture

Herbst et al. (2011) describe the core idea of how people in a particular field employ specific standards to explain their decisions and, conversely, reject other options. These standards, often seen as the values that members of the practice use to justify or reject potential actions, play a crucial role in shaping individuals' choices. Moreover, this concept provides insights into the complexities of teaching while also distinguishing the rational foundations of pedagogy from the implementation of such principles by educators. Values possess the capacity to offer rationalization for an action, and these very values are frequently embraced without the need for validation. The context of their origin is inseparable from values, as what is valued within one community might diverge from that of other societies. In one environment, the possible values are not necessarily the same. Values are formed socially through a long historical process and always find new forms (Totterdell, 2000), which is termed a process of dialectics. In a dialectical scheme, values develop and change continuously from one form to the next, until they find an ideal form. However, this ideal form does not really become the final form, since at some point a value may be recognized as being in deficit, or even containing an anomaly, so that it develops and changes yet again into a new ideal form. This means that a value that is agreed upon by consensus is always transitional and will find a new form in accordance with the current course of history (Gadamer, 1976).

Values that are individual constructions, always occur within a particular social and cultural space (Secombe, 2016). A value, being a product that belongs to an individual, is important and meaningful. The same argument can be made for community values. Berger and Luckmann (1991) suggested that value construction takes place in three stages, namely externalization, objectivation, and internalization. Externalization is a value creation process that involves actors and agents. Negotiation, tension, and conflict are very common in this process. The process of objectivation makes the value that has been agreed upon as an objective reality that requires compliance.

Individuals and communities who have built a consensus about a value are morally and socially bound to carry out and maintain it in the encompassing socio-cultural process. Internalization, the last stage, is where an individual, or for communities every member of society, adopts a value or set of values in their daily life. The agreed values are manifested in the ongoing social and cultural practices.

The manifestations of these community social and cultural practices normally become individual behaviors in the community space. Smith (2001) suggests that individual and group behavior cannot be separated from the culture in which they have been formed. In a community, culture acts as a binding and controlling framework for individual behavior, and in some cases, can even penetrate individuals' thoughts and actions. As suggested by Jasper (2016) who explains cultural expressions at the macro level such as conflicts and political policies can only be fully explained by analyzing the processes that take place in individuals. Everyone has micro-scale forces of expression which are manifested in the arrangement of elements that form perceptions and systems of meaning such as habits of thought, feelings, actions, and value-forming systems that are reflected in their minds.

Seah (2019) refers to values as conative variables that differ from cognition and affection. These values are the link between cognition with affection and behavior. In the context of development and application to mathematics, Seah et al. (2021) emphasize the role of the individual. Choice of values in mathematics education grows and develops by being influenced by other parties. The words or actions of other parties can influence a person's views, decisions and behavior. Closs et al. (2022) states that an individual's views, based on their learning experiences, do not exist in isolation, but rather, are related to the learning environments that they are systematically involved in. For most individuals, views are essentially formed by external factors that are both binding and coercive. Therefore, by looking at values, one can see how individual actions are guided, but also understand that systemically, they cannot be separated from the existence of learning environments. An individual's social relations will affect the values they believe to be a framework for acting (Herbst et al., 2011).

The maturation of a teacher's value system is intricately woven through a prolonged journey shaped by their interactions with the realm of mathematics. Among the key influences on this process is the engagement within mathematics teacher education programs, which provide educators with a dedicated space to refine their pedagogical perspectives. Steffe and Kieren (1994) emphasize the profound impact of an individual's personal mathematical experiences, underscoring their role in crafting distinct cognitive frameworks that in turn shape the interpretation and transmission of mathematical concepts. As educators embark on the journey of mastering mathematics, their aim is to facilitate the construction of students' mathematical knowledge in a manner that aligns with their own beliefs about effective instruction.

Navigating the landscape of mathematics education, particularly within teacher education programs, offers a structured avenue for the evolution of a teacher's value system. These programs provide educators with the opportunity to delve into pedagogical theories and best practices, enriching their repertoire with diverse instructional approaches. Steffe and Kieren's (1994) insights underscore how personal experiences with mathematics become building blocks for creating unique cognitive structures. These cognitive structures, created from personal experiences with mathematics, influence how educators perceive mathematical concepts and communicate them to their students. However, this transmission of values is not a unidirectional process. Chi's (2005) perspective highlights the intricate dynamics within classrooms, where students grapple with integrating new knowledge into their existing mental models. This journey of sense-making sometimes prompts students to revise, reshape, or even discard previous concepts, leading to a diverse array of mathematical pathways among learners. Thus, the interplay between evolving teacher values and the dynamic responses of students forms the crux of the mathematical learning experience within educational settings.

The process of imparting values differs significantly from that of teaching mathematical content, primarily due to the absence of a definitive "correct" answer when it comes to values. It's worth noting that the values chosen by the teacher wield a substantial influence over the values adopted by students, a dynamic highlighted by Bishop (2001), signifying the pivotal role of values in the realm of mathematics education. These values constitute an integral facet of the teaching process and influence the broader ethos of mathematical learning.

Amidst the classroom environment, values integrate seamlessly with the exchange of interpretations of significance, not only between educators and students but also within the student body itself. This interplay between values within the student body is an intrinsic part of the educational dynamic, reflecting the intricate process of value assimilation in learning mathematics. Kalogeropoulos et al. (2021) emphasizes the importance of a strategic approach to inculcating values in mathematical learning. According to their insights, educators must first comprehend their own personal values before effectively selecting appropriate strategies to convey these values to students during their mathematical education. This approach underscores the nuanced and deliberate methodology that underpins the development of values within the domain of learning mathematics.

Hunter (2021) identified that a divergence might exist between the cultural values and the values concerning mathematics education that students uphold. Consequently, this highlights the potential impact of diverse cultural values on classroom interactions. Each society possesses an individualized cultural identity that is manifested within the educational sphere through the materials included in the curriculum. The construction of values within mathematics education transcends the efforts of solely teachers and students, extending into the portrayal of these values within the subject matter itself. As emphasized by Dede et al. (2021), the distinct cultural attributes of each nation can be effectively demonstrated through the articulation of values within mathematical assignments, where the intersection of mathematics and real-world contexts occurs. This intersection allows cultural nuances to be illuminated as values become integrated into mathematical tasks, providing insights into the ways in which cultural perspectives shape the synthesis of mathematics and practical application.

17.3 Mathematical Design of Pedagogic and Local Culture

Shulman (1987) identified pedagogical content knowledge as an important knowledge possessed by teachers and/or educators which is distinct from the knowledge of the subject matter. This is the knowledge needed to teach the subject matter. Ball et al. (2008) suggest mathematics is a unique knowledge thus teaching it requires special knowledge. Thus, knowledge of pedagogical content becomes part of the knowledge of teaching mathematics.

Sociologically, pedagogical practice involves two main subjects, namely teachers and students. These two groups come from different socio-cultural entities so that there is potential for gaps in the learning potential of students if adjustments and synchronization are not considered (Steffe & Kieren, 1994). This situation is always one of inequality, which was criticized by Freire (2005, p. 55) as oppressive education. This issue revolves around inequality in pedagogical dynamics, a concern that resonates with the principles of pedagogical design and the influence of local culture. Freire's perspective highlights the intentional creation of pedagogical disparities by the dominant group—the teachers—leading to a perpetual state of student dependence. This power imbalance enables teachers to adopt the role of investors who reap substantial gains from students' engagement. However, redefining the teacher-student relationship within pedagogical practice emerges as a strategic imperative in the realm of human transformation. When integrating the concept of pedagogical design and local culture, it's evident that an inclusive and culturally sensitive educational framework becomes essential. Acknowledging and adapting to the unique characteristics of local cultures within pedagogical designs can serve as a potent antidote to the oppressive tendencies pointed out by Freire. By considering local cultural nuances, educators can foster an environment where students' diverse backgrounds are not just accommodated but celebrated. This approach not only bridges pedagogical gaps but also aligns with the broader aim of human transformation, emphasizing equitable learning experiences that empower students and catalyze positive societal change.

Trying to change this unbalanced relationship, mathematics educators need to have knowledge about the needs of students as learners and make decisions regarding the appropriate forms of representation, statements, or processes so that these can be accepted and understood by students. Silverman and Thompson (2008) suggest one of the many ways in which mathematical knowledge can be transformed into knowledge of teaching mathematics is through visualization of how students can understand the concepts and identifying the types of activities that can facilitate student learning.

Furthermore, the process of acquiring mathematical knowledge is intrinsically entwined with culture, as underscored by Cairns (2000), given that mathematics is inherently shaped by cultural contexts. Notably, educational institutions exhibit their own unique cultures that encompass embedded values, a notion expounded upon by Lawton (1999). Delving deeper into the intricate nexus between mathematics education and cultural underpinnings, Leung et al. (2006) uncovered a dynamic relationship, revealing that the prevailing educational traditions of a nation exert a considerable influence on the landscape of mathematics education. Simultaneously, mathematics instruction within schools and broader educational practices mirror and are shaped by the cultural tapestry of the country.

As a science, mathematics is produced through a cultural approach that involves the actors in it. The process takes place subjectively, ideologically, and is determined by the power and interests of the dominant actor (Foucault, 1972). In the transmission process, science develops in certain models which are influenced by the habits, patterns, and prepositions of the dominant group. Bourdieu (1993) refers to it as habitus. The existence of habitus as an instrument will support the practice of transmitting knowledge effectively. Habitus in the form of habitual patterns and propositions is present unconsciously by the owner. Thus, pedagogical practices that present habitus encourage the actors within them to be in cohesive and interactive social situations.

Pedagogical design by linking and presenting local culture is useful for encouraging more meaningful mathematics learning (Chronaki, 2000). Mathematics as a scientific discipline is established historically, embedded, and contextualized within culture which is produced as a local connection between teachers and students. The pedagogic process takes place in the social space of teachers and students which necessitates the formation of collective cultural entities. Presenting a collective culture together with the pedagogic process will form the cohesiveness of learning mathematics as an integral part of social practice. Through this pattern, the pedagogic process of mathematics and social production takes place in the same space and is produced collectively by teachers and students.

17.4 Method

Activities to develop local culture-based pedagogical designs were carried out in three student classes of mathematics teacher education programs at State Universities in Indonesia. Participants in this study were 103 pre-service teachers preparing to become primary school teachers who studied in the course "Literacy for the Development of Mathematics Teaching Materials". In this lecture course the topics covered, all attuned to mathematics teaching, include theory of knowledge development, didactic literacy, authentic literacy, functional literacy, culture as a learning resource, development of critical thinking skills and creative thinking using mathematics teaching materials.

Data collection was carried out through an analysis of pedagogical design task documents developed by the pre-service teachers. The pedagogical designs made by individual students. They each drew on their own culture with 5 local cultures represented within the group. The pre-service teachers associated their designs with mathematical concepts. Seah (2003) suggests the use of culture can affect how the same mathematical content can be taught through different approaches and that culture can affect the values embedded in that teaching. We were exploring whether this notion also impacted the materials that these beginning teachers designed.

The data from the pedagogical design assignments were analyzed using thematic analysis of the written documents developed by the pre-service teachers. According to King (2004) thematic analysis can summarize large data sets, it takes a well-structured approach to such data sets, and helps produce a clear and organized final report. During this process for this study the researcher sort to understand the values embedded in the students' text. This analysis method is carried out by grouping ideas in the texts based on semantic meaning (Wæraas, 2022). The three steps in the analysis for this study were first identifying specific text referencing values, second developing codes, and third generating themes which grouped the codes which related specifically to values (Braun & Clarke, 2006, 2020).

In this study, the application of triangulation techniques encompasses two distinct approaches: researcher triangulation and theoretical triangulation, in accordance with Denzin's framework (2006). Within the scope of researcher triangulation, a meticulous assessment was conducted on pedagogical design documents developed by pre-service teachers. This evaluation saw the collaborative participation of two mathematics education experts and one sociologist, each contributing their expertise. While mathematics education specialists directed their scrutiny towards the alignment of the pedagogical structure with mathematical concepts, sociologists focused on evaluating the congruence of the local cultural context as an essential backdrop for the learning process. Mathematicians and sociologist separately analyze the results of such work. After working individually, experts meet to decide on the agreed outcomes of each piece of work done and determine the value contained in the pedagogical design under review. Simultaneously, the values embedded within the pedagogical framework were subjected to meticulous analysis and discourse among the involved experts. This ensured the seamless integration of these values with both the mathematical content and the contextual dimensions of the local culture.

Concurrently, theoretical triangulation entails a comparison of the outcomes stemming from researcher triangulation with an array of theoretical frameworks or emerging perspectives that pertain to the domain of mathematical values. This multifaceted process adds depth to the investigation by juxtaposing the findings with diverse theoretical lenses, enriching the overall understanding of the intricate relationship between pedagogical design, mathematical values, and local cultural nuances.

17.5 Findings

One of the aims of this research is to examine what values are compiled in the design of pre-service teachers' mathematics pedagogy from the local culture in Indonesia. The integration of local cultural values in mathematics pedagogy is expected to cultivate a broader understanding of the interconnectedness of knowledge. In addition, they are expected to see mathematics as a tool that is not only universal but also deeply embedded in their own cultural identity and experiences. The previous lecturer presented an example of a community legacy in the form of an old mosque building. Using these artifacts, pre-service teachers were presented with material on the number of wall motifs and identified the various geometric shapes within the mosque building.

In the lecture process, after the pre-service teachers attended fourteen lecture meetings, they were then given topics regarding culture as a source of learning mathematics. Three of those topics were:

- Representation of values in Indonesian local culture
- Authentic literacy in mathematics learning
- Exploration of local culture in mathematics learning.

The culture preserved by the different ethnic groups in Indonesia contain values and some of these cultural values can be associated with mathematical concepts. The sources of ideas in the pedagogical designs analyzed mostly contain local culture, although some students did not present them as local culture.

Table 17.1 shows that 87% of pre-service teachers used Indonesian local culture in their assignments. Indonesian local culture included traditional dances, cultural heritage buildings, community cultural activities, motifs on traditional cloth (batik) and others. The highest percentage associates' local culture with geometric content. There were also 13% of pre-service teachers who related contextual problems to math content rather than local culture: For example, by presenting camping activities in the form of the tent used.

The main analysis of the students' work, however, was to identify the values embedded in the assignments going beyond what mathematical content was there. Table 17.2 shows examples of text identified as relating to values. These were then assigned a code and finally the codes were grouped into themes. The three themes

Table 17.1 Source of design ideas and content

Source	Number (%)	Algebra (%)	Measurement (%)	Geometry (%)	Statistics (%)
Local culture (87%)	12.6	3.7	3.2	62.6	5.3
Nonlocal culture (13%)	2.6	3.7	1.1	3.7	1.6

Table 17.2 Examples of pedagogical design task analysis

Sample of explicit text	Codes related to values	The Values themes
Art tools function as... Movements in this dance are performed to... ...with this game students will understand... ...in this context teaches...	A benefit, a function of the local culture	Usefulness
... on musical instruments found a variety of mathematical concepts including ... ...there are many local cultures that can be used as a source of learning mathematics	The diversity of mathematical concepts in the local cultural context	Creativity
.. this culture can be used as a context in mathematical concepts to make it more meaningful" ...mathematical concepts can be used in this tradition...	Change	Transformation

identified in this study were namely usefulness, creativity, and transformation. The next sub sections discuss each of these themes in more detail.

17.5.1 Usefulness

The identification of the value of usefulness in these pedagogical designs arises from the pre-service teachers' statements, which provide an explanation of the uses and functions of the given local culture. We draw inspiration from Hernandez-Martinez and Vos (2018), who presented the same category for mathematics content. In these pedagogical designs, pre-service teachers mention the benefits and meanings of local culture. For example, one student created a pedagogy centered around cultural buildings. The motifs on the walls of the building depict cultural meanings within society, and this can be related to mathematical concepts in the learning process (Fig. 17.1).

In Fig. 17.1, the artifact motif can demonstrate the concept of mathematical reflection. In this case, it can help to communicate information more effectively to recipients, enabling them to understand mathematics better. The values contained within these motifs can be identified as beneficial values (Harriyadi, 2020). According to Bishop (2001), artifact motifs, being symbols, can be endowed with value by the community. This useful value of symbols supports the development of objects values in students.

We developed two criteria regarding usefulness as shown in Table 17.3. In category U1, the design of pre-service teachers introduces local culture with its use in

The ornamentation on the temples consists of winged shell shapes, or śaṅkha, positioned at the top of the roof on the four outer sides of the temple's body. The śaṅkha decoration symbolizes liberation, primarily associated with the release of the soul from physical bonds and potential dangers leading to death. These temple ornaments also incorporate the concept of reflection, which is exemplified in the various decorations present on the Pari Temple,utilizing mathematical reflection techniques.

Fig. 17.1 The value of usefulness in pedagogical design of cultural heritage buildings

learning mathematics but is limited in explaining mathematical concepts. Expressions such as "Wayang games have the concept of probability in them" or "Batik motifs have geometric images in them" are followed with an explanation of how the concepts of probability and geometry exist in the local culture. Recognizing the use of local culture in developing students' cognitive abilities attributed to category U2. Expressions in this category such as "through playing kites students can understand the concept of kites..." The design specifically mentions that the local culture presented can help students to understand the concept of a kite's flat shape.

In our exploration, pre-service teachers demonstrated their ability to elucidate mathematical concepts inherent in local cultures. An exemplary instance is provided by Sukron (in this study we used pseudonyms), who proposes that the four cycles of calculations used by Javanese people offer a valuable foundation for introducing numerical concepts. Sukron further showcases this idea by incorporating the counting technique observed in the Sigeh Penguten dance within the arithmetic context. This creative integration falls under the U1 category, where the emphasis lies in the incorporation of local culture into mathematical instruction.

Similarly, Sari takes a distinctive approach by presenting the concept of geometric transformation within the realm of batik motifs. She effectively illustrates how the intricate patterns of batik can serve as a practical tool for helping students grasp the concepts of transformation geometry while simultaneously fostering critical thinking skills. The underlying beauty of this approach is the hidden mathematical patterns waiting to be uncovered within batik designs, providing students with a means to

Table 17.3 Usefulness

Category type	Criterion	Example
U1	Presents local culture with its use in learning mathematics	The Wayang game, if you pay attention, has the concept of probability in the game
U2	Presents the use of local culture related to learning mathematics and other cognitive abilities	Through playing kites students will be able to understand the concept. The concept of a kite can be introduced from how it is made, as shown below...and so on

explore and discover mathematical principles. This comprehensive description aligns with the U2 category, showcasing Sari's proficiency not only in identifying mathematical elements within local culture but also in detailing the cognitive benefits these approaches offer to students.

17.5.2 Creativity

The students' pedagogical design presentations also showed creativity as a value, which is shown by the existence of various mathematical concepts in local culture which are used as learning resources. Levenson (2022) suggests that challenge-based tasks exhibit a quality of flexibility, encompassing diverse methods for teacher responsibilities, which can foster creativity within the realm of mathematics. The diverse mathematical concepts that may be introduced via different local cultures underscores the exploration of these cultures by aspiring educators from multiple angles. The significance of creativity within instructional resources has the capability to reflect this array of ideas. Figure 17.2 has some examples of teaching material designs that incorporate local Indonesian cultural elements, specifically Batik traditional cloth and traditional games. The motifs found in Batik are used as practical resources for the teaching of various mathematical concepts, particularly in transformation geometry. Similarly, the Engklek game can be used to impart understanding of fundamental mathematical concepts, such as numbers, probability, and flat shapes.

	Banyumas batik is a type of batik that originates from the Banyumas region in Indonesia. Each region in Indonesia has its own unique style and meanings for batik. Banyumas batik is characterized by its dark background and prominent motifs of plants and animals. These motifs signify the community's close relationship with nature.
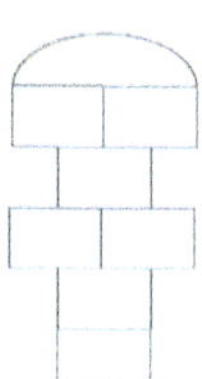	The shape of Engklek, in the form of a square, rectangle, and semicircle. The game is played alternately by multiple children. Each child jumps on one leg while playing. The Engklek involves counting as seen from the order of tiles that need to be traversed in hopscotch. This game is played by several people in turn and can be related to the topic of probability in mathematics.

Fig. 17.2 The value of creativity in the pedagogical design of batik and engklek games

We formulated several criteria for assessing creativity, as outlined in Table 17.4. In the pedagogical designs crafted by pre-service teachers, a significant acknowledgment of diverse mathematical concepts within a single local culture emerges, categorically denoted as C1. This recognition is evidenced through statements embedded in the teaching materials design, underscoring how pre-service teachers correlate a single local culture with multiple mathematical concepts. Phrases like "within batik motifs, students can explore numerous mathematical concepts" or "this game encapsulates various mathematical ideas" succinctly encapsulate this notion. It's worth noting that some pre-service teachers touch upon an array of mathematical concepts within a given local culture, but without providing an intricate description of the material's content.

Furthermore, the pedagogical designs explicitly highlight that there are abundant traditional games that can be intertwined with mathematical concepts, falling within the scope of category C2. Expressions found in this category resemble phrases such as "the Banyumas area boasts an array of local cultures infused with mathematical elements." This emphasis on diverse local cultural influences underpins the narrative. However, the presentation of non-routine problems stemming from local cultural activities stands out as a distinctive aspect of category C3. Expressions belonging to this category include statements like "this tradition can be effectively woven into a classroom story problem." The pedagogical design in this instance specifically outlines that the local culture presented serves as a story problem with the intent of engaging students in critical problem-solving.

We encountered diverse local cultures and a wide array of mathematical concepts within the pedagogical designs of pre-service teachers. For instance, C1, as described by Mufliani, highlighted the utilization of traditional batik motifs to introduce several mathematical concepts, including geometric patterns, comparisons, and geometric transformations. Meanwhile, Dita highlighted that the Lopis Giant Syawalan Tradition incorporates various mathematical concepts, such as comparisons, Pythagoras' theorem, flat shapes, and geometric forms. The incorporation of multiple local cultural concepts was emphasized in C2, with Novi noting that the Central Java region boasts a wealth of local cultures, each lending itself to specific mathematical

Table 17.4 Creativity

Category type	Criterion	Example
C1	Get to know various mathematical concepts in one local culture in a way	The motifs on batik cloth are not only geometric patterns but also calculate the ratio and location of the patterns
C2	Stating that there are many local cultures that can be related to mathematics	This region has many traditional games that can be related to mathematics
C3	Presenting non-routine issues from local cultural activities	The process of seeding rice plants can be used as a contextual problem

concepts. Fasitoh also shared that the Banyumas area hosts a rich diversity of local cultures, many of which encompass mathematical ideas.

In a different perspective, the contribution made by Arinal, categorized as C3, highlighted the Grebek Suran Tradition, where the community constructs tall conical structures adorned with the agricultural products of the community. In this context, Arinal proposed a mathematical challenge involving the calculation of the potential number of agricultural items that could be accommodated within the various possible cone shapes. This diversity of cultural and mathematical integration underlines the multi-faceted nature of pre-service teachers' pedagogical approaches.

17.5.3 Transformation

The transformational value seen in pedagogical design underscores the important role of local culture as a storehouse of valuable tools for learning mathematics. This concept is in line with the views of experts including Bishop who emphasize the importance of the relationship between teacher values and the development of student values. As Bishop (2001) emphasizes, the values upheld by educators significantly affect students, and in this context, the recognition of a teacher's local culture as a powerful medium for shaping students' independent knowledge construction becomes more relevant. Figure 17.3 shows an example of a pre-service teachers' pedagogical design showing gamelan musical instruments.

We then developed two criteria regarding transformation, as shown in Table 17.5. The design of pre-service teachers states that local culture, as a source of learning mathematics, is attributed to category T1. This is expressed through sentences in the design of teaching materials that show how pre-service teachers integrate certain local cultures to assist students in their learning. Expressions such as "shapes on this traditional musical instrument can help students" or "movement beats in dance can be a source of learning numbers" demonstrate this integration. There are also several pre-service teachers who mention mathematics in the local culture, which is attributed to category T2. This is depicted through sentences in the design of pre-service teachers' pedagogy, emphasizing how understanding mathematical concepts can contribute to understanding the workings of local culture. Expressions in this

Gamelan musical instrument is a Javanese culture that resembles geometric shapes. The relationship between musical instruments and mathematical concepts is a form of musical instruments that fulfils the geometrical concept. These shapes can be used as a tool for studying geometric

Fig. 17.3 The value of transformation in the pedagogical design of traditional musical instruments

Table 17.5 Transformation

Category type	Criterion	Example
T1	Presenting local cultural transformation as a learning resource	From here students can understand and learn how to arrange offerings so that they can be arranged like a mountain. This can be applied to the volume of a cone or finding the area of a cone's blanket
T2	Get to know the relationship between mathematical concepts and local culture	When playing Hompimpa we indirectly calculate the chances of winning

category include "if you master the mathematical concepts in it, you can use them in games, increasing your chances of winning".

We found the inherent value of local cultural integration within the mathematics pedagogy designed by pre-service teachers. A prime illustration is provided by T1, where Jumiatin emphasizes the potential of games as a valuable resource for mathematics learning. Fatah, on the other hand, draws attention to the geometric concept of a circle found in the placement of food during traditional Kenduri ceremonies. These instances, categorized under T1, demonstrate a recognition of the hidden educational benefits concealed within local cultural practices, thereby enriching the learning experience.

Furthermore, a noteworthy aspect emerges in the pedagogical designs, represented by T2 as exemplified by Rahma. Rahma points out the application of the volume formula for a cone to determine the amount of rice required for crafting the traditional Tumpeng dish. This perspective goes beyond the immediate mathematical aspect, extending into the culinary domain, and reinforces the notion that local culture carries a wealth of multifaceted insights. Such designs encompassing the underlying value of local culture pave the way for an engaging and holistic approach to mathematics education, promoting a deeper appreciation for both the subject matter and the cultural context.

17.6 Discussion and Conclusion

The diversity of local cultures owned by the Indonesian people is an interesting context to study in learning mathematics. There are three main findings regarding values that appear in the designs of pre-service teachers' pedagogy. The value of usefulness appears in the designs marked by a statement regarding the functions, benefits, and uses of the local cultural identification presented. Parhizgar and Liljedahl (2019) suggest usefulness is related to motivation. Thus, pre-service teachers present this usefulness so that students understand the benefits of local culture which can increase motivation in learning mathematics. Intrinsic motivation according to Ryan and Deci (2000) arises because of the interest in an activity and extrinsic motivation is related to the useful value of the activity itself. The local cultural context which is then used as a model in learning mathematics can increase the value of the usefulness of mathematics (Niss et al., 2007). Usefulness can be related to the benefits of the local culture itself and its use in learning mathematics.

Local culture is born from a process of reflection carried out by individuals and communities on the situation, dynamics, and changes that occur in the surrounding environment. Such responses manifest in diverse forms, molded by the capacities and experiences of both individuals and the entire community. Fang et al. (2022) highlighted the profound influence of community culture on its collective expression, a manifestation that holds significant value for the society that upholds it. This study suggests that the pedagogical designs of the pre-service teachers incorporate usefulness of local culture in mathematics learning, aligning with the discoveries of Hunter (2021), who emphasized the pivotal role of utility in mathematics education. This value underscores the practical application of mathematics in daily life.

Creative values were found in the pedagogical designs of the pre-service teachers. Levenson (2022) suggests that creativity in mathematics is characterized in several ways that use non-algorithmic decision making from different and flexible thinking that allows one to use many ways and perspectives in solving problems. The pedagogical designs contained values of creativity because it was the pre-service teachers who showed that a local culture can have various mathematical concepts as learning resources. Schoevers et al. (2018) explains that creativity can provide a different perspective for someone in analyzing problems by looking at patterns, differences, and similarities, and generating several ideas and choosing the right method to deal with mathematical situations. The local culture around pre-service teachers is an interactive environment which according to Neumann (2007) is a characteristic of an effective environment for fostering creativity.

Culture is creative because it is produced in response to changes in social and environmental situations which can then have implications for deficits or even dysfunction of previous constructions. Creativity can be in the form of new constructions or innovations over previous forms to be more responsive, adaptive, and functional to help subjects give meaning to their surroundings. Cultural creativity is ideological in nature, following the interests of the subjects and will continue to experience parallel dynamics with increasing needs (Geertz, 1973).

Creativity comes from the subject's imagination on the limitations of cultural products that experience deficits and dysfunctions. Individuals and groups have new needs that must be met by cultural instruments. The nature of culture is technological, or the ability to provide solutions through technological approaches (Giddens, 1984). Thus, when a cultural product is in deficit or anomaly, it will be replaced by a new construction through a creative process. Therefore, cultural creativity is essential and will continue to occur together with the increasing needs of individuals and groups. The creativity that emerges within a society eventually influences the learning environment of students.

One particularly influential aspect is mathematical creativity, which plays an essential role in promoting cognitive development among students. Schoevers et al.

(2018) argue that when students proactively search for innovative solutions, they engage their logical, analytical, and problem-solving skills. These skills confer numerous benefits that can be applied to various aspects of life. This finding is consistent with Zhang's (2019) research, which found that Chinese students place a high value on achievement. This is evident in their ability to present a variety of problem-solving methods, which is a key indicator of creativity. This emphasizes the significance of fostering creative thinking as a crucial aspect of education.

The value of transformation in the pedagogical schemes of pre-service teachers shows changes in the function of local culture as a means of learning mathematics that can be utilized by students. In contrast, students master mathematical concepts and can then apply them in culture. As Filho et al. (2018) call it a transformative pedagogy, where in learning, there is a transformational value brought by educators. Transformation leads to continuous learning by making connections with local culture and learning topics. In learning mathematics, this value is realized by connecting mathematical concepts with their application in everyday life and cultural traditions. This aims to enrich the learning experience of students.

The value of transformation within the framework of pre-service teachers' pedagogy signifies a shift in the role of local culture, transforming it into a potent tool for facilitating the learning of mathematics, which students can actively harness. This concept aligns with what Levy and Kerpelman (2010) term as transformative pedagogy, placing a special emphasis on the active involvement of students in the construction of knowledge. This pedagogical approach is characterized by the integration of local cultural elements that resonate with students, providing them with meaningful contexts. Through such design, students gain the opportunity to develop new epistemological perspectives, exploring diverse ways of understanding, and cultivating a heightened awareness for the process of learning (Lopez & Olan, 2018). Incorporating local culture in mathematics instruction, the pedagogical transformation instills a sense of relevance and relatability, enhancing students' engagement and fostering deeper comprehension. This approach recognizes the potency of culture as a dynamic conduit for shaping students' cognitive pathways, encouraging them to view mathematical concepts within the context of their own lived experiences. Consequently, this transformative engagement with local culture not only empowers students as active participants but also broadens their cognitive horizons, reinforcing the multifaceted nature of the learning process and enriching their overall educational experience.

Pre-service teachers' designs of lessons is crucial because it markedly influences the educational experience of students. It is known that the values held by teachers play a significant role in shaping the knowledge and perspectives that students eventually acquire (Bishop, 2001). This is because instilling these values in pre-service teachers is not instantaneous, indeed it requires persistent effort and hands-on experience, particularly in teaching mathematics, a subject known for its challenges and nuances.

Studying the teaching methods suggested by these pre-service teachers in Indonesia, it is evident that they prioritize three key principles. Firstly, they prefer the local culture to render mathematics more practical and beneficial for students, with a strong emphasis on the importance of usefulness. This approach helps students grasp the relevance of mathematical concepts in real-life situations, making learning more meaningful and engaging. Secondly, these pre-service teachers actively explore diverse mathematical ideas embedded within Indonesian culture, those, fostering an environment that stimulates creative thinking among students. This promotes a deeper understanding of both mathematics and our cultural heritage. Lastly, the innovative pedagogical designs employed by these pre-service teachers showcases how our culture can evolve and be a powerful tool for learning, thus highlighting the value of transformation. This multifaceted approach not only enriches students' mathematical cognitive abilities but also demonstrates the dynamic nature of our cultural resources in the educational context, providing an effective and holistic learning experience.

References

Ball, D., Thames, M., & Phelps, G. (2008). Content knowledge for teaching what makes it special? *Journal of Teacher Education, 59*(5), 389–407.

Bedewy, S. E., Lavicza, Z., Haas, B., & Lieban, D. (2022). A steam practice approach to integrate architecture, culture, and history to facilitate mathematical problem-solving. *Education Sciences, 12*(1). https://doi.org/10.3390/educsci12010009

Berger, P. L., & Luckmann, T. (1991). *The social construction of reality*. Penguin Books.

Bishop, A. J. (1988). *Mathematical enculturation: A cultural perspective on mathematics education*. Kluwer Academic Publishers.

Bishop, A. J. (2001). What values do you teach when you teach mathematics? *Teaching Children Mathematics, 7*(6), 346–349.

Bourdieu, P. (1993). *The field of cultural production*. Columbia University Press.

Braun, V., & Clarke, V. (2006). Using thematic analysis in psychology. *Qualitative Research in Psychology, 3*(2), 77–101.

Braun, V., & Clarke, V. (2020). One size fit all? What counts as quality practice in (refexive) thematic analysis? *Qualitative Research in Psychology, 18*(3), 328–352.

Cairns, J. (2000). Morals, ethics, and citizenship in contemporary teaching. In R. Gardner, J. Cairns, & D. Lawton (Eds.), *Education for values* (pp. 6–23). Kogan Page.

Chi, M. T. H. (2005). Common sense conceptions of emergent processes: Why some misconceptions are robust. *The Journal of the Learning Sciences, 14*, 161–199.

Chronaki, A. (2000). Teaching maths through theme-based resources: Pedagogic style, 'theme' and 'maths' in lessons. *Educational Studies in Mathematics, 42*, 141–163. https://doi.org/10.1023/A:1004124328308

Closs, L., Mahat, M., & Imms, W. (2022). Learning environments' influence on students' learning experience in an Australian faculty of business and economics. *Learning Environments Research, 25*, 271–285. https://doi.org/10.1007/s10984-021-09361-2

Daher, W. (2020). Values in the mathematics classroom. *Educational Philosophy and Theory, 52*(3), 284–299. https://doi.org/10.1080/00131857.2019.1618276

Dede, Y. (2013). Examining the underlying values of Turkish and German mathematics teachers' decision-making processes in group studies. *Educational Sciences: Theory and Practice, 13*(1), 690–706.

Dede, Y., Akçakın, V., & Kaya, G. (2021). Mathematical, mathematics educational, and educational values in mathematical modelling tasks. *ECNU Review of Education, 4*(2), 241–260.

Denzin, N. (2006). *Sociological methods: A sourcebook* (5th ed.). Aldine Transaction.

Fang, X., Sauter, D. A., Heerdink, M. W., & van Kleef, G. A. (2022). Culture shapes the distinctiveness of posed and spontaneous facial expressions of anger and disgust. *Journal of Cross—Cultural Psychology, 53*(5), 471–487. https://doi.org/10.1177/00220221221095208

Filho, W. L., Raath, S., Lazzarini, B., Vargas, V. R., et al. (2018). The role of transformation in learning and education for sustainability. *Journal of Cleaner Production, 199*, 286–295. https://doi.org/10.1016/j.jclepro.2018.07.017

Foucault, M. (1972). *The Archaeology of knowledge and the discourse on language*. Pantheon Books.

Freire, P. (2005). *Pedagogy of the oppressed*. Continuum.

Gadamer, H. G. (1976). *Hegel's dialectic, five hermeneutical studies*. Yale University Press.

Geertz, C. (1973). *The interpretation of cultures*. Basic Book.

Giddens, A. (1984). *The constitution of society*. Polity Press.

Haines C., & Crouch R. (2007). Mathematical modelling and applications: Ability and competence frameworks. In W. Blum, P. Galbraith, H. Henn & M. Niss (Eds.), *Modelling and applications in mathematics education* (pp. 417–424). Springer. https://doi.org/10.1007/978-0-387-29822-1_46.

Harriyadi. (2020). The meaning of winged śaṅkha ornaments in Hindu and Buddhist temples in java. *Purbawidya, 9*(20), 113–130. https://doi.org/10.24164/pw.v9i2.377

Herbst, P., Nachlieli, T., & Chazan, D. (2011). Studying the practical rationality of mathematics teaching: What goes into "installing" a theorem in geometry? *Cognition and Instruction, 29*(2), 218–255.

Hernandez-Martinez, P., & Vos, P. (2018). "Why do I have to learn this?" A case study on students' experiences of the relevance of mathematical modelling activities. *ZDM Mathematics Education, 50*, 245–257. https://doi.org/10.1007/s11858-017-0904-2

Hill, J. L., Kern, M. L., Seah, W. T., & van Driel, J. (2021). Feeling good and functioning well in mathematics education: Exploring students' conceptions of mathematical well-being and values. *ECNU Review of Education, 4*(2), 349–375.

Hunter, J. (2021). An intersection of mathematics educational values and cultural values: Pāsifika students' understanding and explanation of their mathematics educational values. *ECNU Review of Education, 4*(2), 307–326.

Iliev, N., & D'Angelo, F. (2014). Teaching mathematics through multicultural literature. *Teaching Children Mathematics, 20*(7), 452–457.

Jasper, J. M. (2016). *Protest: A cultural introduction to social movements*. Polity Press.

Kalogeropoulos, P., Russo, J. A., & Clarkson, P. (2021). Exploring educator values alignment strategies in an intervention context: The emergence of the Beacon strategy. *ECNU Review of Education, 4*(2), 327–348.

King, N. (2004). Using templates in the thematic analysis of text. In C. Cassell & G. Symon (Eds.), *Essential guide to qualitative methods in organizational research* (pp. 257–270). Sage.

Lawton, D. (1999). *Values and the curriculum: A curriculum for the 21st century*. Woburn Press.

Leung, F. K. S., Graf, K. D., & Lopez-Real, F. J. (Eds.) (2006). *Mathematics education in different cultural traditions-A comparative study of East Asia and the West*. Springer. https://doi.org/10.1007/0-387-29723-5_1

Levenson, E. S. (2022). Exploring the relationship between teachers' values and their choice of tasks: The case of occasioning mathematical creativity. *Educational Studies in Mathematics, 109*, 469–489. https://doi.org/10.1007/s10649-021-10101-9

Levy, M. K. H., & Kerpelman, J. L. (2010). Identity process and transformative pedagogy: Teachers as agents of identity formation. *An International Journal of Theory and Research, 10*(1), 76–91. https://doi.org/10.1080/15283481003711684

Lim, C. S., & Ernest, P. (1997). Values in mathematics education: What is planned and what is espoused? In *British Society for Research into Learning Mathematics (BSRLM)* (pp. 37–44). Proceedings of the Day Conference held at the University of Nottingham, 1 March.

Lopez, A. E., & Olan, E. L. (2018). *Transformative pedagogies for teacher education: Moving towards the praxis in an era of change*. Information Age Publishing Inc.

Malloy, C. E., & Malloy, W. W. (1998). Issues of culture in mathematics teaching and learning. *The Urban Review, 30*, 245–257. https://doi.org/10.1023/A:1023261002882

McLean, D. L. (2002). Honouring traditions: Making connections with mathematics through culture. *Teaching Children Mathematics, 9*(3), 184–188.

Neumann, C. J. (2007). Fostering creativity—A model for developing a culture of collective creativity in science. *EMBO Reports, 8*(3), 202–206.

Niss, M., Blum, W., & Galbraith, P. (2007). Introduction. In W. Blum, P. Galbraith, H.-W. Henn, & M. Niss (Eds.), *Modelling and applications in mathematics education* (pp. 1–32). Springer. https://doi.org/10.1007/978-0-387-29822-1_59

Parhizgar, Z., & Liljedahl, P. (2019). Teaching modelling problems and its effects on students' engagement and attitude toward mathematics. In S. Chamberlin & B. Sriraman (Eds.), *Affect in mathematical modeling. Advances in mathematics education* (pp. 235–256). Springer. https://doi.org/10.1007/978-3-030-04432-9_15

Ryan, R. M., & Deci, E. L. (2000). Intrinsic and extrinsic motivations: Classic definitions and new directions. *Contemporary Educational Psychology, 25*, 54–67. https://doi.org/10.1006/ceps.1999.1020

Schoevers, E. M., Kroesbergen, E. H., & Kattou, M. (2018). Mathematical creativity: A combination of domain-general creative and domain-specific mathematical skills. *The Journal of Creative Behaviour, 54*(2), 242–252.https://doi.org/10.1002/jocb.361

Seah, W. T. (2003). The professional socialisation of teachers in transition: A values perspective. *Paper presented at the International Education Research Conference AARE-NZARE*, Auckland, New Zealand.

Seah, W. T. (2019). Values in mathematics education: Its conative nature, and how it can be developed. *Research in Mathematical Education, 22*(2), 99–121.

Seah, W. T., & Bishop, A. J. L. (2003). Values, mathematics, and society: Making the connections. *Prime Number, 18*(3), 4–9.

Seah, W. T., & Wong, N. Y. (2012). What students value in effective mathematics learning: A 'Third Wave Project' research study. *ZDM Mathematics Education, 44*(1), 33–43.

Seah, W. T., Zhang, Q., & Bishop, A. J. (2021). Mentors expressing what they value through their writings: Emphasizing the person in mathematics. *ECNU Review of Education, 4*(2), 230–240.

Secombe, M. J. (2016). Core values and human values in intercultural space. *Politeja, 44*, 265–276. http://www.jstor.org/stable/24920306

Shulman, L. (1987). Knowledge and teaching foundations of the new reform. *Harvard Educational Review, 57*(1), 1–21.

Silverman, J., & Thompson, P. W. (2008). Toward a framework for the development of mathematical knowledge for teaching. *Journal Mathematics Teacher Education, 11*, 499–511. https://doi.org/10.1007/s10857-008-9089-5.

Smith, P. (2001). *Cultural theory*. Blackwell Publishing.

Steffe, L. P., & Kieren, T. (1994). Radical constructivism and mathematics education. *Journal for Research in Mathematics Education, 25*, 711–733.

Totterdell, M. (2000). The moralization of teaching: A relational approach as an ethical framework in the professional preparation and formation of teachers. In G. Gardner, J. Cairns, & D. Lawton (Eds.), *Education for values*. Kogan Page.

Wæraas, A. (2022). Thematic analysis: Making values emerge from texts. In G. Espedal, B. Jelstad Løvaas, S. Sirris, & A. Wæraas (Eds.), *Researching values*, (pp. 153–170). Palgrave Macmillan. https://doi.org/10.1007/978-3-030-90769-3_9

Zhang, Q. (2019). Values in mathematics learning: Perspectives of Chinese mainland primary and secondary students. In P. Clarkson, W. T. Seah, & J. Pang (Eds.), *Values and valuing in mathematics education: Scanning and scoping the territory* (pp. 185–196). Springer. https://doi.org/10.1007/978-3-030-16892-6_13

[illegible] (2008). [illegible]. https://doi.org/10.[illegible]
[illegible] (2018). [illegible] & [illegible]. *[illegible]*.
[illegible] (2012). The [illegible]. https://doi.org/10.[illegible]
[illegible] (2016). [illegible]. *[illegible]*.
[illegible] (2013). [illegible]. *[illegible]*.
[illegible] (2020). [illegible] mathematics [illegible]. *[illegible]*.
[illegible] (2017). [illegible]. https://doi.org/10.[illegible]
[illegible] (2021). [illegible]. *[illegible]*.

Chapter 18
Building on Mathematics Educational Values to Develop Culturally Sustaining Mathematics Pedagogy

Jodie Hunter

18.1 Introduction

Both in New Zealand and internationally, diverse groups of people including Indigenous, migrant, and other minority communities are under-represented in mathematics in higher education settings and employment. In many countries, there is an ongoing narrative of under-achievement and achievement 'gaps' for marginalised students (Faulkner, et al., 2019; Turner et al., 2015). In the case of Aotearoa/New Zealand, there is a growing population of Pacific people who come from the island nations within the Te Moana-nui-a-Kiwa (Pacific Ocean region). This group is composed of multiple generations as well as being comprised of diverse groups of Indigenous people, each with their own language and cultural and social ways of being (Coxon et al., 2002). Despite the heterogenous nature of Pacific people, a communal approach to society shapes all their cultural values and provides a common thread between people from different Pacific nations (Podsiadlowski & Fox, 2011; Uehara et al., 2018).

Equity in schooling can only be achieved when educators explicitly connect to and build on the cultural, social, and linguistic contexts, including values of non-dominant students, their families, and community (Averill & Rimoni, 2019; Sum et al., 2022). While Pacific students in New Zealand enter schooling with deep and rich backgrounds of understandings and experiences, in mathematics classrooms they encounter structural inequities that cause cultural and social dissonance. At the centre of this cultural and social dissonance are the more individualistic approaches and values which have shaped the schooling practices used in New Zealand. These types of approaches are 'assumed as normal for everyone' practices but align with the values of a middle-class Pākehā (European) population while positioning Pacific students in a less favourable way. This has a consequence of disengagement with

J. Hunter (✉)
Massey University, Palmerston North, New Zealand
e-mail: j.hunter1@massey.ac.nz

Y. Dede et al. (eds.), *Values and Valuing in Mathematics Education*,
https://doi.org/10.1007/978-981-99-9454-0_18

mathematics and a view of the subject as having little relevance in their lives, or as something unattainable (Barton, 1995; Hunter, 2021; Hunter & Hunter, 2018; Thaman, 2005).

In the New Zealand setting, key educational policy initiatives have focused on developing pedagogical practices that build on the values of Pacific people as a means to address disparity in education and support the development of a more equitable education system (Ministry of Education (MoE), 2013, 2018). This policy implementation in the context of mathematics education requires both understanding of Pacific values and how these can be integrated into mathematics teaching and learning. A growing body of research both in general education and mathematics education has drawn on the perspectives of Pacific people, including researchers, parents, and students to examine perceptions of values and how they influence teaching and learning (e.g., Armstrong et al., 2021; Hill et al., 2019; Hunter, 2021; Hunter et al., 2016; Rimoni et al., 2022; Uehara et al., 2018). However, there has been limited research which explicitly examines teacher actions that align with the mathematics educational values of marginalised students. The aim of this chapter is to use an exemplary case study to analyse the ways in which an educator used pedagogical approaches that built on mathematics educational values of Pacific students as a strength in the mathematics classroom. Specifically, the chapter will address the gap in the field with the research question: *what pedagogical approaches can be used by educators that build on the mathematics educational values of diverse students to develop culturally sustaining pedagogy in mathematics classrooms?*

18.2 Culturally Sustaining Mathematics Pedagogy

Mathematics as a school subject has been critiqued for using a one size fits all approach with standardised curricula and testing. Compounding this is the typical positioning of mathematics as a value and culture free subject (Leonard et al., 2010; Presmeg, 2007). In contrast, those advocating for culturally sustaining mathematics pedagogy contend that mathematics education is inextricably connected to the sociocultural context of the classroom and that learning is shaped by cultural practices (Hunter & Hunter, 2018; Nasir et al., 2008; Nasresh & Kasmer, 2018). Developing Culturally Sustaining Mathematics Pedagogy (CSMP) is a foundational element in developing more equitable and inclusive mathematical learning environments (Abdulrahim & Orosco, 2020). The definition of CSMP used in this chapter draws on the seminal work of Gay (2010), Ladson-Billings (1995, 2006), Paris (2012), and Paris and Alim (2014) in what has been termed culturally relevant, culturally responsive, and culturally sustaining pedagogy. The following sections will focus on the key aspects of CSMP that inform and guide this study and analysis.

Culturally sustaining mathematics pedagogy involves educators reframing how they think about pedagogical practices and reflecting on their own theories which underpin the practices that they use in the mathematics classroom (Ladson-Billings,

2006). Using CSMP to address equity requires educators to draw upon a multi-dimensional approach and consider access, achievement, identity, and power and to de-centre dominant taken as normal pedagogical practices. A focus on academic achievement means empowering learners to realise and reach their potential for high levels of achievement which also provides learners with access to mathematics (Gay, 2010; Ladson-Billings, 1995, 2006). Educators with cultural competence honour the cultural heritage of their students (identity) and incorporate this into all aspects of the classroom (Gay, 2010). To consider power, educators need to create structures for learners to recognise, understand, and critique social inequalities (Ladson-Billings, 1995). Gutiérrez (2012) contends that in both research and practice, access and achievement (as an indication of equity) have been readily accepted dimensions, however, identity and power have been neglected which potentially creates an overly simplified framing of CSMP. Consequently, a key element of culturally sustaining mathematics pedagogy is broadening the focus of educators to consider identity and power relationships as well as access and achievement in the mathematics classroom.

Using culturally sustaining mathematics pedagogy as a framing means that we acknowledge that students' mathematical experiences in the classroom significantly impact their views of themselves in relation to mathematics (Thomas & Berry, 2019). This type of pedagogy requires educators to understand the contexts of their students as people within a cultural group and community and consider their lives, experiences, and backgrounds (Hunter & Hunter, 2018; Nasir et al., 2008). As Thomas and Berry (2019) contend relationships are at the heart of this and mathematics teachings can be framed as "eliciting shared frames of references to make meaningful connections between teaching and the cultures, lives, and experiences of learners" (p. 22). In this chapter, we also consider mathematics educational values that are embedded within classroom practices as an element to consider when working to develop CSMP.

Mathematics educational values vary depending on the culture of the learner. Previous research has shown that there is a link between cultural values and mathematics educational values (Hunter, 2021; Zhang, 2019). A clash of values or misalignment between a teacher and student values during mathematics teaching and learning potentially causes dissonance and disengagement with mathematics (Kalogeropoulos et al., 2021). In contrast, alignment between what teachers' value and what students' value has the potential to strengthen relationships. This does not mean that all members of the classroom community have to share the same values but instead that the values of all are aligned and in harmony (Seah & Andersson, 2015). This requires teachers to understand student values and what these may look like from the students' point of view in the context of a mathematics classroom.

18.2.1 Pacific People and Mathematics Educational Values

Across different cultural groups, there are varying levels of collectivist or individualistic orientations in relation to ways of being and living (Hofstede, 2011). Hofstede (2011) explains individualist cultures as those where people are expected to look after themselves and their immediate family, in contrast, collectivist cultures integrate people from birth into strong cohesive groups. For Pacific people their cultural values are shaped by a collectivist way of life and society. This means that notions of fono (family as an extended grouping rather than nuclear) are central to how values are considered. In a Pacific framing, fono is important and at the centre of everything that Pacific people do within their lives (Civil & Hunter, 2015; Hunter et al., 2016; Rimoni et al., 2022). Woven within the understanding of what it means to be a member of fono are values including belonging, love, service, spirituality, reciprocal relationships, respect, inclusion, and leadership (MoE, ; Rimoni et al., 2022). While it may be considered that these values would apply to most countries and cultural groups, there are many subtle and obvious differences between how they are enacted by those who have a collectivist or individualistic orientation (Averill & Rimoni, 2019; Hunter, 2021; Zhang, 2019). Importantly, Hunter (2021) found that there were many intersections between Pacific students' cultural values and their ranking and understanding of mathematics educational values.

Mathematics educational values relate specifically to learning and teaching of mathematics as a subject. Only three previous studies (Hill et al., 2019; Hill & Hunter, 2023; Hunter, 2021) have explicitly focused on the mathematics educational values reported by Pacific students themselves in New Zealand schools, however, other studies from New Zealand have examined values through investigating student perspectives of mathematical learning (Anthony, 2013; Sharma et al., 2011) or through analysing how teachers draw on cultural values during mathematics lessons (Hunter & Anthony, 2011; Hunter & Miller, 2022). With fono as a core cultural and mathematics educational value for Pacific people, earlier studies have shown that Pacific students related the importance of help and support from their fono while also striving to achieve to make fono proud (Hunter, 2021). Other studies (Hunter & Anthony, 2011; Hunter & Miller, 2022) have demonstrated how teachers can draw on the wider definition of fono as a community in mathematics lessons and position students to treat other students as fono. Aligned with the concept of fono are the mathematics educational value of collaboration and related aspects of inclusion and belonging. Hill et al. (2019) illustrated that Pacific students linked collaboration with reciprocity and saw group-work as providing reciprocal learning opportunities and were founded on concepts of inclusion and belonging.

Respect is a key cultural value for Pacific people and one identified as a highly ranked mathematics educational value by Pacific learners (Averill & Clark, 2012; Hill & Hunter, 2023; Hunter, 2021). In a study set in secondary classrooms, Averill and Clark (2012) found that Pacific students perceived their teachers as demonstrating respect when they expressed high expectations for the students. Other research literature (Hunter, 2021; Rimoni ct al., 2022) illustrates that for Pacific people, respect is part of a reciprocal relationship so this translates in the classroom to a view that it is important for a Pacific student to respect the teacher and this will in turn lead to respect from the teacher.

18.3 Methodology

This chapter draws on an exemplary case study (Hannula, 2002) from a larger project focused on documenting the mathematical funds of knowledge of Pacific children and fono in New Zealand and exploring how funds of knowledge can be used as a basis for task design and enactment. An exemplary case study was chosen as this is an appropriate methodology to undertake an in-depth examination of the pedagogical actions that teachers can use that build on students' mathematics education values to develop culturally sustaining mathematics pedagogy.

For this chapter, the focus is on one teacher and a lesson involving a group of ten students from her class of Year Five and Six students (aged between eight years and 10 years old). The teacher was an experienced teacher of Pākehā (New Zealand European) heritage while the students were predominantly of Pacific heritage with one student who was Pākehā. The school had also been involved in a transformative school based professional learning and research approach called Developing Mathematical Inquiry Communities for the three years preceding the data collection (for more detail see Hunter, et al., 2018). This initiative is focused on providing more equitable outcomes for Pacific students and is set within a culturally sustaining model (Paris, 2012) while drawing on the tenets of ambitious mathematics (Kazemi, Franke, & Lampert, 2009). The leaders of this work (one an author of this chapter) are New Zealand born of Kuki Airani (Cook Island) heritage. They draw on the Pacific value of fono in leading this work as mother and daughter.

In the lesson presented in this chapter, a task developed from data collected during the photo elicitation interviews undertaken as part of the larger project was used. Specifically, the photo shared with the research team showed children helping out with chores around the house. This was used as the basis of a statistics lesson collaboratively planned with the teacher and researchers. The task was the focus of a 45–60 min lesson that drew on a three phase lesson structure: launch, explore, and summary. At the beginning of the lesson, the teacher arranged the students to work in small groups of three to four to complete the task. During the lesson, the teacher moved between the groups observing their interactions and solution strategies and selecting the contributions to be discussed in the summary phase of the lesson. Although the lesson was predominantly taught in English, both the teacher and the students interchangeably used te reo Māori (the Māori language) in parts of the lesson. The lesson was video recorded both during the whole class and paired work to document the teaching and learning interactions. The video-recordings of the lessons were transcribed and analysed to identify themes.

Data analysis included a mixed deductive and inductive approach. All data was coded by the author of the chapter. Multiple levels of coding were used which focused on both pedagogical practices and student engagement in learning activities. Firstly, the researcher undertook initial deductive coding focused on analysis of the transcript of the video-recorded lesson in relation to mathematics educational values linked to cultural values previously identified in research literature (e.g., Hill et al., 2019; Hill & Hunter, 2023; Hunter, 2021; Hunter & Miller, 2022). This included fono (family),

Table 18.1 Coding indicators for mathematics education values and related teacher actions

Mathematics education value	Teacher actions
Fono (family) (Hill et al., 2019; Hill & Hunter, 2023; Hunter, 2021)	Use of humour Connecting to student experiences in the home Indicating respect for family values
Respect (Hill & Hunter, 2023; Hunter, 2021)	Demonstrating high expectations for students Explicitly identifying helpful ideas from students Affirming questioning and mathematical argumentation
Collaboration and reciprocity (Hill et al., 2019; Hunter & Miller, 2022)	Positioning students to describe norms for collaborative group-work Noticing and affirming collaborative ways of working Asking students to reflect on the mathematical ideas of others Facilitating mathematical argumentation Reinforce expectations of collaborative ways of working
Inclusion and belonging (Hunter & Miller, 2022)	Use of humour Inclusive language such as "we" and "our" Encourage student questioning in a safe learning environment

respect, and the inter-connected values of collaboration, belonging, reciprocity, and inclusion. Following this, the researcher undertook a second layer of both deductive and inductive coding to examine the specific teacher actions that made implicit and explicit connections to the mathematics educational values identified in the first layer of coding. This included analysing the subsequent student responses and ways of participating. Table 18.1 provides examples of the indicators and coding adopted for both the mathematics education values related to Pacific cultural values and the associated teacher actions.

The use of multiple levels of coding supported a focus on the connections that the teacher made to students' cultural values and mathematics educational values, and analysis of how this aligned with the development of CSMP in the classroom. The codes were used to examine, cluster, and integrate emerging themes.

18.4 Finding and Discussion

In this section, a lesson in three parts (launch, explore, summary) is presented as an exemplary case study of a teacher using a pedagogical approach that builds on the mathematics educational values of her predominantly Pacific students to develop culturally sustaining pedagogy in her mathematics classroom. The teacher used a task (see Fig. 18.1) collaboratively developed with the research team.

Nirvana and Nora show aroha (love) at home by helping with the chores. Here is a set of data that shows how many minutes some tamariki (children) in the school spend doing chores in the weekend.

Minutes:
0, 7, 8, 6, 5, 4, 23, 31, 12, 10, 13, 12, 5, 28, 16, 24, 30, 45, 19, 55

Put the data into a stem and leaf graph.

Read the graph, what does it say about tamariki doing chores at home?

Make statements.

Fig. 18.1 Statistics task

18.4.1 Launch of the Lesson

The lesson began with the teacher connecting the students to the overall area of focus in mathematics. A student responded to the initial prompt from the teacher with a humorous response referring to a social focus of collaboration rather than a mathematical focus although it appeared that the student was not intending to be funny. Rather than reprimanding or correcting the student, the teacher herself began to joke with the larger group of students. In this way, she also removed the possibility of the student feeling embarrassed by their unintentionally funny response:

T: "What's our focus in maths this term?"

S1: "Share the pen".

All of the students in the group begin to laugh.

T: (laughing) "What is the special focus? The word I can't say, sta-stick-ticks."

S2: "Sta-stick-ticks".

T: "Now he said it the way I say it because I can't say it".

SS: "Statistics, statistics".

In this initial part of the lesson, the teacher cultivated a sense of belonging and fono (family) in the classroom both through her own use of humour and by allowing the students to express themselves through humour. Previous studies (Civil & Hunter, 2015) have shown how social chat along with good natured humour and laughter are a way of developing a safe learning environment for Pacific learners and defusing tension. In this case, the teacher built on a spontaneous moment of humour from her students to connect with the mathematics educational values of belonging and family in cultivating a connected and safe classroom environment.

The teacher then re-oriented the students to the specific focus of the lesson and engaged them in a discussion about the ways in which they worked in the mathematics classroom:

T: "It's a little bit tricky… it's a new type of graph that we need to learn how to read. What can we do to show aroha (love) to our group today in maths?"

S3: "By including others in our ako (to both teach and learn) but include others that are out of the group".

T: "Absolutely".

S3: "And ask if they understand, if they understand a question".

T: "So that would be a good chance for you to go to Tasa and say "hey Tasa do you.

understand the question?… Lachlan, how can we help everyone succeed?".

S4: "By making sure they are participating and working together".

T: "Does succeeding mean getting everything right?".

S4: "No, it means try your best".

T: "That was beautiful. What did he say?".

S1: "Try your best".

T: "Trying your best, you always try your best".

S5: "I try my hardest".

In this section, the teacher launched the task by acknowledging that there would be challenge in the lesson because it involved new learning for the students. This action aligns with how Pacific students perceive their teachers as demonstrating respect through high expectations (Averill & Clark, 2012). We see the teacher showing the value of respect for the students by framing the task as challenging but as something that they are capable of learning. As the teacher continued to set up the social norms for the students to work together, she also used prompts which provided students with opportunities to bring to the surface their own values of inclusion and reciprocity (Hunter & Miller, 2022; Rimoni et al., 2022). The teacher wove the student values into the launch of the lesson as a way of setting up the norms for how students can engage mathematically. Throughout this interchange, the teacher consistently used inclusive language, for example, "how can **we** help **everyone** succeed?" and in these ways aligned her practice with the values that the students expressed (Seah & Andersson, 2015). This provides an example of how a teacher can engage and align with values and ways of being that relate to student identity and integrate these into classroom practice.

The teacher then began to launch the task by explicitly connecting to the context of the task while also asking the students to share their experiences with doing tasks. In doing this, she also acknowledged differences across families and positioned the students to consider cultural values in relation to chores.

T: "There is so many brilliant ideas but we have to have one person talking at a time. Fogasavaii first".

S6: "I have to take out the trash".

T: "You take out the rubbish, okay, and do you get anything for doing that?"

S6: "No".

T: "No, it is just your way of showing manakitanga (care and love for others).

In this excerpt of the lesson, we see the teacher further building on Pacific understandings of the value of respect and fono by incorporating and explicitly connecting the context of the mathematical task to student experiences in the home and their wider identity (Averill & Clark, 2012).

Following this, the teacher oriented the students to the mathematical aspects of the task and checked their knowledge of a stem and leaf graph. After establishing that this was a new graph for the students, she carefully explained and modelled the process of developing a stem and leaf graph and then reading the data presented on a stem and leaf graph. She finished her explanation with further directions: "I really want you to focus on putting this data into a stem and leaf but then writing statements because remember all numbers and all data tells us stories. What stories can we learn from this data?". The teacher then set the students to work collaboratively in a small group: "I love the aroha and enthusiasm Raevyn has for his group. Tiana is showing aroha, she is waiting patiently". In this way, the teacher reinforced the expectation of working collaboratively with others in an inclusive way.

Throughout this section focusing on the launch of the task, examples have been provided that show how the teacher demonstrated both her care for the students' values and made it a safe environment for students to put their values at the centre while also engaging in collective mathematical learning. In this way, we see the teacher using pedagogical approaches through her prompts and structuring of group-work that built on the mathematics educational values of her students. In turn, these approaches developed CSMP as the teacher showed her understanding of her students as people within a cultural group and community and connected her mathematical teaching to their ways of being as Pacific learners (Nasir et al., 2008; Thomas & Berry, 2019).

18.4.2 Small Group Work During the Lesson

During the small group work, the teacher moved between the groups of students who were working on the mathematical task. At various points during this phase of the lesson, she interacted with the groups of students to support their learning and position them to engage in mathematical discussion.

T: "Okay Sam…why have you chosen six as the largest number" [referring to stem].

S7: "It's five".

T: "Timo, you think it's five. Why do you think it is five?"

S8: "Because 55 is the highest".

T: "Do you guys agree with that? [pause]… Who did zero minutes of chores? That's what.

I want to know!"

S1: "Me sometimes" [laughing].

The teacher moves to the next small group of students and listens to their discussion and then facilitates them to engage in working collectively to understand each other's thinking.

T: "Why did he put a zero?"

S2: "For zero minutes of chores".

S5: "Or if you have a single [digit] number, you can't put it with a ten".

T: "I'm just going to pause you, did you hear what Tiana just said?'

S2: "No".

T: "Okay Tiana, can you repeat that because that was a really fantastic little sentence you just said. She was talking about why we put a zero there".

S5: "Because for the single [digit] numbers, you can't put a ten".

T: "Because what do these represent?"

S5: "Tens".

T: "And the leaves that come off, they represent the?"

S5: "Ones".

As the students continue to work together, the teacher notices that one student looks confused and wants to disagree. The teacher steps in stating: "You've got to trust yourself, I can see you looking at it thinking, "no, actually I think it's this". So this is when you have got to argue and say "no, hold on"". The teacher then moves to a different group of students and at this point notices that one student is predominantly writing and speaking. Again, she scaffolds the participation of students: "So do you maybe want to pass the pen so you can all write stuff. I can see that you've got it, but we want to make sure that everyone is participating and included, don't we?" Following the scaffolding from the teacher, the students worked collectively on the task and produced a range of relevant statements. The teacher concluded the small group work by stating: "Have you figured out how you are going to share back as a group so you are all working together?".

Through the small-group work, the teacher developed a safe and supportive learning environment for her students by building on the mathematics educational values that closely linked with their cultural values. In this way, she continued to develop CSMP by connecting with students' cultural identity and ways of being throughout her mathematics teaching (Gay, 2010). To achieve this, the teacher consistently made links with the values of collaboration, belonging, reciprocity, and inclusion by positioning the students to work collectively and both understand and build from each other's mathematical ideas. She both asked the students to explain their own mathematical thinking and reasoning but also to explain their peers' ideas. This meant that as the lesson progressed the groups of students collectively owned the responsibility for both their own and others' mathematical understanding. When she noticed over-participation from one group member, she re-visited the expectations that built on the values of collaboration and inclusion by framing this as a collective focus on everyone participating and being included. Hill et al. (2019) noted that Pacific students linked the mathematics educational value of collaboration with reciprocity. Similarly, Anthony (2013) found that Pacific secondary students endorsed collaborative values including belonging and reciprocity as supportive for their mathematical learning. This example provides an exemplar of how a teacher can use prompts during small group-work that build on mathematics educational values.

For students to engage in social and intellectual risk-taking during mathematical activity, it is important that they feel safe and secure. In this lesson, it appeared that connecting pedagogy with values was instrumental in developing the supportive culture in the classroom (Binkley et al., 2012; Civil & Hunter, 2015). The teacher built on the value of respect by specifically identifying student ideas that were mathematically helpful, for example, connecting the stem and leaf graph with place value knowledge of tens and ones, and revoicing these ideas to give all students in the group access to the mathematical connections. Previous research (e.g., Fletcher et al., 2009; Hunter & Hunter, 2018) has shown that Pacific students can find questioning and challenging others very difficult because it potentially conflicts with their concept of the value of respect. The teacher in this example instead connected the value of respect to participating in questioning and argumentation as she related this to trust in ones-self. Throughout this part of the lesson, it is evident that the teacher was working through her use of prompts and scaffolding to develop the classroom norms in a way where the values of all students and herself as the teacher were aligned (Seah & Andersson, 2015). Furthermore, with the focus on value alignment, the teacher also facilitated stronger relationships for these students both with herself and with their peers, a key aspect of CSMP (Thomas & Berry, 2019).

18.5 Summary Phase of the Lesson

The teacher began the summary phase of the lesson by again building on the student values to remind them of the expectations for their collective community to show manakitanga (care) to each other. As one of the students began to record the stem and leaf graph on the whiteboard, another student explained their shared solution strategy.

S1: "So first we are writing a stem and leaf to count the minutes the kids do chores. Any questions?

T: "Yes, you can ask if they have any questions, I like that because maybe some people are feeling a bit whakama (shy or shame) and that helps them to get the courage.

to ask. Oh, look at that, straight away [a child has put their hand up].

S7: "Why did you write 60 min?"

The group sharing their solution talk together quietly and then one of them erases the incorrect recording on the stem and leaf graph.

T: "That's okay, can I just follow up? Why did you just go up to five?"

S1: "It's because it only goes up to 55".

As the students who are sharing continue to record their stem and leaf graph, the teacher then prompts them to explain what they are doing. The students sharing explain and once again ask whether their classmates have any questions.

S6: "What are those numbers?".

S1: "Those are the zero, umm, that's the single numbers".

S2: "Yeah, single numbers".

As the students shared their solution, they also self-corrected the stem and leaf graph to further the clarity of their representation. The teacher consistently pressed then to explain this:

T: 'You need to share why you have just changed and put those commas in there".

S2: "It's because it doesn't add up to a bigger number like 55. That's why we put a comma in-between the five so it is not five million-billion.

T: "So how has it changed putting the commas there?".

S2: "So now we can all understand it's five, five, six …".

After the group finished recording, the teacher clarified student understanding of the representation by asking different students to read figures off the graph. She then asked a second group to share the statements that they developed from the stem and leaf graph.

S6: "Lots of people don't spend much time on chores".

T: “Okay, would you agree with that statement based on that data” [indicates stem and leaf graph].

S9: “No”.

T: “No, okay, we need to talk about it. I love an argument, you’ve got to try and persuade agree with the statements or actually go prove it. We need to be able to question and ask questions and be able to prove our thinking and our knowledge. So, if you are shaking your head and going “uh uh” then you need to say “why do you think that?”.

So Tiana?”.

S5: “Why do you think that?”.

S6: “It’s because some, most of the kids don’t have that much time except for that one person that spent like 45 minutes so we think that not much people spend that much time doing chores”.

The students continued to discuss the statement and come to a collective agreement. The group then continued to share other statements about the data.

S10: “No one spent 40 min on chores”.

S1: “I disagree because some people, no, most people spend 40 min doing chores”.

S10: “I know but we are talking about the stem and leaf graph [gestures at their graph], what is on here. We’re not talking about other ones, just on here”.

The teacher then built on this to facilitate the students to reflect on whether the stem and leaf graph was a fair representation of how much time most students would spend doing chores. She concluded the lesson by asking the students to consider the stories that could be told from a graph and the types of graphs that are beneficial to represent different types of data.

The safe and supportive classroom environment created by the teacher through her pedagogical approaches that both centred and built on the student mathematics educational values was evident in the summary phase of the lesson. The students themselves demonstrated the values of collaboration, inclusion, and reciprocity by actively involving the wider group in their explanation and regularly asking whether there were any questions about their solution strategy. This is similar to the examples explained by Hunter and Miller (2022) with younger Pacific students where the teacher actively built on the value of reciprocity. However, with these older students, we begin to see the students themselves independently interact mathematically in ways that build on their values. Connected to CSMP, this can be viewed as clear evidence of the safe power-sharing environment that the teacher created by building on students’ mathematics education values (Gutiérrez, 2012; Seah & Andersson, 2015).

The teacher explicitly connected the act of questioning to respect and reciprocity by framing questioning as an act which takes courage. It was evident also that the questions and interaction from other students supported the group who were sharing their representation to continue to refine and reflect on their solution. Aligned with her pedagogical practices during the launch of the lesson, the teacher maintained the use of inclusive language. For example, in promoting disagreement, she framed this as part of collective learning and collective knowledge building: "**We** need to be able to question and ask questions and be able to prove **our** thinking and **our** knowledge". Towards the conclusion of the lesson, the students demonstrated a sense of normalisation towards discursive interactions that moved beyond agreeing but also involved disagreement as part of the construction of a mathematical explanation. Although the teacher in this classroom and her students came from different cultural backgrounds and value systems, in contrast to Kalogeropoulous et al. (2021) description of potential dissonance, there appeared to be alignment of values during the mathematics lesson which was also evident in the levels of engagement from the students.

18.6 Conclusion and Implications

In conclusion, an exemplary case study of a mathematics lesson has been used to examine the pedagogical approaches that can be used which build on mathematics educational values of diverse students to develop CSMP. By an in-depth examination of one mathematics lesson, we can begin to build a view of mathematics teaching that draws on the mathematics educational values of a marginalised group of students as a strength to support their engagement with mathematics. However, the use of one lesson is also a limitation of this study given that it presents a small snapshot of mathematics teaching.

The chapter illustrates how a teacher from a different cultural group than the predominantly Pacific students in the classroom used prompts and approaches that made both implicit and explicit connections to the students' (and presumably her) values including family, respect, collaboration, inclusion, belonging, and reciprocity in ways that drew on Pacific understandings and reflections of these values. The teacher actions to achieve CSMP through connecting to students' mathematics educational values are shown explicitly throughout the findings. These included prompts to students to position them to reflect on how their values could be used during mathematics learning, the consistent use of inclusive language, demonstrating high expectations for mathematical learning and noticing and affirming student ways of participating in mathematics, and connecting both to students' experiences out of school and their family values. Although Averill and Rimoni (2019) highlighted that non-Pacific heritage teachers may have differing understanding of Pacific values than Pacific heritage teachers, in this example, the teacher actively shaped her practice so that the values of all were aligned and as Seah and Andersson (2015) describe in harmony to strengthen relationships.

New Zealand, similar to many other Western countries, currently has an inequitable schooling system including in relation to mathematics teaching and learning. For too long, Pacific student voices have been silenced within traditional mathematics classrooms by pedagogical practices that draw on the values of the dominant cultural groups in New Zealand with the consequence of disengagement for many Pacific learners (Barton, 1995; Hunter & Hunter, 2018). In contrast to approaches that problematise marginalized groups of students, in this chapter, an illustration is provided of teacher actions which are representative of how Pacific values can be used to scaffold all students into working mathematically. The findings and discussion indicate the potential of such an approach in both supporting mathematical learning and dispositions and the maintenance of a strong cultural identity. Although this chapter specifically focuses on Pacific students and their aligned mathematics educational values, it also provides an example of how teachers can carefully consider the mathematics educational values of their students and engage in pedagogical actions that support value alignment (Kalogeropoulous et al., 2021).

Returning to the focus on CSMP, in this exemplary case study, the teacher demonstrated her understanding of the context of students as people within their cultural group and connected to their lives and experiences through her consistent practice focused on collaboration, inclusivity, reciprocity, and respect. This in turn both supported her relationship with the students and the students' relationship with mathematics. Clear in the findings was how the students during the large group discussion had appropriated productive collaborative practices related to mathematics learning and independently prompted other students to reflect upon and question their mathematical explanation. Using a framing of CSMP, we can see how these pedagogical practices involved addressing access to mathematics, achievement, as well as identity and power relationships in the mathematics classroom (Gay, 2010; Gutiérrez, 2012; Ladson-Billings, 1995, 2006).

In a wider sense, this chapter adds to the current research field in relation to how a teacher can develop pedagogical practices that draw on the strengths such as values and experiences of marginalised students to develop equity. There has been limited research which examines the potential of values alignment in relation to CSMP and this chapter highlights the importance of considering values as a key element of the development of CSMP.

References

Abdulrahim, N. A., & Orosco, M. J. (2020). Culturally responsive mathematics teaching: A research synthesis. *The Urban Review, 52*, 1–25.

Anthony, G. (2013). Student perceptions of the 'good' teacher and 'good' learner in New Zealand classrooms. In G. Anthony, B. Kaur, M. Ohtani, & D. Clarke (Eds.), *Student voice in mathematics classrooms around the world* (pp. 209–225). Sense Publishers.

Armstrong, A. C., Johansson-Fua, S. U., & Armstrong, D. (2021). Reconceptualising inclusive education in the Pacific. *International Journal of Inclusive Education*, 1–14.

Averill, R., & Clark, M. (2012). Respect in teaching and learning mathematics. *Set*, 50–57.

Averill, R., & Rimoni, F. (2019). Policy for enhancing Pasifika learner achievement in New Zealand: Supports and challenges. *Linhas Críticas, 25*, 549–564.

Barton, B. (1995). Cultural issues in New Zealand mathematics education. In J. Neyland (Ed.), *Mathematics education: A handbook for teachers* (Vol. 2, pp. 150–164). Wellington College of Education.

Binkley, M., Erstad, O., Herman, J., Raizen, S., Ripley, M., Miller-Ricci, M., & Rumble, M. (2012). Defining twenty-first century skills. In P. Griffin, B. McGaw, & E. Care (Eds.), *Assessment and teaching of 21st century skills* (pp. 17–66). Springer.

Civil, M., & Hunter, R. (2015). Participation of non-dominant students in argumentation in the mathematics classroom. *Intercultural Journal, 26*(4), 296–312.

Coxon, E., Anae, M., Mara, D., Wendt-Samu, T., & Finau, C. (2002). *Literature Review on Pacific Education Issues*. Ministry of Education.

Faulkner, V. N., Marshall, P. L., & Stiff, L. V. (2019). *The stories we tell: Math, race, bias and opportunity*. Rowman & Littlefield.

Fletcher, J., Parkhill, F., Fa'afoi, A., Taleni, L. T., & O'Regan, B. (2009). Pasifika students: Teachers and parents voice their perceptions of what provides supports and barriers to Pasifika students' achievement in literacy and learning. *Teaching and Teacher Education, 25*, 24–33.

Gay, G. (2010). *Culturally responsive teaching: Theory, research, and practice*. Teachers College Press.

Gutiérrez, R. (2012). Context matters: How should we conceptualize equity in mathematics education? In B. Herbel-Eisenmann, J. Choppin, D. Wagner, & D. Pimm (Eds.), *Equity in discourse for mathematics education: Theories, practices, and policies* (pp. 17–33). Springer.

Hannula, M. S. (2002). Attitude towards mathematics: Emotions, expectations and values. *Educational Studies in Mathematics, 49*, 25–46.

Hill, J., Hunter, J., & Hunter, R. (2019). What do Pasifika students in New Zealand value most for their mathematical learning? In P. Clarkson, W. T. Seah, & J. Pang (Eds.), *Values and valuing in mathematics education: Scanning and scoping the territory* (pp. 103–114). Springer.

Hill, J., & Hunter, J. (2023). Examining the mathematics education values of diverse groups of students. *International Journal of Mathematical Education in Science and Technology, 54*(8), 1614–1633, https://doi.org/10.1080/0020739X.2023.2184280

Hofstede, G. (2011). Dimensionalizing cultures: the Hofstede model in context. *Online Readings in Psychology and Culture, 2*(1). https://doi.org/10.9707/2307-0919.1014

Hunter, R., & Anthony, G. (2011). Forging mathematical relationships in inquiry-based classrooms with Pāsifika students. *Journal of Urban Mathematics Education, 4*(1), 98–119.

Hunter, R., & Hunter, J. (2018). Maintaining a cultural identity while constructing a mathematical disposition as a Pasifika learner. In E. A. McKinley & L. Tuhiwai Smith (Eds.), *Handbook of Indigenous Education*. Springer.

Hunter, J. (2021). An intersection of mathematics educational values and cultural values: Pāsifika students' understanding and explanation of their mathematics educational values. *ENCU Review of Education, 4*(2), 307–326.

Hunter, R., Hunter, J., Anthony, G., & McChesney, K. (2018). Developing mathematical inquiry communities: Enacting culturally responsive, culturally sustaining, ambitious mathematics teaching. *Set, 2*, 25–32.

Hunter, J., Hunter, R., Hannant, B., Kritesh, K., Lachaiya, R., Bills, T., & Cheung, I. (2016). Developing equity for Pasifika learners within a New Zealand context: Attending to culture and values. *New Zealand Journal of Educational Studies, 51*(2), 197–209.

Hunter, J., & Miller, J. (2022). Using a culturally responsive approach to develop early algebraic reasoning with young diverse learners. *International Journal of Science and Mathematic Education, 20*, 111–131. https://doi.org/10.1007/s10763-020-10135-0

Kalogeropoulos, P., Russo, J. A., & Clarkson, P. (2021). Exploring educator values alignment strategies in an intervention context: The emergence of the Beacon strategy. *ECNU Review of Education, 4*(2), 327–348.

Kazemi, E., Franke, M., & Lampert, M. (2009). Developing pedagogies in teacher education to support novice teachers' ability to enact ambitious instruction. In: R. Hunter, B. Bicknell, & T. Burgess (Eds.), *Crossing divides* (*Proceedings of the 32nd Annual Conference of the Mathematics Education Research Group of Australasia*, pp. 11–29). MERGA.

Ladson-Billings, G. (1995). But that's just good teaching! The case for culturally relevant pedagogy. *Theory into Practice, 43*(3), 159–165.

Ladson-Billings, G. (2006). "Yes, but how do we do it? Practicing culturally relevant pedagogy. In J. G. Landsman & C. W. Lewis (Eds.), *White teachers' diverse classrooms: Creating inclusive schools, building on students' diversity, and providing true educational equity* (pp. 33–46). Stylus.

Leonard, J., Brooks, W., Barnes-Johnson, J., Berry, R. Q., & III. (2010). The nuances and complexities of teaching mathematics for cultural relevance and social justice. *Journal of Teacher Education, 61*, 261–270.

Ministry of Education. (2013). *Pasifika education plan 2013–2017*. https://www.education.govt.nz/assets/Documents/Ministry/Strategies-and-policies/PasifikaEdPlan2013To2017V2.pdf

Naresh, N., & Kasmer, L. (2018). Using ethnomathematics perspective to widen the vision of mathematics teacher education curriculum. In T. Bartell (Ed.), *Toward equity and social justice in mathematics education* (pp. 309–326). Springer.

Nasir, N. I. S., Hand, V., & Taylor, E. V. (2008). Culture and mathematics in school: Boundaries between "cultural" and "domain" knowledge in the mathematics classroom and beyond. *Review of Research in Education, 32*(1), 187–240.

New Zealand Ministry of Education. (2018). *Tapasa: Cultural competencies framework for teachers of Pacific learners*. https://pasifika.tki.org.nz/Tapasa

Paris, D. (2012). Culturally sustaining pedagogy: A needed change in stance, terminology, and practice. *Educational Researcher, 41*(3), 93–97.

Paris, D., & Alim, H. S. (2014). What are we seeking to sustain through culturally sustaining pedagogy? A loving critique forward. *Harvard Educational Review, 84*(1), 85–100.

Podsiadlowski & Fox, S. (2011). Collectivist value orientations among four ethnic groups: Collectivism in the New Zealand context. *New Zealand Journal of Psychology, 40*(1), 5–18.

Presmeg, N. (2007). The role of culture in teaching and learning mathematics. In F. K. Lester (Ed.), *Second handbook of research on mathematics teaching and learning* (pp. 435–458). Information Age Publishing.

Rimoni, F., Glasgow, A., & Averill, R. (2022). *Pacific educators speak: Valuing our values*. NZCER Press.

Seah, W. T. (2016). Values in the mathematics classroom: Supporting cognitive and affective pedagogical ideas. *Pedagogical Research, 1*(2), 45–63.

Seah, W. T., & Andersson, A. (2015). Valuing diversity in mathematics pedagogy through the volitional nature and alignment of values. In A. Bishop, Tan, H. & Barkatsas (Eds.), *Diversity in mathematics education: Towards inclusive practices* (pp. 167–183). Springer.

Sharma, S., Young-Loveridge, J., Taylor, M., & Häwera, N. (2011) The views of Pāsifika students in New Zealand about communicating mathematically. *Asia Pacific Journal of Education, 31*(4), 503–519.

Sum, E. S. W., Wong, K. L., & Seah, W. T. (2022). Attributes of mathematics learning valued by ethnic minority students in Hong Kong mainstream school. *Asia Pacific Journal of Education*, 1–5.

Thaman, K. (2005). Whose values and what responsibility? Cultural and cognitive democracy in education. *Pacific Asian Education, 17*(1), 8–19.

Thomas, C. A., & Berry, R. Q., III. (2019). A qualitative metasynthesis of culturally relevant pedagogy & culturally responsive teaching: Unpacking mathematics teaching practices. *Journal of Mathematics Education at Teachers College, 10*(1), 21–30.

Turner, H., Rubie-Davies, C. M., & Webber, M. (2015). Teacher expectations, ethnicity and the achievement gap. *New Zealand Journal of Educational Studies, 50*, 55–69.

Uehara, D. L., Chugen, J., & Staley Raatior, V. (2018). Perceptions of Pacific Islander students in higher education. *Journal of Diversity in Higher Education, 11*(2), 182–191.

Zhang, Q. (2019). Values in mathematics learning: Perspectives of Chinese mainland primary and secondary students. In P. Clarkson, W. T. Seah, & J. Pang (Eds.), *Values and Valuing in Mathematics Education: Scanning and scoping the territory* (pp. 185–196). Springer.

Chapter 19
Exploring Undergraduate Women of Color's Mathematical Values in the United States Through Cultural Values and Mathematical Discourses

Ciera Street

19.1 Introduction

The call for more equitable mathematics education in the USA is apparent; several professional mathematics organizations in the USA have released position statements addressing issues of equity and diversity (e.g., American Mathematical Society, n.d.; Mathematical Association of America, n.d.; National Council of Teachers of Mathematics, 2014). Scholars argue that incorporating and reflecting students' values in the mathematics classroom may be one way to create a more equitable and inclusive environment (Battey & Leyva, 2017; Seah et al., 2016). To this point, research suggests that this incorporation of values into the classroom positively influences students' mathematical learning and affect (Hill et al., 2021; Hunter et al., 2016).

However, undergraduate mathematics in the United States often does not reflect the cultural values of women or students of color (Fong et al., 2019; Leyva, 2021; McGee, 2016). Various scholars recognize how mathematical discourses in the USA often align with, and thus give power to, dominant masculine and white[1] values such as competition, risk, and individualism (Bullock, 2019; Jaremus et al., 2020; Leyva, 2017; Martin, 2006). Leyva (2021) argues that academic mathematical contexts function as white, patriarchal spaces that maintain "inequities at intersections of gender,

[1] I use lowercase white and capitalize all other racial/ethnic groups. This diverges from APA style, but aligns with the Associated Press's extensive research in the matter (Daniszewski, 2020).

This chapter is based on part of the author's thesis (Street, 2023). The thesis has been published on ProQuest.

C. Street (✉)
California Polytechnic State University, San Luis Obispo, California, USA
e-mail: cnstreet@calpoly.edu

Colorado State University, Fort Collins, Colorado, USA

Y. Dede et al. (eds.), *Values and Valuing in Mathematics Education*,
https://doi.org/10.1007/978-981-99-9454-0_19

race, and class" (p. 121). Thus, unique and possibly additional tensions within mathematical spaces emerge at the intersection of race and gender for women of color (Carlone & Johnson, 2007; Leyva, 2021). Supporting and incorporating women of color's cultural values within undergraduate mathematics may help reduce these tensions and counter inequities (Battey & Leyva, 2017; Seah et al., 2016).

While research globally has considered cultural values alongside mathematics education values, minimal work is this area is based in the USA or at the undergraduate level. Given the white, patriarchal values rooted in undergraduate mathematics in the USA, more work needs to consider what women of color value about mathematics and mathematics education and in what ways mathematics spaces can better align with those values. This research engages critical frameworks at the intersections of gender and race to challenge exclusionary adherence to dominant Western values in mathematics. I utilize qualitative data from two women of color STEM majors in the USA to explore what they value about mathematics and mathematics education and to what extent they saw these values reflected in their undergraduate mathematics courses. In particular, I ask:

1. What do two women of color STEM majors in the USA describe as valuable or important about mathematics and undergraduate mathematics education?
2. To what extent do these values align with the values traditionally held in the discipline and in their undergraduate mathematics spaces?

19.2 Values and Valuing

The conceptualization of values and valuing continues to transform and grow with the development of new research and theoretical viewpoints. Bishop et al. (1999) defined mathematics education values as affective qualities fostered through students' experiences in school mathematics. Other researchers describe this fostering of values as more cognitively driven, including a stage of individual choosing of values based on experiences and the options available (Raths et al., 1987). More current research positions values as a combination of affective and cognitive variables, with additional motivational or volitional components (Hannula, 2012; Seah & Andersson, 2015). The addition of a volitional perspective supports values as not only motivating factors, but also a driving factor in an individual's determination to maintain an action even when faced with challenges or obstacles (Seah & Andersson, 2015). For this work, I align with this latter perspective and borrow directly from Seah and Andersson (2015) to define values in mathematics education as follows:

> Values are the convictions which an individual has internalised as being the things of importance and worth. What an individual values defines for her/him a window through which s/he views the world around her/him. Valuing provides the individual with the will and determination to maintain any course of action chosen in the learning and teaching of mathematics. They regulate the ways in which a learner's/teacher's cognitive skills and emotional dispositions are aligned to learning/teaching in any given educational context. (p. 169)

This definition implicitly posits that values are inherently influenced by contextual, social, and cultural factors, such as educational settings, social interactions, and societal norms (Reinholz et al., 2019; Schwartz et al., 2001; Seah, 2018). While values are often considered effectively internalized and stable, this perspective reflects how values can shift or transform based on context and experiences (Seah, 2018). Furthermore, Perez et al. (2014) summarize theory that suggests that personal and collective identities help frame individuals' self-perceptions of their values. On a personal level, values help support notions of uniqueness, while on a collective level, values build connections between oneself and salient social groups.

These various viewpoints support the idea that social identities, such as gender and race, relate to individual values. Schwartz and Rubel (2005) in their research assessing gender differences across 10 basic values show that women tend to value benevolence and universalism more than men, while men value power and achievement more than women. Research also suggests that people of color value collectivism more so than white people (French et al., 2020; Sue et al., 2022). We also see specific values at the intersections of gender and race. For Latina women, *marianismo* represents "a Latino cultural value that describes both positive and negative aspects of traditional Latina femininity… emphasiz[ing] culturally valued qualities such as interpersonal harmony, inner strength, self-sacrifice, and morality" (Da Silva et al., 2021, p. 3755). In Del-Mundo and Quek's (2017) study considering second-generation Filipino American women's meanings of gender identity, multiple participants expressed valuing familial connections and respecting elders. These studies suggest that community and interpersonal relationships are common values for women of color. However, STEM careers are more frequently perceived to align with agentic values (e.g., power, individualism) more so than communal values (e.g. helping others, serving the community) (Diekman et al., 2010). This perception aligns with undergraduate mathematical spaces in the USA and other Western nations that continue to uphold dominant masculine and white values (Leyva, 2017; Martin, 2006; Mendick, 2006).

19.3 Categories of Values and Valuing in Mathematics Education

Stemming from Bishop et al. (1999) work, values and valuing in mathematics education span three subcategories: general education, mathematical, and mathematics education. General education values encapsulate broad moral and ethical values embedded in society's perception of the goals of education (e.g., valuing honesty). Mathematical values include values associated with the discipline itself (e.g. valuing rationalism). Mathematics education values relate to the teaching and learning of mathematics and the norms of the specific mathematics environment (e.g. valuing collaborative learning). These categories are not distinct and their boundaries necessarily blend and inform one another (Seah et al., 2016). This work focuses on the latter two categories, mathematical values and mathematics education values. Similar to more general values, mathematical and mathematics education values are also strongly intertwined with the culture of the learner and societal influences (Hunter, 2021; Lee & Seah, 2015; Zhang, 2019). Research suggests that the mathematical values of Swedish students differ from those of Turkish students which differ from those of Chinese students (Andersson & Österling, 2019; Dede et al., 2022; Tang et al., 2021). Given the USA context of this research, I draw from literature on "Western" mathematical and mathematics education values.

Bishop (1988) developed a well-known framework to describe Western mathematical values in which he categorizes three complementary mathematical value pairs by sentimental, ideological, and sociological components. Each of these value pairs align with some aspect of Western society. For example, the ideological value pair captures Western society's valuing of logic, sensible connections, and making the abstract tangible. Iterations of these values emerge in other research as well. Mendick (2006) posits that Western mathematical values align with those traditionally associated with dominant masculinity, such as logical, competitive, and innate ability. She argues that these values emerge along numerous dichotomous value pairs that position dominant Western masculinity against femininity, such as logical versus emotional and ordered versus creative. Scholars also note how these mathematical values, such as innate ability and individualism, are submerged in notions of whiteness (Leyva et al., 2021; McGee, 2016). This is reflected in Reinholz et al. (2019) work looking at the similarities and differences in cultural components across STEM, where participants perceived mathematical knowledge as largely innate, objective, logical, and individually-driven.

Similar ideas appear when considering Western mathematics education values. Reinholz et al. (2019) describe how the overarching perception of undergraduate mathematics teaching emulates a "sage on the stage," where instructors individually transfer knowledge to the students. Participants explained how this method of teaching related to valuing pedagogical autonomy, teaching as an individual endeavor, and a focus on knowledge acquisition. Ferrare and Hora (2014) as well in their study exploring science and math instructors' views on learning, found that the most prevalent view was that students learn best when they engage in individual, sustained struggle, often outside of the classroom. These pedagogical values conflict with literature around marginalized students' mathematics education values. Voigt et al. (2022) suggest that women may feel more interest and confidence in mathematics courses that include deep engagement with mathematical tasks during class. Espinosa (2011) found that women of color who frequently engaged with peers to discuss course content remained in a STEM major more often than those who did not. These results propose that Western mathematics education values that promote isolated knowledge acquisition imparted solely by the instructor may not align with women of color's values around collaborative and exploratory learning.

19.4 The Gendered-Racialized Mathematics Space

The above description of Western mathematical and mathematics education values helps frame recent theory in mathematics education research that positions the mathematics space in the USA as both gendered and racialized (Leyva, 2017; Martin, 2006; Mendick, 2006). Although focused on British college students, Mendick (2006) provided instrumental work for scholars to detail the gendered nature of undergraduate mathematical spaces in the USA and other Western nations. Mendick positions mathematics as a gendered space by exploring dichotomies within mathematical discourses that structure femininity against masculinity. Mathematical discourses often value and empower dominant masculine characteristics, such as framing mathematical reasoning as ultimate rational thought, mathematical ability as innate, and mathematicians as culturally deviant (Jaremus et al., 2020; Mendick, 2006). In particular, Mendick argues discourses that define mathematics as ultimate rational thought stemmed from long-standing gendered discourses of rationality, which associated logic with masculinity and emotions with femininity. These discourses communicate masculine norms around mathematics participation, e.g. competitive, risk-based, and speed-focused, that often exclude women and compromise their ability to develop robust mathematical identities (Barnes, 2000). With mathematics culturally defined in this way, those whose values more naturally align with these masculinities have easier access to social labels that position them as good at math or a "math person." Thus, women often experience more tension and pressure within mathematical spaces while negotiating societal ideas of femininity with mathematical masculinities (Leyva, 2017; Mendick, 2006).

Scholars also recognize mathematics in the USA as a racialized space (Leyva, 2016, 2021; Martin, 2006). Bullock (2019) argues that reforms for "all" in K-12 mathematics education in the USA continue to define "all" as those who "can embrace the whiteness that standardization represents" (p. 77). In his work deconstructing the Mathematics for All rhetoric, Martin (2003) describes how the goals of mathematics education policy often do not reflect the goals and values held by Black parents for their children's mathematics education. This idea extends to various communities of color within undergraduate mathematics education. The values of students of color often conflict with those expected in STEM departments, such as the disconnect between Indigenous American students' cultural and familial values with a mathematics curriculum detached from community needs and contexts (Abrams et al., 2013; Fong et al., 2019). This conflict of values often goes unchallenged within the racialized discourse of color-evasiveness within mathematics education (McNeill et al., 2022). The color-evasiveness discourse centers "the ideology that mathematics is a body of knowledge independent of culture and race, so there is no need to attend to racism or student race in teaching mathematics" (Leyva et al., 2021, p. 3). Furthermore, this racialized discourse also perpetuates the idea of mathematical ability as innate and communicates a deficit narrative around students of color who struggle with mathematics. Coupled with racial and cultural stereotypes, this discourse imposes a racial hierarchy of ability whereby Black, Latine,[2] and Indigenous students innately lack mathematical ability compared to white students (Abrams et al., 2013; McGee, 2016). Multiple studies also illustrate covert and overt instances of racism in the mathematics classroom and marginalized students' responsive use of coping strategies and stereotype management (Jett, 2019; Leyva, 2021; McGee, 2016). These harmful stereotypes range from placing Asian students as "model minorities" to presuming a lack of ability for Black, Latine, and Indigenous students (Abrams et al., 2013; McGee, 2016; McGee & Martin, 2011; Oppland-Cordell, 2014). These stereotypes in mathematics often impart undue cognitive effort on marginalized students to navigate these assumptions and define mathematics as their own (Martin, 2006; Masten et al., 2011). Thus, students of color often experience more tension and pressure within mathematical spaces while navigating color-evasive and stereotyped mathematical discourses (Leyva et al., 2021; McGee & Martin, 2011).

19.4.1 Values Alignment in Education

Scholars suggest that supporting and incorporating marginalized students' values within undergraduate mathematics can help counter inequities by challenging dominant masculine and white norms (Battey & Leyva, 2016; Hunter, 2021; Seah et al., 2016). Seah and Andersson (2015) borrow from Branson (2008) to highlight the notion of values alignment in education. They describe values alignment as a cooperative and collaborative process whereby the members of an organization work together to clarify common values and develop strategies and systems aimed to support those values and guide actions and norms of behavior in the space. That is to say that the instructor and students collectively negotiate the norms of the classroom to reflect each other's values. Importantly, this process does not propose values inculcation, rather "values alignment facilitates the coexistence of different values as these are held by different people interacting in any given context… [such that] students perceive that their knowledge, skills and dispositions are valued" (Seah & Andersson, 2015, p. 178).

Several studies demonstrate the negative outcomes for marginalized students related to a misalignment of values in the mathematics classroom, including decreased perceptions of ability, sense of belonging, and performance. In their study considering gendered and racialized components of undergraduate Calculus instruction, Leyva et al. (2021) suggest how the value of mathematical knowledge as innate and the valuing of the instructor as the sole mathematical authority displayed in certain instructional instances misaligns with values held by marginalized students. Several of the participants, all undergraduate students of color, perceived these instances to relay "messages of minoritized students lacking ability and not belonging

---

[2] In this work, I use Latine as a gender-neutral term for those who identify with Latin American origins and/or descent (Slemp, 2020).

in STEM" (p. 802). Other studies demonstrate how the particular misalignment between Indigenous students' valuing of family and community with the individualistic and self-promotion values embedded in mathematics can harm these students' mathematical performance, interest, and belonging (Fong et al., 2019; Hunter et al., 2016). Given the prevalence of value misalignment between undergraduate mathematics in the USA and the cultural values of women and students of color, these studies necessitate more work considering how to promote value alignment (Abrams et al., 2013; Fong et al., 2019; Leyva, 2021).

Other studies showcase how an alignment of values can promote positive outcomes for marginalized students in mathematics, such as increased comfort, participation, and sense of belonging. A commonly recognized equity-based pedagogy in the USA, culturally relevant pedagogy, strongly supports this ideology. Culturally relevant pedagogy identifies how current racialized mathematical discourses value certain ways of knowing, mathematical content, and pedagogy that preserve the norms of the dominant group (Allen, 2004; Ladson-Billings, 1995). Teaching practices using a culturally relevant pedagogical lens confront white, masculine discourses of mathematics through "empower[ing] students intellectually, socially, emotionally, and politically using cultural referents to impart knowledge, skills, and attitudes" (Ladson-Billings, 1994, pp. 17–18, Howard & Rodriguez-Scheel, 2017). Acknowledging the cultural capital students of color already bring into the classroom and valuing its role in STEM success can help bolster mathematical identity and support a sense of belonging (Adiredja & Zandieh, 2020; Ortiz et al., 2019). In her study exploring how Latino/a students' participation shifted during a diverse, collaborative Calculus I workshop, Oppland-Cordell (2014) also found that the students participated more and were more comfortable in this space that specifically challenged dominant values around mathematics ability as innate and learning as individualistic and instead supported the students' cultural values. Other studies demonstrate increased student engagement and interest when instructors align their pedagogical practices with their students' values (Hill et al., 2021; Kalogeropoulos & Clarkson, 2019). These various results point toward the promise of values alignment within undergraduate mathematics to counter gendered and racialized inequities for women of color in the USA.

19.5 Theoretical Perspectives

In this work, I utilize two theoretical lenses to frame how values encompass and inform the gendered and racialized nature of undergraduate mathematical spaces. Given the sensitivity of mathematics educational values to the culture of the learner and societal influences (Hunter, 2021; Lee & Seah, 2015; Zhang, 2019), I utilize sociopolitical and intersectional perspectives. Adiredja and Andrews-Larson (2017) draw from social and political contexts in the USA to describe a sociopolitical perspective toward postsecondary mathematics. This perspective emphasizes how students' negotiations of knowledge, power, and identity emerge and are often constrained by the norms and values of the environment and broader social discourses. This viewpoint considers how "accepted" norms and values within mathematics privilege certain forms of knowledge and identity expressions. Identity in this sense captures a dynamic conceptualization of one's beliefs, values, and position relative to others in a certain context, often influenced by one's gender, race, and other social identities (Esmonde et al., 2009). Thus, the valuing of white, masculine knowledge and identities in Western mathematics determines who and what has power in mathematical spaces, including not only formal positions of power, but also the ability for individuals to resist and shape mathematical discourses (Adiredja &Andrews-Larson, 2017). Using this lens situates this work alongside other scholars suggesting that the adherence to white, patriarchal values in undergraduate mathematics excludes, devalues, and disempowers women of color's experiences, knowledge, and place in mathematics (Leyva, 2016, 2021). To counter these inequities, this lens rejects student assimilation and instead promotes transforming systems to consider, support, and afford space to various values and identities.

I also utilize an intersectional lens to ground this work. Intersectionality describes the ways in which interactions between race, gender, class, and other social constructs account for overlapping discrimination imposed on people with multiple marginalized identities within societal systems (Crenshaw, 1991). Specifying race and gender as social constructs is not to say that these categories are insignificant, but rather they provide critical understandings about "the particular values attached to [gender and race] and the way those values foster and create social hierarchies" (Crenshaw, 1991, p. 1297). Often, systems fail women of color because the discourses and narratives around those systems privilege the values and experiences of dominant groups. Thus, in this work I recognize that the values women of color hold are both individually unique as well as shaped by both societal influences and mathematical discourses related to gender, race, and other social constructs. This view helps illuminate varying mathematical experiences across intersecting subgroups of students and mitigate monolithic narratives of marginalized groups within power-laden environments (Esmonde, 2011; Leyva, 2017). Considering women of color's values requires recognizing that identities are interconnected and concurrently influencing how they navigate and perceive the values upheld and empowered within undergraduate mathematical spaces.

Given that undergraduate mathematics in the United States often does not reflect the cultural values of women or students of color, showcasing a variety of both unique and overlapping intersectional narratives outside of dominant groups paints a broader picture of mathematics education and can inform and support more inclusive mathematics education values (Abrams et al., 2013; Fong et al., 2019; Leyva, 2021). Thus, the sociopolitical perspective speaks towards how values and norms help frame power, identity, and knowledge relationships in undergraduate mathematics, while an intersectional lens expands on the nuances of identity, especially when considering a population with multiple marginalized identities in the USA.

19.5.1 Positionality and Reflexivity

As a white woman with degrees in mathematics and experiences with mathematics education as a student, researcher, and instructor, I consider how my positionality permeates throughout this work. My gender identity enabled instances of relatability while simultaneously I recognize my limited perspective on the gendered-racialized experiences ever present for women of color. My background, including my identities and experiences, formed my perception of undergraduate mathematics as a socially exclusionary space and motivates my work toward more equitable mathematics education. However, this gendered exclusion does not directly translate to exclusionary experiences based on other identities, such as race, first-generation status, or sexuality. Throughout this research, I rely on the works of numerous scholars of color, women scholars of color, and activists both inside and outside of the academy to reflect on my perspectives and inform research decisions. I frequently engaged in journaling through my research decisions, exploring differing viewpoints for analysis and discussion, and continued reading through the work of critical scholars to inform my perspectives (Idahosa & Vincent, 2019). This reflexive process promoted more ethical and authentic work, positive relationships with participants, and a consistent centering of the participants' voices and experiences.

19.6 Methods

19.6.1 Data Context and Participants

The data for this study include transcripts of follow-up interviews in Spring 2022 from participants who completed the *S-PIPS-M* survey instrument administered by the Mathematical Association of America's Progress through Calculus (PtC) project in 2017–2018. The survey collected information about undergraduate students' experiences in precalculus and calculus in the USA, including mathematical activities, interactions, and affect (Street et al., 2021). To recruit participants for this follow-up study, I sent out an interest survey in Spring 2022 to all women of color from the survey who consented to future contact ($n = 1121$). Women of color within this dataset includes any participant who selected at least woman from the following select-all response options related to gender: Man, Transgender, Woman, Not listed (please specify) and who selected at least one of the following select-all response options related to race and/or ethnicity: Alaskan Native or Native American, Black or African American, Central Asian, Hispanic or Latinx, Middle Eastern or North African, Native Hawaiian or Pacific Islander, Southeast Asian, South Asian. The recruitment email described the study as an exploration of women of color STEM majors' experiences in undergraduate mathematics and thus participants self-selected as a woman of color to participate in this follow-up study. The interest survey included questions about demographics, experiences and feelings related to undergraduate

Table 19.1 Description of participants

Lyka	Callie
Lyka (she/her) is a Filipino and white cisgender woman. She recently graduated with degrees in Biomedical Engineering and Multidisciplinary Studies from a public university in the Eastern half of the United States. Within the mathematics department at her university, she completed College Algebra, Trigonometry, Calculus I, Calculus II, Multivariable Calculus, and Ordinary Differential Equations	Callie (she/her) is a Hispanic/Latino cisgender woman. She recently graduated with a degree in Biology from a public university on the East coast of the United States. Within the mathematics department at her university, she completed College Algebra, Precalculus, Trigonometry, Statistics, and Calculus I

mathematics, and how those experiences and feelings related to race and gender, if at all.

I selected 12 participants to interview. For this analysis, I focus on two participants, Lyka and Callie. This smaller sample size allows me to convey the depth of their responses and provide nuance in how they view and express values. I selected Lyka and Callie in particular for this analysis because of their differences in demographics, institutions, degrees, and experiences. This afforded more diversity in both academic and personal experiences, thus providing more potential variance in their values. A smaller sample size also limits this research, including less generalizability. The nature of participant recruitment also limited the sample to those who self-selected into the study and the interview structure limited participants' discussion of values in terms of topic and time. Thus, the results presented here represent a subsection of the values and experiences of two women of color and do not reflect all women of color. However, considering prior literature and theory indicates that their values are not discordant with cultural values embedded within gendered and racialized notions, suggesting implications beyond these two participants. Future research will expand this work by analysing the remaining interviews to capture similar, additional, or disparate values. Brief descriptions of Lyka and Callie appear in Table 19.1. The participants self-described their gender and racial identities through an open-ended question on the interest survey.

19.6.2 Data Collection and Analysis

In Spring 2022, I conducted 60–90 min, semi-structured virtual interviews with each participant. The interview protocol included questions about participants' experiences and feelings related to undergraduate mathematics, including their perceptions of themselves as doers of math, what they found to be supportive and valuable in mathematics spaces, and potential connections to gender and race. For example, I asked "what type of teaching style did you feel best supported you in undergraduate mathematics and in what ways?" This protocol intended to bring forth participants' narratives around important and valuable aspects of mathematics and their undergraduate mathematics education as well as their mathematical identity. In this work, I focus on responses related to participants' mathematical and mathematics education values.

Before analyzing the data, I transcribed each interview and replaced identifying names and institutions with pseudonyms. While transcribing, I engaged in pre-coding whereby I noted particularly interesting or powerful statements related to values (Saldaña, 2013). I then continued to use Saldaña's (2013) first cycle and second cycle coding methods. During the first cycle, I open-coded by attending to prior work characterizing students' mathematical, mathematics education, and cultural values (e.g., Hill et al., 2021; Hunter, 2021; Reinholz et al., 2019), but without specific codes in mind. In general, I labeled anything connected to the research questions, including statements related to cultural values, mathematical values, mathematics education values, social identities, and expressions of emotions. These first cycle codes each captured a main idea, a more specific sub-category, and whether the participant described this aspect positively, negatively, or as neutral. For example, I coded the statement "I think there's just so much to learn about [mathematics] and like I had to do in such little time almost that I kind of wish that like I could go... at a better, at my own pace" with *reputation—speed (negative).* This code related the statement to the overall idea that society perceives mathematics and mathematical learning to have certain characteristics (reputation), one of which is the idea that mathematics must be learned and performed quickly (speed), but the participant views this perception negatively. Statements often stretched across multiple codes. For example, the statement "It was definitely, mostly like lecture where it was just like, on your own" was coded as *class structure—lecture* and *reputation—individual.* These codes became the basis of the second cycle coding process which involved grouping the initial codes into theoretical codes related to values. These theoretical codes included career/future, mathematics reputation, real-world meaning, community, accomplishment, pedagogy, and various types of social support. I then split, merged, and rearranged these theoretical codes and subcodes to finalize themes speaking toward the research questions. This resulted in five themes: meaning of mathematics, mathematics as fast-paced, innateness of mathematical ability, mathematics is competitive, mathematics as a gatekeeper.

19.7 Results

19.7.1 Meaning of Mathematics

Hill et al. (2021) describe the dimension of meaning related to mathematics values as "having a sense of direction in mathematics, feeling mathematics is valuable, worthwhile, or has a purpose" (p. 353). This can include perceiving math as valuable for its applications to the real-world, to one's life, to a future career, or in and of itself. Since the 1950s, values associated with college mathematics in the USA tend to emphasize pure or abstract mathematics over applied, including the implementation of major specific calculus courses (e.g. Calculus for Engineers) in mathematics departments to "maintain [their] focus on the beauty and intellectual depth of core mathematical theory" (Tucker, 2013, p. 700). This is seen through pedagogical approaches in mathematics that focus on content knowledge more so than applications and the view that pure mathematical research holds more prestige than applied mathematical research (Reinholz et al., 2019).

Both Callie and Lyka challenge this view by expressing valuing the meaning of mathematics in terms of its applications to the real-world, future careers, and their community. Callie spoke about how she "was always more interested in those real applications… look how it's used, like in a real world application." Lyka as well discussed how excited she was to see how calculus could be applied in a real-world setting through her engineering courses, how "wow, like you can use calculus [in engineering] …so just like actually being able to apply that in the setting. Not only do I know how to do math now, like I know how to apply it." In contrast, Lyka expressed her frustrations with courses that she did not find applicable, such as "ugh, I'm stuck in algebra… I don't need to know this. I just didn't find it very applicable."

Both Callie and Lyka also emphasized value in seeing connections between mathematics and a future career. The importance of a career and the pressure to succeed connected back to their identities. Callie addressed how diverging from a pre-medicine track was challenging given "my family they're like, you need to be a doctor and all that … it's kind of the typical Hispanic like, the doctors are the best … and I did like it for a while but then I realized it wasn't for me." She also described how being a first-generation student and the oldest child in her family there was "another added pressure of you're like, the child that's gonna make it." Lyka talked similarly about her first-generation status when she said "I need to prove myself….So I'm getting two degrees. The first-generation [status], I'm just like, setting the bar high." These experiences align with their perceptions that mathematics is also meaningful when it relates to career success. Lyka mentioned this explicitly, noting how "I didn't just spend all these years in these math courses to not use any of it, like it's very much applicable and prevalent in my career."

Lyka in particular also valued mathematics when it could help others and her community. She expressed how "I feel confident that I'm able to do [math], and…it makes it better, because I know it's going towards something that's going to help someone and improve their quality of life." However, she emphasized how generally disconnected she perceives mathematics is from confronting issues in society at large. For example, in reference to the increase in hate crimes towards Asian people in the USA, she admonished how "this was what I was seeing in my community, like, how am I supposed to want to do [math] or think [math] is valuable in the moment, when all these things are happening, when … I'm needed in my community in some way."

Callie and Lyka's mathematical education values in this regard emerged along similar lines as their mathematical values. They describe experiencing both context-devoid and context-heavy mathematical courses. Both participants explicitly brought up how in their calculus courses, applications "[weren't] necessarily talked about in classes... it was just like the concepts, how you learn it, and that's it." However, both participants also positively spoke of courses in which they could see connections between mathematics and the real-world or future careers. Callie spoke about how "that's why I enjoyed statistics so much, because they were almost like problems that you would see normally, in real life." Lyka described a lesson in high school mathematics where the teacher connected the lesson to a project completed by engineers at a national beverage company, commenting how "I think finding ways to translate math is still important in the undergraduate setting, and just looking for those experiences like, if you know how to do this [math], then ...you can do this in the future." While both Lyka and Callie experienced a variety of mathematics pedagogy, they each valued mathematics courses that included connections between mathematics and real-life contexts or careers.

19.7.2 Mathematics as Fast-Paced

Another long-standing value ingrained in mathematics is the ability to do math quickly and "keep up" with mathematics course content. Multiple scholars report that students across grade levels strongly believe that success in mathematics requires knowing answers and completing problems quickly (Darragh, 2013; Ernest, 2011). This reflects mathematics classrooms that embody a fast-paced curriculum toward the goal of a timed exam (Geist, 2010). In the college setting, this value aligns with standardization within mathematics departments, often including coordinated courses with common syllabi, homework, and tests. This standardization enforces a fast-paced approach to avoid falling behind other sections and cover the required content for the exam (Reinholz et al., 2019).

Callie and Lyka both acknowledged and expressed their frustrations with the value of speed embedded into their undergraduate mathematics courses. Callie recognized that "there's just so much to learn about [math] and I had to do it in such little time that I kind of wish that I could go ... at a better, at my own pace in learning these concepts" and how she wished that there "wasn't so much pressure on the student to learn it immediately." Lyka as well felt like her grade suffered because she "just couldn't learn it as fast as the other [students]." Thus, while the participants successfully navigated the fast-paced mathematics curriculum as STEM majors, they did not value the focus on speed within mathematics learning.

Callie and Lyka both expressed challenges in courses that reflected the fast-paced value of mathematics. Callie described how the majority of her undergraduate mathematics courses involved "the math teacher just kind of spit-vomiting a bunch of math information and then... I would just take notes as fast as I can and then I would have to look over it after class...so I think it was pretty difficult to kind of learn things that way." Lyka similarly expressed that in most of her math courses, if she did not understand a concept, she just accepted that "that's just something I don't know, ... because you really don't have the time or opportunity in class to go over what you missed. It's like, alright, on to the next material, you failed that exam, oh, better study well for the next one." However, Lyka enthusiastically described her experience in her mastery-based upper-level calculus courses where she was required to work through previously missed exam problems and present them to the professor. She emphasized how grateful she was that "I was able to actually go back and learn what I missed" in comparison to her other courses.

19.7.3 Innateness of Mathematical Ability

Another commonly held value in Western mathematics upholds the discipline as a place for those with innate mathematical intellect (Hottinger, 2016; Leyva et al., 2021; Reinholz et al., 2019). This underlines the belief that there are those who are "math people" and those who are not, emphasizing that math ability is something with which someone is born (Ernest, 2011). Scholars suggest that this value permeates into racialized perceptions of mathematical ability. In their study examining student of color's perceptions of racialized and gendered experiences in introductory undergraduate mathematics, Leyva et al. (2021) found that many of the participants who perceived certain classroom instances to imply mathematics ability as innate also perceived these instances to relay messages suggesting minoritized students inherently lack mathematical ability.

Lyka recognized this value explicitly, saying "in our society, they say there are two types of people – math people and not math people... I've always been into math from a very young age." However, rather than adhering to this value, she continued to say that "the reason for that can be just, some people didn't have the... extra attention that they may have needed, so they just grew up with this hatred towards math because it seemed so unattainable." She also described how her mathematics interest faded at the beginning of college and did not return until her mastery-based upper level calculus courses, where "having the opportunity to fail and learn from my mistakes, I could actually take an interest into what I was learning and not just focusing on memorizing the steps for the next quiz." In this regard, Lyka valued the work it took to learn from her mistakes rather than feeling disconcerted over a lack of "innate" ability. Callie expressed in retrospect that she should have given herself more credit for her hard work in her mathematics courses. She described how "I should have been more confident about [math], because they were hard classes, and I should have just been more proud about myself and not so hard on myself." She wished that there was more reassurance that "you are studying a somewhat difficult topic. It's okay and you'll get it eventually."

While Callie did not discuss explicitly how this mathematical value played into what she valued about her mathematics education, Lyka expressed this connection thoroughly when comparing experiences in different mathematics courses. She addressed how messages about mathematical ability interacted with gender and race and her feelings about asking for help. For example, she describes how "when you're a woman, you have to work a little bit harder, or like stand out a little bit more,... because if you mess up, or if you're not as good as your classmates, it's like a reflection not on you, but like on your gender and your race" and how that feeling "just added pressure when it comes to just needing help in math and not like performing well." However, enrolling in the mastery-based mathematics courses "naturally raised my interest and actually doing the math and understanding it, because I wasn't being penalized for not getting it on the first try." This type of course challenged notions about innate mathematical ability and instead valued students' hard work and perseverance to understand mathematics—something Lyka deeply valued from those courses.

19.7.4 Mathematics is Competitive

Another value of the mathematics discipline is competition, or the view that a competitive disposition is necessary to succeed in mathematics (Hottinger, 2016; Mendick, 2006). Mendick (2006) argues that this mathematical value emerges from the dichotomization in Western culture that positions competition as masculine and collaboration as feminine. Studies show women and student of color's preference for collaborative mathematics environments compared to competitive ones (Kogan & Laursen, 2014; Vaughan, 2002). A focus on competition may also undermine marginalized students' mathematical performance. Literature supports the position that men have less aversion to competitive situations and thus women may underperform on mathematics situations posed as more competitive (Cai et al., 2019; Niederle & Vesterlund, 2011). Scholars also suggest that the presence of stereotypes can negatively impact marginalized students' performance during competitive or high-risk mathematics situations (Steele, 2018).

Neither Callie nor Lyka spoke explicitly about their mathematical values related to competition versus collaboration in mathematics, but they described valuing a more collaborative classroom atmosphere. Callie talked about how "in other classes there was a lot of like, peer learning environments, and I did like those in certain classes but...I don't think I ever did math classes like that. It was definitely mostly the lecture, where it was just on your own." Thus, although she did not experience a collaborative mathematics course, she did value that type of learning in other courses. Lyka experienced both mathematics courses that were collaborative and others that were not. In terms of the mathematics courses that did not include collaborative components, Lyka mentions that she "really wasn't a fan" and those courses "didn't really align with my teaching style." However, she described a more collaborative course saying, "I liked the design of that class ... the teacher wouldn't just lecture on the board" and instead "days would just be dedicated to solving problems...and you could work in groups." A collaborative learning environment was also a large part of her deeply valued mastery-based calculus courses.

Regardless of the course style, both Callie and Lyka expressed valuing collaboration with peers outside of class rather than competing with them. Callie expressed that learning in courses that focused on fast-paced lectures was difficult for her, so she "always ended up having to ask people in class, like, how do you do this?" and they would share their understandings back and forth. Callie further talked about a specific friend with whom she had numerous classes with and "it was really great to have her as a friend to study with ... I think she was a really great help in that because she'd [give] all this type of encouragement ... like, I honestly don't know if I would have been able to score high without her." Rather than compete for a higher grade, Callie and her friend supported each other both through mathematical understanding and encouragement. Lyka as well collaborated with her peers frequently, especially in her upper-level calculus courses, describing how "most of our homework was done together ... it really worked out, we all got together and did math and then too, before exams [we'd study] ... it was a really good experience, like I was able to get through the [COVID-19] pandemic because of them." Similarly to Callie, Lyka valued collaboration with her peers both for mathematical and emotional support. While Callie and Lyka did not explicitly address competition in their mathematics courses, they each talk about valuing collaborative atmosphere inside and outside the classroom.

19.7.5 Mathematics as a Gatekeeper

Introductory undergraduate mathematics courses are often labeled as gatekeeper courses – courses structured to "weed out" certain students and prevent them from continuing in a STEM degree (Bryk & Treisman, 2010; Leyva et al., 2020). Mathematics often holds a reputation of being difficult to the point of being unattainable to learn except by those considered most intelligent or potentially "genius" (Leyva et al., 2021; Reinholz et al., 2019). This reputation also posits that mathematical reasoning is purely logical and emotionless, which are values often ingrained in dominant Western masculinity (Mendick, 2006). This value can permeate into distanced relationships between students and instructors. However, women often value having a positive relationship with their mathematics instructors more so than men (Solomon et al., 2011). Unfortunately, Rainey et al. (2019) found that women of color perceive the least amount of instructor care.

Both Callie and Lyka acknowledged society's perception of mathematics as extremely difficult and exclusive. Lyka talked about how some people "just grew up with this hatred towards math because it seemed so unattainable and so hard ... I think more than half of society will be like, yeah, that's not worth it ... either people say eww math or they're terrified of it." Callie expressed similar sentiments when she said, "most of the world... they're just like, oh yeah, [math is] hard, it's like unattainable, it's impossible to do." Lyka reflected on how she felt the "gatekeeping" value in her introductory mathematics courses, expressing how "it was really math holding me back" and in particular, described Calculus II as "the weed out class" and one of her "worst experiences." Callie related this value to the culture of mathematics and how she wishes "just to kind of get that like, culture of harshness out of math...or just not as daunting." However, they did express valuing challenging mathematics for the rewarding feeling of accomplishing a difficult task. Callie emphasized that "math inherently is, it can be hard, but then I remember like, when I would answer stuff, like I'd get really happy." Lyka as well described a similar feeling, saying "[math] is definitely not easy...I'm not gonna downplay the content of math...[but] just completing a problem is, it just feels like you've accomplished something." Overall, it appears that Lyka and Callie both valued mathematics as a challenging discipline, but disagreed that mathematics should continue to be promoted as impossible and exclusionary.

Both Callie and Lyka described situations where this value was reflected in their mathematics courses and the tension they felt in those scenarios. Callie described how one of her classes "was really hard. I remember I was very stressed out with that class." However, she was insecure about asking for help because "if I asked a stupid question, like, that's embarrassing and annoying." Thus, the harshness of mathematics translated to the existence of "stupid questions" even when the class was hard. Lyka illustrated this disconnect of values with a specific exam experience, where "I was the last one to leave and I was just so stressed out and I actually ended up crying after the [test]...I didn't understand the material...and I wasn't seen as a math doer, I was just seen as a failing student." She continued to say how after the test, she was further disappointed that there was no check-in or follow-up about the test and everything "proceeded in class like normal." Lyka and Callie experienced this value differently when it comes to relationships with mathematics instructors. Callie explained how "[going to office hours] was daunting. I did not like doing that...I went in once and this professor, he was just not very welcoming." Interestingly, she specified that "I think the math [office hours] were definitely a bit more intimidating for me, as opposed to the science classes." Thus, there is a specific placement of this value of intimidation in mathematics that Callie does not perceive in science. Lyka described a very positive experience where she felt an instructor went out of his way to support her and challenge this idea of "weeding out" weaker students by offering mathematics help to former students in their future mathematics courses. Overall, Lyka and Callie both experienced situations in their mathematics education that align with or challenge the value that mathematics is a "gatekeeper" and impossible. However, their own values align with the mindset that while math is difficult, students should feel supported and that they can succeed.

19.8 Discussion

The results of this work corroborate other studies showing the misalignment between marginalized students' mathematical values and the values of the mathematics discipline in the USA (Diekman et al., 2010; Fong et al., 2019; Leyva et al., 2021). Callie and Lyka described ways in which they noticed gendered and racialized sociohistorical values embedded in their mathematical experiences and the negative feelings associated with trying to navigate those values. This reflects the role of identity that stretches throughout Callie and Lyka's discussion about their values, suggesting the importance of gender, race, and other identities when considering women of color's values in mathematics. In some ways, they talked about this influence directly. For example, Callie spoke about how sometimes she felt like her actions represented those of her entire race and gender, which added pressure around asking for help. They also noticed how few women, people of color, and women of color faculty they saw in the mathematics department. Callie expressed how "just having like more women of color teaching or like, being in positions of power within the university" would have made her feel more supported and valued. In other ways, this influence was more subliminal. Callie and Lyka also identify as first-generation, which research suggests that cultural mismatch in higher education may impede first-generation students' help-seeking behavior (Chang et al., 2020).

Throughout their interviews, Callie and Lyka also expressed views on how to challenge exclusionary values and create mathematics spaces with values alignment. For example, they both discuss how they value mathematics that applies to the real-world or their lives and suggest integrating more applications into the classroom to support this value. In this way, the distinction between mathematical values and mathematics education values start to blur. This reflects how these categorization of values necessarily overlap and inform one another (Seah et al., 2016). It follows that one way to prompt mathematical values in the USA to better reflect the values of marginalized groups may be to start integrating more undergraduate mathematics pedagogies that reflect the mathematics values of these groups. These results suggest various implications for pedagogical practices to integrate women of color's values into the undergraduate mathematics classroom. One way is to include applications and connections in mathematics curriculum that reflect students' interests and future careers. Callie and Lyka also expressed positive attitudes toward collaborative, in-class activities. Research suggests that encouraging questions and prioritizing understanding over speed during collaborative activities can help challenge gendered and racialized values and support marginalized students' comfort and learning in class (Leyva, 2021). This aligns with Callie and Lyka's valuing of peer and instructor relationships. Encouraging student–student and instructor-student relationships during collaborative activities helps counter mathematics as emotionless and support values around community and care. Lastly, Lyka and Callie also valued learning from their mistakes, which challenges values around speed and innate ability. In practice, this could include test corrections or discussing homework problems in class. Lyka expressed that many of the above practices were interwoven into the very nature of her mastery-based mathematics courses. It was clear how much Lyka valued her experiences in these classes and suggests that utilizing mastery-based course designs in undergraduate mathematics education may be a way to broadly reflect women of color's values in the classroom and help build more equitable learning environments.

19.9 Conclusion

This work supports other results that suggest women of color's values are not reflected in undergraduate mathematics spaces in the USA (Abrams et al., 2013; Leyva, 2021). Lyka and Callie expressed misalignment with each of the five traditional mathematical values presented in the results. From a sociopolitical viewpoint, this suggests that women of color may feel excluded or hindered by a misalignment of values while navigating white, masculine discourses in undergraduate mathematics. However, Callie and Lyka experienced different pedagogical settings that influenced how corresponding mathematics education values emerged in the classroom. Lyka enthusiastically talked about her mathematics courses that challenged traditional mathematical values, while Callie felt like she was just getting through her mathematics courses that strongly reiterated traditional mathematical values. These results speak to the potential for institutions in the USA to better support women of color through values alignment in the mathematics classroom. Implementing pedagogy that portrays mathematical and mathematics education values that focus on collaborative learning, perseverance, and applications to the real-world may better reflect those of women of color in the USA and spur more positive cognitive and affective outcomes. Future work will explore this connection between values alignment and positive outcomes for women of color in the USA in undergraduate mathematics more explicitly. Additional interviews and survey data may add to these results by asking participants more directly their mathematical and mathematics education values and whether they perceive these values imbued in their undergraduate mathematics courses. Supporting and aligning with marginalized students' values in undergraduate mathematics classrooms may be one approach to answer the powerful call for more equitable and inclusive mathematics education in the USA.

References

Abrams, E., Taylor, P. C., & Guo, C.-J. (2013). Contextualizing culturally relevant science and mathematics teaching for indigenous learning. *International Journal of Science and Mathematics Education, 11*(1), 1–21.

Adiredja, A. P., & Andrews-Larson, C. (2017). Taking the sociopolitical turn in postsecondary mathematics education research. *International Journal of Research in Undergraduate Mathematics Education, 3*(3), 444–465.

Adiredja, A. P., & Zandieh, M. (2020). The lived experience of linear algebra: A counter-story about women of color in mathematics. *Educational Studies in Mathematics, 104*(2), 239–260.

Allen, R. L. (2004). Whiteness and critical pedagogy. *Educational Philosophy and Theory, 36*(2), 121–136.

American Mathematical Society (n.d.). *Equity, diversity and inclusion.* www.ams.org/about-us/diversity

Andersson, A., & Österling, L. (2019). Democratic actions in school mathematics and the dilemma of conflicting values. In P. Clarkson, W. T. Seah, & J. Pang (Eds.), *Values and valuing in mathematics education* (pp. 69–88). Springer International Publishing.

Barnes, M. (2000). Effects of dominant and subordinate masculinities on interactions in a collaborative learning classroom. In J. Boaler (Ed.), *Multiple perspectives on mathematics teaching and learning* (pp. 145–169). Ablex.

Battey, D., & Leyva, L. A. (2016). A framework for understanding whiteness in mathematics education. *Journal of Urban Mathematics Education, 9*(2), Article 2.

Bishop, A. J. (1988). *Mathematical enculturation: A cultural perspective on mathematics education.* Kluwer.

Bishop, A. J., FitzSimons, G., Seah, W. T., & Clarkson, P. (1999). *Values in mathematics education: Making values teaching explicit in the mathematics classroom.* Annual Meeting of the Australian Association for Research in Education and the New Zealand Association for Research in Education, Melbourne, Australia.

Branson, C. M. (2008). Achieving organisational change through values alignment. *Journal of Educational Administration, 46*(3), 376–395.

Bryk, A. S., & Treisman, U. (2010). Make math a gateway, not a gatekeeper. *Chronicle of Higher Education, 56*(32).

Bullock, E. C. (2019). Mathematics curriculum reform as racial remediation: A historical counterstory. In J. Davis & C. Jett (Eds.), *Critical race theory in mathematics education* (pp. 75–97). Routledge.

Cai, X., Lu, Y., Pan, J., & Zhong, S. (2019). Gender gap under pressure: Evidence from China's national college entrance examination. *The Review of Economics and Statistics, 101*(2), 249–263.

Carlone, H. B., & Johnson, A. (2007). Understanding the science experiences of successful women of color: Science identity as an analytic lens. *Journal of Research in Science Teaching, 44*(8), 1187–1218.

Chang, J., Wang, S., Mancini, C., McGrath-Mahrer, B., & Orama de Jesus, S. (2020). The complexity of cultural mismatch in higher education: Norms affecting first-generation college students' coping and help-seeking behaviors. *Cultural Diversity and Ethnic Minority Psychology, 26*(3), 280–294.

Crenshaw, K. (1991). Mapping the margins: Intersectionality, identity politics, and violence against women of color. *Stanford Law Review, 43*(6), 1241.

Da Silva, N., Verdejo, T. R., Dillon, F. R., Ertl, M. M., & De La Rosa, M. (2021). Marianismo beliefs, intimate partner violence, and psychological distress among recently immigrated, young adult Latinas. *Journal of Interpersonal Violence, 36*(7–8), 3755–3777.

Daniszewski, J. (2020, July 20). Why we will lowercase white. AP Style Blog. https://www.apstylebook.com/blog_posts/16

Darragh, L. (2013). Sticking with it or doing it quickly: What performances do we encourage in our mathematics learners? In V. Steinle, L. Ball, & C. Bardini (Eds.), *Mathematics Education: Yesterday, Today and Tomorrow: Proceedings of the 36th Annual Conference of the Mathematics Education Research Group of Australasia* (pp. 218–225). Mathematics Education Research Group of Australiasia.

Dede, Y., Akçakın, V., & Kaya, G. (2022). Identifying students' mathematical and mathematics educational values in Turkish culture: A cross-sectional study (Identificación de los valores matemáticos y de la didáctica matemática de los estudiantes en la cultura turca: Un estudio transversal). *Culture and Education, 34*(3), 597–629.

Del-Mundo, J. L., & Quek, K.M.-T. (2017). Balancing the old and the new: The case of second-generation Filipino American women. In K.M.-T. Quek & S.-R.S. Fang (Eds.), *Transition and change in collectivist family life: Strategies for clinical practice with Asian Americans* (pp. 67–77). Springer International Publishing.

Diekman, A. B., Brown, E. R., Johnston, A. M., & Clark, E. K. (2010). Seeking congruity between goals and roles: A new look at why women opt out of science, technology, engineering, and mathematics careers. *Psychological Science, 21*(8), 1051–1057.

Ernest, P. (2011). *The psychology of learning mathematics: The cognitive, affective and contextual domains of mathematics education.* LAP Lambert Academic Publishing.

Esmonde, I. (2011). Snips and snails and puppy dogs' tails: Genderism and mathematics education. *For the Learning of Mathematics, 31*(2), 27–31.
Esmonde, I., Brodie, K., Dookie, L., & Takeuchi, M. (2009). Social identities and opportunities to learn: Student perspectives on group work in an urban mathematics classroom. *Journal of Urban Mathematics Education., 2*(2), 18–45.
Espinosa, L. (2011). Pipelines and pathways: Women of color in undergraduate STEM majors and the college experiences that contribute to persistence. *Harvard Educational Review, 81*(2), 209–241.
Ferrare, J. J., & Hora, M. T. (2014). Cultural models of teaching and learning in math and science: Exploring the intersections of culture, cognition, and pedagogical Situations. *The Journal of Higher Education, 85*(6), 792–825.
Fong, C. J., Alejandro, A. J., Krou, M. R., Segovia, J., & Johnston-Ashton, K. (2019). Ya'at'eeh: Race-reimaged belongingness factors, academic outcomes, and goal pursuits among Indigenous community college students. *Contemporary Educational Psychology, 59*, 101805.
French, B. H., Lewis, J. A., Mosley, D. V., Adames, H. Y., Chavez-Dueñas, N. Y., Chen, G. A., & Neville, H. A. (2020). Toward a psychological framework of radical healing in communities of color. *The Counseling Psychologist, 48*(1), 14–46.
Geist, E. (2010). The anti-anxiety curriculum: Combating math anxiety in the classroom. *Journal of Instructional Psychology, 37*, 24–31.
Hannula, M. S. (2012). Looking at the third wave from the West: Framing values within a broader scope of affective traits. *ZDM Mathematics Education, 44*, 83–90.
Hill, J. L., Kern, M. L., Seah, W. T., & van Driel, J. (2021). Feeling good and functioning well in mathematics education: Exploring students' conceptions of mathematical well-being and values. *ECNU Review of Education, 4*(2), 349–375.
Hottinger, S. (2016). *Inventing the mathematician: Gender, race, and our cultural understanding of mathematics*. SUNY Press.
Howard, T. C., & Rodriguez-Scheel, A. (2017). Culturally relevant pedagogy 20 years later: Progress or pontificating? What have we learned, and where do we go? *Teachers College Record, 119*(1).
Hunter, J. (2021). An intersection of mathematics educational values and cultural values: Pāsifika students' understanding and explanation of their mathematics educational values. *ECNU Review of Education, 4*(2), 307–326.
Hunter, J., Hunter, R., Bills, T., Cheung, I., Hannant, B., Kritesh, K., & Lachaiya, R. (2016). Developing equity for Pāsifika learners within a New Zealand context: Attending to culture and values. *New Zealand Journal of Educational Studies, 51*(2), 197–209.
Idahosa, G. E., & Vincent, L. (2019). Enabling transformation through critical engagement and reflexivity: A case study of South African academics. *Higher Education Research & Development, 38*(4), 780–792.
Jaremus, F., Gore, J., Prieto-Rodriguez, E., & Fray, L. (2020). Girls are still being 'counted out': Teacher expectations of high-level mathematics students. *Educational Studies in Mathematics, 105*(2), 219–236.
Jett, C. (2019). Mathematical persistence among four African American male graduate students: A critical race analysis of their experiences. *Journal for Research in Mathematics Education, 50*, 311.
Kalogeropoulos, P., & Clarkson, P. (2019). The role of value alignment in levels of engagement of mathematics learning. In P. Clarkson, W. T. Seah, & J. Pang (Eds.), *Values and valuing in mathematics education* (pp. 115–127). Springer.
Kogan, M., & Laursen, S. L. (2014). Assessing long-term effects of inquiry-based learning: A case study from college mathematics. *Innovative Higher Education, 39*(3), 183–199.
Ladson-Billings, G. (1994). *The dreamkeepers: successful teachers of African American children*. Jossey-Bass Inc.
Ladson-Billings, G. (1995). Toward a theory of culturally relevant pedagogy. *American Educational Research Journal, 32*(3), 465–491.

Lee, H. F., & Seah, W. T. (2015). "Math is not for us, not an Indigenous thing, you know": Empowering Taiwanese Indigenous learners of mathematics through the values approach. In S. M. B. Greer (Chair), *Symposium conducted at the meeting of the eight international mathematics education and society conference*, Portland, OR, 21st to 26th June 2015.

Leyva, L. (2016). An intersectional analysis of Latin@ college women's counter-stories in mathematics. *Journal of Urban Mathematics Education, 9*(2), 41.

Leyva, L. (2017). Unpacking the male superiority myth and masculinization of mathematics at the intersections: A review of research on gender in mathematics education. *Journal for Research in Mathematics Education, 48*(4), 397–433.

Leyva, L. (2021). Black women's counter-stories of resilience and within-group tensions in the white, patriarchal space of mathematics education. *Journal for Research in Mathematics Education, 52*, 117–151.

Leyva, L. A., McNeill, R. T., Marshall, B. L., & Guzmán, O. A. (2021). "It seems like they purposefully try to make as many kids drop": An analysis of logics and mechanisms of racial-gendered inequality in introductory mathematics instruction. *The Journal of Higher Education, 92*(5), 784–814.

Leyva, L. A., Quea, R., Weber, K., Battey, D., & López, D. (2020). Detailing racialized and gendered mechanisms of undergraduate precalculus and calculus classroom instruction. *Cognition and Instruction, 39*(1), 1–34.

Mathematical Association of America (n.d.). *About MAA*. https://maa.org/about-maa

Martin, D. B. (2003). Hidden assumptions and unaddressed questions in mathematics for all rhetoric. *The Mathematics Educator*, *13*(2), Article 2.

Martin, D. B. (2006). Mathematics learning and participation as racialized forms of experience: African American parents speak on the struggle for mathematics literacy. *Mathematical Thinking and Learning, 8*(3), 197–229.

Masten, C. L., Telzer, E. H., & Eisenberger, N. I. (2011). An fMRI investigation of attributing negative social treatment to racial discrimination. *Journal of Cognitive Neuroscience, 23*(5), 1042–1051.

McGee, E. (2016). Devalued Black and Latino racial identities: A by-product of STEM college culture? *American Educational Research Journal, 53*(6), 1626–1662.

McGee, E., & Martin, D. B. (2011). "You would not believe what I have to go through to prove my intellectual value!" Stereotype management among academically successful Black mathematics and engineering students. *American Educational Research Journal, 48*(6), 1347–1389.

McNeill, R. T., Leyva, L. A., & Marshall, B. (2022). "They're just students. There's no clear distinction": A critical discourse analysis of color-evasive, gender-neutral faculty discourses in undergraduate calculus instruction. *Journal of the Learning Sciences, 31*(4–5), 630–672.

Mendick, H. (2006). *Masculinities in mathematics*. McGraw-Hill Education (UK).

National Council of Teachers of Mathematics. (2014). *Access and equity in mathematics education.* National Council of Teachers of Mathematics.

Niederle, M., & Vesterlund, L. (2011). Gender and competition. *Annual Review of Economics, 3*(1), 601–630.

Oppland-Cordell, S. B. (2014). Urban Latina/o undergraduate students' negotiations of identities and participation in an Emerging Scholars Calculus I workshop. *Journal of Urban Mathematics Education, 7*(1), 19–54.

Ortiz, N. A., Morton, T. R., Miles, M. L., & Roby, R. S. (2019). What about us? Exploring the challenges and sources of support influencing Black students' STEM identity development in postsecondary education. *The Journal of Negro Education, 88*(3), 311–326.

Perez, T., Cromley, J. G., & Kaplan, A. (2014). The role of identity development, values, and costs in college STEM retention. *Journal of Educational Psychology, 106*(1), 315–329.

Rainey, K., Dancy, M., Mickelson, R., Stearns, E., & Moller, S. (2019) A descriptive study of race and gender differences in how instructional style and perceived professor care influence decisions to major in STEM. *International Journal of STEM Education, 6*(6), 1–13.

Raths, L. E., Harmin, M., & Simon, S. B. (1987). Selections from 'values and teaching'. In J. P. F. Carbone (Ed.), *Value theory and education* (pp. 198–214). Malabar: Robert E. Krieger.

Reinholz, D. L., Matz, R. L., Cole, R., & Apkarian, N. (2019). STEM Is not a monolith: A preliminary analysis of variations in STEM disciplinary cultures and implications for change. *CBE—Life Sciences Education, 18*(4), mr4.

Saldaña, J. (2013). *The coding manual for qualitative researchers* (2nd ed). SAGE Publications.

Schwartz, S. H., & Rubel, T. (2005). Sex differences in value priorities: Cross-cultural and multimethod studies. *Journal of Personality and Social Psychology, 89*, 1010–1028.

Schwartz, S. H., Melech, G., Lehmann, A., Burgess, S., Harris, M., & Owens, V. (2001). Extending the crosscultural validity of the theory of basic human values with a different method of measurement. *Journal of Cross-Cultural Psychology, 32*(5), 519–542.

Seah, W. T. (2018). Improving maths pedagogy though student/teacher valuing: Lessons from five continents. In G. Kaiser, H. Forgasz, M. Graven, A. Kuzniak, E. Simmet, & B. Xu (Eds.), *Invited lectures from the 13th international congress on mathematics education* (pp. 561–580). Springer.

Seah, W. T., & Andersson, A. (2015). Valuing diversity in mathematics pedagogy through the volitional nature and alignment of values. In A. Bishop, H. Tan, & T. N. Barkatsas (Eds.), *Diversity in mathematics education* (pp. 167–183). Springer International Publishing.

Seah, W. T., Andersson, A., Bishop, A., & Clarkson, P. (2016). What would the mathematics curriculum look like if values were the focus? *For the Learning of Mathematics, 36*(1), 14–20.

Slemp, K. (2020). *Latino, Latina, Latin@, Latine, and Latinx: Gender inclusive oral expression in Spanish (Order No. 29247791)* [Master's thesis, Western University]. ProQuest Dissertations & Theses Global. https://www.proquest.com/dissertations-theses/latino-latina-latin-latine-latinx-gender/docview/2714874201/se-2

Steele, C. (2018) Stereotype threat and African-American student achievement. In D. B. Grusky (Ed.), *Social stratification: Class, race, and gender in sociological perspective*. Routledge.

Solomon, Y., Lawson, D., & Croft, T. (2011). Dealing with 'fragile identities': Resistance and refiguring in women mathematics students. *Gender and Education, 23*(5), 565–583.

Street, C. (2023). *Exploring women of color's expressions of mathematical identity: The role of institutional resources and mathematical values* (Order No. 30526284) Doctoral dissertation, Colorado State University. ProQuest Dissertations and Theses Global.

Street, C., Apkarian, N., Gehrtz, J., Tremaine, R., Barron, V., Voigt, M., & Hagman, J. E. (2021). *X-PIPS-M Data Summary*. arXiv preprint arXiv:2111.01795.

Sue, D. W., Sue, D., Neville, H. A., & Smith, L. (2022). *Counseling the culturally diverse: Theory and practice* (9th ed.). Wiley.

Tang (唐恒钧), H., Seah (佘伟忠), W. T., Zhang (张侨平), Q., & Zhang (张维忠), W. (2021). The mathematics learning attributes valued by students in Eastern China. *ECNU Review of Education, 4*(2), 261–284.

Tucker, A. (2013). The history of the undergraduate program in mathematics in the United States. *The American Mathematical Monthly, 120*(8), 689–705.

Vaughan, W. (2002). Effects of cooperative learning on achievement and attitude among students of color. *The Journal of Educational Research, 95*(6), 359–364.

Voigt, M., Hagman, J., Street, C., Guglielmo, J., Martinez, A., & Tremaine, R. (2022). A quantitative critical analysis of instructional practices and math affect. In S. S. Karunakaran & A. Higgins (Eds.), *Proceedings of the 24th Conference on Research in Undergraduate Mathematics Education* (pp. 666–674).

Zhang, Q. (2019). Values in mathematics learning: Perspectives of Chinese mainland primary and secondary students. In P. Clarkson, W. T. Seah, & J. Pang (Eds.), *Values and Valuing in Mathematics Education* (pp. 185–196). Springer International Publishing.

[illegible] Dissertation [illegible]. [illegible]
[illegible]
[illegible]
[illegible]

Chapter 20
Turkish High School Students' Mathematics Values in Terms of Feelings About Mathematics

Yüksel Dede, Veysel Akçakın, and Gürcan Kaya

20.1 Introduction

Although the future is difficult to predict, we can adapt by keeping up with some popular trends today, learning to develop and even shape the future, and helping our children learn. Here, we must develop not only students' knowledge and skills but also their attitudes and values that can lead them to ethical and responsible actions (The Organization for Economic Co-operation and Development [OECD], 2022). In this sense, the OECD Learning Compass 2030 report proposed a very conceptual and helpful framework that defines the knowledge, skills, and values (and attitudes) that learners need to fulfill their potential and contribute to the well-being of their societies, cultures and the world (OECD, 2019).

Here, the well-being of communities in general, and individuals in particular, depends on the practical characteristics of the speaker's environment at the time of the judgment (Alexandrova, 2017). This concept of wellbeing has been studied in different dimensions through different definitions in a wide variety of areas (e.g. health, education, economy etc.). Similarly, OECD (2017) has considered well-being at school in terms of psychological, physical, cognitive, and social contexts. When considered within the classroom context, well-being changes depending on the class-room environment and the values adopted in that classroom environment (Hill et al.,

Y. Dede (✉)
Gazi University, Yenimahalle, Türkiye
e-mail: ydede@gazi.edu.tr

V. Akçakın
Uşak University, Uşak, Türkiye
e-mail: veysel.akcakin@usak.edu.tr

G. Kaya
Afyon Kocatepe University, Afyonkarahisar, Türkiye
e-mail: gurcankaya@aku.edu.tr

Y. Dede et al. (eds.), *Values and Valuing in Mathematics Education*,
https://doi.org/10.1007/978-981-99-9454-0_20

2021). Likewise, mathematical well-being [MWB] is closely related to one's ultimate or core values in the context of positive emotions in mathematics education (Hill & Seah, 2022).

Based on these considerations, this chapter aims to determine students' mathematics (mathematical values and mathematics educational values) values in Türkiye according to their feeling good about dealing with mathematics in school. For this aim, the mathematics values of Turkish Grade 9 students who think/do not think that mathematics is taught well in schools and who feel good or, on the contrary, feel bad while dealing with mathematics are only reported here, as an extreme or deviant case sampling, in Turkish schools.

20.1.1 Theoretical Background

This section discusses the relationships between culture, values, mathematics, and student values emerging in the context of well-being in mathematics classroom practices.

20.1.1.1 Relationships Among Culture, Values, and Mathematics

Culture is defined in this chapter as "a way of dividing people into groups according to some features of these people which helps us to understand something about them and how they are different from or similar to other people" (Scollon et al., 2012, p. 3). In this context, culture encompasses communication (language, symbols, and artifacts), interaction (traditions, practices, and patterns of interaction), and values (shared values, beliefs, norms, and expectations) that guide people or groups (Pang, 2005). As can be seen from this, each culture and educational system may have its own culture and different student expectations. Since there has been a clear and transparent link between culture and the quality of the education system (see Williams, 1961), the concepts of culture and education are already intertwined. Therefore, sociocultural factors are crucial in transferring education structure, content, and processes to new generations (Powe, 1993).

On the other hand, although values and valuing ideas are always totemic in education, these ideas bring out some problems when applied to mathematics. Even nowadays, mathematics is considered a value-free discipline (Bishop, 1991) by most mathematics teachers and students. According to them, school mathematics consists only of learning the skills of manipulating numbers, some abstract ideas of algebra, and dealing with sometimes geometry and measurement concepts. Similarly, mathematics learning for students in schools is limited to using and learning pedagogical methods and strategies to provide good marks on tests and exams (Clarkson et al., 2019). Mathematics is therefore seen as the discipline least affected by sociocultural factors (Stigler & Perry, 1988). However, in recent years, mathematical education research and practices have begun to discuss this from a critical perspective against this judged approach to mathematics and learning/teaching mathematics (Clarkson et al., 2019). So, there are more recent claims in the literature that mathematics is value-laden and is, thus, influenced by culture (see Bishop, 1988; Ernest, 2007). From this perspective, it is also noted in the literature that students' high achievement in mathematics is more closely related to cultural values rather than specific mathematics teaching approaches (Askew et al., 2010). In this sense, students' learning and understanding of mathematics may vary in different cultures and societies (see Dede et al., 2020). Hence the same classroom practices may reflect different underlying values across cultures in which a particular value can be espoused differently (Hill & Seah, 2022). Thus, values affect the learning practices and processes of individuals/students in general, and mathematics learning in particular, because values are enacted beliefs and deep affective structures in the minds of individuals (Bishop & Seah, 2008). They have a crucial function in learners' choice of whether to engage in mathematical tasks (Bishop et al., 2006). In this context, Seah and Andersson (2015) definition of value is useful:

> The convictions that an individual has internalized as being the things of importance and worth. What an individual values defines for her/him a window through which s/he views the world around her/him. Valuing provides the individual with the will and determination to maintain any course of action chosen in the learning and teaching of mathematics. They regulate how a learner's/teacher's cognitive skills and emotional dispositions are aligned to learning/teaching in any given educational context. (p. 169)

In general, Bishop argued quite early in his writing that values are one of the critical factors underlying individuals' decision-making processes (see Bishop & Whitfield, 1972). But he, with others, gradually expanded on this general argument. So, values can be viewed as both a sociocultural and conative variable in the learning environment and processes of mathematics (see Goldin, 2019; Seah, 2019). Because while the sociocultural environment shapes our values (Seah, 2019), conative variables "regulate the individual's activation of cognitive skills and affective dispositions in complementary ways" (Seah, 2019, p.107). In this sense, three practical value frameworks have been proposed for mathematics classes, two of which are directly mathematics education and the other specific to cultural values. Firstly, Bishop (1988) proposed three main categories for classifying values taught in mathematics classrooms: general education, mathematical values, and mathematics educational values. These values cannot be completely separated from each other (Seah & Bishop, 2000). Then, Lim and Ernest (1997) proposed three classifications for the values taught in mathematics lessons: epistemological, social and cultural, and personal values. Here, the epistemological values proposed by Lim and Ernest (1997) can, to some extent, be associated with the mathematical values of Bishop (1988). Similarly, the social and cultural values suggested by Lim and Ernest (1997) may be related to Bishop (1988) general educational value category. Finally, Lim and Ernest (1997) personal values category may be evaluated in the category of mathematics education values of Bishop (1988) (see Dede, 2011).

As for cultural values, Hofstede (1980, 2023) determined that cultural values differed between countries and proposed the following six cultural dimensions: power distance index (high versus low), individualism versus collectivism, masculinity versus femininity, uncertainty avoidance index (high versus low), long- versus short-term orientation, and indulgence (see Hofstede, 1980, 2023). Cooper et al. (2007) evaluated the reflection of these cultural values of Hofstede on classroom practices. According to this evaluation, in societies such as Türkiye, with collectivism, a high-power gap, and a high uncertainty avoidance, the teacher-centered teaching approach is at the forefront; individuals are asked to make sacrifices for society, and uncertain situations are not emphasized. On the contrary, in individual, low power range, and low uncertainty avoidance societies such as Germany, student-centered education is at the forefront, learning environments open to technological changes, uncertain, open-ended, and even conflict environments are designed as an element that encourages teaching. However, it should be noted here that Hofstede's cultural dimensions are not the ultimate focus for studying countries and therefore are not taken as absolute data. In this context, these cultural dimensions are used in this section only to assess what some of the findings mean.

20.1.1.2 Student Values Emerging in the Context of Well-Being in Mathematics Classroom Practices

MWB is fulfilling one's ultimate or core values, accompanied by positive feelings and functioning in mathematics education (Hill & Seah, 2022). In this context, a sense of well-being will likely change according to culture, society, school/lesson culture, and the values attributed to that culture and lesson. Also, an individual's well-being in mathematics may be expected to differ from the well-being experienced in other lessons (e.g., music) depending on the degree to which they perceive and value the lesson (Hill et al., 2021). In this sense, Clarkson et al. (2010) proposed a framework of MWB that considered three domains: a cognitive domain as knowledge and skills needed to do mathematics at school, an affective domain as combining values in mathematics education, and finally, an emotional domain as feelings and reactions to mathematics. According to Bishop (2012), one of the important features of MWB, which represents a new way of thinking about value development in mathematics education, is a developmental rather than just a sequential form. In this way, MWB offered valuable data for classroom practices for researchers, especially teachers, and in this context, Bishop (2012) proposed the six stages of the construct of MWB. These stages, from simple to complex, are awareness of mathematical activity, recognition and acceptance of mathematical activity, positively responding to mathematical activity, valuing mathematical activity, having an integrated and conscious value structure for mathematics, and finally, independently competent and confident in mathematical activity (see Bishop, 2012).

On the other hand, Hill and Seah (2022) stated that the MWB framework of Clarkson, Bishop, and Seah has some limitations and claimed that the framework was not fully compatible with the many later contemporary conceptualizations of well-being or human flourishing. They also noted that the framework neglects some social aspects of well-being and learning mathematics and has not been validated for school-age students. They proposed different dimensions of a MWB framework identified within positive psychology linked to positive mathematics outcomes with descriptions of the dimensions and examples of supporting studies. Their seven dimensions are accomplishment, cognitions, engagement, meaning, perseverance, positive emotions, and relationships (see Hill et al., 2021). Considering that the values affecting learning mathematics processes are like the MWB dimensions (Hill & Seah, 2022), these MWB dimensions may affect whether a student feels profoundly good when doing mathematics. Tiberius (2018) discussed values at two levels according to their impact on well-being.

> ... I mentioned that there are ultimate values and instrumental values. We value some things (like friendship, health, and meaningful work) for their own sake and others (money, health-care, and a college diploma) as a means to something more ultimate. I don't think the line between ultimate and instrumental values is always a bright one for people. ... It does matter that some goals we have are so clearly instrumental that it starts to strain the notion of "value" to call them values. For example, I aim to floss my teeth regularly to avoid gum disease and ultimately for the sake of my physical health. It would be odd to say that I value flossing my teeth.... When this happens, I think it's more natural to talk about desiring the means to a more ultimate value… (pp. 42–43)

Hill and Seah (2022) discussed the difference between these two values when directly associating them with classroom practices. For this, they considered values such as teacher-student relations, group work, or respect as instrumental values, and the value of relationships that are tried to be reached along these values as an ultimate value. In addition, they stated that instrumental (mathematical) values sometimes allow for more than one ultimate value through the following example: valuing teacher explanations can provide the valuing of relationships, such as the teacher providing support, cognitions such as facilitating better understanding, and accomplishment, such as explanations enhancing mathematical accuracy.

In recent years, it has been observed that there has been an increase in the number of studies based on different regions and cultural backgrounds to determine student values in mathematics classrooms (some of which appear in this volume). As mentioned before, because different cultures contain different values, mathematics teachers working in other cultures do not teach the same societal or pedagogical values even if they teach the same curriculum (Bishop et al., 2000). Thus, these studies investigating student values in mathematics classrooms naturally reach different results. As for recent studies about student values, for example, Dede (2019) found relevance, practice, rationalism, and fun values for Turkish students and relevance, fun, rationalism, and consolidating values for German students. On the other hand, he also identified three value categories for Turkish immigrant students in Germany: relevance, rationalism, and communication. When these values are considered in the context of MWB, Turkish and German students emphasize the values of rationalism, relevance, and fun. Accordingly, this student group aims at the ultimate values of cognition, meaning, and positive emotions in the context of MWB. On the other hand, immigrant students differ from these two groups and prioritize the ultimate value of relationships through the instrumental value of communication. In a slightly different context, Aktaş et al. (2021) examined the mathematics educational values of students in religious vocational middle schools in Türkiye, where religious content and contexts are taught more intensively. And they determined that the importance given by the students in these schools to instrumental values such as relevance, learning approach, consolidating, and practice decreased as the grade level increased. From this, it can be said that the students in these schools prioritize the ultimate values of meaning, cognition, and engagement (and perseverance), respectively, through the instrumental values of relevance, learning approach, consolidating, and practice.

Moving to different cultures, Hong Kong students stated that fun, achievement, student participation, and teacher support are essential for effective mathematics learning (Law et al., 2012). From this, it can be stated that Hong Kong students emphasize the ultimate values of positive emotions, accomplishment, engagement, and relationships through the instrumental values here. Hunter (2021), on the other hand, found that Pāsifika students in New Zealand prioritized the instrumental values of practice, family, respect, and persistence in their mathematics classrooms. Accordingly, Pāsifika students in New Zealand emphasize the ultimate values of engagement and perseverance, relationships, positive emotions, and perseverance. Finally, Pang and Seah (2021) determined that Korean students emphasized the values of understanding, connections, fun, accuracy, and efficiency in their mathematics learning. For this, Korean students give more importance to the ultimate values of cognition and meaning, positive emotions, accomplishment, and engagement.

20.2 Research Process

The Turkish education system generally has a rationalist philosophy in terms of epistemology, a deductive philosophy in terms of problem-solving approaches, a centrally directed philosophy in terms of management approaches, and finally, a society-centered (i.e., Durkheimian) philosophy in terms of the criterion of correct knowledge (Hesapçıoğlu, 2009). In this sense, Türkiye's current education system structure has shown a growing tendency towards reform-driven student-centered approaches, particularly since the early 2000s when the curriculum was updated four times in 2005, 2009, 2013, and 2018. In this sense, the current mathematics curriculum in Türkiye also gives more attention to developing everyday skills such as using Turkish correctly, effectively, and beautifully, critical thinking, creative thinking, decision-making, and information technologies. (Ministry of National Education of Türkiye [in Turkish: MEB], 2018). Also, ten moral and cultural values, such as justice, friendship, honesty, self-control, patience, respect, love, responsibility, patriotism, and benevolence, have been included in the curriculum since 2018.

Compulsory education in Türkiye is 12 years. The first four years are called primary school (Grades 1–4), the second four are called middle school (Grades 5–8), and the last four are called high school (Grades 9–12). The Turkish education system generally focuses on getting a promising career on high-stake exams with multiple-choice tests, which puts pressure on Turkish students and their families (Dede & Tasos, 2019). In this context, this chapter aimed to determine Turkish

Grade 9 students' mathematics values in the context of the MWB. These Grade 9 students in Türkiye are first-year high school students and can be a good resource for assessing the values conveyed from middle school to high school. As for determining values, qualitative research designs are used to measure values because of the nature of values (Seah, 2008). In recent years, there has been a trend toward research on the determining and developing values in mathematics classrooms, and as a result of this interest, measuring values in the classrooms have been assessed using valid and reliable Likert-type measurement tools (see Akyıldız et al., 2021; Dede, 2011; Dede & Tasos, 2019; Seah et al., 2017; Zhang et al., 2016 for Likert-type scale and, Dede et al., 2022 for bipolar scale). In this context, this chapter planned to measure values in mathematics classrooms using a different measurement tool than the ones mentioned above.

In this explanatory sequential mixed design study, firstly, quantitative data were collected. Then qualitative data were collected and analyzed to explain and deepen the understanding of the quantitative data (see Creswell & Plano Clark, 2018). In this context, quantitative data in this study were obtained through two different questions. The first was acquired by a Likert-type question form, including the statement, "I learn mathematics well at school." This form was given to 343 Grade 9 students selected according to the deviant or extreme sampling method, one of the purposive sampling methods (see Cohen et al., 2011). Accordingly, students who marked the numbers 1 (if they did not learn mathematics well at school) and 5 (if they learned mathematics well at school) for the Likert-form question "I learn mathematics well at school" were determined. In this chapter, only the answers given by these students were analyzed and reported.

The second question was, "Imagine that we are going to produce a magic pill. Anyone who takes this magic pill will become very good at mathematics! What will you choose to be the top 3 main ingredients of this magic pill?" It was derived from a semi-open-ended question expressed as "Be imaginative; this main ingredient can be something we can touch and see or something we can feel but cannot see, for example." In this semi-open-ended question, students were asked to write three components of importance and justify their reasons for choosing them. By trying to measure students' values through such a semi-open-ended question, it is considered that the closed measure of quantitative approaches and incomplete assessment of qualitative approaches are avoided, allowing researchers, especially participants, to focus more on the phenomenon under investigation. The study's qualitative data were obtained from semi-structured interviews with the students to determine the details and the underlying reasons for the answers to the above questions.

Examination of the students' responses showed that 79 of the 343 (23%) Grade 9 students marked 1, and 68 of the students (nearly 20%) marked 5. As can be seen, the percentages and numbers of both groups are close to each other, hence it is legitimate for comparisons between both groups of students. Thus, the data of a total of 147 students were analyzed and reported.

Interestingly the 147 students whose data were analyzed gave different answers to the magic pill question. Although they were asked to state three main ingredients, which most did, some wrote 4, and some wrote just one main ingredient. For this reason, the number of students differ for the value category frequencies. The data were analyzed using the constant comparative analysis method. In this context, firstly, some codes were created according to the open coding of the data. These codes were developed by examining the students' answers line by line, dividing them into parts, and assigning words with similar meanings to the same code. After the codes were created, they were examined in detail, and appropriate categories were developed which encompassed groups of codes (see Strauss & Corbin, 1998). In addition, the data were quantified by giving the frequencies of these values categories.

The study's qualitative data were obtained from semi-structured interviews with ten Grade 9 students (students selected according to the deviant or extreme sampling) to determine the possible reasons underlying these emerging categories. For example, one of the value categories that emerged was "ability." During the interview, students were asked, "Can ability be one of these main ingredients? Please explain." The value categories obtained in the study are discussed and reported together with the reasons given by the students. An example of the coding process for two value categories is shown in Table 20.1.

Different reliability methods have been used to ensure the reliability of the study data. Member checking (see Birt et al., 2016, for details) was used first. Transcripts of student answers in the interviews were made for this purpose and submitted for students' approval without any changes to the transcripts (see Creswell, 1998). Following this, the answers given by the students were first read a few times independently by the three researchers (authors of this chapter) and categorized through open coding. Then, these categories, which the researchers created independently, were discussed together by the researchers. These codes and categories, which the researchers developed together, were sent to two experts with a doctorate in mathematics education for review. To conclude this process, a consensus was sought on the codes and categories, taking into account the external experts' comments. Hence peer review was used for the second reliability of the data analysis (Lincoln & Guba, 1985). Also, the participants' statements were directly included in the text, and attention was paid to a thick description and richness in reporting the findings (see Creswell, 2012). Finally, the values emerging in the current study were assessed in the context of Bishop's (1988) and Lim and Ernest (1997) mathematics values,

Table 20.1 Sample analysis of student statements

Expressions for ability value	**Reasons in an interview for inclusion in this category**
Being good at maths	Statements "being good at mathematics" and "mathematical intelligence" indicate ability
Mathematical intelligence	
Expressions for teacher value	**Reasons in an interview for inclusion in this category**
Listening to the teacher	These statements refer to an activity that is directly related to the teacher
Teaching method	
Teacher	

Hofstede's (1980, 2023) cultural values, and the MWB of Hill et al. (2021) frameworks. Thus, theoretical triangulation was implemented on the values (see Cohen et al., 2011).

20.3 Findings

As a result of the analysis of the answers given to the "magic pill" question by the students who marked the 1 and 5 Likert categories for the "I learn math well at school." Six value categories were obtained: Ability, effort, mathematics concepts, fun, teacher, and materials that enhance thinking. The percentage frequencies related to these values are given in Table 20.2. Although the number of students in the two groups is similar, the number of responses for the groups is quite different: 153 compared to 127. Hence in the analysis, percentage frequencies are compared in the Table.

The explanations for each of the six values given in Table 20.2 are as follows:

Ability: With this value, it is stated that ability is essential in mathematics and mathematics learning. When the students' answers were examined, it was found that the students who thought they learned mathematics well at school attributed more importance to ability while learning mathematics (in percentage) compared to those who did not think that they learned mathematics well. However, as can be seen from Table 20.2, the percentages of both groups of students are close to each other. Therefore, it can be said that both groups of students attach more or less equal importance to their ability while learning mathematics. In addition, as seen in Table 20.2, both groups of students state that ability is the most critical component in mathematics learning. The sample statements of the students in both groups about the ability category are given in Table 20.3.

In the semi-structured interviews conducted with the students in these two groups (who marked the 1 and 5 options for the relevant question), they were asked to explain

Table 20.2 Student choices for the "I learn mathematics well at school" and value category frequencies of "magic pill."

Value categories	1	5
Ability	27.45%	33.07%
Effort	18.95%	16.54%
Mathematical Concepts	18.95%	13.39%
Fun	16.34%	13.39%
Teacher	11.11%	14.17%
Materials Enhance Thinking	7.19%	9.45%
Totals (f)	153	127

1: Students who think that they did not learn mathematics well at school

5: Students who think they learned mathematics well at school

Table 20.3 Sample student statements of ability value

Learned mathematics well (5)	Didn't learn mathematics well (1)
Maths is certainly about intelligence and attention Ability Intelligence, ability to understand Being good at maths	Intelligence is very important Ability to understand Ability to solve questions Without talent, people don't want to do anything

why the ability value is an important component in mathematics learning. Since the S1-coded student, who believes that he does not learn mathematics well at school, and the S5-coded student, who believes that he has learned mathematics well at school, made explanations that more clearly reflect the characteristics of their own categories, it was preferred to use the expressions of these students while exemplifying the semi-structured interviews. In this context, the following is an excerpt from an interview regarding the ability value of a student who stated that he learned mathematics well at school and marked himself as 5, not a typical student of this group. (including the effort value) (R: Researcher, S5: a student who stated that he learned mathematics well at school):

R: Could ability be one of the components of learning mathematics?

S5: Yes. Ability matters the most.

R: Why? Can you explain a little more?

S5: For example, two people love football. One of these two people is Ronaldo, and the other is another football player. Both football players love football. The other football player can't be a Ronaldo, he can go up to a level, but he can't be a Ronaldo. Such ability matters.

R: Is putting in the effort a part of learning math? What do you think?

S5: Of course, it's important to put in the effort. A little effort gives rational value. But as I said in the previous example, he cannot reach the level of those with high ability and cannot be a Ronaldo even if he puts in the effort.

Effort: This value emphasizes the importance of study, effort, persistence, etc. while learning mathematics. As seen in Table 20.2, the percentage of the effort value are close to each other in both groups of students who think they have learned mathematics well at school and those who do not. In addition, students who do not think they learned mathematics well at school emphasized this value more in learning mathematics than those who did. Some sample student statements are given in Table 20.4.

Table 20.4 Sample student statements of effort value

Learned mathematics well (5)	Didn't learn mathematics well (1)
Effort	Being hardworking
Studying yourself	Studying hard
Hardworking	Make a person hardworking
Regular study	Study

Here is an excerpt from an interview regarding the ability value of a student who stated that he did not learn mathematics well at school and marked himself as 1, not a typical student of this group. (including the ability value) (S1: a student who stated that he did not learn mathematics well at school and marked 1):

R: Could ability be one of the components of learning mathematics?

S1: Yes.

R: Why? Can you explain a little more?

S1: Ability and intelligence are necessary to learn mathematics very well.

R: How are intelligence and ability required in mathematics? Can you explain?

S1: Yes, it is required. Because if you have a talent for mathematics, you will learn the subject more easily.

R: If you need the ability for math?

S1: Those who don't have the talent can make up for it by putting in the effort because you can do the math by studying.

R: Then you say that you can learn mathematics by studying?

S1: Yes. Those without ability do not learn as easily as those with ability. For example, those who have the ability can understand by studying for a short time. Those who don't have the ability have to spend a long time. As I said, if we don't have the ability, we can reach a certain level by working.

R: Here, you specified the level. Can you explain the level you stated in terms of ability and effort?

S1: Those with talent pass this level. Those who put in the effort do not go as far as those with talent. Because with talent, they learn in a short time.

Mathematical Concepts: This value indicates the importance of knowing and understanding mathematical concepts, definitions, processes, etc., in mathematics learning. As seen in Table 20.2, both groups of students rated this value relatively highly than the other values. However, it can be said that students, who do not think they learned mathematics well at school, emphasized mathematical concepts somewhat lower in percentage than the other group. However, as seen from the interview excerpts below, it has also been revealed that these students lack understanding regarding the role and place of mathematical concepts in mathematics learning. In addition, although students in both groups emphasized different mathematical concepts for mathematical concept value, they generally emphasized mathematical formulas, operations, and the use of these formulas and functions (see Table 20.5).

Here is an excerpt from an interview with a student who stated that he did not learn mathematics well at school regarding this value:

Table 20.5 Sample student statements of mathematical concepts value

Learned mathematics well (5)	Didn't learn mathematics well (1)
Formulas	Four operations
Four operations	Formulas
Algebraic expressions	Angles
Shapes	Ratio theorem

R: Are mathematical concepts important in learning mathematics? Can you explain?

S1: Knowing some formulas might help.

R: Which formulas? Can you give an example?

S1: For example, algebraic expressions.

On the other hand, below is an excerpt from an interview with a student who stated that he learned mathematics well at school:

R: Are mathematical concepts important in learning mathematics? Can you explain?

S5: Yes. Since some mathematical concepts are included in many other mathematics subjects, we learn mathematics better if we know these concepts related to other mathematical subjects. However, I don't think that knowing these mathematical concepts that are related to other mathematics subjects will have much effect on our learning of mathematics because one can learn any subject later by studying it.

Fun: This value expresses having fun, being happy, loving mathematics, and having self-confidence while doing or learning mathematics. As seen in Table 20.2, both groups of students emphasized the value of fun in mathematics, especially those who did not think they learned mathematics well in school (see Table 20.6).

Two short excerpts from the interviewed students' opinions about the fun value are given below:

S1: Because doing a lesson you love is always easy and fun.

S5: We want to do more if we have fun at what we do.

Teacher: This value includes the teachers' teaching method in mathematics learning, the teachers' characteristics and behaviors, and the teachers' communication with students. In this context, students in both groups emphasized the teacher factor while learning mathematics (see Table 20.2). Some sample student statements regarding this value are given in Table 20.7.

Here is an excerpt from an interview with a student who stated that he did not learn mathematics well at school and about teacher value:

Table 20.6 Sample student statements of fun value

Learned mathematics well (5)	Didn't learn mathematics well (1)
To get pleasure from maths	Love the numbers
It is a hobby to get pleasure	Fun math
Math should be fun	Make people laugh
Self-confidence	Love of math

Table 20.7 Sample student statements of teacher value

Learned mathematics well (5)	Didn't learn mathematics well (1)
Teacher	Listening to the teacher
If a teacher teaches math well	A teacher should be friendly
Teacher to be like our fellow	Get on well with maths teacher
Medicine to make maths teacher amusing	Teacher's way of teaching

R: Could teachers be one of these components?

S1: Yes, the teacher is important.

R: Can you open a little more?

S1: Because the teacher should know how I learn better. And he must make me love the lesson.

R: You talked about making the lesson loved. Can you explain? What does it mean to make the lesson loved?

S1: The teacher can either make the lesson loved or hated. If I love the lesson, I will want to learn.

R: Are you saying that after the teacher, loving the lesson is also important in learning mathematics?

S1: Yes. I said if I love, I will work better and more willingly.

On the other hand, below is an excerpt from an interview with a student who stated that he learned mathematics well at school about the teacher's value:

R: Could teachers be one of these components?

S5: It could be. If a person's math teacher is good, his life will improve. For example, I found my middle school math teacher close to me as a friend. I would study and listen to the lesson so there would be no disrespect to the teacher. He made me love maths.

R: You talked about loving math class. So, can loving math be a component of learning math?

S5: Yes, it is important. Because if you don't like math, you'll have some issues learning it. But you still learn, but you struggle.

Materials Enhance Thinking: This value emphasizes some foods, materials, and concepts such as enzymes, honey, green tea, almonds, and sleep, which are thought to improve students' thinking skills and capacities while learning mathematics. In this context, both groups of students emphasized these foods and materials, which they thought developed thinking while learning mathematics (see Table 20.2). Some sample student statements regarding this value are also given in Table 20.8.

The interviews conducted with the students regarding this value determined that the students in both groups generally put more emphasis on some foodstuff. Although students in both groups give the highest importance to the value of talent, they relatively less emphasize materials that can increase their thinking capacity and, therefore, their abilities (the value they attach the most importance to) (see Table 20.2). Two short excerpts from the interviewed students' views on this value are given below:

Table 20.8 Sample student statements of materials enhance thinking value

Learned mathematics well (5)	Didn't learn mathematics well (1)
Enzymes to make the brain operate	Sleeping regularly
Nutrition	Hazelnut and walnut
Fish oil	Red meat
Lime tea, lemon, honey	Candy for learning and comprehending

Table 20.9 Category comparison of Hill et al. (2021) and the current study

Current study	Hill et al. (2021)
Ability	Cognitions
Effort	Perseverance
Mathematical concepts	Cognitions
Fun	Positive emotions
Teacher	Relationships
Materials enhance thinking	Cognitions

S1: It could be. For example, reading a book is one of them; it increases your intelligence and thinking.

S5: Yes, because substances that increase your intelligence and memory also increase your learning.

20.3.1 Discussion

The results of this study suggested that there were six values that this group of Grade 9 students in Türkiye consider the most important in learning mathematics: ability, effort, mathematical concepts, fun, teacher, and materials to enhance thinking.

These six values were evaluated within the scope of instrumental values. In contrast, the MWB categories specified in the study of Hill et al. (2021) were evaluated within the scope of ultimate values (see Tiberius, 2018). In this context, the MWB equivalents of 6 instrumental values in the study of Hill et al. (2021) are presented in Table 20.9. It should be noted here that some values that emerged in the current study may be considered under more than one MWB category/dimension at the same time (for example, the value effort may be considered in the context of perseverance and accomplishment MWB categories).

In the current study, unlike Hill et al. (2001), it was determined that the values that both student groups attach importance to in the learning of mathematics are generally similar, even if the students' levels of feeling good in mathematics (MWB) were different. One reason may be that students see ability, effort, and teachers as the most important factors for success in mathematics, as evidenced by the semi-structured interviews. The fact that teaching in Türkiye, in general, and mathematics teaching, in particular, is exam-oriented (Dede & Tasos, 2019) may be why Turkish students consider these three values more important in learning mathematics. This is because, in Türkiye, mathematics questions (together with Turkish) play a leading role in the exams held for the selection of high school and especially university students. Besides math ability, this situation highlights that Turkish students should prepare for these exams by solving as many mathematics questions as possible and clearly their mathematics teachers and their teaching practices are crucial for this while they are learning mathematics. Similar results were obtained in other studies conducted with Turkish students (see Aktaş, Akyıldız & Dede, 2021; Dede & Tasos, 2019; Dede et al., 2022; Dede et al., 2023).

In addition, it can be said that these results coincide with the mathematical values prominent in the study of Hill et al. (2021). This is because cognitions and accomplishment ultimate values emerged more in the study of Hill et al. (2021). Similarly, in the current study, the ability value of cognitions ultimate values, which can be considered an instrumental value, has come to the fore as the value students see as most important while learning mathematics. One reason for this may be that, as Hill et al. (2021) stated, the questions in the questionnaire led students to think of mathematics as more cognitive and achievement oriented. As mentioned before, Turkish students are likely to have developed an achievement-oriented perspective in mathematics due to the central high-stakes exams in Türkiye. For this reason, Turkish students can see the ability value as the cornerstone of mathematical success, as can be understood from semi-structured interviews.

In addition, the results of this study are like Weiner's (1985) causal attribution theory. In Weiner theory (1985), effort and ability are stated as internal causal attribution, and luck and task difficulty as external causal attribution. On the other hand, in this study, the categories of ability, effort, and fun, which the students emphasized most, overlap with Weiner's work as internal causal attribution. The other categories of this study, mathematical concepts, teacher, and materials enhancing thinking, can be considered external causal attribution.

In the current study, the effort value is another value obtained for students in both extreme groups. The effort value considered an instrumental value, can be linked to the perseverance's ultimate value of Hill et al. (2021). Unlike the study of Hill et al. (2021), in the current study, effort emerged as the value that Turkish students attach the most importance to in learning mathematics after ability. In the study of Hill et al. (2021), it can be said that although the value of perseverance is among the values that students give less importance to in learning mathematics, the value of perseverance ultimately supports and complements the values of cognition and accomplishment. In this context, it can be said that the results of the current study are similar at this point. Still, due to different cultural understandings and insights (Hill & Seah, 2022), Turkish students can consider this value differently. Additionally, Dede et al. (2022) determined the values of Turkish students with the bipolar scale; the ability-effort values were positioned as the two extreme values of the bipolar scale. In Dede et al. (2022) study, students preferred the effort value more than the ability value. In the current study, the ability value was emphasized more than the effort value of Turkish students in learning mathematics. However, it was determined that the students answered the questions in an open-ended manner. Considering the structure of the bipolar scale (choosing one of the two extreme values), Turkish students still attach more importance to these two extreme values (ability-effort) than the other values in the learning of mathematics.

Another value that students chose in this study is the fun value. This fun value can be evaluated in the context of an ultimate value of positive emotions by Hill et al. (2021). In a study where Dede et al. (2022) determined the values of Turkish students with the bipolar scale, it was determined that the students whose pleasure (fun)-hardship values were positioned as the two extreme values of the bipolar scale, they gave more importance to the pleasure (fun) value than the hardship value. The fun value emerged at a frequency close to other mathematical values in the current study. Semi-structured interviews also revealed that students liked mathematics as a component of learning mathematics. At the same time, the value of fun emerged in Turkish and German students representing two different cultures in the study of Dede (2019), in which he examined the values of Turkish, German, and Turkish immigrant students living in Germany in mathematics classes, while I did not emerge for Turkish immigrant students (in acculturation process). On the other hand, Hill et al. (2022) state that when students understand mathematics, positive emotions occur, and positive emotions trigger engagement. In this context, in the current study, the values of ability, effort, and fun can be considered three values that interact.

On the other hand, although the peer instrumental value is more prominent in the relationships of an ultimate value in the study of Hill et al. (2021), it is remarkable that in the current study, only the teacher-instrumental value is emphasized by the students in the context of this ultimate value. One of the reasons for this situation can be shown as the fact that Turkish society is a collectivist society (see Hofstede, 1980, 2023), which means that the teacher is seen as a dominant figure and plays a role in classroom practices (see Cooper et al., 2007). In addition, the competitive structure of the education system in Türkiye, which is exam-oriented, causes teachers to make their teaching exam-oriented. This situation causes students to prefer teachers who prepare the best for the exam, solve the most questions while learning mathematics, and keep the interaction and sharing between peers at a limited level in this competitive environment. A similar situation manifests itself mainly in Asian students, and the idea of "practice makes perfect" puts much pressure on these students (Zhang et al., 2016).

Finally, the materials enhance thinking and mathematical concepts instrumental values obtained in the current study can be evaluated in the context of cognitions ultimate value of Hill et al. (2021). The cognitions ultimate value includes values related to the scientific discipline of mathematics (for example, rationality) (see Atweh & Seah, 2008). It represents the common values that emerge in mathematics classrooms in all cultures. For example, it is frequently noted in the literature that the common value that Turkish and German students (see Dede, 2019) and Asian students (see Zhang et al., 2016) give the most importance while learning mathematics is the value of rationality.

20.4 Moving On

The six values identified by students in the current study suggest that they see mathematics as a worthwhile and doable subject, as it has complementary properties in mathematics learning and causes students to feel good about themselves (MWB). This is because students approach mathematics in ways that can be more harmful than many other school lessons. It is essential to understand the well-being of students

(MWB) in mathematics education and to prepare supportive teaching environments and processes based on this understanding to change and improve the way mathematics is perceived and experienced by students (see Hill et al., 2021). This is because well-being depends in part on one's values and generally on the context (Hill et al., 2021), as determined by the characteristics of one's environment (Alexandrova, 2017). For this reason, values (and attitudes) are positioned in the OECD Learning Compass 2030 among the things that students should gain alongside knowledge and skills within the framework it offers to reveal students' potential and contribute to the well-being of societies. In this context, this study may contribute significantly to the literature on students' MWB. The sampling used in this study does suggest that these results may well generalize to the wider population of students.

There are also implications for teacher practice. Teachers were shown to be an essential value for mathematics learning in the current study. It follows that they can increase students' interest in mathematics and their success if they prepare environments and processes that support students' MWBs. This notion is supported by an argument from Waters et al. (2019) who determined that when the well-being of students is supported, in particular, the value of perseverance increases, leading to students' academic success. In this context, as revealed in the semi-structured interviews, it is also essential for teachers to establish good relations with students and support their positive emotions towards mathematics. Thus, with the development of students' MWBs, they become more engaged in mathematics. Ensuring teacher and student values alignment is another factor that supports students' MWBs. Kalogeropoulos and Clarkson (2019) found that when teachers align their classroom pedagogical values with their students' values, student engagement in the learning of mathematics increases. In other words, student engagement can be increased by supporting students' positive emotions.

This study is limited to the answers given by the students to the magic pill question. Building on this restricted context, it may be useful to conduct further research focused on classroom observations to reveal the reflections of the student values obtained in classroom practices. In addition, even though both Türkiye and many countries have recently made massive investments in the technological integration of mathematics classes, Turkish students never emphasized the value of Information and Communication Technology [ICT] for the value of mathematics education. Similar results were found in studies in other cultures (see Hill et al., 2021; Dede, 2019; Zhang et al., 2016). In this context, further research may be conducted on why students do not mention ICT value.

Acknowledgements The study involving students from Türkiye was supported, in part, by grants from the Scientific and Technological Research Council of Türkiye (TÜBİTAK, Grant No: 1059B191401095). Any opinions expressed herein are those of the author and do not necessarily represent the views of TÜBİTAK.

References

Aktaş, F. N., Akyıldız, P., & Dede, Y. (2021). Students' mathematics educational values at religious vocational middle schools: A cross-sectional study. *International Journal of Primary, Elementary and Early Years Education 3–13, 51*(3), 480–492.

Akyıldız, P., Aktaş, F. N., Dede, Y., & Hacıömeroğlu, G. (2021). Mathematics teachers' values about teaching mathematics. *Studies in Educational Evaluation, 68*, 1–11.

Alexandrova, A. (2017). *A philosophy for the science of well-being*. Oxford University Press.

Askew, M., Hodgen, J., Hossain, S., & Bretscher, N. (2010). *Values and variables: Mathematics education in high-performing countries*. Nuffield Foundation

Atweh, B., & Seah, W. T. (2008). *Theorising values and their study in mathematics education*. AARE Annual Conference, Fremantle, Australia. Retrieved February 8, 2020 https://www.aare.edu.au/data/publications/2007/atw07578.pdf

Birt, L., Scott, S., Cavers, D., Campbell, C., & Walter, F. (2016). Member checking: A tool to enhance trustworthiness or merely a nod to validation? *Qualitative Health Research, 26*(13), 1802–1811.

Bishop, A. J. (1988). *Mathematical enculturation: A cultural perspective on mathematics education*. Kluwer Academic.

Bishop A. (1991). *Mathematical enculturation: A cultural perspective on mathematics education*. Kluwer Academic.

Bishop, A. J. (2012). From culture to well-being: A partial story of values in mathematics education. *ZDM, 44*, 3–8.

Bishop, A. J., & Seah, W. T. (2008). Educating values through mathematics teaching: Possibilities and challenges. In M. H. Chau & K. Trevor (Eds.), *International perspectives on education* (pp. 118–138). Continuum.

Bishop, A. J., & Whitfield, R. C. (1972). *Situations in teaching*. McGraw-Hill.

Bishop, A. J., Gunstone, D., Clarke, B., & Corrigan, D. (2006). Values in mathematics and science education: Researchers' and teachers' views on the similarities and differences. *For the Learning of Mathematics, 26*(1), 7–11.

Bishop, A., Clarkson, P., FitzSimons, G., & Seah, W. T. (2000). Why study values in mathematics teaching: Contextualising the VAMP Project. In W. Horng & F Lin (Eds.), *History in mathematics education: Challenges for a new millennium* (pp. 147–154). Taipei, Taiwan: National Taiwan Normal University. https://www.researchgate.net/publication/228600609_Why_study_values_in_mathematics_teaching_Contextualising_the_VAMP_project

Clarkson, P., Bishop, A. J., & Seah, W. T. (2010). Mathematics education and student values: The cultivation of mathematical wellbeing. In T. Lovat, R. Toomey, & N. Clement (Eds.), *International research handbook on values education and student wellbeing* (pp. 111–135). Springer.

Clarkson, P., Seah, W. T., & Pang, J. (2019). Scanning and scoping of values and valuing in mathematics education. In P. Clarkson, W. T. Seah, & J. Pang (Eds.), *Values and valuing in mathematics education: Scanning and scoping the territory* (pp. 1–10). Springer.

Cohen, L., Manion, L., & Morrison, K. (2011). *Research methods in education* (7th ed.). Routledge Falmer.

Cooper, P. J., Calloway-Thomas, C., & Simonds, C. (2007). *Intercultural communication: A text with readings* (1st ed.). Allyn and Bacon.

Creswell, J. W. (1998). *Qualitative inquiry and research design: Choosing among five traditions*. Sage.

Creswell, J. W. (2012). *Research design: Qualitative, quantitative, and mixed methods approaches* (4th ed.). Sage.

Creswell, J. W., & Plano Clark, V. L. (2018). *Designing and conducting mixed methods research*. Sage.

Dede, Y. (2011). Mathematics education values questionnaire for Turkish preservice mathematics teachers: Design, validation, and results. *International Journal of Science and Mathematics Education, 9*(3), 603–626.

Dede, Y. (2019). Why mathematics is valuable for Turkish, Turkish immigrant and German students? A cross-cultural study. In P. Clarkson, W. T. Seah, & J. Pang (Eds.), *Values and valuing in mathematics education* (pp. 143–156). Springer.

Dede, Y., Akçakın, V., & Kaya, G. (2020). Mathematical, mathematics educational, and educational values in mathematical modeling tasks. *ECNU Review of Education, 4*(2), 241–260.

Dede, Y., & Tasos, B. (2019). Developing a questionnaire to evaluate Turkish students' mathematics values and preferences. *Journal of Theoretical Educational Science, 12*(4), 1142–1163.

Dede, Y., Akçakın, V., & Kaya, G. (2022). Identifying students' mathematical and mathematics educational values in Turkish culture: A cross-sectional study. *Culture and Education, 34*(3), 597–629. https://doi.org/10.1080/11356405.2022.2058795

Dede, Y., Aktaş, F. N., & Akyıldız, P. (2023). Mathematics educational values of students and teachers in Imam-Hatip high schools. *International Journal of Mathematical Education in Science and Technology*. https://doi.org/10.1080/0020739X.2023.2172621

Ernest, P. (2007). The philosophy of mathematics, values and keralese mathematics. *Montana Mathematics Enthusiast, 4*(2), 174–187.

Goldin, G. A. (2019). Exploring a conative perspective on mathematical engagement. In S. A. Chamberlin & B. Sriraman (Eds.), *Affect in mathematical modeling* (pp. 111–129). Springer International Publishing.

Hesapçıoğlu, M. (2009). Türkiye'de Cumhuriyet Döneminde Eğitim Politikası ve Felsefesi. *Marmara Üniversitesi. Atatürk Eğitim Fakültesi Eğitim Bilimleri Dergisi, 29*, 121–138.

Hill, J. L., Kern, M. L., Seah, W. T., & van Driel, J. (2021). Feeling good and functioning well in mathematics education: Exploring students' conceptions of mathematical wellbeing and values. *ECNU Review of Education, 4*(2), 349–375. https://doi.org/10.1177/2096531120928084.

Hill, J. L., & Seah, W. T. (2022). Student values and wellbeing in mathematics education: perspectives of Chinese primary students. *ZDM–Mathematics Education, 55*, 385–398. https://doi.org/10.1007/s11858-022-01418-7.

Hofstede, G. H. (2023). Hofstede Insights. https://www.hofstede-insights.com/country-comparison-tool?countries=turkey.

Hofstede, G. H. (1980). *Culture consequences: International differences in work-related values.* Sage.

Hunter, J. (2021). An intersection of mathematics educational values and cultural values: Pāsifika students' understanding and explanation of their mathematics educational values. *ECNU Review of Education, 4*(2), 307–326.

Kalogeropoulos, P., & Clarkson, P. C. (2019). The role of value alignment in levels of engagement of mathematics learning. In P. C. Clarkson, W. T. Seah, & J. Pang (Eds.), *Values and valuing in mathematics education: Scanning and scoping the territory* (pp. 115–127). Springer.

Law, H. Y., Wong, N. Y., & Lee, N. Y. L. (2012). A study of espoused values in Hong Kong's mathematics classrooms. *ZDM, 44*, 45–57.

Lim, C. S., & Ernest, P. (1997). *Values in mathematics education: What is planned and what is espoused?* British Society for Research into Learning Mathematics day conferences.

Lincoln, Y. S., & Guba, E. G. (1985). *Naturalistic inquiry*. Sage.

Ministry of National Education. (2018a). *Matematik dersi öğretim programı (ilkokul ve ortaokul 1, 2, 3, 4, 5, 6, 7 ve 8. sınıflar) [Mathematics curriculum (primary and middle school, Grade 1, 2, 3, 4, 5, 6, 7 and 8)]*. Author. Ankara, Türkiye.

Organisation for Economic Co-operation and Development (OECD). (2022). *Trends shaping education*. OECD Publishing, Paris, Retrieved January 10, 2023 from https://doi.org/10.1787/6ae8771a-en.

Organisation for Economic Co-operation and Development. (OECD). (2019). *OECD Future of education and skills 2030: What is the Learning Compass?* Retrieved December 30, 2022 from http://www.oecd.org/education/2030-project/teaching-and-learning/learning/.

Organisation for Economic Co-operation and Development OECD. (2017). *PISA 2015 results: Students' well-being* (Volume III). PISA, OECD Publishing.
Pang, J. S., & Seah, W. T. (2021). Excellent mathematical performance despite "negative" affect of students in Korea: The values perspective. *ECNU Review of Education, 4*(2), 285–306.
Pang, V.O. (2005). *Multicultural education: A caring-centered reflective approach.* McGraw-Hill.
Powe, K. W. (1993). Values education and the local school board. *Updating School Board Policies, 24*(6), 1–4.
Scollon, R., Scollon, S. B. K., & Jones, R. H. (2012). *Intercultural communication: A discourse approach* (3rd ed.). Wiley-Blackwell.
Seah, W. T., & Bishop, A.J. (2000). *Values in mathematics textbooks: A view through the Australasian regions.* Paper Presented at the Annual Meeting of the American Educational Research Association, New Orleans, LA.
Seah, W. T. (2008). Valuing values in mathematics education. In P. Clarkson & N. Presmeg (Eds.), *Critical issues in mathematics education: Major contributions of Alan Bishop* (pp. 239–252). Springer.
Seah, W. T. (2019). Values in mathematics education: Its conative nature, and how it can be developed. *Journal of Korean Society for Mathematics Education, 22*(2), 99–121.
Seah, W. T., Baba, T., & Zhang, Q. (2017). The WIFI study: Students' valuing of mathematics learning in Hong Kong and Japan. In J.-W. Son, T. Watanabe, & J.-J. Lo (Eds.), *What matters? Research trends in international comparative studies in mathematics education* (pp. 333–354). Springer.
Seah, W. T., & Andersson, A. (2015). Valuing diversity in mathematics pedagogy through the volitional nature and alignment of values. In A. Bishop, H. Tan, & T. Barkatsas (Eds.), *Diversity in mathematics education: Towards inclusive practices* (pp. 167–183). Springer.
Stigler, J. W., & Perry, M. (1988). Cross-cultural studies of mathematics teaching and learning: Recent findings and new directions. In D. A. Grouws & T. J. Cooney (Eds.), *Perspectives on research on effective mathematics teaching* (pp. 194–223). Lawrence Erlbaum.
Strauss, A., & Corbin, J. (1998). *Basics of qualitative research: Techniques and procedures for developing grounded theory* (2nd ed.). Sage.
Tiberius, V. (2018). *Well-being as value fulfillment: How we can help each other to live well.* Oxford University Press.
Waters, L., Loton, D. J., Grace, D., Jacques-Hamilton, R., & Zyphur, M. J. (2019). Observing change over time in strength-based parenting and subjective wellbeing for pre-teens and teens. *Frontiers in Psychology, 10*, 2273.
Weiner, B. (1985). *Human motivation.* Springer.
Williams, R. (1961). *The long revolution.* Pelican.
Zhang, Q., Barkatsas, T., Law, H.-Y., Leu, Y.-C., Seah, W. T., & Wong, N.-Y. (2016). What primary students in the Chinese Mainland, Hong Kong and Taiwan value in mathematics learning: A comparative analysis. *International Journal of Science and Mathematics Education, 14*(5), 907–924.

Chapter 21
Transforming Powers in a Magic Pill that Makes Anyone Good at Mathematics

Lisa Österling, Anette De Ron, and Annica Andersson

21.1 Introduction

The mathematics and the mathematical values in schools are rarely adapted to the mathematical wellbeing of children (Hill et al., 2021). Mathematics has been used for classifying children as able or disabled (Popkewitz, 2018) rather than taking what students find important as the starting point. Where curriculum development is often based on what adults in a historic and cultural time and place desire for future generations to know (Bishop, 1991), we hope that this chapter inspires teachers to listen to what students value, and thereby contribute to more inclusive classroom practices.

School mathematics holds several contradicting messages about what students need to know or what they need to be. Looking at the history of Swedish school mathematics, the message in curriculum is that working thoroughly is more important than being fast, students are supposed to demonstrate engagement, curiosity and joy when solving problems, and good students are depicted as those who are thinking logically, using reasoning and being rational (de Ron, 2022). However, in a Swedish values-survey (Andersson & Österling, 2019), students were found to reproduce a discourse about the importance of teachers' explanations and knowing the times-tables. Hence, when Swedish students are asked, they envision mathematics as fixed rules and formulas to memorize effectively, in contrast to the messages in curriculum.

L. Österling (✉) · A. De Ron
Stockholm University, Stockholm, Sweden
e-mail: lisa.osterling@su.se

A. De Ron
e-mail: anette.de-ron@su.se

A. Andersson
University of South-Eastern Norway, Notodden, Norway
e-mail: annica.andersson@usn.no

Y. Dede et al. (eds.), *Values and Valuing in Mathematics Education*,
https://doi.org/10.1007/978-981-99-9454-0_21

The gap between direct and dialogic instruction (Munter et al., 2015), has been a long-lasting discussion in mathematics education. Direct instruction refers to teacher-centred models of pre-defined goals and teacher support to reach them, whereas dialogic models an approach where students are granted more agency in the classroom, such as reform-teaching or inquiry-based teaching (Lampert et al., 2011; Stein et al., 2008). In mathematics teacher professional development, the dialogic model has been strongly emphasized (see Weingarden & Heyd-Metzuyanim, 2023). A focus on rules and formulas are questioned by becoming teachers (Andersson et al, 2023). Still, their importance in mathematics is repeated, by teachers (Horn, 2007; Pansell & Björklund Boistrup, 2018), in news media storylines (Abtahi & Barwell, 2019, Andersson et al., 2023), but also among students (Seah et al., 2016). Many researchers acknowledge how the gap between the dominance of dialogic models in teacher professional development, and the dominant discourses of valuing fixed rules and formulas in mathematics classrooms, poses a challenge to overcome. Lampert et al (2011) and Weingarden and Heyd-Metzuyanim (2023), each propose a middle ground to be established in teaching.

Teachers who are committed to challenge educational values need to resolve several contradictions in the classroom (James & Pedder, 2006), where an awareness of what students find important or value is a crucial point of departure. Hoping to reach a more nuanced understanding of what students' do value in mathematics, we asked students across Sweden to each propose three ingredients for a magic pill to make anyone "very good at mathematics". The magic pill-question was part of the international WIFI-study (see Clarkson et al., 2019), a questionnaire in three parts, targeting mathematical values of 10- and 15-year-old students. The assumption was that this particular question opens up for learners to suggest ingredients they would require to make mathematics more accessible, relevant or engaging. However, a concern that grew during the analysis of students' responses was that the task to suggest ingredients for transforming a person has an underlying implication that there is something which need improvement in that person. Therefore, we reflect on how taking a pill implies that something must be fixed in the person taking the pill, with impact on who feels included in mathematics.

An aim of this chapter is therefore to understand the transforming powers in a magic pill. In particular we explore what mathematical values students consider as part of being "very good at mathematics". The research question we pose is: What ingredients could Swedish mathematics students put in a pill that makes anyone who takes the pill very good at mathematics? From this, we engage in a discussion on what these pill-ingredients reveal about what students value, their attitudes towards mathematics, and what they wish the pill to transform in the person taking the pill.

21.1.1 Mathematical Enculturation, Wellbeing and In(ex)clusion

To identify gaps between values and classroom practices facilitates values-alignment (Kalogeropoulos & Clarkson, 2019). Thus, to ask students to suggest what 'good at mathematics' means can contribute to the development of values-informed classroom practices. We use Bishops (1991) mathematical values and the concepts of MWB (Hill et al., 2021) to make sense of the suggested ingredients in the pill. In addition, some responses made us turn to the political dimensions of the transforming powers in a pill for making anyone "very good at mathematics".

Mathematics as a societal construct has particular meanings, where individuals need to "recreate Mathematical culture in relation to the other meanings and knowledge they possess" (Bishop, 1991, p. 153). The claim is that learning mathematics entails a process of enculturation into this Mathematics culture. Recent research includes students' perspectives, as in Aspects of Mathematical Wellbeing (MWB), which connect students' values to "include emotions, social aspects, and a sense of engagement, several of which have been associated with various positive mathematics outcomes" (Hill et al., 2021, p. 351). MWB is defined as a context-dependent rather than static construct (Kern et al., 2019), and goes beyond a simplistic deterministic relation between performance and affect. Similarly, the complex relations involved between mathematics, affect and context is acknowledged in a range of studies. In Canada, love is found to be important in student group work on problem solving (Andersson & Wagner, 2019), and in Sweden, Nyman and Sumpter (2019) show how the intrinsic motivation to study mathematics include both cultural motivations as well as personal, and emotional as well as cognitive aspects.

Earlier reported results of the magic-pill-question show how Chinese students commonly include effort, diligence, formula smartness and memory in the pill, whereas ability, wisdom or thinking, were less common (Zhang, 2019). Few ingredients related to affects and emotions were reported in that study. These findings are in line with Stigler and Hiebert (2009) observations that Eastern Asian education tend to focus on effort rather than ability. As a contrast, Boaler (2015) claims how ability is strongly emphasized in western classrooms, and suggest teachers make the processes and efforts of learning the focus in teaching and learning mathematics. The Turkish results from the magic-pill question also show that students there gave both ability and effort the highest importance (Dede, Akcakin & Kaya, chapter in this volume), but they also rated the value of fun highly. Little variance between students who categorise themselves as learning mathematics well or not was found in the Turkish context. In contrast, Hill et al. (2021) argue MWB is related to achievement in mathematics. The context dependency of mathematical values and MWB makes it relevant in reporting our results from Sweden, as a means for increasing the awareness of what makes mathematics more relevant or engaging for students in diverse countries and contexts. When mathematics is used for classifying children as able or disabled (Popkewitz, 2018), the boarders for the mathematics culture becomes part of construing what is desired as normal for a child, but at the same time what becomes the non-desired or even pathological type of child. Despite the ambition to enable change and inclusion, the same move establishes exclusion: "Yet, the recognition of difference establishes difference" (Popkewitz, 2018, p. 148). This double gesture about establishing who the mathematical student is, at the same time imposes a need for the deviating student to change and adapt, or to be excluded.

Bishop (1991) outlines principles for an enculturation curriculum for overcoming the asymmetry between the outsiders, the students who are newcomers to the culture of mathematics, and the insiders, the teachers or mathematicians. Mathematical enculturation establishes shared values for members in the mathematics culture, values which also become the boarder for who is part of mathematics culture, and who is not. Questions of "Who is in and who is out" (Goodin, 1996, p. 343) is central for democratic theorists, where inclusion aims for moving the excluded "over the line" (p. 348). Thereby, processes of inclusion become part of construing borders for participation. In mathematics enculturation, children are expected to move over the boarder to become included in the mathematics culture. From this, we will now look at the ingredients of a pill that makes anyone "very good at mathematics" as the values which are part of the boarder for participation in mathematics, in a particular time and place.

As part of the processes of enculturation, learners are expected to share a particular set of values for becoming part of the mathematics culture. The claim is that "Enculturation must be for all - Mathematical enculturation should be for all" (Bishop, 1991, p. 96). Bishop describes three complementary pairs of values; the ideological values of objectism and rationalism, the sentimental values of control and progress, and finally the sociological values of openness and mystery. Bishop (1991) uses these pairs for a critique of the mathematics curriculum. The argument is that objectism (the valuing of static mathematical abstractions) and control (a sense of certainty) contribute to deepen the mystery of mathematics by making the objects more abstract, and through constituting an exclusiveness in who is doing mathematics. To enable a more democratic enculturation, Bishop (1991) therefore suggests an emphasis on the opposing values of rationalism, progress and openness in mathematics curriculum. Rationalism foregrounds the logic and argument, progress the sense of development and growth, and Bishop claims how rationalism and progress enable openness in who can do mathematics and in what ways. Consequently, rationalism, progress and openness are central values for a democratic inclusion in mathematics under the assumption that they open the boarders for many ways of being mathematical. As a contrast, the values of objectism and control imposes a singular correct way of being mathematical, and mystery when the process of becoming mathematical is covert and inaccessible for outsiders.

21.2 Methods

The magic pill-question (Fig. 1) is part of the international WIFI-survey in which Sweden participated (Andersson & Österling, 2013, 2019). The WIFI study was originally developed and piloted in English in an Australian-Asian context, and has been implemented in 18 countries (Clarkson et al., 2019).

Mathematical values are methodologically challenging to capture (Chan & Wong, 2019), and the approach of asking students to imagine a pill, and anyone who takes this magic pill will become very good at mathematics [den som tar pillret blir väldigt bra på matematik] is aimed at opening up spaces for students to freely suggest what they consider important.

21.2.1 Survey Sample and Data Collection

The survey was provided as an online-survey, although a small proportion of teachers preferred a paper version. We engaged teachers to help us to distribute the survey to obtain a geographic spread. The participating teachers received a letter of instruction which stressed how participation was voluntary, together with the web-link for the online-survey, or a file from which they could print the paper version. We received 850 completed surveys and checked the sampling for relatively equal proportions for gender and between years five and eight students. To avoid non-serious answers, we removed respondents with too many responses missing (Bryman, 2016). By setting the limit low, at 10% of missing answers, we eliminated students who did not complete the last pages of the survey, as well as those who skipped a whole page. This left 742 students. Seven did not report their gender, and were excluded as a last step, leaving us with 735 respondents, and a total of 2125 suggested ingredients to a magic pill to make you "very good at mathematics".

21.2.2 Data-Analysis

The analysis began with inductive coding of the ingredients the students had suggested in their answers in the questionnaire, with 27 categories as a result (see Table 21.1). Spreadsheets were used for coding and statistics. The categories were merged in six themes. Inspired by the idea of a "magic pill", we gave pharmaceutically inspired names to the different themes: targeting mathematics, targeting the person, for best effect, excipients, experimental agents and bulk.

The first two themes are composed of suggested ingredients for the pill targeting something that can be related to being "very good at mathematics". The first theme, 'targeting mathematics', includes both mathematical content and mathematics learning. The second theme, 'targeting the person', included the mind or body of the

Section C

Imagine that we are going to produce a magic pill.
Anyone who takes this magic pill will become very good at mathematics!

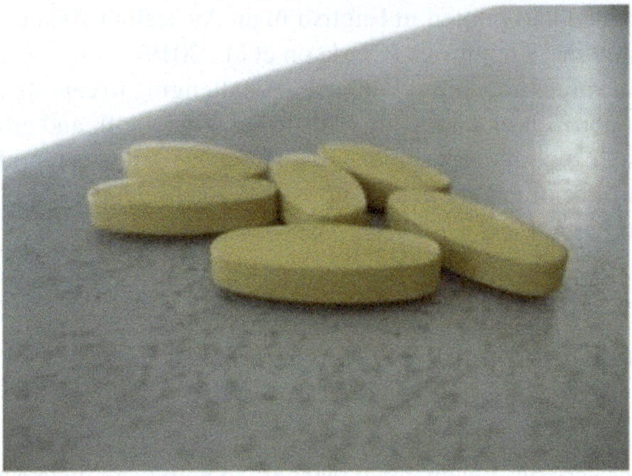

What will you choose to be the **top 3 main ingredients** of this magic pill? (Be imaginative, this main ingredient can be something we can touch and see, or something we can feel but cannot see, for examples)

77. Most important ingredient:

78. Second most important ingredient:

79. Third most important ingredient:

80. I selected these ingredients because:

Fig. 1 The magic pill-question, the WIFI-questionnaire. English version

Table 21.1 Table of categories of the students' ingredients for a magic pill for becoming very good at mathematics (Percentages refer to N = 2125, the number of ingredients suggested by students)

Table of categories[a]	
Targeting mathematics: 30%	
14%	Abstract mathematical objects
2%	Mathematical rationalism
5%	Physical mathematical tools
4%	Know and understand
2%	Connections
1%	More knowledgeable
1%	Read, write and explain
Targeting the person: 27%	
8%	Affect and feeling
7%	Intelligence
3%	Effort, energy, endurance
2%	Concentration, commitment
2%	Creativity, aesthetics
2%	Efficiency, being fast
2%	Physical body
0%	Thoroughness
For best effect: 7%	
4%	Environment
2%	Teachers or mathematical friends
0%	Tools, non-mathematical
0%	Subject integration
0%	Future opportunities
Excipients: 19%	
10%	Nutrition: vitamins, proteins, Omega 3
4%	Carbohydrates
4%	Pharmaceutics, Narcotics, alcohol
Experimental agents: 8%	
8%	Magic
0%	Being lucky
Bulk: 10%	
7%	Binding substances, colouring
3%	Flavour

[a]In this list, the percentage of each ingredient is rounded to the closest integer. A dose less than 0,5% are shown as 0%, and the sum is the rounded sum of exact numbers, which gives a more accurate value than the sum of integers.

person taking the pill; for example, "better memory" or "Einstein DNA". We analysed the relation between ingredients and mathematical values within these two themes:

- **objectism**, the valuing of static mathematical abstractions, or **rationalism**, which foregrounds logic and argument,
- **control** as sense of certainty, or **progress** the sense of development and growth, and
- **mystery** of mathematics through constituting an exclusiveness in who is doing mathematics, or **openness** in who can do mathematics and in what ways.

Not all ingredients grouped into these themes were possible to understand in terms of mathematical values. We used inductive categories to further structure results.

The remaining four themes show how students draw on earlier experiences of taking pills. The third theme, 'for best effect', we understood as improving the effect of the pill, which included the environment where the pill is used. For the usual pharmaceutical pill, this would be to take the pill before or after a meal. For this imagined pill, the theme for best effect, also includes suggested ingredients as good teachers, friends and a calm classroom environment. A fourth theme is 'excipients'; that is helper-ingredients which improves the effect. These can help to improve performance or well-being, in mathematics, and includes vegetables, vitamins, but also drugs or pain-killers. The fifth theme is what we choose to refer to as 'experimental'. The ingredients here have a very uncertain effect for becoming good at mathematics. We included many magic ingredients, as well as of "being lucky". And finally, 'bulk-ingredients' were included as a theme, for ingredients which make the pill look and taste like a pill, as binding substances, colouring or taste.

After the categorisations and thematisations, the relative frequencies were calculated. Our research question focuses on values among Swedish students, and therefore we explore the ingredients suggested for the whole group of students. In addition, we explore differences between boys and girls, and the two different age-groups. We also made a qualitative reading of ingredients suggested by a small group of students ($N = 21$) who rated themselves as "not good at maths", as the basis for our discussion of in-/exclusion in mathematics enculturation.

21.2.3 Limitations and Methodological Challenges

The data was collected in a Swedish context only, and due to the contextual variation for mathematical values, the transferability beyond a Scandinavian context cannot be assumed. A part of this chapter engages the construct validity of the pill-items, where we question if it is only values that the ingredients reflect. The fact that students suggested many eatable ingredients can be explained from them envisioning a pill to eat, and the question for a "magic pill" probably explains all the magic ingredients. In the analysis, we had to make several assumptions with a substantial effect on results. One of the most common ingredients was "sugar", an ingredient which adds both flavour and carbohydrates, and therefore energy, to the pill. We decided to categorise sugar as carbohydrates, and since this was a big category, it has an effect on the results. And finally, when respondents were divided into sub-groups (see appendix), some categories become too small to substantiate any generalisability.

21.3 Results

We stay true to the quest to actually find a pill for becoming "very good in mathematics" (Fig. 2), and present results as a table of contents from the etiquette of the jar for magic pills (Table 21.1).

The six themes are described and exemplified below. The first two themes, 'targeting mathematics' and 'targeting the person', s consist of ingredients targeting what is needed to become good in mathematics, whereas the remaining four themes contain what we termed helper-ingredients. The themes are described and exemplified below by first listing the categories they consist of, and then by taking each category and giving examples of the ingredient's students suggested which fit into each category. In the final section of results, we attune the dose based on the ingredients for the different gender- and age-groups, as a basis for a discussion of who is included and excluded in mathematics.

21.3.1 Ingredients Targeting Mathematics or Mathematics Learning

In this section, ingredients targeting mathematics content together with those targeting mathematics learning are included. All examples are translations from Swedish, and we use double quotation-signs to indicate when an example is a translated quote from students.

Fig. 2 A pill-jar for magic pills

The first two categories in the first theme, *abstract mathematical objects* and *mathematical rationalism,* consist of mathematics content. Among the 'abstract mathematical objects', numbers and operations on numbers (de fyra räknesätten) were the most common ingredients in year five. Algebra, equations and related content enter in year eight. This category corresponds to the ideological values of objectism, where the static fixed mathematical abstractions are core. Thus, the lists of ingredients such as numbers, operations and concepts in this category reflect how the pill targets primarily abstract objects and not the relation between the student and the mathematical object.

In the 'mathematical rationalism' category, "problem-solving", "logical thinking" or "reasoning" are all included. This category corresponds to the value of rationalism, as it includes the valuing of logic and reasoning. These are ingredients that are still particular for mathematics, but with some more openness for students to participate, to think and discuss, compared to the category of abstract mathematical objects. As shown in Table 21.1, mathematical rationalism was not as common as abstract mathematical objects among the suggested ingredients (14% compared to 2%). We interpret this as the enculturation they need to take part in is to ingest the abstract objects and ideas of mathematics, spiced with rare occurrences of participation and rationalism.

The remaining five categories in this theme target mathematics learning. In the 'physical mathematical tools' category, calculators or textbooks were commonly mentioned ingredients. Putting them into the pill is a way of internalising either calculations or all the mathematics in the textbooks, and having it with you at all times. In the category of 'know and understand', these two words were often put together by students; "to know and understand". The most common ingredients were knowledge, thinking or understanding, one at a time. One student wrote "knowledge – thinking", connecting the two, and another wrote "The ability to tell right from wrong", which we saw as a way of expressing thinking or understanding. Some student suggestions gave more precisions, as "to think easily or quickly". In the cases where some explanation is provided, the pill makes understanding come instantly, magically. From this, when students' write "understand", it might well imply the understanding of facts as abstract objects or the textbook. Therefore, their pill includes a large dose of control, the sense of certainty of secure knowledge and quick calculations. There are small doses of progress, the sense of struggle and growth.

The 'connection' category connects understanding to mathematics in particular, as well as connections within mathematics. An example from one student was "The ability to connect things to a pattern to be able to solve problems in a more flexible way", which says both that to connect things is a useful ability for doing mathematics, and that it is necessary for becoming good at problem solving. Some students described an intuitive or automatic process, "you just see it", or "to find the equations". Others include ingredients such as to "recognise strategies", as "knowing which is the easiest method", or that the pill gives the "steps for solving a task as soon as you see it". Altogether, the ingredients in this category give an insightful description of mathematics learning as learning to connect the objects and methods, not just knowing them one by one. Still, the envisioned learning from taking the pill is having intuition or a gaze for strategies or connections. The dose of openness in the process of learning is low, and mystery of the origin of mathematical connections comes in a stronger dose. The pill gives direct access to connections and makes it possible to escape the struggle to develop them.

The' more knowledgeable' category includes how "good at mathematics" is related not only to your own knowledge, but how much you know in relation to others. This is manifest in ingredients as explaining to others, to think outside the box, or "to be able to calculate stuff that no 14-year-old should know of".

Finally, in the 'read, write and explain' category, some of the students' proposed ingredients were "reading tasks", "presenting solutions", and "using the math-words". Several of their answers echoed criteria about presenting clear solutions, emphasised in both national tests and in grading criteria, and signalling how mathematics is mainly a written language.

To summarise, students include abstract mathematical objects as numbers or equations more often than problem solving or reasoning. Hence, objectism came in a stronger dose than rationalism. For the categories targeting mathematics learning, there is a large dose of control through ingesting calculators and textbooks, to write and explain, and to know and understand. There are few traces of valuing progress, where students see the meaning of mathematics for their own life and the surrounding society, and the processes of how mathematics develops are still covert in mystery. Hence, the effect of the pill is to effortlessly and quickly, perhaps magically, provide understanding of mathematical objects and language.

21.3.2 Ingredients Targeting the Person

One student wished for an ingredient as "A key to open my brains to think of calculations more easily". This quote captures what part of the body that needs to change to become "good at mathematics": mathematics is taking place inside the brain, through thinking of calculations or other mathematical processes, and this should ideally happen easily and quickly. As a contrast, one student instead includes "time for doing mathematics at one's own pace", which is an example of how context and frames for learning are targeted, instead of the body of the person learning mathematics. Almost a third of the students' suggested ingredients targeted the person taking the pill and hence were included under this theme. Eight categories were used to give meaning to the theme Targeting the Person: feelings or affect, intelligence, effort, concentration, creativity, efficiency, the body or thoroughness.

The first category, 'affect and feelings', includes ingredients that are mainly positive sensations, as love, joy, engagement and friendship, as when one student suggested "an ounce of love and joy" in the pill. What the pill should do is to make mathematics learning fun and bring feelings of love and engagement. Putting this in the pill can be understood either how students experience mathematics, or how they would like to experience mathematics.

The second category targets the person's 'intelligence'. This category included "braincells", "DNA from the smartest mathematician in the world", "Albert Einstein", "IQ" and "good memory". These ingredients establish mathematics as mainly situated in the brain, and that requires intelligence. Under this cultural understanding, mathematics is only accessible for some people: those born with good brains, perhaps a particular "Einstein-DNA". The physical body besides brains was mentioned only by a few students; for example, "sleep", "hunger", or "superpowers". Consequently, the pill includes a strong dose for changing the biology, but mainly targeting the brain.

Other ingredients target attitudes which can be trained or learned, as 'effort, energy and endurance'; this category captured students' answers such as "perseverance to practice a lot", "patience", "endurance to struggle and to not give up". This category also included "interest", "will" or "motivation for doing mathematics".

Five more categories round out this theme: concentration, creativity, efficiency, the physical body, and very few suggestions of thoroughness. A set of ingredients relates to the category 'concentration and commitment'; "to be able to listen", "to focus", and "to pay attention". A small dose of 'creativity/aesthetics' was suggested, including fantasy, but also music as well as their favourite music or artists. We cannot be sure whether ingredients as K-pop-idols are to be interpreted as a nonsense response, or actually a way of making mathematics more motivating or joyful! A small dose of being 'efficient or being fast' is suggested; for example, "To be able to count as quickly as possible, that is to almost be able to see how to count and what the right answer is". Thus, effort, commitment, efficiency or motivation are all suggested as being incorporated into the pill, but in smaller doses compared to intelligence or memory. Ingredients related to the physical body were suggested most commonly by boys, which we comment on below (see Fig. 5).

Intelligence, effort, concentration and efficiency come in small doses, but together, they would form the person taking the pill into someone who is well aligned with present discourses of learning or knowing mathematics as being fast at calculations (Andersson & Österling, 2019; Horn, 2007). To challenge this discourse, a substantial dose of joy and happiness is included. This makes us wonder if this is something important for mathematics learning, or if it is rather what students see as missing in mathematics classrooms at present.

21.3.3 The Helper-Ingredients

As helper-ingredients, we describe ingredients which students consider necessary for the pill, but which have no direct relation to being "very good at mathematics". Rather, they improve the effect or facilitate the function of the pill.

The 'For best effect' theme contains the ingredients we interpreted as being necessary for ensuring the effect of the pill. Students include a supporting context, the category of 'environment' including "peace and quiet so I don't make mistakes", or "an invisible friend which only I can see and who is always next to me to help me with those things I am poor at". In these examples, students acknowledge the importance of the context, but also how they shouldn't make mistakes, and how they feel they are poor at some things in math. The next most common responses included a "good math teacher", and to have "calm and silence in the classroom". Some included to have parents who knows mathematics, or to have friends to ask. This indicates how taking a pill is not enough, the environment as well as knowledgeable teachers or others are important for being good at mathematics. This result reasons with how explanations by the teacher were valued as the most important item in previously reported results from WIFI-study (see Andersson & Österling, 2019). These pill-ingredients add to our understanding of how teachers are important not only for knowledge, but also for feelings of well-being in mathematics.

The next theme is a set of 'excipients' included in the pill to increase health in general of the person taking the pill, or to ensure the effect of the active ingredients. The most common ingredients were categorized as 'Nutrition: vitamins, proteins, Omega 3' with a category of 'carbohydrates' giving energy as next. Parents in Sweden often say how fish is good for your brains, and Omega 3 is known as healthy fat in fish. Thus, energy, together with food that improves your health and your brains were common ingredients. In this theme, there was also a category of 'pharmaceutics, narcotics, alcohol' which also included ingredients such as tranquilisers or painkillers. We will discuss the destructive nature of these ingredients later in this chapter.

Under the 'experimental agents' theme, we consider a 'magic' category which grouped ingredients such as "unicorn horns" or "troll snot". We also included "being lucky" in this theme, which we consider as an experimental suggestion for mathematics learning. We believe this theme emerged from the question of suggesting a "magic pill". Still, we understand the wish to include a magic-wand-ingredient to effortlessly learn mathematics without having to do mathematics. In addition, several of the magical ingredients, as "rainbows" or "unicorn horns", are widespread symbols in child- and youth culture (Weida & Bradbury, 2020). There, they hold a range of meanings, as creativity, purity, healing, diversity or uniqueness. Hence, the magic of the pill includes a dose of cultural meaningfulness in mathematics learning.

The 'bulk' and flavor theme contains ingredients that are the things included to make the pill look like a pill, and to taste good. The ingredients were sometimes possible to relate to the image of the pill included in the question (see Fig. 1), as yellow coloring. The theme also included flour or concrete, presumably used to bake a pill. As suggested flavoring ingredients, sweets are the most common, as candy, soda or ice-cream.

To summarise, many students took the question to suggest a magic pill seriously, and suggested things to make it look like a pill, taste good and to include magic. Still, we learn about the importance of a supportive context for best effect of the pill. These excipients reinforce the messages about good brains and energy needed to do mathematics. And, a dose of magic reveals the things that makes mathematics more attainable but also more relevant.

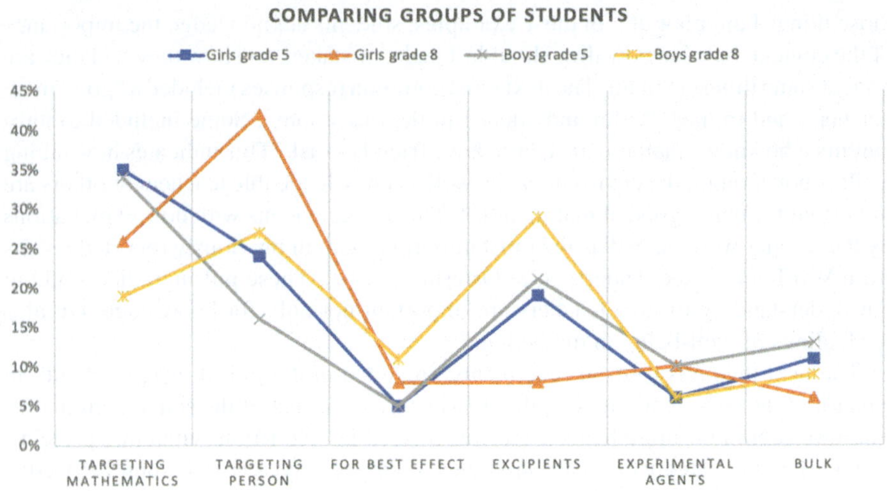

Fig. 3 The relative frequency by themes for how often ingredients were suggested in each group of students

21.3.4 Attuning the Dose

The Appendix shows how proportions of some of the ingredients varied between boys and girls, and also between the participating age-groups. The relative proportion of ingredients were calculated for the different groups to attune the dose for a particular group of students. Figure 3 show the distribution between the themes, and allows a comparison of grade 5 and 8, as well as of boys and girls.

For three themes, *for best effect*, *experimental agents* and *bulk*, there is not a significant difference between groups at the overarching level. For the other three themes, differences[1] are found. The three themes where there were significant differences will be explored in more detail below.

In the *targeting mathematics* theme, there is little difference between boys and girls, but the dose for abstract objects is stronger in grade five than in grade eight (Fig. 4).

The diagram in Fig. 4 show how in grade five, students tended to put a lot of *abstract mathematical objects*, as numbers and arithmetic operations, in the pill. And even though it comes in a small dose, it is among girls in grade eight that the *mathematical connections* were most common. As a contrast, the dose of textbooks and calculators, the *physical tools*, is stronger for boys in grade five. Hence, we see a slight shift away from suggesting mathematics to be a range of discrete objects in grade five, to how girls in grade eight starts talking about connections, as recognising patterns, finding paths through a solution or find connections between parts of a problem.

[1] For all comparisons between groups, a chi-square-test was used to check that statistic significance of difference was at least at a 95% level.

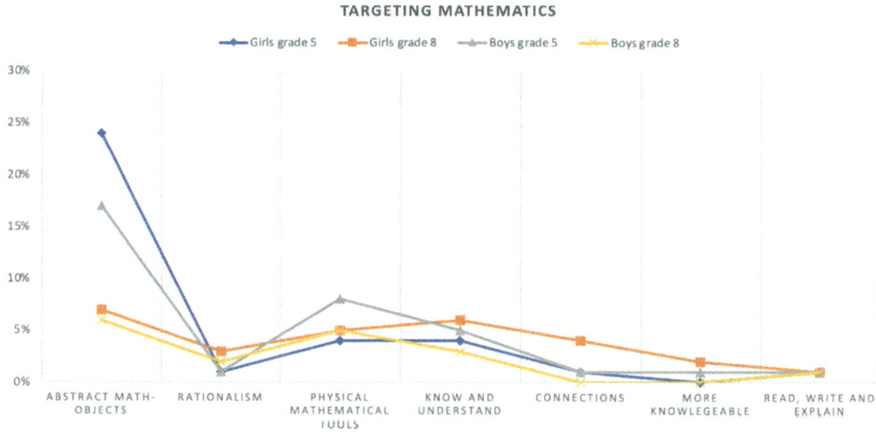

Fig. 4 Comparing the preferences for categories in the targeting mathematics theme for different groups

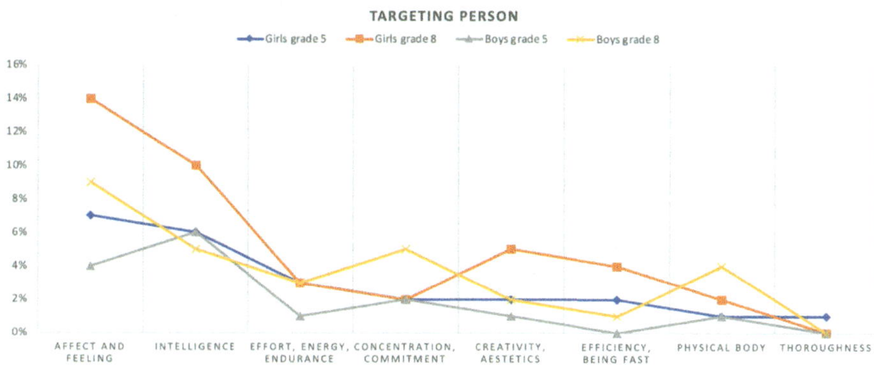

Fig. 5 Comparing the preferences for categories in the targeting person theme for different groups

The decreased dose of mathematical objects from year five to year eight correspond to an increased dose of ingredients targeting person, a difference bigger for girls than for boys.

Although the dose for affect and intelligence is high in all groups, girls privileged this more often (see Fig. 5). For affect, girls often suggest joy and happiness. The categories with small differences between boys and girls are effort and endurance, properties connected to hard work in mathematics. Still, the dose is quite low for both groups. Boys in grade 8 instead add more ingredients of concentration and commitment, as well as a higher dose of ingredients related to the body, mostly concerning rest or sleep. Thus, both boys and girls lower the dose of mathematics and increase the dose of affect from year five to eight. However, for girls, the dose for improving intelligence and efficiency, increase in grade eight. For boys, there is instead an increase in the dose for both the physical body, together with an increase

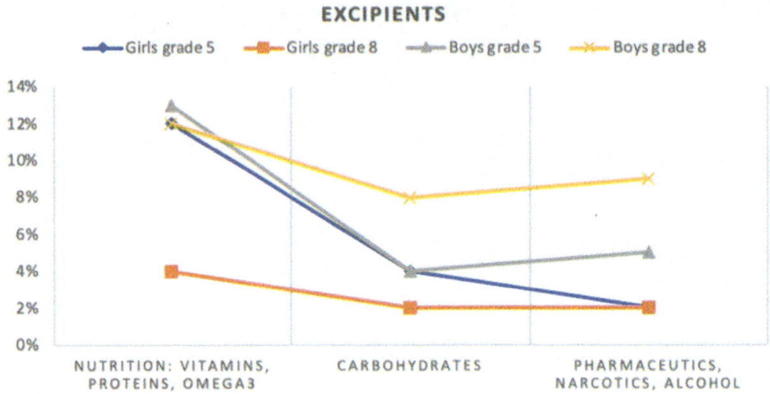

Fig. 6 Comparing the preferences for categories in the excipient theme for different groups

of energy in the pill. It seems girls increasingly see a need to improve the mind, whereas more boys see a need for bodily strength or energy. This is in line with how boys prefer more carbohydrates in the excipients theme, shown in Fig. 6.

Among the categories in the *excipients* theme, boys and girls follow the same pattern in their preferences in grade five (Fig. 6). A divide emerges in grade eight, where girls in general mention excipients less, and boys instead increase the dose of carbohydrates, and pharmaceutics, alcohol or drugs. Some ingredients listed in the pharmaceutics category were quite destructive, as steroids or drugs, which can be interpreted as non-serious responses. However, these destructive ingredients might tell us a story about how mathematics makes young people feel or perceive of themselves.

21.3.5 What a Pill Can Do for Students Feeling They Are "Not Good at Math"

In the WIFI-survey, we asked respondents to rate themselves as students in mathematics, on a scale from "good at math" to "not good at math", and we looked in depth at the 21 students who responded "not good at math" to better understand what they wish from the pill. In this group, eight students suggest ingredients as IQ or being smart, being fast, do calculations, and the textbook, which is a similar message about including mathematical objectism and intelligence in the pill as found in the big group. Affects or attitudes were not so common, only one suggests joy, one to include interest, another confidence, knowledge and concentration. One student suggests "a good teacher, a better teacher, and a good teacher", with a lot of emphasis on the need to include a good teacher, which might reflect some frustration of the current teacher. Among the experimental ingredients, one student includes all members of a K-pop-band, three included some food and water, and four included rainbows, a cat palate, or eternal luck. The sad part is that five students in this group includes narcotic drugs or alcohol in their pill. No one includes ingredients that can be related to rationalism or openness in mathematics, as a way of enabling a more inclusive participation.

In this group of students who rate themselves as "not good at math", none of the students challenged the idea of mathematics as objectism. Also, intelligence, interest, confidence, knowledge or concentration was included in the pill to make someone "very good at mathematics". Some in this group put their trust in magic help from the pill. But when this group of students include drugs in the pill, it can be a sign of a will to escape mathematics, or a will to self-destruct due to the sense of not fulfilling the expectations of being someone intelligent, fast and confident as they perceive is required for learning mathematics.

21.4 Conclusions and Implications for Practice

In line with earlier reported results from the Swedish WIFI-survey (Andersson & Österling, 2019), the dose of objectism and control is high in the students' pills. We interpret ingredients as numbers and equations as objectism, and textbooks and calculators as control with their focus on facts and correct answers. This can be contrasted to the earlier mentioned focus on rationality in mathematics education research (de Ron, 2022). Hence, teachers who aim for developing rationalism or openness in their classrooms face strong opposing values among students. Values are often difficult to challenge (Kalogeropoulos & Clarkson, 2019), and we fear that teachers who emphasise problem solving, reasoning and dialogic classrooms (de Ron, 2022) face a contrasting view of what it means to be good at mathematics from their students.

We are concerned about how many students, and girls in particular, include ingredients to improve themselves. The most common were intelligence, good memory or being fast at doing mathematics. This result is different from the low importance Chinese students attributed ability (Zhang, 2019), and confirms the western focus on fixed mindset, that Boaler (2015) critiques for being an unsuccessful approach to teaching and learning mathematics. The focus on how to improve intelligence or efficiency is not primarily about enculturation, instead, it becomes an individual often impossible boarder to cross for mathematics which excludes those who do not fulfil the requirements. The pill-question risk to reinforce messages that connects mathematics to who you are, not what you know. This seems to be particularly strong among girls. For teachers, we believe this is a reminder of the importance of decreasing the emphasis on the exclusionary practices, as ability groupings or a focus on being fast, in the mathematics classroom. Such practices have the potential of imposing involuntary transformations on students, or else, leaving them with a sense of being excluded.

Some novel findings are that students include helpful friends, parents, good mathematics teachers, sleep and the whole classroom environment in the pill. One interpretation of these ingredients is that mathematics is not enough; that the pill will only be effective and efficient in a supportive environment with helpful and knowledgeable teachers, a calm and supportive classroom milieu, and enough time to sleep, eat, for pleasures, for emotions, and for properly understanding mathematics. Hill et al (2021) points to the benefit of mathematical wellbeing as important for mathematics outcomes. We add that for some students, wellbeing is the desired outcome from taking the pill, not mathematics.

Some of the unexpected answers, as unicorns and rainbows, but also pain-killers, we have had a struggle to make sense of. Our attempted interpretations range from that these might be nonsensical answers, or to unicorns as a means for making mathematics more beautiful, or rainbows for making it more diverse and allowing access to mathematical ideas, as for the painkillers, we wonder whether students included them for supporting or reducing the pain of doing mathematics. But we cannot know, and therefore, we wish we could go back and ask a new group of children to elaborate on what these ingredients could do for mathematics learning.

When students suggested emotions or affect as ingredients, some foregrounded the importance of love and joy. Considering the emphasis on students to be engaged, curious and joyful when solving problems and the focus in mathematics education of promoting possibilities for developing those attributes (de Ron, 2022) this also gives us a glimpse of how mathematics classrooms can be perceived as harmful to the wellbeing of some students. Again, the importance of the affective and emotional aspects in MWB is reinforced. The implication is given by one girl, to include "an ounce of joy and happiness" in mathematics teaching.

Since our study is based on Swedish data only, we propose the magic pill-question to be used by teachers or in teacher education to get perspectives on what is valued in their particular contexts. Still, this needs to be done with caution and in an inclusive setting to avoid imposing transformations on students. Instead, we hope this is the magic pill to spark engaging discussions on mathematical values.

Appendix

The proportion of ingredients in the pill for different groups of students.

	Girls grade 5 *(n = 614)*	Girls grade 8 *(n = 525)*	Boys grade 5 *(n = 568)*	Boys grade 8 *(n = 418)*	Total *(n = 2125)*
Targeting mathematics	**35%**	**26%**	**34%**	**19%**	**30%**
Abstract math-objects	24%	7%	17%	6%	**14%**
Rationalism	1%	3%	1%	2%	**2%**
Physical mathematical tools	4%	5%	8%	5%	**5%**

(continued)

(continued)

	Girls grade 5 (*n = 614)*	Girls grade 8 (*n = 525)*	Boys grade 5 (*n = 568)*	Boys grade 8 (*n = 418)*	Total *(n = 2125)*
Know and understand	4%	6%	5%	3%	**4%**
Connections	1%	4%	1%	0%	**2%**
More knowlegeable	0%	2%	1%	0%	**1%**
Read, write and explain	1%	1%	1%	1%	**1%**
Targeting person	**24%**	**42%**	**16%**	**27%**	**27%**
Affect and feeling	7%	14%	4%	9%	**8%**
Intelligence	6%	10%	6%	5%	**7%**
Effort, energy, endurance	3%	3%	1%	3%	**3%**
Concentration, commitment	2%	2%	2%	5%	**2%**
Creativity, aestetics	2%	5%	1%	2%	**2%**
Efficiency, being fast	2%	4%	0%	1%	**2%**
Physical body	1%	2%	1%	4%	**2%**
Thoroughness	1%	0%	0%	0%	**0%**
For best effect	**5%**	**8%**	**5%**	**11%**	**7%**
Environment	3%	5%	4%	5%	**4%**
Teachers or mathematical friends	2%	3%	1%	4%	**2%**
Tools, non-maths	0%	0%	0%	0%	**0%**
Subject integration	0%	0%	0%	1%	**0%**
Future opportunities	0%	0%	1%	1%	**0%**
Excipients	**19%**	**8%**	**21%**	**29%**	**19%**
Nutrition: vitamins, proteins, Omega3	12%	4%	13%	12%	**10%**
Carbohydrates	4%	2%	4%	8%	**4%**
Pharmaceutics, narcotics, alcohol	2%	2%	5%	9%	**4%**
Experimental agents	**6%**	**10%**	**10%**	**6%**	**8%**
Magic	6%	9%	10%	6%	**8%**
Being lucky	0%	0%	0%	1%	**0%**
Bulk	**11%**	**6%**	**13%**	**9%**	**10%**
Binding substances, colouring	3%	2%	4%	4%	**7%**
Flavour	8%	4%	9%	5%	**3%**

References

Abtahi, Y., & Barwell, R. (2019). Mathematical morality tales: Mathematics education in Canadian newspapers. *Canadian Journal of Science, Mathematics and Technology Education, 19*(1), 48–60. https://doi.org/10.1007/s42330-019-00042-0

Andersson, A., Foyn, T., Simensen, A. M., & Wagner, D. (2023). Storylines in voices of frustration: Implications for mathematics teacher education in changing times. *Education Sciences, 13*(8), 816. https://doi.org/10.3390/educsci13080816

Andersson, A., & Österling, L. (2013). Measuring immeasurable values. In A.M. Lindmeier & A. Heinze (Eds.), *Proceedings of the 37th Conference of the International Group for the Psychology of Mathematics Education (PME)* (Vol. 2, pp. 17–24). International Group for the Psychology of Mathematics Education (PME). https://www.igpme.org/publications/current-proceedings/

Andersson, A., & Österling, L. (2019). Democratic actions in school mathematics and the dilemma of conflicting values. In P. Clarkson, W. T. Seah & J. S. Pang (Eds.), *Values and valuing in mathematics education: Scanning and scoping the territory* (pp. 197–208). Springer. https://doi.org/10.1007/978-3-030-16892-6

Andersson, A., & Wagner. D. (2019). Respond or dismiss: Interactions that may support loving, bullying and solitude in mathematics. *Journal of the Canadian Association for Curriculum Studies, 17*(1), 47–74. https://jcacs.journals.yorku.ca/index.php/jcacs/article/view/40399

Bishop, A. (1991). Mathematical enculturation: A cultural perspective on mathematics education. *Kluwer Academic Publishers*. https://doi.org/10.1007/978-94-009-2657-8

Boaler, J. (2015). *Mathematical mindsets: Unleashing students' potential through creative math, inspiring messages and innovative teaching*. Wiley.

Bryman, A. (2016). *Social research methods* (3rd ed). Oxford University Press.

Chan, Y. C., & Wong, N. Y. (2019). Methodological issues in the investigation of values in mathematics. In P. Clarkson, W. T. Seah & J. S. Pang (Eds.), *Values and valuing in mathematics education: Scanning and scoping the territory* (pp. 197–208). Springer. https://doi.org/10.1007/978-3-030-16892-6

Clarkson, P., Seah, W. T., & Pang, J. (2019). Scanning and scoping of values and valuing in mathematics education. In *Values and valuing in mathematics education: Scanning and scoping the territory* (pp. 1–10). Springer. https://doi.org/10.1007/978-3-030-16892-6

de Ron, A. (2022). Problem i matematiken—Diskursiva sanningar om matematikundervisningens varför och hur. In P. Valero, L. Björklund Boistrup, I. M. Christiansen, & E. Norén (Eds.), *Matematikundervisningens sociopolitiska utmaningar*. (pp. 69–100). Stockholm University Press. https://doi.org/10.16993/bcc

Goodin, R. E. (1996). Inclusion and exclusion. *European Journal of Sociology/Archives Européennes de Sociologie*, 37(2), 343–371. https://www-jstor-org.ezp.sub.su.se/stable/23997023

Hill, J. L., Kern, M. L., Seah, W. T., & van Driel, J. (2021). Feeling good and functioning well in mathematics education: Exploring students' conceptions of mathematical well-being and values. *ECNU Review of Education, 4*(2), 349–375. https://doi.org/10.1177/2096531120928084

Horn, I. S. (2007). Fast kids, slow kids, lazy kids: Framing the mismatch problem in mathematics teachers' conversations. *The Journal of the Learning Sciences, 16*(1), 37–79. https://www.learntechlib.org/p/166142/

Kalogeropoulos, P., & Clarkson, P. (2019). The role of value alignment in levels of engagement of mathematics learning. In P. Clarkson, W. T. Seah & J. Pang (Eds.), *Values and valuing in mathematics education* (pp. 115–127). Springer International Publishing. https://doi.org/10.1007/978-3-030-16892-6_8

Kern, M. L., Zeng, G., Hou, H., & Peng, K. (2019). The Chinese version of the EPOCH measure of adolescent wellbeing: Testing cross-cultural measurement invariance. *Journal of Psychoeducational Assessment, 37*, 757–769. https://doi.org/10.1177/0734282918789561

Lampert, M., Boerst, T., & Graziani, F. (2011). Organizational resources in the service of school-wide ambitious teaching practice. *Teachers College Record, 113*(7), 1361–1400. https://doi.org/10.1177/016146811111300706

Munter, C., Stein, M. K., & Smith, M. S. (2015). Dialogic and direct instruction: Two distinct models of mathematics instruction and the debate(s) surrounding them. *Teachers College Record, 117*(11), 1–32. https://doi.org/10.1177/016146811511701102

Nyman, M., & Sumpter, L. (2019). The issue of 'proudliness': Primary students' motivation towards mathematics. *Lumat: International Journal of Math, Science and Technology Education, 7*(2), 80–96. https://doi.org/10.31129/LUMAT.7.2.331

James, M., & Pedder, D. (2006). Beyond method: Assessment and learning practices and values. *The Curriculum Journal, 17*(2), 109–138. https://doi.org/10.1080/09585170600792712

Pansell, A., & Bjorklund Boistrup, L. (2018). Mathematics teachers' teaching practices in relation to textbooks: Exploring praxeologies. *The Mathematics Enthusiast, 15*(3), 541–562. https://doi.org/10.54870/1551-3440.1444

Popkewitz, T. S. (2018). Reform and making human kinds: The double gestures of inclusion and exclusion in the practice of schooling. In: E. Hultqvist, S. Lindblad & T. Popkewitz (Eds.), *Critical analyses of educational reforms in an era of transnational governance. Educational governance research* (Vol 7, pp. 133–150). Springer. https://doi.org/10.1007/978-3-319-61971-2_8

Seah, W. T., Andersson, A., Bishop, A., & Clarkson, P. (2016). What would the mathematics curriculum look like if values were the focus? *For the Learning of Mathematics, 36*(1), 14–20. https://www.jstor.org/stable/44382694

Stein, M. K., Engle, R. A., Smith, M. S., & Hughes, E. K. (2008). Orchestrating productive mathematical discussions: Five practices for helping teachers move beyond show and tell. *Mathematical Thinking and Learning, 10*(4), 313–340. https://doi.org/10.1080/10986060802229675

Stigler, J. W., & Hiebert, J. (2009). *The teaching gap.* Free Press.

Weida, C. L., & Bradbury, C. (2020). The unicorn as lifelong companion: Remixing inclusive, intergenerational art education journeys with the freedom of Froebel and wildness of Waldorf. *International Journal of Lifelong Learning in Art Education, 3*(1), 41–53. https://doi.org/10.25889/417w-2r47

Weingarden, M., & Heyd-Metzuyanim, E. (2023). What can the realization tree assessment tool reveal about explorative classroom discussions? *Journal for Research in Mathematics Education, 54*(2), 97–117. https://doi.org/10.5951/jresematheduc-2020-0084

Zhang, Q. (2019). Values in mathematics learning: Perspectives of Chinese mainland primary and secondary students. In P. Clarkson, W. T. Seah, & J. S. Pang (Eds.), *Values and valuing in mathematics education: Scanning and scoping the territory* (pp. 185–196). Springer. https://doi.org/10.1007/978-3-030-16892-6

The manufacturer's authorised representative in the EU is Springer Nature Customer Service Centre GmbH, Europaplatz 3, 69115 Heidelberg, Germany. If you have any concerns regarding our products, please contact ProductSafety@springernature.com

Printed and bound by CPI Group (UK) Ltd, Croydon, CR0 4YY
30/09/2026
02225319-0019